# PRINCIPLES OF PHYSICS

BY

## M. NELKON

M.Sc. (Lond.), A.Inst.P., A.K.C.

FORMERLY HEAD OF THE SCIENCE DEPARTMENT
WILLIAM ELLIS SCHOOL, LONDON

1970

Published by
Chatto & Windus (Educational) Ltd
42 William IV Street
London WC2

*

Clarke Irwin and Co. Ltd
Toronto

SBN   7010   0140   2

| | | | |
|---|---|---|---|
| *First Published* | . | . | 1951 |
| *Reprinted* | . | . | 1953 |
| *Reprinted* | . | . | 1954 |
| *Reprinted* | . | . | 1956 |
| *Revised Edition* | . | . | 1956 |
| *Reprinted* | . | . | 1957 |
| *Reprinted* | . | . | 1957 |
| *Reprinted* | . | . | 1958 |
| *Reprinted* | . | . | 1959 |
| *Reprinted* | . | . | 1960 |
| *Revised and Reset* | . | . | 1961 |
| *Reprinted* | . | . | 1962 |
| *Revised Edition* | . | . | 1964 |
| *Reprinted* | . | . | 1965 |
| *Reprinted* | . | . | 1967 |
| *Reprinted twice* | . | . | 1968 |
| *Reprinted* | . | . | 1970 |

Printed by T. & A. Constable Ltd., Edinburgh

# PREFACE

THIS book deals with the fundamental principles of Physics in all its branches, up to the "Ordinary" level of the General Certificate of Education. Since experiment is the very stuff of science, many experiments are referred to in the text; but, in order that each teacher may exercise individual discretion in laboratory procedure, detailed experimental arrangements have, for the most part, not been described. Physical science is essentially quantitative in nature: care has therefore been taken to present clearly the essential mathematics of each topic and to define carefully the units of quantities encountered. Further, in order to illustrate the subject-matter and to help pupils to acquire facility in tackling problems, numerically worked examples from past examination papers have been incorporated in the text. Since we live in a world whose material basis is in debt to science, stress has been laid on the contribution of the physicist to the welfare of the community; while, in order to help the pupil to gain some appreciation of the great historical sweep of scientific achievement, regard has been paid to the history of physics and the outstanding contributions of its great pioneers.

Throughout the book, the idea of energy and the principle of its conservation have been kept in the foreground. The Mechanics section begins with a consideration of moving bodies and their energy, and then proceeds to a general discussion of statics, this order of treatment having been found to be particularly successful with fifth-formers. The gas laws, Joule's equivalent, and the concept of temperature have been stressed in the Heat section. After careful consideration of the relative merits of the alternative sign-conventions recommended in the Physical Society's *Report on the Teaching of Geometrical Optics* it has been

v

decided that both the "real is positive, virtual is negative" and the "New Cartesian" conventions should be available for the convenience of teachers. There are therefore two editions of the book. To assist pupils who have difficulty with lens and mirror calculations, a sufficient number have been included in the text to illustrate the technique. Plane-progressive and stationary waves have been fully treated in the section on Sound. In the Electricity section, the concept of electric potential has been associated fundamentally with electrical energy, although analogies have also been used. To permit the logical development of the subject, electrostatics has been discussed before current electricity, and the nature of the carriers of the current through metals, liquids and gases has been emphasized. In view of their importance, simple accounts of the valve, the cathode-ray tube, the photo-electric cell and atomic structure have been included.

From the presentation of the topics, it is hoped that the book will provide a useful background for pupils proceeding to a more advanced course in the subject. The questions at the end of each chapter have been graded in order of difficulty, so that the book should provide useful for a two-year course in Physics.

My thanks are due to the late P. Parker, M.Sc., formerly Senior Lecturer, Northampton College of Advanced Technology, London, for his valuable assistance with the book as a whole, and to my former colleagues J. Duffey, B.Sc., Ph.D., and C. W. Othen, M.Sc., of Cardiff University, for their suggestions relating to parts of it. I am also very much indebted to Dr. Cyril Bibby, M.A., M.Sc., F.L.S., Principal, Hull Training College for Teachers, for many editorial suggestions at all stages of its preparation.

## PREFACE TO 1964 EDITION

In this Edition, an account of the principles of atomic structure and a selection of some experiments on modern physics have been added. This is in accordance with developments in the new physics syllabus. Recent examination questions have also been added.

# ACKNOWLEDGMENTS

The author acknowledges with thanks the permission to reprint questions set in Physics papers by the following examining bodies: Matriculation and Schools Examination Council, London University (L.); Northern Universities, Joint Matriculation Board (N); Cambridge Local Examinations Syndicate (C); Oxford and Cambridge Joint Board (O. & C.); Central Welsh Board (C.W.B.).

He is also much indebted to the following (whose names are given, for convenience of reference, in the sequence of the numbers of the *first* picture acknowledged in each case) for allowing the use without fee of photographs and blocks; Philips Electrical, Ltd. (*Frontispiece*); British Steel Piling Co., Ltd. (24); English Electric Co., Ltd. (27, 427); Griffin & George, Ltd., (55, 386, 401); London Transport Executive (60); Wellman, Smith Owen Engineering Co. (70 (*a*) ); Siebe, Gorman & Co. (83); Davy & United Engineering Co., Ltd. (90); Goodyear Tyre & Rubber Co., Ltd. (113); Negretti & Zambra, Ltd. (115 (*a*) ); Elliott Bros., Ltd. (115 (*b*) ); A. Lewis & Co. (161); Heath & Co. (186, 369); C. A. Parsons & Co., Ltd. (196); U.S. Information Service (242); Ilford, Ltd. (250); Weston, Ltd., Enfield (256); E.M.I. Sales & Service, Ltd. (263, 278); John Compton Organ Co., Ltd. (275); Massachusetts Institute of Technology (295); Muirhead & Co., Ltd. (304 (*b*) ); British Electricity Authority (312, 316); Mather & Crowther, Ltd. (326); General Electric Co. (328, 393, 437 (*b*) ); Rhokana Corporation, Ltd. (334 (*a*) ); W. Canning & Co., Ltd. (334 (*b*) ); Postmaster General (344); The British Thermostat Co., Ltd. (351 (*a*) ); The Hawker Siddeley Group, Ltd. (351 (*b*) ); Automatic Coil Winder & Electrical Equipment Co., Ltd. (353); Kelvin & Hughes (379); Ericsson Telephones, Ltd. (397); British Thomson-Houston Co., Ltd. (411); A. C. Cossor, Ltd. (435 (*b*), 436 (*a*) ); Science Museum, South Kensington (By courtesy of the Royal Society) (440); University of California and U.S. Atomic Energy Commission (441); United Kingdom Atomic Energy Authority (456).

Acknowledgment for permission to reproduce photographs is also made to : The National Physical Laboratory (Crown Copyright) (34) ; Exclusive News Agency (27 (*a*) ); Keystone Press Agency, Ltd. (37 (*a*), 128, 436 (*b*) ); Imperial War Museum (37 (*c*), 70 (*b*) ; Science Museum (Crown Copyright) (133); Air Ministry (Crown Copyright) (157) ; News Chronical (223) ; C.O.I. (252) ; B.B.C. (265, 431).

# ACKNOWLEDGMENTS

The author wishes to thank the permission to reprint questions set in the past papers by the following examining bodies: Metropolitan and Schools Examination Council, London University (L.U. Northern Universities Joint Matriculation Board (N.), Cambridge Local Examinations Syndicate (C.), Oxford and Cambridge Joint Board (O. & C.), Central Welsh Board (C.W.B.).

He is also indebted to the following (whose names are given for convenience of reference) in the sequence of the numbers above for the are acknowledged in each case for permission by whose reproduction for photographs and blocks: Philips Electrical Ltd. (Frontispiece, British Std Plate 1 C. 2 etc. 66); English Electric Co. Ltd. (V. 20); Globe & George Ltd. (33, 38, 40); Linotype Corporation Machine (40); William Smith Owen Instrument Co. 170 (61); Staine, Curtis & Co. (53); Davy & United Engineering Co. Ltd. (28); Goodyear Tyre & Rubber Co. Ltd. (138; Maguire & Zumbal Ltd. (113 (b)); Elliott Bros. Ltd (115) (D); A. Lock & Co. (Tel.); Heath & Co (Eng. 604); G.A. Parsons & Co. Ltd. (156); 113; Information Service (242); Ibo (2 Ltd. (250); Watson Ltd. London Photos; E.M.I. Sales & Service Ltd. (250); Army Council (Organ Cot Ltd. (250); Massachusetts Institute of Technology (255); Mumford & Co. Ltd. (204 (b)) British Electricity Authority (212 (b)); Mather & Crossley Ltd. (130); General Electric Co. (130, 203, 417 (b)); Rheostat Corporation (114 (a)) A. W. Chapman & Co., Ltd. (431 (a) 1); Ferranti General Ltd.); The Royal Technical Co., Ltd (471 (a)); The Electric Sin Co.; Chron, Ltd. (251) (a); H. Armstrong (Coll Winder & Electrical Engineers etc. Ltd.) (258); Selwin & Hughes Ltd. (Electric Telephone Ltd. (457)); British Thomson-Houston Ltd. Laboratory; R.C. Glenn, Ltd. (455 (b), 36 (a)); Science Museum, South Kensington (B) Courtesy of The Royal Society (440); University of California and U.S. Atomic Energy Commission (441); Unitech English Atomic Energy Authority (522).

Acknowledgment for permission to reproduce photographs is also made to: The National Physical Laboratory, Teddington (Crown Copyright) (431, Frontispiece No. 3; Appx 67 (a)); Kodansha Press (etc.); Ltd. 132 (b); 133; 136 (b) ); Imperial War Museum IV (a); 70 are Science & Instrument Crown Copyright (131); Air Ministry (Crown Copyright) (131) News Chronicle (214); C.O.I. (255); B.B.C. (235, 424).

# CONTENTS

## INTRODUCTION

## MECHANICAL ENERGY

## HEAT ENERGY

## LIGHT ENERGY

1*

# CONTENTS

## SOUND ENERGY

## ELECTRICAL ENERGY

## MODERN PHYSICS

# INTRODUCTION

# CHAPTER I

## THE TREND OF SCIENCE

"WHEN someone is ill, the doctor is called by telephone, visits his patient by automobile, measures his temperature with a thermometer and his pulse with a watch, examines his heart and lungs with a stethoscope and his throat with a light reflector. Every one of these operations uses a tool and technique supplied by the physicist." This quotation, from the American scientist K. T. COMPTON, illustrates very clearly the dependence of modern man upon scientific method and invention. It is not merely that science has produced a few gadgets and mechanical devices. Science has changed (and is still changing) our way of life, bringing many benefits to mankind.

**Scientific method.** From the very early days, when our remote ancestors fashioned the first tools of wood and stone, man has sought to achieve mastery of his environment. Later, he learned to make implements of copper and bronze and iron, to make pottery and spin threads, to grow crops and keep herds, and even to erect great monuments such as the pyramids of the Pharaohs. While developing these skills and mastering these techniques, man also sought to devise some rational explanation of the phenomena of nature, and the history of science is very largely the history of an ever-closer approximation to truth. This combination of hand and head, of practical experiment and theoretical reasoning, is still perhaps the outstanding characteristic of the scientific method. The results of an experiment suggest a theory; the development of the theory suggests other experiments; and so theory and practice perpetually fertilize each other.

An essential feature of this scientific method is a willingness to abandon previously held theories in the light of newly discovered facts. Very often, in the past, there has been a tendency to believe something merely because an eminent person stated it; and this

1

is the very reverse of the scientific spirit. A good example of this tendency is to be found in the physics of falling bodies. The famous Greek philosopher and scientist ARISTOTLE had stated more than three centuries before Christ, that heavy objects always fell to the ground faster than light ones and, so great was the reverence in which Aristotle was held, it was not until nearly two thousand years later that this belief was shown to be false. In the early part of the seventeenth century, someone (not Galileo, on the authority of Sir Charles Singer) performed a simple experiment to test the truth of this theory. He dropped a heavy and a light object simultaneously from a high building and observed that they reached the ground almost together. The previously held theory was thus in conflict with the observed fact; and it was the theory which had to be altered, for facts cannot be altered.

The really great advances in science have been made by those scientific workers whose bold imagination and careful experimentation have led them into new fields. SIR ISAAC NEWTON, who in the seventeenth century first stated the main laws of mechanics in a form which has lasted until our own time; MICHAEL FARADAY, who in the early nineteenth century discovered how to convert magnetism into electricity; LORD RUTHERFORD, who during the present century made clear the basic structure of the atom: all these were great seekers after truth. They did not always know what material benefits might follow from their discoveries (Faraday can certainly have had no conception of the vast electrical industry of to-day), but they pushed ahead with their fundamental work.

**Units of measurement.** As science has developed through the centuries, the importance of accurate measurement has become increasingly clear; and, in order that scientists working in different parts of the world may agree in and compare their measurements, it is necessary for certain basic units of measurement to be generally accepted. The most important of these internationally accepted units are as follows—

A *metre* is the distance between two fixed marks on a certain platinum-iridium rod kept under specified conditions in a vault in Paris.

A *kilogram* is the mass of a certain piece of platinum kept in Paris.

A *mean solar day* is the average period, taken over twelve months, between successive transits of the sun across the meridian at any part of the earth's surface.

In practice, these units are inconveniently large, and the following smaller units are generally used in scientific work:

The *centimetre*, which is 1/100th part of a metre.

The *gram*, which is 1/1000th part of a kilogram.

The *second*, which is 1/86,400th part of a mean solar day.

This *centimetre-gram-second* system of units is commonly abbreviated to *c.g.s. units* and is used not only by scientists but also by most peoples of the world. In Great Britain, however, different units of length and mass are normally used:

A *foot* is one-third of the distance (a yard) between two fixed marks on a certain bronze rod kept under specified conditions in London.

A *pound* is the mass of a certain piece of platinum kept in London.

This *foot-pound-second* system of units (abbreviated to *f.p.s. units*) has the disadvantage of inconvenience in calculation, as the following contrast shows—

| | |
|---|---|
| 100 cm. = 1 metre | 36 in. = 1 yard |
| 1000 m. = 1 kilometre | 1760 yd. = 1 mile |
| 1000 gm. = 1 kilogram | 16 oz. = 1 pound |
| 1000 kg. = 1 tonne | 2240 lb. = 1 ton |

In the different branches of physics, other units of measurement are used—*degrees* for temperature, *amperes* for electric current, *foot-candles* for illumination for example.

**Accurate measurement of length.** One of the marks of a good scientist is the care he takes to measure things really accurately. Even such a simple operation as the measurement of the length of a straight bar requires careful attention.

When it is necessary to measure small distances accurately, a special type of scale constructed by PIERRE VERNIER is often used. In Fig. 1 (*a*), suppose that the main scale *M* is graduated in centimetres and millimetres. The vernier scale *V* is 0·9 cm. long and is graduated in ten equal parts. Thus the vernier graduations are 0·09 cm. apart, which is 0·01 cm. less than the distance between graduations on the main scale.

Now suppose that, in order to measure the length of an object A, the vernier V is moved from the position shown in Fig. 1 (a) to that shown in Fig. 1 (b). The length of A clearly lies between 10·2 cm. and 10·3 cm. but to obtain the second decimal place it is necessary to examine the vernier scale and *note which division on it coincides with a division on the main scale*.

In this case, the coincident vernier division is the fourth, marked X. Thus the distance from the end of A to X is 4 × 0·09 cm., i.e. 0·36 cm. But the distance from the 0·2 cm. mark on

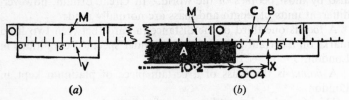

FIG. 1. Vernier principle

the main scale to X is 4 × 0·1 cm., i.e. 0·4 cm. Therefore the distance from the 0·2 cm. mark to the end of A is 0·04 cm. (0·4 − 0·36); and the length of A is 10·24 cm.

The argument would be precisely the same wherever on the two scales the coincidence had been: the second decimal place in the measurement is given by the number of the vernier scale marking which coincides with a main scale marking.

The same principle is applied in the vernier slide *calipers* (Fig. 2). The object A, whose length is required, is placed between a fixed jaw X and a movable jaw Y. The zeros of the main scale M and the vernier are originally coincident, and thus the length of the object illustrated is 7·76 cm. For measuring angles accurately, circular vernier scales are made, graduated in degrees or half-degrees.

Many other devices have been invented to measure accurately not only length, but also temperature, weight, time and all the other quantities with which the scientist deals. No matter how delicate or ingenious the device may be, however, it cannot produce accurate measurements without the utmost care on the part of the person using it.

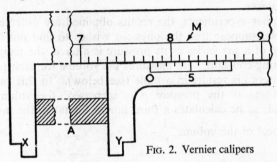

FIG. 2. Vernier calipers

**Graphical methods.** Very often, the physicist finds that it is convenient and helpful to examine the results of his experiments with the aid of a graph. For example, suppose that he wishes to find out how the stretching of a spiral spring varies as different weights are suspended from it. He measures the extension for each weight, and enters his results in a table thus—

| Weight (gm.) . . | 5 | 10 | 20 | 30 | 40 | 50 | 60 |
|---|---|---|---|---|---|---|---|
| Extension (cm.) . | 0·9 | 1·8 | 3·6 | 5·4 | 7·2 | 9·0 | 10·8 |

He now plots a graph between the weight $W$ and the extension $l$,

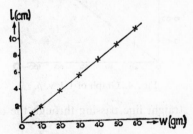

FIG. 3. Graph of weight ($W$) v. extension ($l$)

as in Fig. 3. The seven points of the graph lie on a *straight line passing through the origin*. This shows that, up to weights of

60 gm., $l$ is *directly* proportional to $W$; or, in mathematical notation, $l \propto W$. The same result could have been obtained by calculation, but the graphical method is often much simpler and quicker.

In other experiments, the results obtained are different. For example, suppose that the physicist wishes to find out how the volume of a gas varies as its pressure is altered, the temperature remaining constant. He measures the volume for each pressure, and enters his results in a table (see below). In this case, it is clear that, as the pressure $p$ is increased, the volume $V$ is reduced, so he calculates a third line in his table, the inverse or reciprocal of the volume, $\dfrac{1}{V}$ —

| Pressure $p$ . | 60 | 30 | 40 | 20 | 50 | 100 |
|---|---|---|---|---|---|---|
| Volume $V$ . | 40 | 80 | 60 | 120 | 48 | 24 |
| $\dfrac{1}{V}$ . . | 0·025 | 0·0125 | 0·0167 | 0·0084 | 0·0208 | 0·0417 |

He now plots a graph between $p$ and $\dfrac{1}{V}$, as in Fig. 4. The six

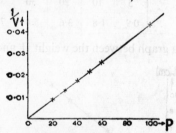

FIG. 4. Graph of $1/V$ v. $p$.

points lie on a straight line passing through the origin, showing that

$$p \propto \frac{1}{V},$$

i.e. pressure is inversely proportional to volume.

**The conservation of energy.** Energy, which may be defined as the capacity for doing work, exists in many different forms. Thus, a moving train possesses *mechanical energy*, and does work in overcoming the frictional resistance to its movements. A fire possesses *heat energy*, and does work in heating a room or perhaps in converting water to steam. An accumulator possesses *chemical energy*, and does work in sending an electric current round a

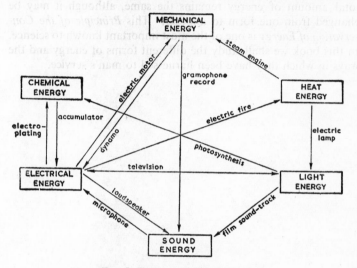

FIG. 5. Energy and transformation

circuit. This *electrical energy* may then do work in ringing a bell or starting up a motor-car engine. Similarly, *sound energy* does work in setting the air in vibration and eventually causing the ear to hear; while *light energy* causes the eye to see.

As a result of the careful work of physicists over the centuries it has gradually become clear that energy can be transformed from one form to another. Thus, when coal is burned in the furnace of a steam locomotive, chemical energy is converted into heat energy; then the heat energy vaporizes the water into steam and drives the engine, thus being converted into mechanical energy. In the case of trolley buses, electrical energy is converted into mechanical energy; while dynamos at power stations convert mechanical energy into electrical energy. Similarly, light energy from the sun is changed by the green chlorophyll of plants into chemical energy, television receivers change electrical energy into light energy, radio receivers change electrical energy into sound energy, and so on (see Fig. 5).

In all these transformations, *energy is never destroyed*. The total amount of energy remains the same, although it may be changed from one form to another. This *Principle of the Conservation of Energy* is one of the most important known to science. In this book we shall study the different forms of energy and the ways in which they have been harnessed to man's service.

# MECHANICAL ENERGY

# CHAPTER II

# ELEMENTS OF DYNAMICS

IF an object is pushed or pulled, a **force** is said to be exerted on it, for example, a force is exerted on a football when it is kicked, and on a moving train by its engine. **Mechanics** is the name given to the study of forces, and a knowledge of its principles is required by the engineer, as in the construction of a motor-car, a building, or an aeroplane.

The subject of Mechanics divides into two parts—

(1) **Dynamics,** the study of motion and the forces which keep an object in motion (e.g. the forces which keep an aeroplane in flight);

(2) **Statics,** the study of forces which keep an object in equilibrium (e.g. the stresses in the various parts of a bridge).

In this chapter we begin with the fundamental principles of Dynamics.

## DYNAMICS

**Velocity.** If a car travels 100 miles in 4 hours, its average speed is 25 miles per hour. This result does not depend on the car's direction; it might travel the 100 miles along a winding road.

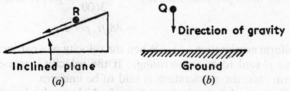

FIG. 6. Rectilinear (straight line) motion

When the movement of a body is specified as being in a definite *direction*, we do not use the term "speed"; we refer to its **velocity** in that direction. In this book we shall deal with the motion

11

of bodies moving in straight lines (**rectilinear motion**); for example, in Fig. 6 (*a*), *R* is a car, moving in a straight line down a hill; in Fig. 6 (*b*), *Q* is a ball falling vertically towards the ground under the action of gravity.

**Uniform velocity.** If an object travels in a straight line and moves the same distance every second, we say that its velocity is *uniform*. We should feel even more justified in calling the velocity uniform if we knew that it moved the same distance every 1/1000th of a second.

**Uniform velocity is the motion of a body which travels in a straight line and moves over equal distances in equal times, no matter how small the latter may be.**

If *s* is the distance travelled in time *t* by a body moving with uniform velocity *v*, it can be seen that, since

$$velocity = \frac{distance}{time},$$

$$v = \frac{s}{t} \quad . \quad . \quad . \quad . \quad . \quad (1)$$

and

$$s = vt \quad . \quad . \quad . \quad . \quad . \quad (2)$$

The units employed must be carefully noted. In equations (1) and (2), if *s* is in feet and *t* is in seconds, *v* is in feet per second; if *s* is in centimetres and *t* in seconds, *v* is in centimetres per second.

It is useful to remember that

$$60 \; miles \; per \; hour \; (m.p.h.) = \frac{60 \times 5280}{3600} \; ft. \; per \; sec.$$

$$= 88 \; ft. \; per \; sec.$$

**Uniform acceleration ( *f* ).** When the velocity of a car increases, the car is said to be *accelerating*. If the velocity increases at a uniform rate, the acceleration is said to be uniform.

**Uniform acceleration is the motion of a body moving in a straight line whose velocity increases by equal amounts in equal times, no matter how small the latter may be.**

Suppose that a train is moving at 25 m.p.h. and is then uniformly accelerated to 40 m.p.h. in 10 secs. Then its

$$acceleration = \frac{change\ in\ velocity}{time\ taken\ to\ make\ the\ change}$$

$$= \frac{(40 - 25)\ m.p.h.}{10\ sec.} = 1\cdot5\ m.p.h.\ per\ sec.$$

When the units of velocity are expressed in "ft. per sec." or "cm. per sec." the corresponding units of acceleration are "ft. per sec. per sec." or "cm. per sec. per sec.", the time element occurring twice. The units of acceleration are therefore frequently abbreviated to "ft. per sec.$^2$" (or "cm. per sec.$^2$")

The magnitude, $f$, of the acceleration is given by

$$f = \frac{increase\ of\ velocity}{time}.$$

When an object falls to the ground under the action of gravity, its velocity increases at the constant rate of about 32 ft. per sec. for every second of time it is dropping (p. 25), i.e. the acceleration of a falling body at a given place on the earth's surface is constant.

The acceleration of a body due to gravity is commonly denoted by the letter $g$; its magnitude in c.g.s. units is about 980 cm. per sec.$^2$, and in f.p.s. units about 32 ft. per sec.$^2$

*Formulae.* If the velocity of a body is initially $u$ and this increases to $v$ in a time $t$ with uniform acceleration, the acceleration of the body, $f$,

$$= \frac{change\ in\ velocity}{time\ to\ make\ change} = \frac{v - u}{t}$$

$$\therefore f = \frac{v - u}{t} \qquad . \qquad . \qquad . \qquad . \qquad . \qquad (3)$$

On some occasions, the final velocity ($v$) of a moving object is required when it accelerates uniformly from a velocity $u$ for a time $t$. If the acceleration is $f$, it follows from equation (3) that

$$v = u + ft. \qquad . \qquad . \qquad . \qquad (4)$$

It can be seen from the definition of acceleration in (3) that when a body is moving with a uniform velocity its acceleration is zero.

**Retardation.** A body is said to be *retarded* if its velocity is decreasing. Thus a train is retarded when it slows down on approaching a station. A ball thrown straight up into the air is retarded by the attraction of the earth, and its velocity decreases, becoming zero at the highest point of its flight.

*Retardation* is measured in the same way as acceleration, i.e. $\dfrac{\text{change in velocity}}{\text{time}}$, and has the same units. It can be regarded as "negative" acceleration; and by paying attention to the signs of the quantities in the formula $v = u + ft$, one can apply the formula to cases of retardation.

Consider a car which is moving with a velocity of 88 ft. per sec. and which, when the brakes are applied, undergoes a uniform retardation of 10 ft. per sec.$^2$ The initial velocity ($u$) of the car = 88 ft. per sec., and the retardation ($f$) = $-$ 10 ft. per sec.$^2$, as the car is slowed down. Thus, after 5 secs., the velocity ($v$) of the car is given by

$$v = u + ft = 88 - 10t = 88 - 50 = 38 \text{ ft. per sec.}$$

The velocity of the car becomes zero when $v = 0$. Hence the time $t$ for this to occur, from the instant the brakes are applied, is given by

$$v = u + ft, \text{ i.e. } 0 = 88 - 10t.$$

Thus $\qquad\qquad\qquad t = 8\cdot8 \text{ secs.}$

When a body is dropped, it falls with an acceleration of 32 ft. per sec.$^2$ (p. 13). When a body is thrown vertically upwards with an initial velocity ($u$) of 96 ft. per sec. it undergoes a retardation of 32 ft. per sec.$^2$; and thus the velocity ($v$) 2 secs. after the instant it is thrown up is given by

$$v = u + ft = 96 - 32 \times 2 = 32 \text{ ft. per sec.}$$

The time taken to reach the top of its flight is given by making $v = 0$ in the formula $v = u + ft$, as the body is then stationary. In this case

$$0 = 96 - 32t,$$

from which $t = 3$ secs.

The following examples on acceleration and retardation should be studied carefully:

1. *A car travelling at* 30 *m.p.h. undergoes an acceleration of* 2 *ft. per sec.*$^2$ *Calculate its velocity in m.p.h. after* $\frac{1}{4}$ *minute.*

Here $u$ = 30 m.p.h. = 44 ft. per sec., $f$ = 2 ft. per sec.$^2$, $t$ = 15 sec.

∴   final velocity, $v = u + ft = 44 + 2 \times 15 = 74$ ft. per sec.

But 88 ft. per sec. = 60 m.p.h.

∴   $v = \dfrac{74}{88} \times 60 = 50.5$ m.p.h.

(Note that attention must always be paid to the *units* in a problem before substituting in formulae.)

2. *A train is moving at* 60 *m.p.h. when the brakes are applied. The retardation is* 4 *ft. per sec.*$^2$ *Find (a) the velocity after* 10 *secs., (b) the time taken for the train to come to rest.*

We have $u$ = 60 m.p.h. = 88 ft. per sec., $f = -4$ ft. per sec.$^2$

(*a*) After $t$ = 10 sec., $v = u + ft = 88 - 4 \times 10 = 48$ ft. per sec.

(*b*) When the train comes to rest the final velocity $v = 0$.

Now                              $v = u + ft,$

∴      $0 = 88 - 4t$

∴      $t = 22$ sec.

This is the time taken for the train to come to rest.

**Graphs in dynamics.** When the velocity of a moving object is plotted against the time, the resulting graph is known as the "velocity-time curve" of the motion. Velocity-time curves give useful information about the movement of the object, and are much used by engineers in their researches.

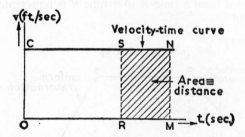

FIG. 7 (*a*). Uniform velocity graph

**Uniform velocity.** Fig. 7 (*a*) illustrates the simple velocity-time curve *CSN* of a body moving with a uniform velocity of *OC* ft. per sec. As the velocity is constant, *CSN* is a straight line parallel to the time-axis *OM*. The distance travelled by the body in *OR* secs. = velocity × time = $OC × OR$. But $OC × OR$ represents the *area* of the rectangle *OCSR*; hence the distance travelled is represented by the area enclosed between the velocity-time curve and the time-axis. In the same way, the distance travelled in *OM* secs. is represented by the area *OCNM* and that travelled in the time-interval *RM* is represented by the area *RSNM*, shown shaded in Fig. 7 (*a*).

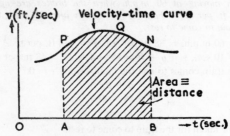

FIG. 7 (*b*). Variable velocity graph

It can be shown that however irregular the shape of the velocity-time curve may be, *the distance travelled is represented by the area between the curve and the time-axis*. Thus if *PQN* is the velocity-time curve of a train on a journey (Fig. 7 (*b*) ), the distance travelled by it from a time *A* to a time *B* is represented by the area *PQNBA*.

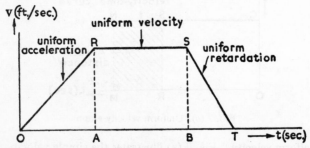

FIG. 8. Uniform acceleration and retardation

**Uniform acceleration.** Fig. 8 illustrates the velocity-time curve *OR* of a car moving with uniform acceleration for a time *OA*.

In this case the velocity increases by equal amounts in equal times (p. 12), and hence the velocity-time graph is a straight line inclined to the time-axis. If the velocity of the car becomes constant at a time $A$ and remains so until a time $B$, the velocity-time curve follows a straight line $RS$ parallel to the time-axis. Suppose the car's brakes are applied at the instant $B$, so that the car undergoes a uniform retardation and comes gradually to rest at a time corresponding to $T$. Since the decrease in velocity in equal times is the same, the velocity-time curve follows a straight sloping line $ST$, until the velocity of the car becomes zero at $T$.

The acceleration $f$ of the car in the time-interval $OA$ can easily be deduced from the graph. If the velocity at the instant $O$ is zero,

$$\text{the acceleration, } f = \frac{\text{change in velocity}}{\text{time}} = \frac{RA}{OA}.$$

This ratio $\dfrac{RA}{OA}$ is known as the **gradient** of the line $OR$. Similarly, the retardation in the time-interval $BT$ is given by the gradient $\left(\dfrac{SB}{BT}\right)$ of the line $ST$. The gradient of the line $RS$ is zero.

The distance travelled by the car in the total time $OT$ secs. is represented by the area between the velocity-time curve $ORST$ and the time-axis, as stated on p. 16. Since $ORST$ is a trapezium with $RS$ parallel to $OT$, the distance travelled, $s$, can be calculated from $s = \frac{1}{2}(RS + OT)RA$.

**Distance-time curve.** When the distance $s$ of a body is plotted against the time $t$, the resulting graph is known as a "distance-time curve". When an object moves with uniform velocity from a place at $O$, the distance-time curve is a straight line $OPQ$, since equal distances are then travelled in equal times (Fig. 9 (a) ). Suppose the distance travelled in $OA$ secs. is $AP$ ft., and in $OB$ secs. is $BQ$ ft. Then, considering the time $AB$ secs., the velocity $= \dfrac{\text{distance travelled}}{\text{time}} = \dfrac{BQ - AP}{AB} = \dfrac{QT}{PT}$, where $PT$ is drawn perpendicular to $BQ$. But $\dfrac{QT}{PT}$ is the gradient of the straight line $OPQ$. Hence

*the velocity is given by the gradient of the line.*

Fig. 9 (*b*) illustrates the distance-time curve, *OCD*, for the height of a cricket-ball thrown into the air, or the height of a

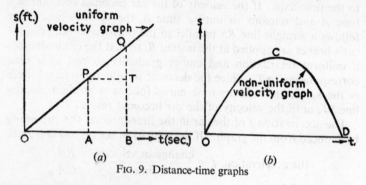

FIG. 9. Distance-time graphs

projectile fired upwards from a gun. The height travelled is a maximum at *C*, which corresponds to the top of the flight, and is zero at *D* when the flight is ended. In this case the velocity is not uniform.

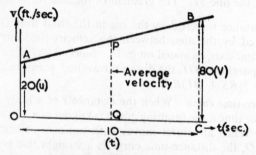

FIG. 10. Uniform acceleration

**The distance travelled by a body moving with uniform acceleration.** Suppose that a train moving with a velocity of 20 ft. per sec. is accelerated at the rate of 6 ft. per sec.² for 10 seconds. The velocity increases uniformly from *OA* to *CB* along the straight-line velocity-time curve *AB* (Fig. 10), and the velocity after 10 secs. $= u + ft = 20 + 6 \times 10 = 80$ ft. per sec.

The average velocity of the train during the 10 seconds is

$\frac{1}{2}(20 + 80)$, or 50 ft. per sec. This corresponds to $PQ$ in Fig. 10. The distance travelled by the train during the 10 seconds is therefore $50 \times 10$, or 500 ft.

We can now obtain a useful formula for the distance, $s$, travelled by an accelerating object. Suppose $u$ and $v$ are the initial and final velocities respectively of a moving object having a constant acceleration $f$ for a time $t$. The average velocity during the time is $\frac{1}{2}(u + v)$.

$$\therefore \quad s = \frac{1}{2}(u + v)t.$$

Now $v = u + ft$. By substitution,

$$s = \frac{1}{2}(u + u + ft)t$$
$$\therefore \quad s = ut + \frac{1}{2}ft^2$$

**The formula for $v$ in terms of $u, f, s$. Equations of motion.** The equations of motion we have deduced so far when a body moves with uniform acceleration are

$$v = u + ft \qquad . \qquad . \qquad . \qquad . \quad (5)$$
$$s = ut + \tfrac{1}{2}ft^2 \qquad . \qquad . \qquad . \quad (6)$$

Often a relation between $v, u, f, s$, is required, and this can be obtained by eliminating $t$ from (5) and (6). Thus, from (5),

$$t = \frac{v - u}{f}.$$

$$\therefore \quad s = \frac{1}{2}(u + v)t = \frac{1}{2}(u + v)\left(\frac{v - u}{f}\right)$$
$$= \frac{v^2 - u^2}{2f},$$

from which, after simplifying, we obtain

$$v^2 = u^2 + 2fs \qquad . \qquad . \qquad . \qquad . \quad (7)$$

The equations (5), (6), (7) are the standard equations for rectilinear motion when an object is accelerating uniformly. They should be memorized.

**Representation of velocity and acceleration.** Velocity and acceleration are called **vector** quantities, because they have direction as well as magnitude; up to now we have only been concerned with their magnitude. If we wish to represent a velocity or an acceleration by a straight line, we can do so by making the length of the line proportional to the *magnitude* of the velocity or acceleration, and drawing the line in the *direction* of the velocity or acceleration. The direction is denoted by an arrow on the line.

Imagine a car moving with a uniform velocity of 30 m.p.h. along a road *OA* at an angle 40° N. of E. of a road *OX* (Fig. 11 (*a*) ). Its velocity is then completely represented by the line *OA* drawn 6 cm. long at an angle of 40° to *OX*, where 1 cm. represents a velocity of 5 m.p.h. Similarly the acceleration of 10 ft. per sec.² of a bicycle down a road inclined at 15° to the horizontal is represented by *PQ* in Fig. 11 (*b*), where *PQ* is 5 cm. long and 1 cm.

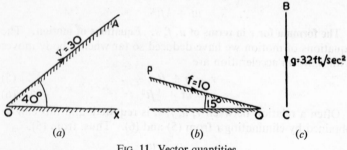

FIG. 11. Vector quantities

represents an acceleration of 2 ft. per sec.² Again, an object thrown into the air is subjected to a vertical retardation (*g*) of 32 ft. per sec.² as it moves upwards, and to a vertical acceleration of 32 ft. per sec.² as it starts to move downwards towards the earth. In *either* case this effect of gravity is represented by a vertical line *BC* drawn to represent 32 ft. per sec.², with an arrow on it pointing downwards, as in Fig. 11 (*c*).

**Parallelogram of velocities.** Suppose that a boy is running with a velocity of 8 m.p.h. in the direction *OB* (Fig. 12) across the deck of a ship, and that the ship itself is travelling with a velocity of 15 m.p.h. in the direction *OA* inclined at an angle of 70° to *OB*.

If the boy were running with the speed of 8 m.p.h. in the direction *OA*, his velocity relative to the sea would be (8 + 15), or 23 m.p.h. When the boy's velocity and that of the ship are not in the same direction, their sum or **resultant** can be found by drawing a parallelogram whose sides *OA*, *OB* represent 15 m.p.h. and 8 m.p.h. in magnitude and direction. Thus if 1 cm. represents 2 m.p.h., *OA* = $7\frac{1}{2}$ cm. and *OB* = 4 cm. (Fig. 12). The *diagonal OC through O* of the completed parallelogram *OACB* then represents the magnitude and direction of the boy's resultant velocity. Suppose *OC* is 10 cm. long and angle *COA* is 25°; the resultant velocity of the boy is then 20 m.p.h., and is in a direction inclined at 25° to the direction of the ship. The parallelogram *OACB* is known as the **parallelogram of velocities** for the motion

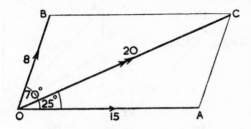

Fig. 12. Parallelogram of velocities

of the boy (see also p. 54). Vector quantities other than velocities, such as accelerations and forces, can always be added by the parallelogram method.

It will be noted that the velocity of 8 m.p.h. is also represented by the line *AC* in Fig. 12, since *AC* is parallel and equal to *OB*. Thus, if *OA* is drawn first to represent 15 m.p.h., and *AC* is then drawn to represent 8 m.p.h., the side *OC* of the completed triangle *OAC* represents the resultant of the two velocities. This is a very useful method of finding the resultant, which is used shortly.

**Relative velocity.** If two ships, *X*, *Y*, are travelling in the same direction with velocities of 15 and 30 m.p.h. respectively, the velocity of *Y* relative to *X* = velocity of *Y* − velocity of *X* = 30 − 15 = 15 m.p.h. If the ships are moving in opposite directions, the velocity of *Y* relative to *X* = 30 − ( − 15) = 45 m.p.h.

Suppose, however, that the ships *X*, *Y* are moving with

velocities of 15 m.p.h., 30 m.p.h., in directions north and north-west respectively (Fig. 13 (*a*) ). The velocity of *Y* relative to *X* = velocity of *Y* − velocity of *X* = velocity of *Y* + (− velocity of *X*); in other words, the velocity of *Y* relative to *X* is the velocity of *Y* *plus a velocity equal and opposite* to that of *X*.

In Fig. 13 (*b*), *OA* is drawn to represent 30 m.p.h., the velocity of *Y*, in magnitude and direction; *AB* is drawn from *A* to represent a velocity of 15 m.p.h. in a *southward* direction, i.e. *AB* represents a velocity equal and opposite to that of *X*. Now it was shown on p. 21 that the resultant of the two vector quantities

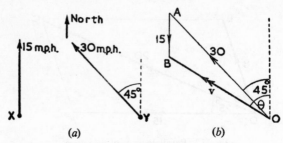

Fig. 13. Relative velocity (*not to scale*)

*OA* and *AB* is represented by the side *OB* of the completed triangle. Measurement of *OB* and the angle $\theta$ show that the velocity *v* of *Y* relative to *X* is 22 m.p.h. in a direction 73·5° W. of N.

**Components.** Consider an object *O* sliding down a smooth incline *YX* (Fig. 14). The effect of gravity is less than if the object fell freely downwards, and the acceleration of *O* is thus less than *g*, the acceleration due to gravity (p. 13).

The effective part of *g* along the plane *YX* is known as the *resolved component* of *g* in this direction. Its magnitude can be found as follows: Draw *OR* to represent *g* in magnitude and direction. Draw a parallelogram *OBRC* which has *OR* as its diagonal and its sides *OB*, *OC* respectively along and perpendicular to the plane. *OB* then represents the component, or effective part, of *g* along the plane. It can be seen, from the parallelogram of accelerations, that *OR* (*g*) is the resultant of the accelerations represented by *OB* and *OC*.

Since the angle *BOC* is 90°, the triangle *OBR* is a right-angled triangle. Hence, if angle *BOR* is θ, *OB* = *OR* cos θ = *g* cos θ. *Thus the component of g in a direction inclined at an angle θ to it is given by g cos θ.* In a later chapter we shall show that this result

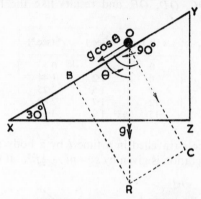

Fig. 14. Component of *g*

can be extended to components of other vectors, besides acceleration, which are represented by a straight line in magnitude and direction (see p. 57).

If the angle of inclination *YXZ* of the plane in Fig. 14 is 30°, then θ = 60°. Hence a body sliding down the plane has an acceleration of *g* cos 60°, or 32 cos 60° ft. per sec.² down the plane. Since cos 60° = 0·5, the acceleration is 16 ft. per sec.² This assumes there is no friction between the body and plane.

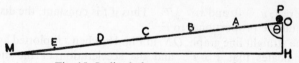

Fig. 15. Inclined plane measurement of *g*

**Experiment to show uniform acceleration. Measurement of *g*.**
A rough experiment to show uniform acceleration, due originally to Galileo, can be performed with a long smooth inclined plane *OE*, graduated from *O* in known lengths *OA*, *AB*, *BC*, *CD*, *DE* (Fig. 15).

A small sphere $P$ can travel between rails on the plane, starting from rest at $O$. With a stop-watch graduated in tenths of a second, the time taken by the sphere to travel from $O$ to $A$ can be determined. The experiment is repeated in turn for the distances $OB$, $OC$, $OD$, $OE$, and results like the following may be obtained—

| $s$ (ft.) | $t$ (sec.) | $t^2$ (sec.²) |
|---|---|---|
| 0 | 0 | 0 |
| 6 | 1·2 | 1·44 |
| 9 | 1·5 | 2·25 |
| 12 | 1·7 | 2·89 |
| 15 | 1·9 | 3·61 |

The distance $s$ travelled in a time $t$ by a body moving with a uniform acceleration is given by $s = ut + \frac{1}{2}ft^2$. If the body starts

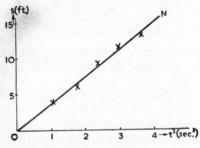

FIG. 16. Graph of $s$ v. $t^2$

from rest, $u = 0$, and $s = \frac{1}{2}ft^2$. Thus if $f$ is constant, the distance $s$ should be proportional to $t^2$.

The straight line graph, $ON$, in Fig. 16, when $s$ is plotted against $t^2$, indicates that $s \propto t^2$; and the acceleration of the moving sphere down the plane is therefore constant. See also p. 157.

This experiment can also be adapted to obtain a value for $g$, the acceleration due to gravity. Suppose $t$ is the time taken by the sphere to fall from rest at $O$ to $E$ a distance $s$, say. Then, since $s = \frac{1}{2}ft^2$ from above, the acceleration $f$ down the plane $= \dfrac{2s}{t^2}$, and can hence be calculated. But $f = g \cos \theta$, the component of $g$ down the plane, where angle $EOH = \theta$. Thus $g = f/\cos \theta$; $\cos \theta$ can be found, since it is equal to $OH/OM$, and hence $g$ can be calculated. This method gives only an approximate value.*

**Determination of $g$ by simple pendulum method.** Study of the motion of the pendulum was made by GALILEO, who was struck by the regular to-and-fro motion of a swinging lantern suspended from a roof. By using the beats of his pulse, Galileo discovered that the time of a complete swing was always the same; he was probably the first person to consider the value of a pendulum in designing accurate clocks.

The **period** ($T$) of a pendulum is the time taken to make a complete (i.e. to-and-fro) swing. Thus if a small sphere, suspended from a fixed point $O$, is oscillating, the period is the time taken for the sphere to go from $B$ to $C$ and back to $B$; or from $A$ to $B$ across to $C$, and back to $A$ (Fig. 17). The period of swing, $T$, depends in general on the length, $l$, of the pendulum, and a calculation, beyond the scope of this book, shows that

$$T = 2\pi\sqrt{\frac{l}{g}} \qquad . \qquad . \qquad . \quad (8)$$

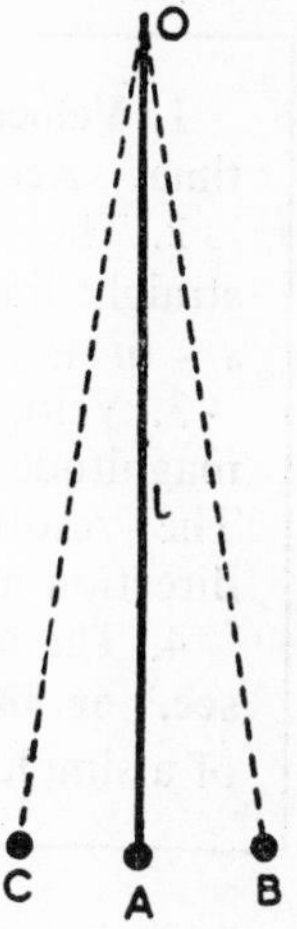

where $g$ is the acceleration due to gravity at the place where the pendulum is used. This formula for $T$ is true only if the pendulum swings to-and-fro through a *very small* angle. In (8), $T$ is in seconds, $l$ in feet (or cm.), and $g$ in feet (or cm.) per second[2].

Fig. 17. Determination of $g$

The formula for $T$ can be used to obtain a fairly accurate value for $g$, the acceleration due to gravity. A small sphere is attached to one end of a very long thread, whose other end is fixed to a point $O$. The sphere is displaced slightly so that it swings through a small angle, and the time taken for 50 complete oscillations is observed. Suppose it is 162 sec.; then the period, $T = 162/50 = 3\cdot24$ sec.

* Using a rolling cylinder instead of a sphere, then, by advanced mechanics, $g = 1\frac{1}{2}f/\cos \theta$ accurately. This relation gives good values for $g$.

The length, $l$, from $O$ to the middle of the sphere is then obtained; suppose it is 261·5 cm. Substituting for $T$ and $l$ in the equation (8), we have

$$3·24 = 2\pi\sqrt{261·5/g}$$

Squaring both sides,     $3·24^2 = 4\pi^2 \times 261·5/g$

$$\therefore \quad g = \frac{4\pi^2 \times 261·5}{3·24^2} = 983.$$

Thus the acceleration due to gravity, $g$, according to this experiment, is 983 cm. per sec.$^2$

---

## Summary

1. **Velocity = distance travelled in constant direction time. Acceleration = velocity change/time.**

2. The equations of motion for an object moving in a straight line with uniform acceleration $f$ are: $v = u + ft$; $s = ut + \frac{1}{2}ft^2$; $v^2 = u^2 + 2fs$.

3. Velocity and acceleration can be represented in magnitude and direction by a straight line drawn to scale. The resolved component of an acceleration, $f$, in a direction inclined at an angle $\theta$, is $f\cos\theta$.

4. **The acceleration due to gravity, $g$, is about 32 ft. per sec.$^2$ or 980 cm. per sec.$^2$** It can be measured by means of a simple pendulum whose period, $T, = 2\pi\sqrt{l/g}$.

---

## WORKED EXAMPLES

1. *Define uniform acceleration and describe an experiment to find g, the acceleration due to gravity. A body is projected vertically upwards from the ground and reaches a height of 625 ft. Find the velocity of projection and the time to rise to the highest point.* (L.)

Let $u$ = velocity of projection in ft. per sec.

From               $v^2 = u^2 + 2fs$

we have          $0 = u^2 - 2 \times 32 \times 625,$

since $v = 0$ at the highest point and $f = -32$ ft. per sec.$^2$

$$\therefore \quad u^2 = 2 \times 32 \times 625$$

$$\therefore \quad u = \sqrt{2 \times 32 \times 625} = 200 \text{ ft. per sec.}$$

Let $t$ = the time in sec. to rise to the highest point.

Then, since         $v = u + ft$

$$0 = 200 - 32t$$

$$\therefore \quad t = \frac{200}{32} = 6\tfrac{1}{4} \text{ sec.}$$

2. (a) *What is meant by an acceleration of* 20 *m.p.h. per min.?* *What is its value in ft. per sec.²?* (b) *A body with an initial velocity of* 2 *ft. per sec. is uniformly accelerated at* 2 *ft. per sec.² for* 3 *sec. It then travels for* 3 *sec. with constant velocity and finally is uniformly retarded to rest in another* 3 *sec. Construct a velocity-time graph of the motion, plotting velocity as ordinate and time as abscissa . Find from it the distance moved by the body in each phase of its motion. What is the value of the retardation?* (L.)

(a) The acceleration in ft. per sec.², $f$,

$$= \frac{\text{change in velocity in ft. per sec.}}{\text{time in sec.}}$$

But         20 m.p.h. $= \frac{88}{3}$ ft. per sec., and 1 min. = 60 sec.

$$\therefore f = \frac{88/3}{60} = \frac{88}{3 \times 60} = \frac{22}{45} \text{ ft. per sec.}^2$$

(b) Fig. 18 illustrates the velocity-time curve $ABCD$ for the 9 sec. of the motion. The initial velocity $OA$ is 2 ft. per sec.; the velocity after 3 sec. $= 2 + 3 \times 2 = 8$ ft. per sec. and is represented by $BQ$. Thus $AB$ is the

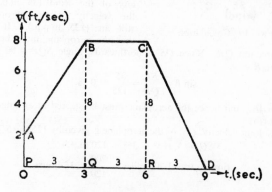

Fig. 18. Velocity-time graph

straight line representing the velocity-time graph during the uniform acceleration. $BC$ represents the uniform velocity. The uniform retardation is illustrated by the straight line $CD$.

The distance moved in each phase of the motion is equal to the area between the corresponding graph and the time-axis (p. 16). The area of the trapezium $PABQ = \frac{1}{2}(AP + BQ) \times PQ = \frac{1}{2}(2 + 8)3 = 15$ ft.; the area of the rectangle $QBCR = QB \times QR = 8 \times 3 = 24$ ft.; the area of the triangle $CRD = \frac{1}{2}CR \times RD = \frac{1}{2} \times 8 \times 3 = 12$ ft.

The retardation $= \dfrac{\text{decrease in velocity}}{\text{time}} = \dfrac{CR}{RD}$

$= \dfrac{8}{3} = 2\frac{2}{3}$ ft. per sec.$^2$

3. *Explain (a) resultant velocity, (b) a component velocity, (c) the parallelogram of velocities. An aeroplane is to travel from A to another place B,* 200 *miles*

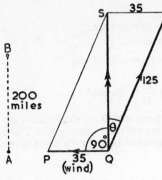

FIG. 19. Calculation

*due north of it. Assuming the wind to remain constant at* 35 *m.p.h. directly from the east, and the aeroplane to have a speed of* 125 *m.p.h. in still air determine (i) the direction in which the aeroplane has to be steered, (ii) the time taken to travel the* 200 *miles.* (C.W.B.)

The *resultant* velocity ($QS$) of the aeroplane in a northward direction is represented by the diagonal of a parallelogram (Fig. 19), whose sides, $QP$, $QR$, represent the velocity of the wind (35 m.p.h.) and the velocity of the aeroplane in still air (125 m.p.h.). It follows that the aeroplane must be steered in the direction $QR$. Since $QS$ is northwards, angle $SQP = 90°$. Thus, if angle $SQR = \theta$,

$$\sin \theta = \frac{SR}{QR} = \frac{35}{125} = 0{\cdot}28.$$

Thus $\theta = 16°$, and hence the aeroplane must be steered at an angle of 16° east of north.

The resultant velocity, $QS$, of the aeroplane is given by $125^2 = SQ^2 + 35^2$. Hence $SQ = \sqrt{125^2 - 35^2} = 120$ m.p.h.

$\therefore$ time to travel 200 miles $= \dfrac{200}{120} = 1\frac{2}{3}$ hours.

# EXERCISES ON CHAPTER II

**1.** Name three applications of the subject of Dynamics in everyday life.

**2.** A car travels with a uniform velocity of 30 m.p.h. for 10 sec. What is the distance travelled?

**3.** A train, moving with a velocity of 30 ft. per sec., starts to accelerate for $\frac{1}{4}$ minute at the rate of 3 ft. per sec.$^2$ Calculate the final speed of the train, and the distance travelled, at the end of the time.

**4.** A ball is thrown vertically upwards with a velocity of 64 ft. per sec. What is the retardation ($f$) of the ball ? What distance does it travel before it comes to rest at the top of its motion ? (Use $v^2 = u^2 + 2fs$.)

**5.** In question 4, find the time taken for the ball to reach its highest point. What is the time taken by the ball to return to the thrower ?

**6.** A box slides down a smooth inclined plane inclined at 30° to the horizontal. Neglecting friction, what is the acceleration of the box in cm. per sec.²? What is the acceleration when the angle of inclination of the plane is 60° ?

**7.** In question 6, how far down the plane does the box travel from rest in ¼ minute, when the angle of inclination is 30° ?

**8.** A car accelerates from rest at the rate of 4 ft. per sec.² for 15 sec., travels at a uniform velocity for the next 30 sec., and then comes to rest after 10 sec. more. Draw a velocity-time graph of the motion. From the graph, find the total distance travelled and the retardation of the car. Check by formula.

**9.** Give an example of an object moving with (i) uniform velocity, (*ii*) uniform acceleration, and explain your answer.

**10.** An aeroplane is travelling due north at 200 m.p.h. A wind of 80 m.p.h. starts to blow from the east. Find the resultant velocity of the aeroplane and its direction of travel (*i*) by drawing, (*ii*) by calculation.

**11.** A ship $X$ is travelling due west with a velocity of 20 m.p.h.; another ship $Y$ is travelling north-east with a velocity of 15 m.p.h. Find the relative velocity of $X$ with respect to $Y$ by drawing.

**12.** Make a *list* of the apparatus you would use, and a *list* of the observations you would make, in order to determine a value for $g$. (The method of calculation is *not* required.) (*N.*)

**13.** Explain the meaning of *uniform acceleration* and deduce expressions for (*a*) the velocity acquired in a time $t$, (*b*) the velocity acquired in a distance $s$, by a body which starts from rest and is subject to a uniform acceleration $a$. A train moves from rest with an acceleration of 1·65 ft. per sec.² Find the speed, in m.p.h., which it reaches in moving through its own length, which is 110 yd. (*L.*)

**14.** What is meant by *uniform acceleration*? Sketch a typical velocity-time graph and a distance-(time)² graph for a body starting from rest and moving with uniform acceleration.

A train, starting from rest, accelerates uniformly so that it attains a speed of 30 m.p.h. in 2 min. It travels at this speed for 5 min., and is

**2\***

then brought to rest with uniform retardation in 3 min.  Find the total distance travelled by the train.  (*C.W.B.*)

**15.** An aeroplane has an air-speed of 198 miles per hour at right-angles to a strong wind.  The velocity of the machine is 202 miles per hour relative to the earth.  What is the velocity of the wind ?  (*N.*)

**16.** Explain, with the aid of a diagram, the theorem of the *parallelogram of velocities*.  A man sets out in a boat to cross a river from a point *A* on one bank.  He steers at an angle of 60° to this bank and reaches the far bank at *B*, a point exactly opposite *A*, the speed of the river being 4 miles per hour.  Determine (*a*) the actual speed at which the boat crosses the river, (*b*) the time taken to cross, *B* being half a mile from *A*.  (*C.W.B.*)

**17.** Find expressions for (*a*) the velocity *v*, (*b*) the displacement *s* of a body *t* seconds after its velocity is *u* ft. per sec., if the acceleration is constant and equal to *a* ft. per sec.$^2$  Show clearly the steps in your argument.  A body is projected up a smooth plane, 72 ft. long and inclined at 30° to the horizontal, and just reaches the top of the plane.  Find the velocity of projection and the time taken to reach the top. (*L.*)

**18.** Show, *from first principles*, that in time *t*, a body moving from rest with uniform acceleration *a*, covers a distance given by $s = \frac{1}{2}at^2$.  Describe an experiment to determine the acceleration due to gravity, pointing out the measurements which are necessary and showing how the result is obtained.  (*L.*)

**19.** Define *uniform velocity*, *uniform acceleration*.  Show how, from the velocity-time graph for a body (*a*) the acceleration at any instant, (*b*) the average velocity can be calculated.

A car, starting from rest, is uniformly accelerated at $\frac{1}{2}$ ft. per sec.$^2$ until it reaches a speed of 50 ft. per sec.  It travels at this speed for 3 min. and is then uniformly retarded so as to come to rest 8 min. after starting.  Plot the velocity-time graph of the motion and use the graph to determine the average speed of the car.  (*O. & C.*)

**20.** Explain the terms: *uniform acceleration, acceleration due to gravity*.  Describe a method of determining the acceleration due to gravity at a given place.  A tennis ball is hit vertically upwards and returns 6 sec. later.  Calculate (*a*) the greatest height reached by the ball, (*b*) the initial velocity of the ball.  (*O. & C.*)

# CHAPTER III

## FORCE AND ENERGY

ONE of the great scientific geniuses of the world was SIR ISAAC NEWTON (1642-1727). He developed mathematically the theory that all the planets, including the earth, move round the sun under the influence of attractive forces. He also discovered that white light consists of many different colours; and he designed the forerunner of the largest telescope in the world (p. 342). He made numerous other fundamental discoveries and researches in science and mathematics, too many to list here. At the age of 23 he became a professor of mathematics at Cambridge, and devoted himself to a search for truth. In later life he was appointed Master of the Mint, in which capacity he reorganized the country's coinage system. But it is for the Laws of Motion which he formulated that he is best known.

**Newton's laws of motion.** In 1687 Newton completed a monumental work on Mechanics, in which, for the first time, "velocity" and "acceleration" were accurately related to the idea of "force." His *Laws of Motion* are as follows—

**Law I. Every body continues in its state of rest or uniform motion in a straight line, unless impressed forces act upon it.**

**Law II. The impressed force is proportional to the change of momentum (p. 32) per unit time which it produces, and the change of momentum takes place in the direction of the straight line along which the force acts.**

**Law III. Action and reaction are always equal and opposite.**

The first law recognizes that a body has a certain amount of **inertia,** which has to be overcome to make the body move from rest. This book, for example, will not move unless it is pushed, pulled, or carried. When a bus or train starts suddenly, standing passengers are jerked backwards, as they tend to remain in their state of rest. Similarly, if an object is moving with a uniform velocity, it will continue to move with this velocity unless a force acts on it. Thus an object hurled along the ice on a frozen pond will continue to move for some distance with undiminished velocity, owing to the very small frictional force acting on it; and when a bus or train comes to rest suddenly, standing passengers are jerked forwards, as they tend to retain the velocity they had possessed.

**Force and acceleration.** When a bus begins to move, we know that a force, due to the engine, is acting on it. When the bus slows down, we know that a force, due to the brakes, is acting on it. If a cricket ball is caught, its speed is changed suddenly to zero, as a result of the retarding force acting on it. In general, a force acts on an object when its velocity increases or decreases, and Newton's second law provides a definition of the magnitude of the force.

When a body of mass $m$ (see p. 34) is moving with a uniform velocity $u$, it is said to have a **momentum** equal to $mu$ units, the product of its mass and velocity; and as long as it continues to move with the same velocity the momentum is constant. If a force now acts on the body and increases its velocity to $v$, the momentum of the body increases from $mu$ to $mv$ units. Newton stated that the magnitude of the force was proportional to the change of momentum that it produced in one second (Law II).

Hence, if the force $P$ acts for a time $t$,

$$P \propto \frac{mv - mu}{t}, \text{ or } P \propto \frac{m(v - u)}{t}.$$

Now $v = u + ft$, where $f$ is the uniform acceleration of the body as the result of the action of the force $P$, and therefore $f = \frac{v - u}{t}$ (p. 13). Substituting in the expression for $P$ above, we have

$$P \propto mf \quad . \quad . \quad . \quad . \quad (1)$$

Thus, if the mass of a body is constant, the acceleration $f$ is proportional to the force $P$ acting on it. See Experiment, p. 155.

**Units of force.** The unit of force in the foot-pound-second (f.p.s.) system is the **poundal**, which is defined as *that force which gives an acceleration of* 1 *ft. per sec.*$^2$ *to a mass of* 1 *lb.* The force acting on a mass of 1 lb. which gives it an acceleration of 8 ft. per sec.$^2$ is hence 8 poundals, and the force ($P$) acting on a mass ($m$) of 5 lb. which gives it an acceleration ($f$) of 8 ft. per sec.$^2$ is 40 poundals. In general,

$$P = mf \qquad . \qquad . \qquad . \qquad . \quad (2)$$

where $P$ is in *poundals* when $m$ is in *lbs.* and $f$ is in *ft. per sec.*$^2$ It should be noted that if $P$ is required in pounds weight (lb. wt.), the number of poundals must be divided by 32 (see p. 35).

The unit of force in the centimetre-gram-second (c.g.s.) system is known as a **dyne**, and is defined as *that force which gives an acceleration of* 1 *cm. per sec.*$^2$ *to a mass of* 1 *gm.* The equation $P = mf$ thus also applies to cases in which $P$ is measured in dynes, $m$ in grams, and $f$ in cm. per sec.$^2$ From (2), it follows that: (*a*) the force acting on a body of mass 5 lb. which gives it an acceleration of 20 ft. per sec.$^2$, is 100 poundals; (*b*) a force of 200 dynes acting on a mass of 8 gm. gives it an acceleration of $\dfrac{200}{8} = 25$ cm. per sec.$^2$

Suppose that a cricket ball of $5\frac{1}{2}$ oz. (i.e. $5\frac{1}{2}/16$ lb.), travelling with a velocity of 45 ft. per sec., is in contact with the bat for $\frac{1}{20}$ sec., and then travels in the *opposite* direction with a velocity of 25 ft. per sec. The change in the velocity in $\frac{1}{20}$ sec. is then (45 + 25), or 70 ft. per sec., since the velocities before and after hitting are opposite. Hence $f$, the acceleration due to the force exerted by the bat, is given by $f = 70/\frac{1}{20} = 1400$ ft. per sec.$^2$ Consequently the force, $P$, on the ball when it meets the bat is given by

$$P = mf = \frac{5\frac{1}{2}}{16} \times 1400 = 481\frac{1}{4} \text{ poundals.}$$

**Distinction between weight and mass.** The *weight* of a body is defined as the *force acting on it due to the attraction of the earth.* Now a body of mass $m$ lb. has a uniform acceleration of $g$ ft. per sec.$^2$ when it falls under gravity. Hence, since $P = mf$, the force acting on the body is given by $P = mg$ poundals. Thus

the weight of the body $= mg$ poundals   .   . (3)

Of the two quantities ($m$ and $g$) concerned, experiment shows that $g$, the acceleration due to gravity, varies over the surface of the earth. This is because the earth is not a perfect sphere, but is flattened at its poles, and the attraction of the earth at a point on its surface depends on the distance of the point from the centre of the earth. Thus $g$ has the value of 32·19 ft. per sec.[2] in London, and 32·09 ft. per sec.[2] at the Equator.

The *mass* of a body is difficult to define exactly. Newton stated that it was "the quantity of matter in a body". A given piece of chocolate contains the same "quantity of matter" in England as at the equator or north pole (or even in interstellar space, if we could put it there) and we consider that *the mass of a body is constant wherever it may be*. It follows from equation (3), however, that the weight of a body varies over the surface of the earth, since $g$ varies.

In 1905 the great scientist ALBERT EINSTEIN showed mathematically that the mass of a body is a measure of the total energy of its atoms. This was proved to be true from observations in radioactivity, in which the mass of an element diminished and an equivalent amount of energy was produced. Einstein's deduction is utilized in the subject of Atomic Energy (p. 633), and it has also resulted in a recasting of two important scientific laws. These were: (1) *The law of conservation of mass*, which states that the total mass of a given system of bodies remains constant, and (2) *The law of conservation of energy*, which states that the energy of a given system of bodies remains constant (p. 8). Since, according to Einstein, the mass of a body is a measure of the total energy of its atoms, these two laws can now be combined into *one* law, the law of conservation of energy.

**The balance and spring-balance.** Suppose that a body of unknown mass $m$ is placed on a scale-pan of a balance and counterpoised by a known mass $m_1$ on the other scale-pan. The lever of the balance is acted upon by forces, which are the weights $W$, $W_1$ of the respective masses; and since $W = mg$ and $W_1 = m_1g$, where $g$ is the acceleration due to gravity at the place on the earth where the balance is situated, it follows that $\dfrac{W}{W_1} = \dfrac{m}{m_1}$. If the arms of the balance are exactly equal, $W = W_1$ (p. 77); and hence $m = m_1$. It can thus be seen that, although a balance functions as a result of the attraction of gravity on the objects on its scale-pans, the balance actually compares the *masses* of the two bodies. If the same balance is taken to any other place on the earth and the same two masses are used, it will again be found that $m = m_1$.

The position is different, however, when a spring-balance is used. The extension of the spring is due to the force acting at the end of it, and is therefore a measure of the *weight* of the object attached to the hook of the balance. If the same object and spring-balance are taken to different parts of the earth, the weight registered on the balance will be slightly different, as $g$ varies with latitude.

**The relation between 1 lb. wt. and 1 poundal; and 1 gram wt. and 1 dyne.** A common unit of force is that exerted by gravity on a mass of 1 lb. It is known as a *pound-weight* (*lb. wt.*), and varies over the surface of the earth. This unit should not be confused with the *mass* of a pound, which is a constant quantity.

Since $P = mf$, the force due to gravity on a 1 lb. mass is given by $P = 1 \times g = g$ poundals = 1 lb. wt. Thus if $g = 32$ ft. per sec.$^2$ in England, then

$$32 \; poundals = 1 \; lb. \; wt. \qquad . \qquad . \qquad . \quad (4)$$

The distinction we have drawn between 1 lb. and 1 lb. wt. applies to 1 gm. (mass) and 1 gm. wt. (force). In the c.g.s. system $g$ is approximately 980 cm. per sec.$^2$; hence the force on 1 gm. due to gravity = $mf = 1 \times 980$ dynes. Thus

$$980 \; dynes = 1 \; gm. \; wt. \qquad . \qquad . \qquad . \quad (5)$$

As they are frequently required, the relations (4) and (5) should be memorized.

**Action and reaction are equal and opposite.** When a box rests on the ground, it is in equilibrium under the action of two forces acting on it: (i) The weight of the box; (ii) the force exerted by the ground on the box, which is known as the *reaction* of the ground on the box (Fig. 20 (*a*) ). Since the box is in equilibrium, the reaction of the ground must be equal and opposite to its weight, and the latter may be termed the *action* on the ground. In the case of a body resting on a smooth surface, experiment shows that the reaction is always *perpendicular* to the surface.

The force of reaction is always called into play when two bodies are in contact. Fig. 20 (*b*) illustrates an object $N$, of 100 lb. wt., held on a smooth plane inclined at 30° to the horizontal. The reaction $R$ of the plane is perpendicular to the plane; the weight acts vertically downwards. The component of the weight perpendicular to the plane is 100 cos 30° lb. wt. (p. 23), and this is the "action" of the weight on the surface, which is equal and opposite to $R$. Consequently $R = 100$ cos 30° = 87 lb. wt. (approx.).

The law of action and reaction applies when the buffers at a terminal railway station bring a train to rest: the buffers provide reaction in the opposite direction to the force exerted on it by the train. For starting a race, many athletes use "holes" in the

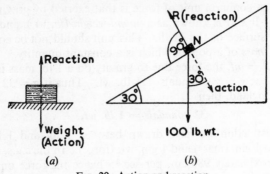

FIG. 20. Action and reaction

track, and obtain a force of reaction at their feet which assists them in a quick take-off.

The jet aeroplane, which has revolutionized flying, utilizes the principle of action and reaction. In brief, the gas is exploded so that a backward jet of it is produced behind the power units, and the force of the explosion urges the plane in a forward direction, according to the law of action and reaction. Rockets have been fired backward from aeroplanes and speedway motor-cycles to obtain additional forward speed, and ships operating by jet propulsion are promised for the future.

The following examples on force and acceleration or retardation should be carefully studied.

1. *A motor-car, weighing ¾ ton, is travelling at 30 m.p.h. Calculate the retarding force acting on the car when the brakes bring it to rest (i) in ½ min., (ii) in 60 yds., and express your answer in lb. wt.*

(i) Since the car is brought to rest, its final velocity, $v$, is zero. The initial velocity, $u = 30$ m.p.h. $= 44$ ft. per sec., and the time, $t = 30$ sec. Applying $v = u + ft$, we have

$$0 = 44 + 30f$$

i.e.
$$f = -\frac{44}{30} \text{ ft. per sec.}^2$$

Now the mass, $m$, of the car $= \frac{3}{4} \times 2240$ lb. $= 1680$ lb.

$$\therefore \quad \text{force, } P = mf = 1680 \times \frac{44}{30} = 2464 \text{ poundals}$$

$$\therefore \quad \text{force in lb. wt.} = \frac{2464}{32} = 77 \text{ lb. wt., since 32 poundals} = 1 \text{ lb. wt.}$$

(ii) The most suitable equation to find the retardation, $f$, in this case is $v^2 = u^2 + 2fs$. We have $v = 0$, $u = 44$ ft. per sec., $s = 180$ ft. Substituting,

$$0 = 44^2 + 2 \cdot f \cdot 180$$

i.e.
$$f = -\frac{44^2}{360}$$

$$\therefore \quad \text{force, } P = mf = 1680 \times \frac{44^2}{360} \text{ poundals}$$

$$= \frac{1680 \times 44^2}{32 \times 360} \text{ lb. wt.}$$

$$= 282\tfrac{1}{3} \text{ lb. wt.}$$

2. *A boy is free-wheeling on a bicycle down a road inclined at 20° to the horizontal. The total weight of the boy and the bicycle is 10 stone, and the frictional force is 1 lb. wt. per stone. Find the acceleration of the boy.*

The weight of the boy and bicycle is 140 lb. wt., and acts vertically downwards at $A$ on the road (Fig. 21). The "effective part," or "component," of this force down the incline is 140 cos 70° lb. wt. (see p. 23), and the frictional force acts in the opposite direction. Since the latter is 1 lb. wt. $\times$ 10, or 10 lb. wt., the resultant force $P$ down the road

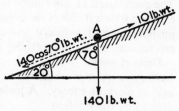

FIG. 21. Calculation

$$= (140 \cos 70° - 10) \text{ lb. wt.}$$

$$= (140 \cos 70° - 10) \times 32 \text{ poundals} = 1212 \text{ poundals}$$

But
$$P = mf$$

where $m =$ mass of boy and bicycle $= 140$ lb., and $P$ is in *poundals*.

$$\therefore \quad 1212 = 140f$$

$$\therefore \quad f = \frac{1212}{140} = 8 \cdot 7 \text{ ft. per sec.}^2$$

## WORK, ENERGY, AND POWER

**How work is measured in Mechanics.** We have already mentioned the different forms which energy can take (p. 7); in this section we shall be concerned only with *mechanical energy* and how it is measured.

When an engine moves a train along the railway line it is said to be doing work; a horse pulling a barge along a canal is also doing work. A boy climbing the stairs, and a mountaineer ascending a height, are doing work in moving themselves. If an engine exerts a constant force of 500 lb. wt. and moves a truck 80 ft., the work it does is greater than if it moved the truck 10 ft. On the other hand, if a different engine exerts a constant force of 2000 lb. wt. in moving a different truck 80 ft., it does a greater amount of work than the first engine in moving its truck through the same distance. It can now be seen that the work done by a force increases with the magnitude ($P$) of the force and with the distance ($s$) concerned, and the amount of work done is defined as $P \times s$ units *if the distance moved is in the direction of the force*. With this last condition fulfilled, then, it can be stated that

$$\textbf{work} = \textbf{force} \times \textbf{distance} \qquad . \qquad . \qquad . \quad (6)$$

The calculation of work done in Mechanics takes no note of any exertion unless it moves the object on which it acts. Thus a porter pushing against a heavy trolley loaded with luggage does no mechanical work on the trolley if he cannot move it.

**Units of work.** When the force is a dyne and it moves an object 1 cm., one **erg** of work is said to be done. The erg is the scientific unit of work or energy. A **joule** is a larger unit of energy, and is defined by the relation

$$1 \text{ joule} = 10 \text{ million } (10^7) \text{ ergs.}$$

When the force is a poundal and it moves an object 1 foot, one **foot poundal** of work is said to be done. A larger unit of work or energy used in this country is the **foot pound weight** (ft. lb. wt., or, more briefly but erroneously, ft.-lb.); this is the work done when a force of 1 lb. wt. moves an object a distance of 1 ft. Since 1 lb. wt. = 32 poundals (p. 35), it follows that a number of ft.-poundals of work must be divided by 32 to bring it to ft. lb. wt.

**Some calculations on work.** Consider a barge moved 150 ft. upstream in the direction $BC$ by a force $P$ of 400 lb. wt. acting along a rope inclined at an angle, $\theta$, of 30° to $BC$ (Fig. 22). $P$ has a component $P \cos \theta$ along $BC$ and a component $P \cos (90° - \theta)$, or $P \sin \theta$, in a direction perpendicular to $BC$. We know from experience that a force cannot have any effect in a direction

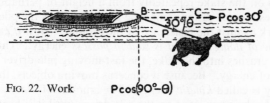

FIG. 22. Work

perpendicular to itself; consequently the latter force does no work as the barge moves from $B$ upstream. From equation (6), the component $P \cos \theta$ does an amount of work = $P \cos \theta \times 150 = 400 \cos 30° \times 150$ ft. lb. wt. Hence the total work done by the force $P = 400 \cos 30° \times 150 = 51,960$ ft. lb. wt. It can be seen that, in general, the angle between the direction of a force and the direction of movement considerably affects the magnitude of the work done; the work done is not calculated by "force × distance" unless the distance moved is in the direction of the force.

Consider a rider and a bicycle of total weight 75 kgm. wt., moving down a

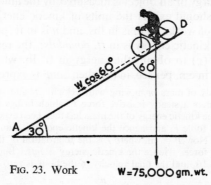

FIG. 23. Work $W = 75,000$ gm. wt.

road $OA$ inclined at an angle of 30° to the horizontal (Fig. 23). The component of the weight down the road = $75,000 \cos 60° = 37,500$ gm. wt. = $37,500 \times 980$ dynes. If a distance of 40 metres is travelled, the work done by the weight (a *force*)

$$= (37,500 \times 980 \times 4000) \text{ ergs}$$

since 40 m. = 4000 cm. The work in joules = $\dfrac{37,500 \times 980 \times 4000}{10^7} = 14,700$.

**Kinetic energy.** After a site is chosen for a new pier or building, the engineers employ a *pile-driver* for driving timber or concrete stakes into the ground, for the foundations (Fig. 24). The pile-driver is a massive block, sometimes 2 tons in weight, which descends on the stake repeatedly from a height of perhaps 6 feet, and drives it some distance into the ground. Thus the pile-driver does work against the resistance of the ground. Now *if an object is capable of doing work, it is said to possess* **energy**. Thus, just before it crashes into the stake, the fast-moving pile-driver has an amount of energy. Because it concerns moving objects, this kind of energy is called *kinetic* energy. The amount of kinetic energy is equal to the work done by the pile-driver until it is brought to rest, when all its energy is spent.

Any moving object, such as the rotating wheel of an engine or a hammer just about to drive a nail into a box, is said to possess kinetic energy. If $m$ is the mass of a moving object and $u$ is its velocity, it is proved below that its

$$\textbf{kinetic energy} = \tfrac{1}{2}\boldsymbol{mu^2} \qquad . \qquad . \qquad . \quad (7)$$

Since the energy of an object is measured by the amount of work it can do, we should expect the units of kinetic energy to be the same as those of work. If $m$ is in lbs. and $u$ is in ft. per sec. in (7) the calculated kinetic energy is in *ft. poundals*: the result must be divided by 32 ($g$) to obtain the energy in ft. lb. wt. If $m$ is in grams and $u$ is in cm. per sec. in (7), then $\tfrac{1}{2}mu^2$ is expressed in *ergs*.

Consider a body of mass $m$ moving with velocity $u$, and suppose that it suddenly encounters a steady resistive force $P$ which brings it to rest in a distance $s$. All the kinetic energy of the mass has then been expended in doing work against the force $P$, and hence the kinetic energy is numerically equal to $P \times s$ units. Now $P = mf$, where $f$ is the retardation of the body when it encounters the force. Hence the kinetic energy $= mfs$. But $0 = u^2 - 2fs$ (since $v^2 = u^2 + 2fs$, and the final velocity of the body is zero). Thus

$$fs = \frac{u^2}{2}.$$

∴    Kinetic energy $= mfs = \tfrac{1}{2}mu^2$, as stated in (7).

**Potential energy.** When a watch is wound up, some work is done, and the spring inside it is coiled more tightly. In this position the spring is said to have an amount of *potential energy*, to distinguish it from the energy of a moving object. As the

FIG. 24. A PILE DRIVER POISED FOR ACTION

It is raised by a steam engine (*in background*) and falls under gravity

spring is slowly released, the potential energy is converted into kinetic energy, and the wheels in the watch are turned. A swimmer poised on the edge of a diving-board has an amount of potential energy, which is converted into kinetic enegy when he dives into the water; while a person in a swing has potential energy at the top of the motion, where he is momentarily stationary. As in kinetic energy, the potential energy of an object is measured by the work it can do; and the potential energy of a body may be defined as *the energy it possesses by virtue of its position, level, or arrangement.*

Consider a body $X$ of mass $m$ held stationary at a height of $h$ above the ground. If the body is released, a constant force of $mg$ units (its weight) acts on it as it falls; and, since work done = force × distance, the work done by the weight of the body = $mg \times h$ when it reaches the ground. Thus, relative to the ground position,

$$\text{potential energy at height } h = mgh \qquad . \qquad . \qquad (8)$$

When $m$ is in grams, $g = 980$ numerically, and $h$ is in cm., then $mgh$ is expressed in ergs. When $m$ is in lbs., $g = 32$ numerically, and $h$ is in ft., then $mgh$ is expressed in ft. poundals.

Divided by $g$ to convert ft. poundals to ft. lb. wt., the potential energy may also be expressed as $mh$ ft. lb. wt.

**Changes of energy.** The potential energy of a body of mass $m$ held at a point $A$ on an inclined plane is $mgh$ units, where $h$ is

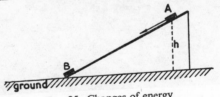

FIG. 25. Changes of energy

the height of $A$ above the ground (Fig. 25). When the body is released and slides down the plane, part of the potential energy is converted into kinetic energy. At the bottom $B$, the kinetic energy would be equal to $mgh$ units, if no energy were lost in moving from $A$ to $B$. In practice the kinetic energy is less than $mgh$ units, since some energy is converted to heat by the frictional forces between the moving body and the plane. As the body moves from *B* along the ground, all its kinetic energy is gradually converted into heat energy by the frictional forces, and it comes to rest. Thus, as pointed out on p. 8, energy can change from one form to another, but the total amount of energy remains constant.

Consider a pile-driver of mass 2000 lb., poised 4 ft. above a timber stake. The pile driver then has a potential energy of 2000 × 4, or 8000 ft. lb. wt. more than when it strikes the stake; and the amount of *kinetic* energy in the latter case is hence 8000 ft. lb. wt., assuming no loss of mechanical energy. Suppose *R* is the net resistance in lb. wt. to the stake when it enters the ground. If the stake penetrates 4 in., the work done against *R* = force × distance = *R* × ⅓ ft. lb. wt., as 4 in. is ⅓ ft. But the energy of the pile-driver has all been spent in doing this work. Hence

$$R \times \tfrac{1}{3} = 8000$$
$$\therefore \quad R = 24{,}000 \text{ lb. wt.}$$

**Power.** An engine which lifts a weight of 200 lb. wt. a distance of 100 ft. in ½ minute is more powerful than one which does the same work in 2 minutes. In general, the *power* of an engine is defined as *the rate at which it does work or expends energy*. Thus

$$\text{power} = \text{work done per sec.}$$
$$= \frac{\text{work done (or energy expended)}}{\text{time taken}}.$$

The unit of power in the f.p.s. system is "ft. lb. wt. per sec." A practical unit is the **horse-power (h.p.)**, which is the rate of working at 550 ft. lb. wt. per sec. In the case of the more powerful of the two engines mentioned above,

$$\text{the power} = \frac{200 \times 100}{30} \text{ ft. lb. wt. per sec.}$$
$$= \frac{200 \times 100}{550 \times 30} \text{ h.p.} = 1{\cdot}2 \text{ h.p. (approx.).}$$

In the c.g.s. system the unit of power is the "erg per second." A practical unit is the **watt**, which is the rate of working at 1 joule ($10^7$ ergs) per sec. An electric lamp of 120 watts consumes energy at the rate of 120 joules per sec. and converts a percentage of it to light energy: a lamp of 60 watts gives out less light energy per sec. than a 120-watt lamp.

As an illustration of the calculation of power, consider a boy weighing 9 stone who walks from the bottom to the top of a building 80 ft. high in

3 min. The work done per second by the boy in moving his weight a vertical distance of 80 ft.

$$= \frac{9 \times 14 \times 80}{180} \text{ ft. lb. wt. per sec.}$$

Thus the average power at which he works is given by

$$\frac{9 \times 14 \times 80}{180 \times 550} \text{ h.p.} = 0.10 \text{ h.p. (to two significant figures).}$$

**Water power.** In many countries the potential energy of water in high rivers and lakes is converted into electric power by *hydroelectric plants*. For example, the water at the top of Niagara Falls

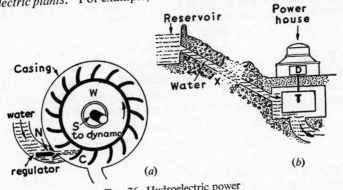

FIG. 26. Hydroelectric power

drops 160 ft., and the energy released is used to drive huge water wheels or turbines. (See Figs. 26 and 27.)

One form of wheel is the *Pelton wheel*, $W$, which has cup-shaped metal discs, $C$, round its circumference (Fig. 26 (*a*).) The wheel is enclosed in a casing, into which the fast-moving water flows, and is driven by playing a jet of water on to the metal cups, $C$, as shown. Two wheels are often arranged on the same shaft, $S$, and drive a dynamo, which, as explained on p. 603, produces electrical energy. A hydroelectric installation is shown diagrammatically in Fig. 26 (*b*). Water flows from a reservoir through a long pipe $X$, and turns a wheel $T$, which is coupled to the dynamo $D$ in the power house.

Canada and America obtain more than a million horse-power from hydroelectric plants. Similar plants in Scotland depend upon the hilly country and lochs: Galloway, for instance, has a large installation, and the British Aluminium Company is supplied with power from a hydroelectric plant. In England and Wales, however, we utilize steam turbines, driven by the energy obtained from burning coal. Many hundreds of tons of coal must be burned every hour to produce all the energy obtained in the same time from hydroelectric plants in Canada and America. It is estimated that about 95 per cent of the energy for heating and lighting purposes in this country comes from coal.

FIG. 27. WATER TURBINE
80-ton Rotor of Generator erected in India

**The sun's energy.** We all welcome the sun after a cold winter. We are warmed by the sun's energy which reaches the earth. The energy is also absorbed by plants, and much of the energy in the food we eat comes initially from the sun. Coal has come from decayed vegetable matter which, with the passage of time, has sunk into the earth; so the energy we derive from coal is stored energy, supplied by the sun millions of years ago. The sun partly evaporates lakes and seas and, when the vapour condenses to droplets, gravity causes them to fall as rain on mountains and hills, forming water supplies for hydroelectric power plants. Thus the electrical energy derived from such plants also comes initially from the sun.

**Conservation of momentum.** There is an important law in Mechanics which concerns the *momentum* (the product of mass and velocity, p. 32) of objects exerting a force on each other.

Suppose that a metal ball $A$ of mass 3 lb., moving with a velocity of 8 ft. per sec., overtakes and collides with a 2 lb. ball, $B$, travelling in the same direction with a velocity of 6 ft. per sec. (Fig. 28). The velocity of $B$ then increases, while $A$ is slowed down. Now Newton's second law states that the *change of momentum* per second of an object is proportional to the force acting on it (p. 31). The force exerted by $A$ on $B$ when the balls collide is equal and opposite to the force exerted by $B$ on $A$, since action and reaction are equal and opposite. Moreover, the forces last for the same

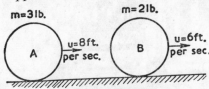

FIG. 28.    Conservation of momentum

time. Since the force and the time are equal, it can be seen that, whatever the change of momentum of $B$ when the collision occurs, an equal and opposite change in momentum occurs in $A$. Consequently no change takes place in the *total* momentum of $A$ and $B$ when the collision occurs, i.e. the sum of the momenta of $A$ and $B$ before collision is equal to the sum of the momenta of $A$ and $B$ after collision. This result, known as the *principle of the conservation of momentum*, may be stated as follows:

If a number of objects collide or interact with each other, the sum of their momenta is constant in the direction along which no external forces act. ("Momentum" is a vector quantity, i.e. it has magnitude and direction (p. 19).)

Since the momentum of an object is the product of its mass and velocity, the sum of the momenta of $A$ and $B$ before collision is $(3 \times 8 + 2 \times 6)$ or 36 units (Fig. 28). After collision, the sum of their momenta is $(3u + 2v)$ units, where $u$, $v$, are the new velocities of $A$ and $B$ respectively. Thus, from the principle of the conservation of momentum, $3u + 2v = 36$.

Suppose that a bullet of mass $\frac{1}{2}$ oz. is shot out of a rifle with a velocity of 300 ft. per sec. The momentum of the bullet is then $\frac{1}{32} \times 300$ units (as $\frac{1}{2}$ oz. $= \frac{1}{32}$ lb.) and the explosion imparts an equal momentum to the rifle, which jerks backwards. Suppose that the rifle has a mass of 6 lb. and moves back with a velocity $v$. Then $6v = \frac{1}{32} \times 300$, i.e. $v = 1 \cdot 6$ ft. per sec. The total momentum of the bullet and gun before the explosion is zero. The total momentum is still zero after the explosion, as the momentum of the rifle is equal and *opposite* to that of the bullet, according to the principle of the conservation of momentum.

If the bullet reaches a target of 3 lb. with its velocity undiminished and is brought to rest in the target, the momentum of the target and bullet is $3\frac{1}{32}v$, where $v$ is their common velocity as the target moves. The original momentum of the bullet is $\frac{1}{32} \times 300$ units, and that of the target is zero as its original velocity is zero; hence the total momentum before the bullet hits the target is $\frac{1}{32} \times 300$ units. By the principle of the conservation of momentum it follows that

$$3\tfrac{1}{32}v = \tfrac{1}{32} \times 300$$

i.e. $$v = 3 \cdot 1 \text{ ft. per sec.}$$

---

## Summary

The following table is a summary of some of the quantities encountered in Dynamics, and their units:

| Quantity | Formula | Units |
| --- | --- | --- |
| Velocity ($v$) | — | ft. per sec.; cm. per sec. |
| Acceleration ($f$) | $\dfrac{v-u}{t}$ | ft. per sec.²; cm. per sec.² |
| Acceleration due to gravity ($g$) | — | 32 ft. per sec.²; 980 cm. per sec.² |
| Force ($P$) | $mf$ | poundal, lb. wt.; dyne, gm. wt. |
| Weight | $mg$ | (32 poundals = 1 lb. wt.; 980 dynes = 1 gm. wt.) |
| Mass ($m$) | — | lb.; gm. |
| Work | $P \times s$ | ft. poundal, ft. lb. wt.; erg, or joule |
| Kinetic energy | $\frac{1}{2}mu^2$ | (Same units as "Work") |
| Potential energy | $mgh$ | (Same units as "Work") |
| Power | Work per sec. | ft. lb. wt. per sec.; erg per sec. watt (1 h.p. = 550 ft. lb. wt. per sec.) |
| Momentum | $mv$ | poundal sec.; dyne sec. |

## WORKED EXAMPLES

1. *Define potential energy. What is the potential energy of a 28 lb. weight held 6 ft. above the ground?* (*N*)

Potential energy = weight × height above ground
= 28 × 6 = 168 ft. lb. wt.

*Note.* The potential energy in foot poundals = $mgh$ = 28 × 32 × 6.

2. *A 4000 lb. bomb is dropped from a height of 5000 ft. Calculate its kinetic energy on reaching the ground if the average effect of air resistance is equivalent to 200 lb. wt.* (*N.*)

The potential energy of the bomb at a height of 5000 ft. = 4000 × 5000
= 20,000,000 ft. lb. wt.

The work expended against the air resistance = force × distance
= 200 × 5000 = 1,000,000 ft. lb. wt.

Hence kinetic energy at ground
= 20,000,000 − 1,000,000 = 19,000,000 ft. lb. wt.

3. *Explain the terms energy and power. Discuss the changes of energy of a carriage travelling along a switchback railway. Why is the carriage unable to rise to the same height from which it started? A motor-car travelling along a level road at 30 m.p.h. develops 4 h.p. What is the total frictional force overcome?* (1 h.p. = 550 ft.-lb. per sec.) (*L.*)

30 m.p.h. = 44 ft. per sec.; 4 h.p. = 2200 ft.-lb. per sec.

Since power = work done per sec., and work done = force × distance
power = force × distance per sec.
$$\therefore \quad 2200 = P \times 44,$$
where $P$ is the force in lb. wt. due to the engine.
$$\therefore \quad P = \frac{2200}{44} = 50 \text{ lb. wt.}$$

Since the car is moving with a constant velocity, its acceleration is zero. Thus the net force on the car is zero. Hence the force $P$ due to the engine = the *opposite* force due to the frictional force.
$$\therefore \quad \text{frictional force} = 50 \text{ lb. wt.}$$

4. *Describe a method of measuring the acceleration due to gravity* (*g*). *A bullet weighing* $\frac{1}{2}$ *oz. is fired vertically upwards from the ground and reaches a height of 900 ft. Neglecting air resistance, calculate* (a) *the initial velocity of the bullet,* (b) *the time that elapses before the bullet reaches the ground again,* (c) *the energy (potential and kinetic) of the bullet at ground level, at the highest point reached, and half-way up.* (*C.W.B.*)

(*a*) Let $u$ = the initial velocity in ft. per sec. At the highest point, the final velocity $v = 0$, and $s = 900$. Substituting in
$$v^2 = u^2 + 2fs,$$

$$\therefore \quad 0 = u^2 - 2 \cdot 32 \cdot 900$$

$$\therefore \quad u = \sqrt{2 \cdot 32 \cdot 900} = 240 \text{ ft. per sec.}$$

(b) Let $t$ = the time to reach the highest point. Then, from $v = u + ft$,

$$0 = 240 - 32t$$

$$\therefore \quad t = \frac{240}{32} = 7\tfrac{1}{2} \text{ sec.}$$

$$\therefore \quad \text{time to reach ground again} = 2 \times 7\tfrac{1}{2} = 15 \text{ sec.}$$

(c) At ground level, the energy is only kinetic and the velocity, $u$, of the bullet is 240 ft. per sec.

$$\therefore \quad \text{energy} = \tfrac{1}{2}mu^2 = \tfrac{1}{2} \times \tfrac{1}{32} \times 240^2 \text{ ft. poundals, as } m = \tfrac{1}{32} \text{ lb.}$$

$$\therefore \quad \text{energy} = 900 \text{ ft. poundals}$$

At the highest point, the energy is only potential energy (P.E.), and given by

$$\text{P.E.} = mgh = \tfrac{1}{32} \times 32 \times 900 \text{ ft. poundals}$$
$$= 900 \text{ ft. poundals}$$

Half-way up, the energy is partly potential and partly kinetic. Since the air resistance is neglected, the total energy is conserved and given by 900 ft. poundals; consequently the potential energy = 450 ft. poundals and the kinetic energy = 450 ft. poundals.

## EXERCISES ON CHAPTER III

**1.** A car of mass 1200 lb. is moving with an acceleration of 20 ft. per sec.$^2$ Find the force acting on it (*i*) in poundals, (*ii*) in lb. wt., (*iii*) in tons wt.

**2.** A force of 12,000 dynes acts on an object of mass 200 gm. What is the acceleration of the object? How far does it travel from rest in $\tfrac{1}{4}$ min.?

**3.** A boy pulls a sledge 20 ft. with a force of 15 lb. wt. Calculate the work done by the boy.

**4.** A man weighing 12 stone is running with a speed of 20 ft. per sec. What is the kinetic energy (*i*) in ft. poundals, (*ii*) in ft. lb. wt.?

**5.** Niagara Falls are 150 ft. high. Calculate the potential energy of 10 lb. of water at the top, relative to the bottom. What is the kinetic energy of this water just before it reaches the bottom, and what happens to the energy after the water reaches the bottom?

**6.** What is the work done when a force of 200 dynes moves an object 300 cm. in its direction? Calculate the work done if the force is 10 gm. wt.

**7.** A boy is free-wheeling on a bicycle down a plane inclined at 20° to the horizontal. The total mass of boy and bicycle is 120 lb. What is the force down the plane? If the frictional force at the ground is 2 lb. wt., calculate (i) the net force down the plane, (ii) the acceleration of the bicycle.

**8.** A car developing 10 h.p. is moving with a uniform velocity of 44 ft. per sec. What energy per second is expended by the engine, and what is the frictional force overcome?

**9.** A bullet of mass 15 gm. has a speed of 400 metres per sec. What is its kinetic energy? If the bullet strikes a thick target and is brought to rest in 2 cm., calculate the average net force acting on the bullet. What happens to the kinetic energy originally in the bullet?

**10.** A rope is attached to a barge so that it makes an angle of 30° with the stream. If the force in the rope is constant at 200 lb. wt., find the work done when the barge moves 100 ft. upstream. What is the component of the force at right angles to the stream?

**11.** Describe an experiment to determine the value of the acceleration due to gravity. A ball is thrown vertically upwards from the ground with an initial velocity of 64 ft. per sec. Determine (a) the height to which it rises, (b) the potential and kinetic energies of the ball, if its mass is 0·5 lbs., (i) when at its maximum height, (ii) when half-way up, showing clearly how you obtain the values and stating the units in which they are expressed. (L.)

**12.** Calculate the acceleration of a falling mass of 20 lb. when there is a steady force of 15 lb. wt. opposing the force of gravity ($g = 32$ ft. per sec.²). (N.)

**13.** The muzzle velocity of a rifle bullet is 2400 ft. per sec. and its mass is half an ounce. Calculate its kinetic energy. (N.)

**14.** Explain the terms: 1 lb. wt., 1 dyne, 1 ft. lb., 1 horse-power. What force acting along a smooth plane inclined at 45° to the horizontal will keep a mass of 10 lb. on the plane in equilibrium? A cyclist rides at 12 m.p.h. along a level road. If he expends energy at the rate of 0·1 horse-power, what is the average value of the frictional force? (L.)

**15.** "To every action there is an equal and opposite reaction." What is meant by this statement? Use it to explain the changes in the horizontal motion of a dinghy when a man in the bows walks to the stern, stands motionless, and then dives into the water.

A dinghy originally at rest, is pulled for 3 sec. with a steady force of 25 lb. wt. Calculate its velocity at the end of this time and the work done in pulling it, assuming the water resistance to be negligible. (Mass of dinghy = 600 lb., $g = 32$ ft. per sec.$^2$) (*N.*)

**16.** Explain clearly the meaning of *potential energy* and *kinetic energy*. Discuss the energy changes which occur when a pendulum-bob, hung from a fixed point by a thread, is allowed to vibrate. A rubber ball of mass 1 lb. is allowed to fall vertically from a height of 16 ft. on to a horizontal table, and it rebounds to a height of 12 ft. Determine (*a*) the velocity of the ball on hitting the table, (*b*) its velocity on leaving the table, (*c*) the potential and kinetic energy of the ball (*i*) on hitting the table, (*ii*) at the maximum height of rebound. (*C.W.B.*)

**17.** Describe how you would determine the average power developed by a boy running up a flight of stairs. The engine of a light aeroplane develops a power of 176,000 ft. lb. wt. per sec. when the machine is cruising at 120 m.p.h. Assuming that the resistance to motion is proportional to the square of the speed, find the power developed when it is travelling at 150 m.p.h. under the same conditions. (*N.*)

**18.** A *force*, *F*, acts on a *mass*, *M*, and gives it an *acceleration*, *A*. Define the three terms in italics, and state the relationship between *F*, *M*, and *A*.

An object weighing 1000 lb. is released from an aircraft flying at 10,000 ft. and falls freely. Neglecting resistance due to the air, calculate (*a*) the gain in kinetic energy, (*b*) its vertical velocity, when it reaches the ground. What is meant by *terminal* velocity ? (*C.*)

**19.** State the *principle of the conservation of energy*. Give *briefly* three examples of the conversion of one form of energy into another. A motor car weighing 1 ton, moving on a horizontal road at 30 m.p.h., is brought to rest by uniform application of its brakes in 50 yards. Calculate (*a*) the braking force in lb. wt., (*b*) the deceleration. (*C.*)

**20.** Describe how you would determine experimentally a value for the acceleration due to gravity. A body of mass 3 lb., initially at rest, falls from a height of 4 ft. on to a hard surface and rebounds to a height of 1 ft. Give an account of the energy changes which occur and calculate (*a*) the velocity of the body just before impact, (*b*) the loss of mechanical energy due to the impact. (Assume *g* to be 32 ft. per sec.$^2$) (*N.*)

**21.** Explain the terms: *force*, *work*, *power*. In what units are these quantities measured, both on the C.G.S. and British systems ? The mass of a train is 300 tons and the total resistance to its motion is 20 lb. wt. per ton. Calculate (*a*) the h.p. of the engine which can maintain a uniform speed, on the level, of 30 miles per hour, (*b*) the distance travelled on the level before coming to rest after shutting off steam and without applying the brakes. (1 h.p. = 550 ft. lb. per sec.) (*O. & C.*)

22. Distinguish between the *mass* and the *weight* of a body. How would you show by experiment that, at a given place, mass and weight are strictly proportional ? A lift has a mass of 2000 lb. Find the tension in the cable supporting the lift when (*a*) the lift is descending at a uniform speed, (*b*) the lift is descending with an acceleration of 3 ft. per sec.$^2$, (*c*) the lift is ascending with a retardation of 3 ft. per sec.$^2$ State in what unit the tension is calculated. (*O. & C.*)

23. Define *acceleration* and *force*. How are they related in any particular instance ?

A motor car of mass 15 cwt. has its speed reduced from 80 m.p.h. to 20 m.p.h. in 11 sec., by the application of its brakes. Determine (*a*) the retardation (assumed constant), (*b*) the force which produces it. Determine also, from the instant the brakes are applied, (*c*) the time taken to come to rest and (*d*) the distance the car travels before coming to rest. (*L.*)

24. Define *dyne* and *erg*, and explain the relationship between the dyne and the gram-weight.

A mass of 6 kgm. is pulled by a force of 30 gm.-wt. along a frictionless, horizontal plane. Calculate (*a*) the acceleration, (*b*) the velocity 3 sec. after starting from rest, (*c*) the distance gone and the work done in those 3 sec. (*L.*)

25. Distinguish between the *mass* and the *weight* of a body. Explain why the variation in the weight of a body at different places on the earth's surface can be detected by means of a very sensitive spring balance but not by a beam balance.

A metal block, mass 20 lb., comes to rest in 2·5 sec., after sliding 25 ft. in a straight line across a horizontal floor. Assuming that the block is uniformly retarded, find the force in lb. wt. opposing its motion and the work done against this force in the first 2 seconds of its motion.

Trace the changes of energy that occur during the motion of the block. (*L.*)

# CHAPTER IV

## STATICS. PARALLELOGRAM AND TRIANGLE OF FORCES

STATICS concerns the forces that keep an object in equilibrium. As these forces affect the stability of houses and bridges, civil and mechanical engineers must have a thorough knowledge of the principles of Statics. When designing a bridge, for example, the engineer applies his knowledge to estimate, among other things, the relation between the strength of the supports, the forces in the girders, and the weight of the bridge.

FIG. 29. MODEL OF NEW SEVERN BRIDGE

High winds produce oscillations in a bridge, and these may cause the bridge to break. When a new suspension bridge over the Severn was proposed, a model of it was constructed at the

National Physical Laboratory; it has been tested in the wind-tunnel for stability at various wind-speeds, and the design has been altered according to the results obtained. (Fig. 29.)

Fig. 30 (*a*) illustrates a ladder *AB*, whose weight acts at a point *G*, resting against a smooth wall. At *B*, on the ground, a force of reaction *S* acts vertically upwards, and a frictional force

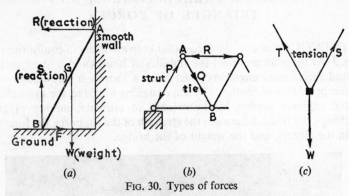

FIG. 30. Types of forces

*F*, which prevents the ladder from slipping, acts along the ground as shown. The reaction *R* of the smooth wall on the ladder is normal to this surface (p. 35). Fig. 30 (*b*) illustrates a section of part of a bridge. A joint at *A* is acted upon by forces *P*, *Q*, *R*, in the girders meeting there. The forces *P*, *R*, are called **thrusts**. The force *Q* is known as a **tension** or **tie** in the girder *AB*; the same name is used for the force in a rope or a string. Fig. 30 (*c*) illustrates the two tensions *T*, *S*, in ropes supporting a heavy weight *W*.

To depict a force on paper, two factors must be taken into account: (1) the magnitude of the force; (2) its direction. A straight line, representing by its length the magnitude of the force, is drawn in the direction in which the force acts, and an arrow is placed on the line to show the actual direction. (see p. 20.)

## THE PARALLELOGRAM AND TRIANGLE OF FORCES

**The parallelogram of forces.** In many cases in Statics the sum, or **resultant**, of two forces is required. The force *P* in the girder at

*A*, for example (Fig. 30 (*b*) ), should just balance the sum of the other two forces *Q, R* acting at *A*.

If two boys are pulling a sledge by ropes inclined to each other at 65°, and the forces *P, Q*, exerted on the sledge are 15 and 20 lb. wt. respectively, these forces have a resultant, *R*, on the sledge, which urges it forward. Experiment (p. 55) shows that the magnitude of *R* and its direction can be found by drawing a certain diagram, known as the "parallelogram of forces." A line *OB* is first drawn in the direction of the force, *Q*, of 20 lb. wt., to represent it in magnitude: for example, if 1 in. represents 5 lb. wt., *OB* is drawn 4 in. long to represent *Q*. A line *OA*, 3 in. long, is then drawn at

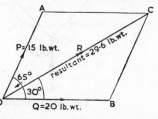

FIG. 31. Parallelogram of forces

an angle of 65° to represent the magnitude of the force *P* of 15 lb. wt. (Fig. 31).

The parallelogram *OACB*, having *OA, OB* as its two sides, is now completed by drawing the lines *BC* and *AC* through *B, A*, respectively, and the diagonal *OC* through *O* is drawn. *OC* *represents the magnitude and direction of the resultant, R, of the* *forces P and Q.* Suppose *OC* = 5·92 in. and angle *BOC* = 30° (by measurement). Then *R* = 5·92 × 5 = 29·6 lb. wt., acting in a direction of 30° to the force *Q*.

**Verification of parallelogram of forces.** The apparatus illustrated in Fig. 32 (*a*) can be used to verify the parallelogram of forces. Two spring balances are suspended from fixed points *M, N*. Two light strings, *OL, OK*, are knotted at *O* and each attached to one end of a spring-balance, as shown. A weight, *W*, of 100 gm. wt. is then suspended from *O* by a light string, so that there are three forces in equilibrium at *O*: namely, *W* and the tensions, *T, P*, in the strings. The magnitudes of *T* and *P* are indicated on the respective balances; suppose they are 58 and 74 gm. wt. respectively. Then, since the resultant of the forces *T* and *P* must balance the force *W* for equilibrium, 100 gm. wt. is the resultant of the forces 58 and 74 gm. wt. acting in the directions *OL, OK* respectively.

To verify this from the parallelogram of forces for the 58 and 74 gm wt., a sheet of paper is placed behind the strings and their directions, *OL, OK*, are marked carefully on the paper with a pencil. The paper is then taken away, and the two forces 58 and 74 gm. wt. are drawn accurately to scale along the directions obtained. Suppose they are represented by *AB*, *AC* respectively (Fig. 32 (*b*).) The parallelogram *ABDC* is then completed, and the diagonal *AD* is measured and converted to gm. wt. *It will be found that, to*

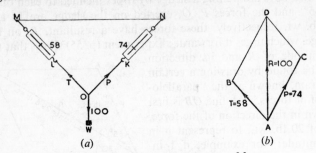

FIG. 32. Verification of parallelogram of forces

*a close examination, AD represents* 100 *gm. wt.* This confirms that the resultant of two inclined forces is represented by the diagonal of the parallelogram of forces. The experiment can be repeated by altering the positions of *M* and *N*, and by using other weights in place of *W*, the 100 gm. wt.

**Resultant of several inclined forces.** When a ship is accidentally grounded on mud-banks or sands, tugs are sent to pull the ship clear and to refloat it. The resultant force on the ship can be found by using the parallelogram of forces, even when there are more than two ropes tugging at the ship. Thus, for example,

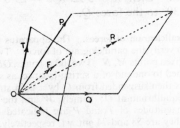

FIG. 33. Resultant of several forces

suppose there are four forces, *T*, *P*, *Q*, *S*, acting at a point *O* (Fig. 33). The resultant, *R*, of the two forces *P*, *Q*, is found by completing the parallelogram of forces to scale, as explained above. The resultant, *F*, of the remaining two forces *S*, *T*, is found in a similar way. The sum of the four forces is hence equal to the sum of the forces $R$, $F$, and the final resultant can be obtained by completing the parallelogram of forces (not shown) for these two forces. By combining two forces at a time in this way, the resultant of any number of forces meeting at a point can be found.

**Components of forces.** A barge moves along a canal in a direction inclined to that of the rope pulling it, and so the force acting on the barge (the tension in the rope) moves it in a direction inclined to the force. Similarly, a lawn-mower is moved by a force inclined to the horizontal.

Suppose that a lawn-mower is pushed with a force of 50 lb. wt. acting at an angle of 30° to the ground. The effective part of the force in a horizontal direction is known as the *resolved component* in that direction, and can be found from a parallelogram of forces. Let $CO$ represent 50 lb. wt. in magnitude and direction (Fig. 34). From $C$ draw a line $CA$ perpendicular to the horizontal $OA$, and

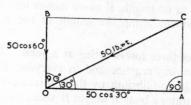

FIG. 34. Components of forces

a line $CB$ perpendicular to the vertical through $O$. Then, since $OACB$ is a parallelogram (strictly, a rectangle), the forces represented by the sides $AO$, $BO$, are together equal to the force represented by $CO$. Conversely, the force represented by $CO$ is equivalent to a force represented by $AO$ acting along the horizontal as shown, together with a force represented by $BO$ acting along the vertical into the ground. $AO$, $BO$, respectively, thus represent the resolved components, or effective parts, of the 50 lb. wt. force along the ground and the vertical.

Since $OAC$ is a right-angled triangle, $\dfrac{OA}{OC} = \cos 30°$. Consequently, $OA = OC \cos 30° = 50 \cos 30°$ lb. wt. The result can be generalized as follows: The component of a force $P$ in a direction inclined at an angle $\theta$ to it is given by

$$P \cos \theta.$$

The component along the vertical direction $BO$ is thus $50 \cos 60°$ lb. wt. (Fig. 34), since angle $COB$ is $60°$, and this downward force is balanced by the upward reaction of the ground. See also p. 22.

The "cosine law" for the resolved component of a force is also illustrated by the case of a nail about to be driven into wood. If it is held upright and then hit by a vertical blow of 6 lb. wt. with a hammer, the nail receives the full effect of the blow along its length, which drives it into the wood. If the blow of 6 lb. wt. is delivered carelessly, so that it strikes the nail at an angle of $60°$ to it, the nail is driven into the wood by an effective force $= 6 \cos 60° = 6 \times 0.5 = 3$ lb. wt. This force is the resolved component of the blow along the nail. If the nail is struck at right angles to its length, it never moves into the wood; the resolved component along the length of the nail $= 6 \cos 90° = 6 \times 0 = 0$.

**Equilibrium of three forces acting at a point.** There are many cases in engineering practice and in everyday life in which an object is kept in equilibrium by three forces acting at a point. Fig. 35 (a) illustrates a boat of weight $W$ when it is swung out from the side of a ship. In this position the pulley $B$ is in equilibrium

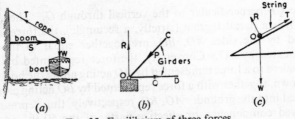

FIG. 35. Equilibrium of three forces

under the action of the weight $W$, the tension $T$ in the rope, and the thrust $S$ in the boom. Fig. 35 (b) illustrates the hinge, $O$, of the corner of a bridge resting on a support. The equilibrium of $O$ is maintained by the forces $P$, $Q$, in the girders $CO$, $OD$, and the reaction $R$ of the support. In Fig. 35 ($c$) an object is held on a smooth inclined plane at $O$ by a string along the plane. The forces maintaining the equilibrium at $O$ are the weight $W$ of the object, the tension $T$ in the string, and the reaction $R$ acting normally to the surface.

Suppose the object resting on the inclined plane in Fig. 35 ($c$) has a weight $W$ of 100 lb. wt. Since it is in equilibrium, the sum (resultant) of the two forces $R$, $T$, must also be 100 lb. wt. and must act in an *opposite* direction to $W$; if either of these two conditions is not obeyed, $O$ cannot be in equilibrium. For the same reason, the resultant of the forces $T$ and $S$ in Fig. 35 ($a$) must equal $W$. The name **equilibrant** is given to the single force which maintains two forces in equilibrium.

**The triangle of forces.** Consider three forces $P$, $Q$, $S$, keeping a point $O$ in equilibrium Fig. 36 ($a$). The resultant $R$ of $P$ and $Q$ is represented by $OC$, the diagonal of the parallelogram of forces

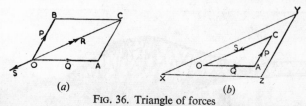

FIG. 36. Triangle of forces

$OACB$, which represents $P$ and $Q$. As $O$ is in equilibrium, the third force $S$ must be equal to $R$ in magnitude and act in the direction $CO$. Thus if the magnitude and direction of *two* of the three of the forces keeping a point in equilibrium are known, the unknown third force can always be found by using the parallelogram of forces.

If the magnitude of *one* and the direction of all the three forces are known, the other two can still be found, but a new method is adopted. In the parallelogram of forces $OACB$, the force $P$ may be represented by $AC$ (which is equivalent to $OB$ in magnitude and direction). We can therefore represent the three forces $P$, $Q$, $S$, by the figure $ACO$ (Fig. 36 ($b$) ). This is the *triangle of forces* for $P$, $Q$, $S$. Any other triangle such as $XYZ$, drawn with the same values for its angles (that is, having its sides parallel to the directions of the forces $P$, $Q$, $S$), will have its sides proportional to the magnitudes of $P$, $Q$, $S$. If the magnitude of one of these

FIG. 37 (*a*). SYDNEY HARBOUR BRIDGE

The bridge floor is suspended by steel cables from the arch, which has a span of 550 yards

FIG. 37 (*b*). THE FORTH BRIDGE

$1\frac{1}{2}$ miles in length, it consists of three double cantilevers joined by two short trusses or open girder systems

forces is known, a scale-drawing of such a triangle will give us the magnitudes of the other two forces by measurement. For example, if $XZ$ is 5 cm. long and represents the known force $Q$ of 100 lb. wt., and the length of $ZY$ is found to be 6 cm., the force $P$ (represented by $ZY$) = $100 \times 6/5$ = 120 lb. wt. Similarly $YX$, found to be 8 cm. long, represents the force $S$ and is $100 \times 8/5$ = 160 lb. wt.

We may now state the following general result: *If three forces acting at a point maintain it in equilibrium, then any triangle with its three sides drawn parallel to the direction of the forces has the length of its sides proportional to the forces.*

FIG. 37 (c). A BAILEY BRIDGE

This type of girder bridge, invented by Sir Donald Bailey for Army use, is speedily constructed of interchangeable units yet takes heavy traffic.

**An application of triangle of forces.** Suppose that an electric lamp of 50 gm. wt. is pulled by an attached horizontal string so that the flex makes

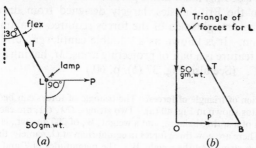

FIG. 38. Problem on triangle of forces

an angle of 30° to the vertical (Fig. 38 (a) ). To find the tensions $P$, $T$ in the string and flex, draw a line $AO$ parallel to the *known* force of 50 gm. wt. to

**3\***

represent it in magnitude on some scale, e.g. 1 c.m. ≡ 10 gm. wt. (Fig. 38 (b) ). Then *AO* is 5 cm. long. To obtain the triangle of forces, draw lines from *A* and from *O* which are parallel to the other two forces *T*, *P*, and let them intersect at *B* (Fig 38 (b) ). Then triangle *OAB* is a triangle of forces for *T*, *P*, and the 50 gm. wt., and *OB* represents *P*, the force to which it is parallel, while *BA* represents *T*. Suppose *OB* is 2·9 cm. long and *BA* is 5·8 cm. long by measurement. Then, since 10 gm. wt. is represented by 1 cm., *P* = 29 gm. wt., and *T* = 58 gm. wt.

**Equilibrium in bridges.** Fig. 39 (a) illustrates a *Warren bridge*, which consists of light girders forming triangles of 60°. If the reaction *R* at the support *a* is known, the forces *T*, *S* in the two girders meeting at *a* can be found by drawing the triangle of

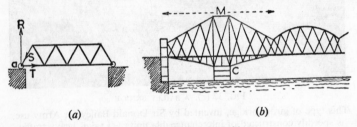

*(a)*                                        *(b)*

FIG. 39. Equilibrium in bridges

forces. Fig. 39 (b) illustrates a part of the *Forth bridge*, the second longest in the British Isles, largely designed from drawings by engineering draughtsmen of the forces required to maintain it in equilibrium. It is known as a "double cantilever" bridge, as its principal feature is a pair of projecting arms, *M*, having a common support *C*. (See also Fig. 37 (b), p. 60.)

**Verification of triangle of forces.** The triangle of forces can be verified by the apparatus shown in Fig. 40 (a). Two strings, *OX*, *OY*, are each attached to the ends of a spring-balance, and a weight, *W*, of 200 gm. wt., is suspended from *O*. There are now three forces in equilibrium at *O*, namely, the tensions, *T*, *P*, in each string and the weight *W*. The magnitudes of *T* and *P* are read from the spring-balances as 164, 152 gm. wt. respectively.

A sheet of paper is now held behind the strings, and the directions *OX*, *OY*, *OZ* of the three forces are drawn on the paper with a pencil. The three directions are then continued so that a triangle *ABC* is obtained (Fig. 40 (b) ), having its sides parallel to the three forces at O. AB, BC, CA are now measured, and are found to be 5, 4·1, 3·8 cm. respectively. Thus

$$AB : BC : CA = 5 : 4·1 : 3·8 = 1 : 0·82 : 0·76$$

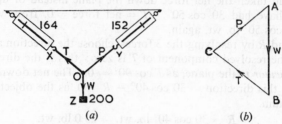

(a)                    (b)

Fig. 40. Verification of triangle of forces.

But   $W : T : P = 200 : 164 : 152 = 1 : 0·82 : 0·76$.   Consequently   the three sides of the triangle ABC represent to scale the magnitudes and direction of the three forces in equilibrium at O, and the triangle of forces is verified.

The experiment can be repeated by altering the weight W or the position of the spring-balances.

**Equilibrium, and the relation between resolved forces.** Suppose that a number of forces act at a point and are in equilibrium; for example, in Fig. 41 an object O of 30 lb. wt. is supported in equilibrium on a smooth plane by a string of unknown tension T, and R is the unknown reaction of the plane acting at right angles to it. Since the object is station-ary, it follows that the resultant force acting on it in any direction is *zero*. We may therefore re-solve the three forces at O in any direction we desire, and equate the net force to zero. In this way, we shall be able to obtain equations to calculate the un-known forces T and R.

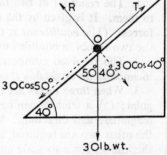

Fig. 41. Components

Since the resolved component of R along the plane is $R \cos 90°$, or zero, as $\cos 90°$ is zero, we choose to resolve the three forces *along the plane*; R is then eliminated from the equation. Thus, if the angle of inclination of the plane to the horizontal is 40°,

$T - 30 \cos 50° =$ net force upward along plane $= 0$

$\therefore \quad T = 30 \cos 50°$ lb. wt. $= 19\cdot3$ lb. wt.

If we had taken the net force down the plane instead of up, we should have had $30 \cos 50° - T =$ net force $= 0$, from which $T = 30 \cos 50°$ lb. wt. again.

To find $R$ by resolving the 3 forces, choose the direction along which the resolved component of $T$ is zero; this is the direction *perpendicular to* the plane, as $T \cos 90° = 0$. The net downward force in this direction $= 30 \cos 40° - R = 0$, as the object is in equilibrium.

$\therefore \quad R = 30 \cos 40°$ lb. wt. $= 23\cdot0$ lb. wt.

If we had chosen to resolve the three forces in a vertical direction, then $R \cos 40° + T \cos 50° - 30 = 0$. We should need another equation to find $R$ and $T$. Resolving the three forces horizontally, we should have $R \cos 50° - T \cos 40° = 0$, as the resolved component of 30 lb. wt. in this direction is zero. Solving the two simultaneous equations for $R$ and $T$ would enable these forces to be calculated; but it should be noted by the reader how much more easily the problem is solved by choosing to resolve the forces in a direction perpendicular to one of the unknown forces.

---

## Summary

1. **The resultant of two forces is a single force equivalent to them. It is given by the diagonal of the parallelogram of forces.** (The *equilibrant* is the single force which maintains the two forces in equilibrium.)

2. The resolved component of a force $P$ in a direction inclined at an angle $\theta$ is $P \cos \theta$.

3. **When three forces are in equilibrium, (a) they meet at a point; (b) a triangle can be drawn to represent the forces in magnitude and direction.** When one force is known, and the other two are required, the triangle is started by drawing the known force to scale and then drawing lines parallel to the remaining forces.

4. When three forces are in equilibrium, the sums of the resolved components in two perpendicular directions are each zero.

## WORKED EXAMPLES

1. *The diagram represents the radio aerial of an aeroplane. BC and AD are rigid struts and ABD is a wire under tension. Find, either by calculation or by drawing on the diagram, the thrust in BC if the tension in each part of the wire is 20 lb. wt. (N.)*

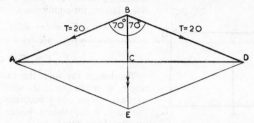

FIG. 42. Resultant

The tensions, *T*, in *AB* and *BD* are equal and act at *B* (Fig. 42). The thrust, *R*, in *BC* is thus equal and opposite to the resultant of the two equal tensions. The parallelogram *ABDE* is now completed by drawing lines from *D* and *A* parallel to *BA* and *BD* to meet at *E*, and the diagonal *BE* is measured. As *BA* represents 20 lb. wt. to scale, the magnitude of the resultant, *BE*, can be calculated; and this value is equal to the thrust in *BC*. The reader can verify that the thrust is 13·7 lb. wt.

2. *What do you understand by (a) the resultant, (b) the equilibrant of two forces? State the parallelogram rule for finding the resultant of two forces and describe an experiment to verify the truth of the rule. Forces of 50, 40, and 30 gm. wt. act at a point in directions north, east, and south respectively. Find their resultant in magnitude and direction. (L.)*

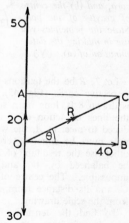

FIG. 43. Resultant

Suppose the three forces are as shown in Fig. 43. The resultant of the 50 gm. wt. and 30 gm wt. is 20 gm. wt. in the direction of the greater force, and is represented to scale by the line *OA*. On the same scale *OB* represents 40 gm. wt. The parallelogram (rectangle) *OACB* is then completed, and the resultant, *R*, is represented by *OC*. The direction of the resultant is obtained by measuring the angle *θ*. Measurements show that *R* = 44·8 gm. wt. and acts at 27° *N*. of *E*.

3. *A picture is hung from a nail by a wire fastened near the two top corners*

*of the frame. Show that there is less tension in the wire when it is long than when it is short.* (*L.*)

Suppose *ABCD* is the frame, and that the weight of the picture is *W*. The weight is balanced by the components of the tensions, *T*, at *A,B* in a *vertical* direction, which are each $T \cos \theta$ (Fig. 44). Hence, for equilibrium,

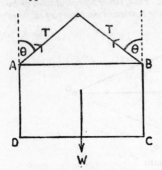

$$T \cos \theta + T \cos \theta = W$$
$$\therefore \quad 2T \cos \theta = W$$
$$\therefore \quad T = W/(2 \cos \theta) \quad (1)$$

The longer the wire, the smaller is the angle $\theta$ made by the wire with the vertical. Now $\cos \theta$ becomes *greater* as the angle $\theta$ diminishes; hence *T* becomes smaller, from equation (1).

FIG. 44. Components

4. *A tapering pole, 18 ft. long and weighing 150 lb., is suspended by a cord at each end, as shown in the following diagram, the centre line of the pole being horizontal; the cords make angles of 45° and 30° with the horizontal. By means of a scale diagram (drawn in your answer book), or by calculation, determine (a) the tension in each cord, and (b) the centre of gravity of the pole. State the principles you use in making the determination of (a).* (*N.*)

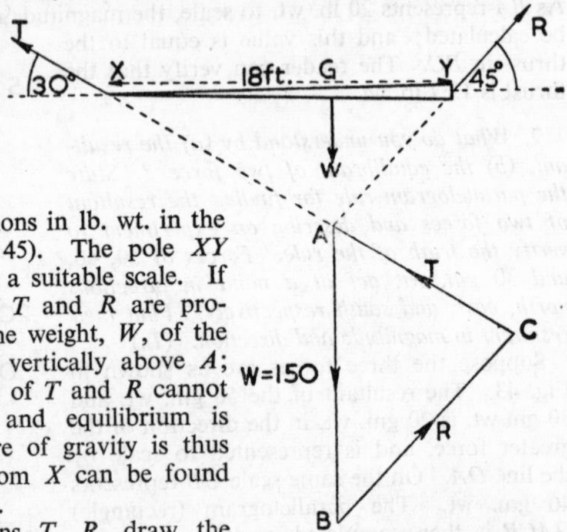

Let *T*, *R* be the tensions in lb. wt. in the respective cords (Fig. 45). The pole *XY* is drawn 8 ft. long on a suitable scale. If the lines of action of *T* and *R* are produced to meet at *A*, the weight, *W*, of the pole must be situated vertically above *A*; otherwise the resultant of *T* and *R* cannot be balanced by *W*, and equilibrium is impossible. The centre of gravity is thus *G*, and its distance from *X* can be found from the scale drawing.

To find the tensions *T*, *R*, draw the triangle of forces for the three forces *W* (150 lb. wt.), *T*, *R*. The line *AB* is drawn

FIG. 45. Triangle of forces

to represent 150 lb. wt. according to some scale; and then the lines *BC* and *AB* are drawn parallel to the forces *R* and *T* respectively to meet at *C*. By measuring *CA* and *CB*, and converting the lengths to lb. wt. from the chosen scale, the forces $T$ and $R$ are obtained. (The accurate scale diagram is left to the reader.)

## EXERCISES ON CHAPTER IV

**1.** Name three practical applications of the subject of Statics.

**2.** Find the resultant of forces of 50 lb. wt. and 80 lb. wt. acting at an angle of (*i*) 60°, (*ii*) 120°, (*iii*) 90° to each other. Give one practical example of each case where the forces act at these angles.

**3.** A sledge is pulled by three ropes, the middle rope making angles of 30° with each of the others. The forces in each of the outer ropes is 20 lb. wt., and the force in the middle rope is 15 lb. wt. Find the resultant of the three forces.

**4.** A barrel of 20 lb. wt. is held on a smooth inclined plane by a rope pulling up the plane. The inclination of the plane to the horizontal is 30°. Find the force in the rope and the reaction of the plane on the barrel.

**5.** A lamp is pulled by a horizontal force of 120 gm. wt. so that the flex makes an angle of 20° with the vertical. Find the weight of the lamp and the tension of the flex.

**6.** Two girders of a bridge, inclined at 60° to each other, rest on a smooth support (see Fig. 35 (*b*) ). If the reaction at this support is 500 lb. wt., find the force in each girder.

**7.** A picture is supported on a wall by two strings attached to a nail. Each string is inclined at 30° to the horizontal, and the weight of the picture is 1000 gm. wt. Find the tension in each string by resolving the forces vertically, and by one other method.

**8.** A car of 1000 lb. wt. is resting on a road inclined at 20° to the horizontal. Calculate the component of the car down the plane. What is the frictional force on the car at the ground, and the normal reaction of the road on it ?

**9.** What is meant by (*a*) the resultant of a number of forces acting at a point, (*b*) the component of a force in a given direction ? How would you find experimentally, the resultant of two given forces acting in different directions at a given point ?

An electric lamp, of mass 2 lb., hangs on a flex which is pulled aside by a horizontal string until the flex makes an angle of 30° with the vertical. Find the tension in the string and in the flex. (*L.*)

**10.** What is the principle of the *parallelogram of forces* ? Describe an experiment to verify it. An unknown mass is suspended by a light cord and is acted upon by a horizontal force of 100 gm. wt. In the position of equilibrium the cord makes an angle of 30° with the vertical. Draw a force diagram and find the value of the mass. (L.)

**11.** Explain why less force is needed to roll a barrel up an inclined plane than to lift it vertically. (N.)

**12.** Explain why a weathercock sets with its head towards the direction of the wind. (N.)

**13.** What conditions have to be satisfied for a body to be in equilibrium under the action of three non-parallel forces ? Describe an experiment to verify your statement.

A string AB, carrying a load of 5 lb. at the lower end B hangs from a fixed point A. A horizontal force P is applied to the lower end B so that AB is inclined at 30° to the vertical. Determine *graphically* the value of P and of the tension in AB. (C.W.B.)

**14.** Explain the conditions that have to be fulfilled for three forces acting at a point to be in equilibrium. A crate of mass 2 tons is supported by a rope from a crane. A horizontal force is applied by a slip rope just above the crate, and the rope from the crane is observed to be at 15° to the vertical. Determine *graphically* (a) the horizontal force. (b) the tension in the rope from the crane. (C.W.B.)

**15.** State the conditions under which three forces passing through a point maintain a body in equilibrium. Describe an experiment to confirm your statement. A uniform girder 13 ft. long and weighing 12 cwt. stands with its upper end against a smooth vertical wall and with its lower end on the ground 5 ft. from the base of the wall. Find the force which the girder exerts on the wall. (This force is perpendicular to the wall.) (L.)

**16.** State the theorem known as the parallelogram of forces. Describe how you would verify it experimentally. A glider weighing 6000 lb. is towed at a constant speed and at a constant height by a tow-rope which sloped downwards from the glider and which makes an angle of 18° with the horizontal at the point at which it joins the glider. If the tension in the tow-rope is 500 lb. wt., find, graphically or otherwise, (a) the force exerted on the glider by the resistance of the air, (b) the upward "lift" acting on the glider. (N.)

**17.** Explain the terms *moment of a force*, *work*. State the conditions under which a rigid body remains in equilibrium under the action of three coplanar forces which are not parallel. A uniform ladder of mass

200 lb. rests on a pavement against a smooth wall at an angle of 60° to the horizontal. A boy of mass 100 lb. climbs up the ladder, which is prevented from slipping by a force of 100 lb. applied horizontally at the foot of the ladder. What fraction of the ladder's length can the boy ascend before the ladder slips ? (*O. & C.*)

**18.** Explain the terms: *rigid body, equilibrium.* State the conditions of equilibrium of a rigid body acted on by three non-parallel forces in a plane. A smooth uniform rigid sphere of radius 6 in. and mass 2 lb. is supported in contact with a smooth vertical wall by a string fastened to a point on its surface, the other end being attached to a point on the wall. If the length of the string is 6 in., find (*a*) the tension in the string, (*b*) the force exerted by the sphere on the wall. (*O. & C.*)

**19.** State the parallelogram law of forces. Describe an experiment to illustrate it, and show how the law may be applied to resolve a force into two perpendicular components.

A nail projects horizontally from a vertical wall and a cord attached to its head is pulled at an angle of 30° to the wall, with a force of 12 lb. wt. By a scale drawing, or otherwise, find (*a*) the force tending to bend the nail, (*b*) the force tending to pull it out of the wall. (*L.*)

**20.** State the triangle law for forces and describe an experiment to illustrate it.

A ventilating window, 2 ft. from top to bottom, has a mass of 2 lb. and has a smooth hinge along its lower, horizontal edge. A horizontal cord, attached to its upper edge, holds the window at an angle of 30° with the vertical. Make a scale diagram, indicating a triangle of forces. Hence, or otherwise, determine (*a*) the direction and magnitude of the reaction of the hinge, (*b*) the tension in the cord. (Assume the centre of gravity of the window to be at its middle point.) (*L.*)

**21.** What are the conditions for a body to be in equilibrium under the action of three non-parallel coplanar forces ?

Describe an experiment to find the resultant of forces of 3 lb. wt. and 5 lb. wt. acting at a point in directions inclined at an angle of 60°. Explain why your method gives the required result.

A uniform cylinder of mass 200 gm. rests with its axis horizontal in the right angle formed by two smooth planes inclined at 30° and 60° respectively to the horizontal so that its curved surface is in contact with the planes. What force does the cylinder exert on each plane? (*L.*)

# CHAPTER V

## MOMENTS. PARALLEL FORCES
## CENTRE OF GRAVITY

THERE are many cases in everyday life in which forces exert a *turning effect*; for example, when a door is opened, or a screwdriver is used, or a bicycle is ridden. Vehicles and most of the machines in industry (such as the dynamo) are driven by the turning-effects of forces on wheels of various sizes.

**The moment of a force.** The turning-effect of a force about a given point is known in Mechanics as its *moment* about that point, and is measured by the product

*force × perpendicular distance from*
*point to line of action of force* . . (1)

We can understand why this definition of moment was chosen if we cast our minds back to our experiences on a see-saw. The further away we were from the turning point of the see-saw, the greater was the turning-effect about that point; thus the moment increased when the distance from the point increased. A heavy child placed on the see-saw had a greater moment than a light child at the same place, showing that the moment increased as the magnitude of the weight (or force) increased. The force and the distance are both taken into account in the definition of moment in (1) above.

When the moment of a force is calculated, the answer is in *lb. wt. ft.* if the distance concerned is in feet, and the force is in lb. wt. The moment is expressed in *gm. wt. cm.* (or *dyne cm.*) when the distance concerned is in centimetres and the force is in gm. wt. (or dynes).

Thus, in Fig. 46 (*a*), which represents weights of 40 lb. wt. and 100 lb. wt. on a light horizontal beam with a turning-point *O*, the moment of the former weight about *O* is $40 \times 3 = 120$ lb. wt. ft., and the moment of the latter is $100x$ lb. wt. ft.   Fig. 46 (*b*) represents a box of 20 lb. wt. pulled by a horizontal force $P$ of 10 lb. wt.

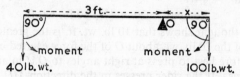

FIG. 46 (*a*).  Moment of force

at the top $B$;  the moment of the force about $O$ is 80 lb. wt. ft., and the moment of the weight of the box about $O$ is 40 lb. wt. ft. Consequently the box will tilt about $O$ in the direction of $P$.

**How the "perpendicular" affects the magnitude of moment.** Suppose that a boy riding a bicycle presses on the upper pedal, $M$, with a *downward* force of 20 lb. wt. when the crank $OM$ (6 in. long) is inclined at 30° to the vertical (Fig. 47).   The moment of the

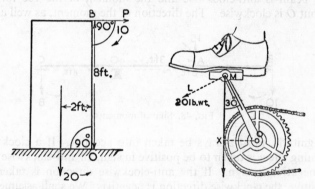

FIG. 46 (*b*).  Moments          FIG. 47.  Moment

force about the centre, $O$, is given by $20 \times OX$, where $OX$ is the perpendicular from $O$ to the direction of the force.   But $OX$ $= OM \sin 30° = 6 \times 0.5 = 3$ in. $= \frac{1}{4}$ ft.

$\therefore$   moment about $O = 20 \times \frac{1}{4} = 5$ lb. wt. ft.

Suppose, however, that the force of 20 lb. wt. is directed *perpendicular* to $OM$, along $ML$, in a downward direction (Fig. 47).

The moment of the force about $O$ is now given by $20 \times OM$, since $OM$ is perpendicular to the direction of the force. As $OM = 6$ in. $= \frac{1}{2}$ ft.

moment $= 20 \times \frac{1}{2} = 10$ lb. wt. ft.

A little thought shows that 10 lb. wt. ft. is the greatest possible moment of the 20 lb. wt. about $O$ of the force exerted by the rider. It is therefore best to press at right angles to $OM$ in any position of the latter. If the rider presses in the direction $MO$, there is *no* turning-effect about $O$, since the perpendicular from $O$ to the direction of the force is zero in this case.

It should be noted that it is meaningless to discuss, or calculate, the moment of a force unless the point about which the moment is taken is specified or known.

**The sign of moment.** The turning-effect of a force may cause rotation in either a clockwise or an anti-clockwise direction. In Fig. 46 (*a*), for example, the moment of the 40 lb. wt. about $O$ on the beam is anti-clockwise and the moment of the 100 lb. wt. about $O$ is clockwise. The direction of the moment, as well as its

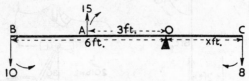

FIG. 48. Sign of moments

magnitude, must always be taken into account. If a clockwise turning effect is taken to be positive in sign, an anti-clockwise one is negative in sign. If the anti-clockwise direction is taken as positive, the clockwise direction is negative. We shall assume the latter convention in this book.

As an illustration of the sign convention, consider a light beam pivoted at $O$ and acted upon by the forces shown (Fig. 48). If $OB = 6$ ft., $OA = 3$ ft., $OC = x$ ft., the sum of the moments about $O$

$$= + 10 \times 6 - 15 \times 3 - 8x = (15 - 8x) \text{ lb. wt. ft.}$$

**Equilibrium under parallel forces.** In a number of cases in mechanics a body is in equilibrium under the action of *parallel* forces. For example, Fig. 49 illustrates a bridge, resting on supports at *a* and *b* which have vertical reactions *R*, *S*, respectively.

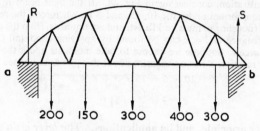

FIG. 49. Parallel forces

If the bridge has loads on it of weights 200, 150, 300, 400, 300 lb. wt. acting as shown, then

$$R + S = 200 + 150 + 300 + 400 + 300 = 1350 \text{ lb. wt.}$$

as the upward forces must balance the downward forces.

When parallel forces maintain an object in equilibrium, another condition must be fulfilled, besides the condition that the sum of the forces in one direction must balance the forces in the opposite direction. This can be seen from Fig. 50, in which a wheel, capable of turning about an axle at *O*, is acted upon by two parallel forces of 40 lb. wt. acting tangentially to the wheel at the rim. The force in one direction is equal and opposite to the other force in this case; but the wheel is *not* in equilibrium, since the two moments act in the same direction and hence cause rotation of the wheel.

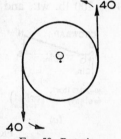

FIG. 50. Rotation

The following are therefore the necessary conditions for equilibrium between parallel forces—

(1) *The sum of the forces in one direction must equal the sum in the opposite direction.*

(2) *The sum of the moments of the forces about any point in an*

*anti-clockwise direction must be equal to the sum of the clockwise moments of the forces about the same point. See Experiment, p. 158a.*

As an illustration, consider again Fig. 48. If the light beam $BC$ pivoted at $O$ is in equilibrium under the 10, 15, and 8 lb. wt. forces, the reaction $R$ at the pivot (not shown) must be 3 lb. wt. and act upwards, from the condition (1). The distance $OC = x$ can be found by applying condition (2) to the system of forces. Suppose we choose to take moments about the pivot $O$. Then the reaction $R$ has no moment about this point, and hence

$$10 \times 6 = 15 \times 3 + 8 \times x$$

Thus $$x = \frac{15}{8} = 1\tfrac{7}{8} \text{ ft.}$$

**The lever principle, and its applications.** The *lever* is an arrangement which enables a small force to overcome a large weight or resistance. The scissors, nut-crackers, pliers and tongs are examples of appliances which utilize a lever principle, and all these appliances can be divided into three types or classes of levers.

**First class of levers.** A crowbar, $DON$, is an example of the "first class" of levers (Fig. 51 (*a*)). A force $P$ is applied at one end $N$ to overcome some resistance or weight $W$ at the other end $D$. The bend, $O$, of the crowbar is the point about which the leverage takes place, and is called the **fulcrum**. Suppose the weight $W$ is 200 lb. wt., and the distances $OA$, $OB$, are 2 and 28 in.

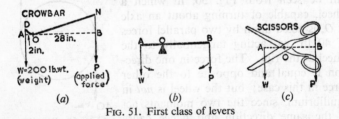

FIG. 51. First class of levers

respectively. Then, taking moments about $O$ to eliminate the moment of the reaction there, $P \times 28 = 200 \times 2$, where $P$ is the force which just overcomes the weight. Thus $P = 14\tfrac{2}{7}$ lb. wt., and by means of the lever this small force overcomes one of 200 lb. wt. The first class of levers, then, is one in which the load ($W$) and effort ($P$) are on opposite sides of the fulcrum (Fig. 51 (*b*)).

It is interesting to note that the head is moved to-and-fro by means of a lever action similar to that just described. A muscle at the back of the neck exerts a force, $P$, which overcomes the weight, $W$, of the head, the fulcrum, $O$, being at the base of the skull.

**Mechanical advantage of levers.** The lever is the simplest example of a "machine", which is a device for exerting a force at one place by applying a force at another place. The *mechanical advantage* (*M.A.*) of a machine is defined as the ratio

$$\frac{\text{weight or resistance overcome}}{\text{force applied}},$$

or $\dfrac{W}{P}$ in the notation applied with the lever (see also p. 90). The crowbar has thus a mechanical advantage of 14, as $W = 200$ lb. wt. and $P = 14\frac{2}{7}$ lb. wt. (Fig. 51 (*a*) ). The magnitude of the mechanical advantage of the lever depends, in general, on the relative lengths of the "arms" of the lever. Thus, taking moments about the fulcrum $O$ (Fig. 51 (*b*) ), we have $W \cdot OA = P \cdot OB$. Hence $\dfrac{W}{P} = \dfrac{OB}{OA} =$ the mechanical advantage. Since $OB$ is always greater than $OA$ in this class of lever, the mechanical advantage is always greater than 1.

Fig. 51 (*c*) illustrates the scissors as a first class of lever. To tear cloth directly usually requires a considerable force, but the resistance $W$ of the fabric can be overcome by a much smaller force $P$ applied at the end of the scissors. The action of a pair of pliers in cutting wire is very similar to that of the scissors.

**Second class of levers.** A nut-cracker with a hinge at $O$ is an example of the "second class" of lever (Fig. 52 (*a*) ); here the load $W$ acts on the same side of the fulcrum as the effort $P$ but is nearer to the fulcrum than $P$. The resistance, $W$, of the nut-shell at $A$ can be overcome by a much smaller force, $P$, applied at the end $B$; and if $W = 40$ lb. wt., $OA = 1$ in., $OB = 8$ in., the force is given by $P \times 8 = 40 \times 1$, taking moments about the hinge $O$. Hence $P = 5$ lb. wt., and the mechanical advantage of the lever,

$\dfrac{W}{P} = 8$. Fig. 52 (b) illustrates the conventional appearance of a lever of the second class, from which, by taking moments about O, $\dfrac{W}{P} = \dfrac{OB}{OA}$. As OB is always greater than OA, the mechanical

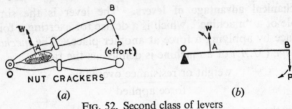

NUT CRACKERS

(a)                            (b)

FIG. 52. Second class of levers

advantage is always greater than 1. A further example of the second class of levers is the wheelbarrow, with its wheel and axle at one end, and the weight and the lifting force at the other. Similarly, when one is walking, the muscles at the heel of the foot exert an upward force P which overcomes the weight W of the body, the ball of the foot acting as the fulcrum O.

**Third class of levers.** We come now to the last class of levers, of which coal tongs and sugar tongs, are examples (Fig. 53 (a) ). Suppose the coal at A has a weight W of 8 oz. wt., and OA, OB, are 18 and 9 in. respectively. Taking moments about the hinge

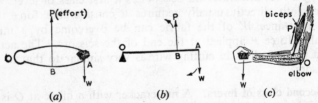

(a)                       (b)                       (c)

FIG. 53. Third class of levers

at O, we have $P \times 9 = 8 \times 18$; thus $P = 16$ oz. wt., i.e. a *greater* force than the weight is required. The mechanical advantage, $\dfrac{W}{P}$, is $\frac{1}{2}$, which, interpreted physically, means a mechanical *disadvantage*. A conventional diagram of a lever of the third class is shown in Fig. 53 (*b*). It will be noted that *P* is nearer to the fulcrum *O* than *W*, which is the essential difference between the third and second classes of levers. By taking moments about *O*,

$$\frac{W}{P} = \frac{OB}{OA};$$ and as *OB* is always less than *OA* the mechanical advantage is always less than 1.

The forearm, with the elbow *O* as a fulcrum, is an example of the third class of lever; the muscular force *P* exerted by the biceps is nearer to *O* than is a weight *W* picked up by the hand (Fig. 53 (*c*) ). Fig. 54 illustrates a *lever safety valve*, used on stationary boilers, which has a valve *V* fitting into the boiler, and a lever *OB*

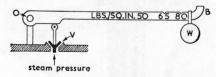

FIG. 54. Safety valve

carrying a weight *W*. When the steam pressure exceeds the dangerous value, the pressure raises *V* and allows the steam to escape until the boiler is relieved. It can be seen that this safety valve acts as a lever of the third class, with the fulcrum at *O*.

**The common balance** is a lever of the first class, and the two arms are usually of equal length. Fig. 55 illustrates a common balance used in laboratories. When the beam at the top is raised, it is pivoted at its centre on an agate knife-edge, pointing downwards, which rests on agate at the top of the central pillar. The scale-pans are suspended from agate knife-edges, pointing upwards, at the ends of the beam. The three edges lie in a horizontal line. When the beam is horizontal, the point of a plumb-line bob should be directly above the point of the metal attached to the central pillar; if this is not the case, the screws beneath the base are adjusted.

If the two arms of the balance are unequal, and of lengths *a*, *b*, respectively, an object can be weighed accurately by placing it in each scale-pan in turn and weighing it. This method is known as *double weighing*.

Suppose $W_1$, $W_2$, are the slightly different weights to give a balance in each case. Taking moments about the point of suspension of the beam, $W \cdot a = W_1 \cdot b$, where $W$ is the true weight

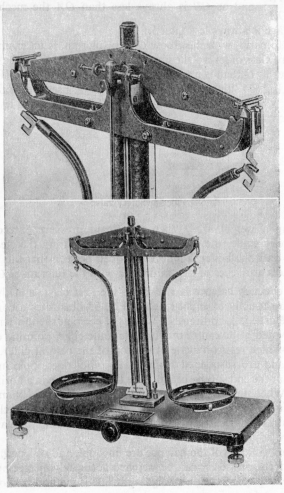

FIG. 55. COMMON BALANCE

of the object; if the object is weighed on the other scale-pan, $W \cdot b = W_2 \cdot a$. Thus $\dfrac{W}{W_1} = \dfrac{b}{a}$, and $\dfrac{W_2}{W} = \dfrac{b}{a}$. Hence $\dfrac{W}{W_1} = \dfrac{W_2}{W}$, or $W^2 = W_1 \, W_2$. Thus $W = \sqrt{W_1 \, W_2} =$ the true weight.

Another method of weighing accurately when the two arms of a balance are unequal consists of counterbalancing the unknown weight $W$ with sand. $W$ is then removed and weights are placed in the *same* scale-pan until a balance is again obtained. Then $W$ is equal to the total weight on the pan.

## CENTRE OF GRAVITY

A cricket ball, hit into the air, begins to descend to earth after a short time; a footballer, knocked off his balance, falls downwards. These are effects of gravity, and all objects, no matter how small, or whether gaseous, liquid, or solid, are attracted towards the centre of the earth by a force which is proportional to their mass (p. 33).

Consider a rod of wood, *AB*, of a constant cross-sectional area, i.e. a *uniform* rod (Fig. 56). Each particle in the rod is attracted towards the centre of the earth by a force which is the particle's weight and, since the centre of the earth is a very long way off, the weights of the various sections of the rod, such as *c*, *d*, act downwards in parallel directions. We thus have a large number of parallel forces, and their resultant, which is the *total weight* of the rod *AB*, acts at some

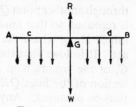

FIG. 56. C.G. of rod

point in the rod. The *centre of gravity* of a body is the name given to *the point through which its total weight appears to act*. A flat object can be balanced at its centre of gravity (C.G.), *G*; the reaction *R* then counteracts the weight *W* (Fig. 56).

**Positions of the C.G.** The C.G. of a *uniform rod* is at its mid-point. The C.G. of a *circular disc or ring* is at the centre of the circle. A thin sheet (lamina) in the form of a *parallelogram, rectangle, or square* has its C.G. at the point of intersection of the diagonals. A *triangular lamina* has its C.G. at the point of intersection of the medians (the lines joining the vertices to the respective mid-points of the opposite sides). The centre of gravity, $G$, can be shown by geometry to be two-thirds of the way along the median from the vertex, i.e. $AG = \frac{2}{3} AM$ in Fig. 57.

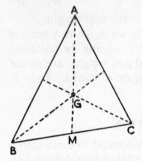

FIG. 57. C.G. of triangle

**Experiment to find the C.G. of an irregularly shaped lamina.** The C.G. of an irregularly shaped lamina cannot be found very precisely by balancing it on a point, as anyone who has tried it knows. A better way is to suspend the lamina $L$ from a point $A$ so that it can move freely about a pin $P$ forming a horizontal axis through $A$ (Fig. 58). When the lamina is at rest its weight will be somewhere vertically below $A$, the support. A small metal ball $B$ at the end of a long thread (a plumb-line) is then suspended from $A$, and the vertical line $AM$ is drawn on the lamina. The latter is then suspended from a horizontal axis through another point $Q$, and the procedure is repeated, so that another line $QR$ is obtained along which the weight appears to act. It follows that the centre of gravity, $G$, of the lamina is at the point of intersection of the lines $QR$ and $AM$, and can thus be obtained. Any other vertical line drawn through a support will pass through $G$ when the lamina is hanging freely.

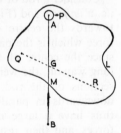

FIG. 58.
Experiment to find C.G.

**C.G. of regular combined objects.** Consider a lamina cut in the form of a shape $ABHCD$, which can be considered as a rectangle $ABCD$ joined to an isosceles triangle $BCH$ of the same base. Suppose $AB = 10$ in., $AD = 6$ in., and the height $HN$ of the triangle is 12 in. (Fig. 59). The C.G. of the rectangle is at $G_1$, which is at a distance $G_1H$ from $H$. But $G_1N = 5$ in., $NH = 12$ in., and hence $G_1H = 17$ in. The C.G. of the triangle $BCH$ is two-thirds from $H$ along the median, and hence $G_2H = 8$ in. Further, the area of $ABCD$ is 60 sq. in., and if 1 sq. in. of the material of the lamina weighs $w$ lb. wt. the weight of $ABCD$ is $60w$ lb. wt. The triangle $BHC$ has an area $= \frac{1}{2}$ base $\times$ height $= \frac{1}{2} \times 6 \times 12 = 36$ sq. in., and has therefore a weight of $36w$ lb. wt. The total weight of $ABHCD$ is thus $36w + 60w = 96w$ lb. wt.

We are now able to find the position of the C.G. of *ABHCD*.  Suppose its distance from *H* is *x*.  Let us tabulate our information—

|  | Rect. *ABCD* | △ *HBC* | *ABHCD* |
|---|---|---|---|
| Weight in lb. wt. . | 60w | 36w | 96w |
| Distance of C.G. from *H* | 17 in. | 8 in. | x |

Now the weight of 60w lb. wt. at $G_1$ and the weight of 36w at $G_2$ are *parallel* forces, and their resultant weight 96w acts at a point *G* which is the C.G. of

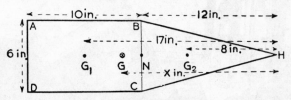

Fig. 59. C.G. of object

the whole lamina.  Since the moment of the resultant about any point = the sum of the moments of the individual forces about the same point (p. 72), we have, taking moments about *H*,

$$96w \times x = 60w \times 17 + 36w \times 8$$
$$\therefore \quad 96x = 1308$$

dividing through by *w*.  Thus $x = 13\frac{5}{8}$, and hence *G* is $13\frac{5}{8}$ in. from *H* along *HN*.

**Stability and centre of gravity.**  In designing a car or a ship, the engineer must take into account its *stability*, especially at high speeds.  Years ago, when omnibuses first appeared, tests for their stability on an incline were carried out by loading the top deck with sandbags representing an equivalent weight of passengers on this part of the bus.  (See Fig. 60.)  Ships, which are liable to considerable rolling and lurching in heavy seas, must be designed to be in stable equilibrium even in rough weather.

The stability of an object is intimately connected with the position of its centre of gravity.  As a simple example, consider an oil-drum, *C*, held on an inclined plane as shown in Fig. 61 (*a*). The vertical through its centre of gravity, *G*, just passes through the edge of the base $OA$, and if the drum is slightly displaced about $O$ to the left so that the vertical through $G$ falls to the left of $O$, the reaction $R$ of the plane has a moment about $O$ in the *same* direction as the moment of the weight $W$ of the drum about $O$. Equilibrium is therefore impossible, and if it is now released

FIG. 60. BUS UNDERGOING STABILITY TEST

A sandbag "load" is distributed over the upper deck equivalent to a full complement of passengers. The scales show the tilts of chassis and body respectively, the difference being due to the flexibility of the springs and tyres.

the drum will topple over about $O$. The drum is hence said to be in **unstable equilibrium.**

Suppose that a cylinder of the same diameter, but a smaller height, is placed on the inclined plane (Fig. 61 (*b*) ). The centre of gravity, $G_1$, is then lower than that of the previous cylinder, and the vertical through $G_1$ may then pass through a point on the base $OA$, as shown. Unlike the case of unstable equilibrium (Fig. 61 (*a*) ), the moment of the weight $W_1$ about $O$ is in the opposite direction to the moment of the reaction $R_1$ about the same point, and equilibrium is therefore possible. If the cylinder is very slightly displaced about $O$ and then released, the weight $W_1$ will exert a moment about $O$ which will restore the cylinder to its former position, and the cylinder is therefore said to be in **stable equilibrium.**

If the cylinder is taken and placed with its curved surface on a horizontal plane, it will be in equilibrium no matter what position it assumes when it is rolled. This is due to the fact that the reaction $R$ of the plane always acts in the same vertical line as the weight

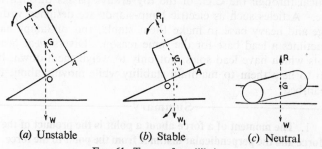

(a) Unstable      (b) Stable      (c) Neutral

FIG. 61. Types of equilibrium

$W$ (Fig. 61 (c) ). The cylinder is therefore said to be in **neutral equilibrium** in this case.

A criterion can be given for the three types of equilibrium just discussed. If the vertical line through the centre of gravity of an object passes through its base when the object is slightly displaced, the object is in *stable* equilibrium; if it passes outside the base when the object is slightly displaced, the object is in *unstable* equilibrium; and if the vertical always passes through the base, no matter to what position the object is displaced, the equilibrium is *neutral*.

**Applications.** It can be seen from the above that the risk of unstable equilibrium is increased as the height of the centre of gravity of an object is increased; the vertical through the C.G. is then more liable to fall outside the base when the object is slightly tilted. The nearer the centre of gravity is to the ground, the more stable is the equilibrium likely to be. For this reason, extra passengers are allowed on the lower deck of a crowded bus, but not on the upper deck; the centre of gravity, $G$, of bus and passengers must be low, so that the vertical through $G$ falls between the wheels even when the bus is rounding a corner. Racing cars are built low for the same reason. The base of the stand of a punchball is a very heavy piece of metal; the C.G. of the whole arrangement is then so low that it is in stable equilibrium, and the ball returns to the boxer however powerfully it is struck. Toys which spring up to the vertical, no matter how they are laid on the table, have a rounded heavy lead base, so that the vertical through the C.G. of the toy always passes through the base. Articles such as electric lamp-stands are designed with a large and heavy base to make them stable, and oil-lamps have sometimes a lead base for the same reason. Divers wear heavy boots which have lead soles, not only to weigh them down, but also to help them to maintain stability while moving about the ocean-bed.

## Summary

1. **The moment of a force about a point is the product of the force and the perpendicular distance from the point to the force.** It is expressed in lb. wt. ft., or gm. wt. cm. (or dyne cm.).

2. The conditions of equilibrium for parallel forces are: **(i) the sum of the forces in one direction = the sum of the forces in the opposite direction ; (ii) the sum of the clockwise moments about any point = the sum of the anti-clockwise moments about that point.**

3. In the first class of levers, the load and effort are on opposite sides of the fulcrum. In the second and third classes, the load and effort are on the same side of the fulcrum. In the third class, the effort is nearer the fulcrum than the load is.

4. The centre of gravity of a rectangle or square is the point of intersection of the diagonals. **The centre of gravity of a triangle is two-thirds along the median from any apex.**

5. The equilibrium of an object is stable or unstable according as the vertical falls inside or outside the base when the object is slightly displaced. The equilibrium is neutral if the vertical always passes through the base, however the object is displaced.

## WORKED EXAMPLES

**1.** *In the diagram, AB is a light rod loaded as shown.  Find the point at which it must be supported in order to balance horizontally.* (*N.*)

Let $O$ be the point, distant $x$ ft. from $A$.  Then, since $AC = 10 - 2 = 8$ ft.
$OC = (8 - x)$ ft.: and $OB = (10 - x)$ ft.  (See Fig. 62.)

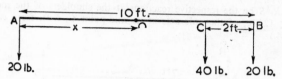

FIG. 62. Moments

The sum of the clockwise moments about $O$ = the sum of the anti-clockwise moments about $O$.

$$\therefore \quad 40(8 - x) + 20(10 - x) = 20x$$
$$\therefore \quad 320 - 40x + 200 - 20x = 20x$$
$$\therefore \quad 80x = 520, \text{ i.e. } x = 6\tfrac{1}{2} \text{ ft.}$$

**2.** *Define moment of a force about a point.  State the principle of moments and describe an experiment to verify it.  A uniform rod AB, length 10 ft. and mass 40 lb., rests on two supports P and Q such that AP = 1 ft. and BQ = 2 ft. A weight of 200 lb. rests on the plank, between P and Q, 1·5 ft. from Q.  What is the least force necessary to raise the loaded rod off the support Q and where must it act?* (*C.W.B.*)

When the rod is taken off the support $Q$, the rod tilts about $P$.  The least force, $R$, to just raise the rod is that which acts as far away from $P$ as possible, and is perpendicular to $AB$.  Thus $R$ acts upwards at $B$.

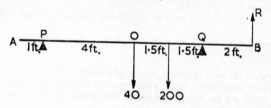

FIG. 63. Parallel forces

The weight, 40 lb., of the rod acts at the mid-point $O$.  Taking moments about $P$, then, from the distances shown in Fig. 63,

$$R \times 9 = 200 \times 5.5 + 40 \times 4$$
$$\therefore \quad 9R = 1260, \text{ i.e. } R = 140 \text{ lb. wt.}$$

4

3. *Describe three different methods of using a lever and in each instance give a practical application of the method. Two men carry a load of 120 lb. wt. by suspending it from a 12 lb. pole and supporting the ends of the pole on their shoulders. If the load is 7 ft. from one man and 5 ft. from the other, what fraction is supported by each man? What work is done by each man when lifting the load through a distance of 3 ft. on raising the ends of the pole to their shoulders? (L.)*

Let $R$, $S$ be the respective reactions at the shoulders of the men. The weight, 12 lb., of the pole acts at its mid-point $G$, 6 ft. from each man (Fig. 64.)

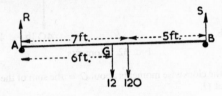

FIG. 64. Parallel forces

To calculate the magnitude of $S$, take moments about $A$, the shoulder of the other man. Then, since the moment of $R$ about $A$ is zero.

$$S \times 12 = 12 \times 6 + 120 \times 7$$
$$\therefore \quad 12S = 912, \text{ i.e. } S = 76 \text{ lb. wt.}$$

Since the upward forces, $R$, $S$, balance the downward forces,

$$\therefore \quad R + S = 120 + 12$$
$$\therefore \quad R = 132 - S = 132 - 76 = 56 \text{ lb. wt.}$$

Thus the fraction of the load supported at $A = \frac{56}{132} = \frac{14}{33}$; the fraction supported at $B = \frac{76}{132} = \frac{19}{33}$.

The work done by the man at $A$ = force × distance = $56 \times 3 = 168$ ft. lb. wt.

The work done by the man at $B$ = force × distance = $76 \times 3 = 228$ ft. lb. wt.

4. *Explain the term centre of gravity of a body. Distinguish between stable and unstable equilibrium, using some common object to illustrate your statements. Why are the engine and petrol tank placed as low as possible in the chassis of a motor-car?*

*A composite cube of 6 in. side consists of a block of wood 2 in. high and of specific gravity 0·6 with a block of glass 4 in. high and of specific gravity 2·4 above it. How high above the base is the centre of gravity of the cube? (C.W.B.)*

Suppose $ADCB$ is the glass, and $BCFE$ is the wood (Fig. 65). The volume of $ADCB = 6 \times 6 \times 4 = 144$ cu. in.; hence the weight = $144 \times 2 \cdot 4 = 345 \cdot 6w$, where $w$ is the wt. of 1 cu. in. of water.

The volume of $BCFE = 6 \times 6 \times 2 = 72$ cu. in.; hence the weight

$$= 72 \times 0.6 = 43.2 \text{ gm. wt.}$$

$$\therefore \quad \text{total weight} = 345.6 + 43.2$$

$$= 388.8 \text{ gm. wt.}$$

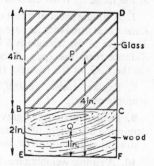

The C.G. of $ADCB$ is at $P$, where the distance of $P$ from the base $EF$ is 4 in.; the C.G. of $BCFE$ is at $Q$, where the distance of $Q$ from the base is 1 in. Suppose the C.G. of the whole block is at a distance of $x$ in. from the base. Then, taking moments about the base,

$$388.8 \times x = 43.2 \times 1 + 345.6 \times 4$$

$$\therefore \quad 388.8x = 1425.6$$

$$\therefore \quad x = \frac{1425.6}{388.8} = 3.7 \text{ in.}$$

FIG. 65. Centre of gravity

## EXERCISES ON CHAPTER V

**1.** The perpendicular distance from a door-knob to the edge $D$ of the door containing the hinges is 2 ft. The door is opened by a force of 4 lb. wt. applied perpendicularly to the door. Calculate the moment about $D$.

**2.** Calculate the moment about $D$ in question 1 if the door is opened with a force of 10 lb. wt. acting at an angle of (i) 60°, (ii) 30° to the door.

**3.** A boy riding a bicycle applies a vertical downward force of 20 lb. wt. to one pedal when the crank is 45° to the horizontal. If the length of the crank is 9 in., find the moment about the axle of the crank. Why is the force best applied perpendicularly to the crank?

**4.** A uniformly packed wooden case weighs 60 lb. wt., and rests on the ground; its horizontal dimensions are 4 ft. by 4 ft., and its height is 6 ft. A rope is attached to the middle of the upper edge and is pulled with a horizontal force $P$ which just makes the case tilt. Calculate the moment of the weight of the case about the turning edge and the magnitude of $P$.

**5.** A light horizontal beam is 6 ft. long and is pivoted 2 ft. from one end. The beam is kept in equilibrium by a 20 lb. wt. at the end nearer the pivot, a 5 lb. wt. at the middle, and a weight $W$ at the other end. Find the magnitude of $W$. What is the force on the pivot itself?

**6.** A nut-cracker has arms 9 in. long, and a nut is placed $\frac{1}{2}$ in. from the hinge. Calculate the resistance of the shell if it is broken by a force of 4 lb. wt. applied at the end of the nut-cracker. What is the reaction at the hinge ? What class of lever is the nut-cracker ?

7. A square sheet of metal of 8 in. side has an isosceles triangle of the same metal fitted with its base on one side so that there is no overlapping. The base of the triangle is 8 in. and the height is 9 in. Calculate the distance from the apex of the triangle of (i) the C.G. of the square, (ii) the C.G. of the triangle, (iii) the C.G. of the whole figure.

8. Explain *stable*, *neutral*, and *unstable equilibrium* and give *one* example of each. (*L.*)

9. Define the centre of gravity of a body, and explain how you would determine the weight of a closed umbrella if you were given a foot rule, a straight edge and a 2 oz. weight. (*L.*)

10. A simple weighing machine is made of a uniform bar 50 in. long, weighing 10 lb., and pivoted 1 in. from one end. Find the weight that must be suspended at the end of the long arm so as to balance a load of 730 lb. suspended at the end of the short arm. (*N.*)

11. What are the conditions necessary for three parallel forces to be in equilibrium ? Describe an experiment to demonstrate the truth of your statements. A non-uniform plank $AB$, of length 12 ft. and mass 40 lb., is supported on two trestles, $C$ and $D$, so that $AC = 1$ ft. and $DB = 3$ ft. The force exerted on $C$ is three times that exerted on $D$. Find the position of the centre of gravity of the plank. (*C.W.B.*)

12. What do you understand by the *centre of gravity* of a body ? How would you determine experimentally the centre of gravity of a uniform sheet of cardboard of irregular shape ? Explain how the stability of a body is affected by the position of the centre of gravity. (*L.*)

13. Describe the common beam balance and explain its action. A uniform metal rod $AB$, 24 in. long and of mass 4 lb., is fastened at $A$ to the surface of a metal ball 4 in. in diameter and of mass 21 lb., the centre of the ball being on the line of the axis of $AB$. The arrangement is suspended by a string attached to the rod at a point $C$, 2 in. from $A$. What mass must be suspended from $B$ for the rod to be horizontal ? (*L.*)

14. Describe briefly the common laboratory balance, pointing out the most important parts and their purpose. How would you use a balance whose arms are of unequal length to weigh accurately ? A certain balance was badly constructed, so that the centre of gravity of the beam was *above* the line joining the knife-edges instead of below it. When equal weights were placed in the two pans, the equilibrium of the beam

was unstable. Explain this, pointing out in your answer the meaning of "unstable equilibrium." (C.)

**15.** What do you understand by the *principle of moments* ? Describe how you would verify the principle experimentally for parallel forces. A uniform rod of length 50 cm. and mass 100 gm. is pivoted 8 cm. from one end A. Loads are weighed by attaching them to A, and moving a weight of 500 gm. along the rod, on the opposite side to the pivot, till the rod balances horizontally. Find the position of the weight when a load of 2·4 kgm. is attached to A and calculate the total upthrust at the pivot. (N.)

**16.** Explain what is meant by the moment of a force about an axis. Describe an experiment to illustrate the law of moments. A uniform ladder 16 ft. long and weighing 100 lb. rests on a horizontal floor. A man lifts one end of the ladder until it is inclined at an angle of 30° to the floor, the other end remaining at rest. Calculate (a) the force necessary to maintain the ladder in this position, assuming that the force is exerted at right angles to the ladder, (b) the work done in lifting the ladder to this position. (O. & C.)

**17.** Define *centre of gravity* and describe how you would find it experimentally for an irregularly shaped sheet of cardboard. It is proposed to order from a glazier a glass disc 18 in. in diameter with a 2 in. diameter circular hole with its centre 4 in. from the edge of the disc. This disc is to have a small hole bored by the glazier accurately through its centre of gravity. Where should the hole be bored ? (O.&C.)

**18.** Define the term *centre of gravity of a rigid body*, and explain with illustrative force diagrams why its position must be considered in building (a) a ship, (b) an omnibus.

If masses of 1, 2, 3, and 4 lb. are placed at the corners of a square of 10 in. side, taken in order, calculate the position of the centre of gravity. (O. & C.)

**19.** Define *centre of gravity*.

A uniform, plane sheet of metal ABCD of negligible thickness has AC = 5 in., AD = AB = 3 in., and CB = CD = 4 in. Draw the piece of metal to scale and mark the positions of the centres of gravity of the triangles ABC and ADC. Hence find the centre of gravity of the whole figure, explaining your method. How could you find your result experimentally ? (L.)

## CHAPTER VI

## MACHINES

WE turn next to a class of machines known as *pulleys*. Like all machines, such as the lever, pulleys are normally used to overcome a resistance or raise a weight by the application of a much smaller force, or by pulling in a more convenient direction. Pulleys are found in large factories and in railway stations; they are also used by builders for hauling heavy loads to higher floors, and at docks in the form of cranes for lifting heavy cargoes into and out of ships. (Fig. 70 (*a*) ).

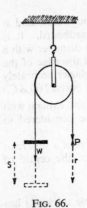

FIG. 66.
Simple pulley

**The simple pulley.** A simple pulley consists essentially of a wheel with a rope round it which is contained in a suspended framework (Fig. 66). Suppose $W$ is the load or weight in lb. wt. attached to one end of the rope, and $P$ is the effort in lb. wt. applied downwards at the other end which can just raise $W$. If the rope is very light and there is no friction round the wheel, the tension in the rope is $W$ lb. wt., and $P$ must hence equal $W$. However, there is considerably less strain in pulling the rope of a pulley downward to lift heavy loads than in raising them to the same height by a direct upward force, and the hauler can also use his own weight in this case.

The **mechanical advantage (M.A.)** of a machine is defined as the ratio $\dfrac{load}{effort}$, $\dfrac{W}{P}$. We shall see later that the load raised in a practical system of pulleys is much greater than the effort, so that the mechanical advantage is much greater than 1. In the case of the

90

simple pulley, however, $W = P$, neglecting friction; hence the mechanical advantage, $\dfrac{W}{P}$, is 1.

The **velocity ratio (V.R.)** of a machine is defined as the ratio $\dfrac{r}{s}$, where $r$, $s$ are the distances moved by the effort and load respectively in the same time. In general, the *effort* (applied force) moves a much greater distance than the load when the latter is raised, and the velocity ratio is thus much greater than 1. In the simple pulley (Fig. 66), the distance ($r$) moved by the effort is equal to the distance ($s$) moved by the load, since the rope which raises the weight is also used to apply the effort. Thus the velocity ratio, $\dfrac{r}{s}$, is 1 in this case. It should be carefully noted that the magnitude of the velocity ratio, unlike the mechanical advantage, is not affected by friction in the pulley system.

**The block and tackle.** In practice, pulleys are designed to provide a large mechanical advantage, so that a large load or weight can be raised with a small effort. Fig. 67 illustrates a useful system of pulleys, known as a *block and tackle* system, which contains two sets of pulleys with one continuous rope round them. The load of weight $W$ is attached to the lower set of pulleys, which is movable, while the upper set is supported from a beam and is fixed in position. Assuming that the pulleys and rope are light, and that no friction is present, the tension in every part of the rope is equal to $P$, the applied effort. Hence, since there are four portions of rope round the *lower* set of pulleys, which support $W$, it follows that $4P = W$. Thus the mechanical advantage, $\dfrac{W}{P}$, is 4 in this case. In practice the

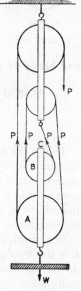

FIG. 67.

Block and tackle

mechanical advantage is less than this figure, since there is friction between the rope and the pulleys, and, moreover, the pulleys have weight. For example, if the two lower pulleys in Fig. 67 have a total weight of 50 lb. wt., then $4P = W + 50$, considering the equilibrium of the two lower pulleys. Consequently $4P$ is greater in magnitude than $W$, so that $\dfrac{W}{P}$ is *less* than 4.

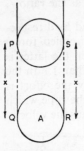

FIG. 68.

Velocity ratio

To find the velocity ratio (V.R.) of the system of pulleys, suppose that the weight $W$ is raised a distance $x$ when the effort $P$ is applied. If we imagine the lowest pulley $A$ raised this distance from $QR$ to $PS$ (Fig. 68), the length of rope made available $= PQ + SR = x + x = 2x$. The upper pulley, $B$ (Fig. 67), also rises a distance $x$, so that the movement of this pulley also makes a length $2x$ of rope available. Thus a total length, $4x$, of rope slips round the pulleys when the load is raised a distance $x$, and hence $4x$ is the distance moved by the effort.

$$\therefore \quad \text{velocity-ratio} = \frac{\text{distance moved by effort}}{\text{distance moved by load in same time}}$$

$$= \frac{4x}{x} = 4.$$

This is the magnitude of the velocity ratio obtained even when friction and the weights of the pulleys are taken into account, since the distance moved by the effort must always be $4x$ when the load is raised a distance $x$. It should again be noted that the magnitude of the mechanical advantage *is* affected by friction and the weights of the pulleys.

In general, the velocity ratio of this system of pulleys is $n$, where $n$ is the *total* number of pulleys in the system. The mechanical advantage is also equal to $n$ when friction and the weights of the pulleys are neglected. In the case of an odd number of pulleys, the upper fixed block has one more pulley in it than the lower movable. Thus, if there were 5 pulleys, the upper fixed block in Fig. 67 would have 3 pulleys, and the string would be connected to $C$ at the lower block. This system of pulleys is used at railway stations and engineering works for hauling heavy loads.

**Archimedean system of pulleys.** The basic form of another pulley system, sometimes known as the *Archimedean or first system of pulleys*, is shown in Fig. 69 (*a*). The load *W* is attached to a pulley $A_1$, which has a rope passing round it. One end of the rope is attached to a fixed point on a beam, while the downward effort *P* is applied to the other end of the rope with the aid of a small fixed pulley *F*.

Suppose that the pulleys are light and smooth, and that the rope is light. The tension in the rope is then equal to *P* lb. wt. The

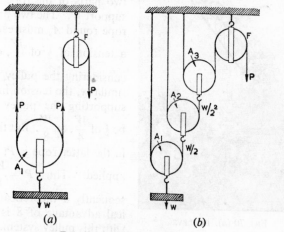

FIG. 69. Archimedean system

upward tensions in the two portions of the rope on either side of the pulley $A_1$ support the load *W*, and hence *W = 2P*. The *mechanical advantage* is thus ideally 2, although in practice it is less. The *velocity ratio* can be deduced by imagining *W* raised a distance *y*, when a length 2*y* of rope moves round the pulley *A* (see also Fig. 68). The effort, *P*, thus moves a distance 2*y*, and hence the velocity ratio is $\dfrac{2y}{y}$, or 2. It will be noted that the fixed pulley *F* offers a convenient means of applying the force *P* in a downward direction; as it is fixed, *F* is not a part of the pulley system, which comprises only $A_1$ in this case.

**4\***

A more practical form of the same pulley system, often used by builders, is shown in Fig. 69 (*b*). In this case there are three movable pulleys, $A_1$, $A_2$, $A_3$, with a load of weight $W$ attached to the lowest pulley. We shall assume the pulleys are very light and that friction is absent. The relation between the effort $P$ and $W$ can be found by noting that the tension in the rope round $A_1$ is $\frac{W}{2}$, since the tensions in the two parts of the rope round $A_1$ support $W$. The two parts of the rope round $A_2$ must each supply a tension of $\frac{1}{2}$ of $\frac{W}{2}$, or $\frac{W}{4}$, by considering the pulley $A_2$; and, similarly, the tension in the rope supporting the pulley $A_3$ must be $\frac{1}{2}$ of $\frac{W}{4}$, or $\frac{W}{8}$. But the tension in the latter rope is $P$, the effort applied. Thus $P = \frac{W}{8}$. Consequently a theoretical mechanical advantage of 8 is obtained with this pulley system.

FIG. 70 (*a*). PULLEYS
Large crane in operation at
Le Havre docks

By considering the lowest pulley $A_1$ and the attached weight $W$ to be raised a vertical distance $x$, and then following the lengths of rope moving round $A_1$, $A_2$, $A_3$, it can be shown that the effort $P$ moves a distance $8x$. This is left as an exercise to the reader. Thus the velocity ratio is 8.

In general, the velocity ratio is $2^n$, where $n$ is the number of pulleys; thus if $n = 3$, as in Fig. 69 (*b*), the velocity is $2^3$, or 8. The mechanical advantage is $2^n$ only if friction and the weights of the pulleys are neglected.

**Efficiency of a machine.** When a machine is said to be very efficient, we mean that nearly as much energy or work can be obtained from the machine as is supplied to the machine. When there is much friction in the moving parts of the machine, a good deal of the work supplied to the latter is used in overcoming the friction, with the result that a smaller amount of useful work is obtained from the machine.

The *useful work obtained* from a pulley system is that due to the

FIG. 70 (*b*). INCLINED PLANE

14-ton tank being loaded on Transporter with the aid of ramps

raising of the load of weight, $W$ lb. wt., say. If the latter moves upward a distance $s$ ft., then

$$\text{work obtained from machine} = \text{force} \times \text{distance}$$
$$= W \times s \text{ ft. lb. wt.}$$

The work done *on* the machine $= P \times r$ ft. lb. wt., where $P$ is the effort in lb. wt., and $r$ is the distance in feet it moves when the load is raised a distance $s$ ft.

The *efficiency* of any machine, mechanical or otherwise, is defined by the ratio

$$efficiency = \frac{work\ obtained\ from\ machine}{work\ done\ on\ machine}.$$

Thus, in the case of the pulley system,

$$\text{efficiency} = \frac{W \times s}{P \times r} = \frac{W/P}{r/s}.$$

But the mechanical advantage (M.A.) of a machine is the ratio $\dfrac{W}{P}$, and the velocity ratio (V.R.) is $\dfrac{r}{s}$, by definition.

Hence                    efficiency $= \dfrac{\text{M.A.}}{\text{V.R.}}$.

In practice, owing to friction and the weights of the pulleys, the mechanical advantage of a pulley system is less than its velocity ratio. Thus the efficiency is less than 1. If the machine were perfect, the efficiency would be 1, and the mechanical advantage would then be equal to the velocity ratio.

**Measurement of mechanical advantage, velocity ratio, efficiency.**
1. The *mechanical advantage* of a pulley system can be measured by attaching the end of the rope, where the effort is applied, to a scale pan, and using a known weight, e.g. 500 gm., as the load $W$, Fig. 71. The scale pan is now loaded until the weight is just raised, and the total weight on the scale pan together with the latter's weight is the magnitude of the effort $P$. If $P = 140$ gm., the mechanical advantage $= \dfrac{W}{P} = \dfrac{500}{140} = 3\cdot56$.

2. The *velocity ratio* of a pulley system is measured by observing the distance $s$ the load ($W$) moves when the free end of the string, where the effort ($P$) is applied, is moved down a distance $r$. Thus if $r = 18$ in., and $s = 4\cdot5$ in., the velocity ratio $= \dfrac{r}{s} = \dfrac{18}{4\cdot5} = 4$.

3. The *efficiency* of a pulley system can be measured by attaching the end of the rope, where the effort is applied, to a scale pan, and then adding weights as before until the effort $P$ raises the load $W$ steadily through a distance $s$, measured by a ruler. If the distance moved by the effort is $r$, then, from the definition of efficiency (p. 95),

$$\text{efficiency of pulley system} = \dfrac{Ws}{Pr}.$$

Alternatively, the efficiency of the system can be calculated after the mechanical advantage (M.A.) and velocity ratio (V.R.) have been measured separately, since the efficiency $=$ M.A./V.R.

**The Weston (differential) pulley.** The Weston pulley, which is capable of providing a large mechanical advantage and velocity ratio, is illustrated in Fig. 72. It consists of a wheel at the top, with two grooves of slightly different diameter, and a smaller wheel at the bottom. An endless chain, which is gripped by the

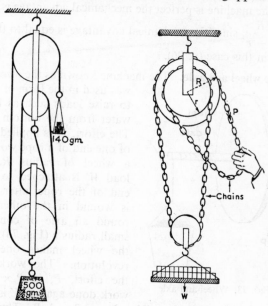

FIG. 71. Experiment on pulleys     FIG. 72. Weston pulley

notched wheels so that it does not slip, passes round the upper and lower wheels as shown. The load of weight $W$ is attached to the lower wheel, and a downward effort $P$ is applied to the chain round the outer groove of the upper wheel.

Suppose the chain is pulled so that the upper wheel is rotated through a complete revolution. The chain on the left of the wheel then moves up a distance $2\pi R$, where $R$ is the radius of the outer groove, but the chain round the right of the inner groove moves *down* a distance $2\pi r$, where $r$ is the radius of the inner groove. Consequently the chain round the lower wheel moves up a distance $2\pi R - 2\pi r$, and the load $W$ then ascends a distance $\frac{1}{2}(2\pi R - 2\pi r)$, or $\pi(R - r)$. Since the distance moved down by the force $P$ has been $2\pi R$, the velocity ratio $= \dfrac{2\pi R}{\pi(R - r)} = \dfrac{2R}{R - r}$. Thus if $(R - r)$ is made small, the velocity ratio is large.

If the machine is perfect the mechanical advantage is also given by $\dfrac{2R}{R - r}$, since the mechanical advantage is equal to the velocity ratio in this case (p. 96).

**The wheel and axle.** The machine known as the *wheel and axle*

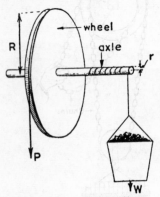

FIG. 73. Wheel and axle

was used in the form of a windlass to raise loads such as buckets of water from the bottom of a well. The effort, $P$, is applied by means of one end of a rope wound round a wheel of radius $R$, and the load $W$ is attached to the other end of the rope, after part of it is wound in the other direction round an axle of comparatively small radius $r$ (Fig. 73). Suppose the wheel makes one complete revolution. The work done by the effort, $P$, is $P \times 2\pi R$; the work done against $W$ is $W \times 2\pi r$.

If the machine is perfect, $W \times 2\pi r = P \times 2\pi R$. Thus

$$\frac{W}{P} = \frac{R}{r} = \text{mechanical advantage};$$

and consequently the larger the radius of the wheel, and the smaller the radius of the axle, the greater is the mechanical advantage.

**The principle of the screwdriver.** The *pitch* of a screw is the distance its point moves when the screw is rotated through a complete revolution. Suppose a screwdriver, $B$, is turned through a complete revolution so that a screw $A$ penetrates into wood, $W$, a distance equal to its pitch (Fig. 74). If the circumference of the handle $H$ of the screwdriver is $1\frac{1}{2}$ in. and the force exerted is $F$ lb. wt., the work done is $(F \times 1\frac{1}{2})$ in. lb. wt. If the resistance of the wood is 200 lb. wt. and the pitch of the screw is $\frac{1}{24}$ in., the work done against the resistance is $200 \times \frac{1}{24}$ in. lb. wt. Consequently, $F \times 1\frac{1}{2} = 200 \times \frac{1}{24}$, from which $F = 5\frac{5}{9}$ lb. wt.

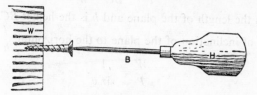

FIG. 74. Screwdriver

Thus the screwdriver is a machine which enables a large resistance to be overcome by the application of a small force; in general, the larger the diameter of the handle $H$, the greater is the mechanical advantage of the screwdriver.

**The inclined plane.** Loads, such as large drums, can often be more easily rolled up inclined planes to a lorry than lifted

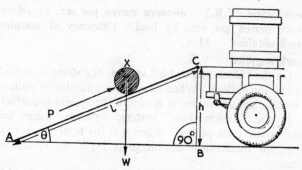

FIG. 75. Inclined plane

directly to the same height. Thus an inclined plane can act as a machine. (See Fig. 70 (b).)

To find the mechanical advantage, consider a barrel, $X$, rolled up an inclined plane from $A$ to $C$ by an effort $P$ applied along $AC$. (Fig. 75). The work done is then $P \times AC$. If the load is lifted vertically from $B$ to $C$, the work done is $W \times BC$, where $W$ is the weight of the barrel. If there is no friction, it follows that

$$W \times BC = P \times AC.$$

$$\therefore \quad \frac{W}{P} = \frac{AC}{BC} = \frac{l}{h},$$

where $l$ is the length of the plane and $h$ is the height $BC$. If $\theta$ is the angle of inclination of the plane to the horizontal, $\dfrac{l}{h} = \dfrac{1}{\sin \theta}$.

$$\therefore \quad \frac{W}{P} = \frac{1}{\sin \theta}.$$

The mechanical advantage is thus $\dfrac{1}{\sin \theta}$. Hence, the smaller the angle of inclination, $\theta$, the larger the mechanical advantage.

---

## Summary

1. **Mechanical advantage (M.A.) = load/effort** $\left( = \dfrac{W}{P} \right)$.

**Velocity ratio (V.R.) = distance moved per sec. by effort/ distance moved per sec. by load. Efficiency of machine** $= \dfrac{\text{work obtained}}{\text{work supplied}} = \dfrac{\text{M.A.}}{\text{V.R.}}$.

2. On the **"block and tackle" system**, one string is wound round all the pulleys. When $n$ is the total number of pulleys the mechanical advantage is also $n$, if the system is perfect.

3. In the **"Archimedean" system**, separate strings are wound round each pulley. When $n$ is the total number of pulleys, the mechanical advantage is $2^n$, if the system is perfect.

4. The Weston pulley has a large and small wheel, with an endless chain round it. When $R$, $r$ are the respective radii of the large and small wheel the mechanical advantage is $2R/(R - r)$. **The wheel and axle, or windlass, has a mechanical advantage $R/r$.** The inclined plane has a mechanical advantage $1/\sin \theta$, where $\theta$ is the inclination to the horizontal.

## WORKED EXAMPLES

**1.** *A screw with 6 threads to the inch has a circular head of circumference 15 in. It is turned by applying a force of $\frac{1}{2}$ lb. tangentially to the circular head and it raises a load of 20 lb. Find (a) the mechanical advantage of the screw, (b) the velocity ratio of the screw. (N.)*

(a) The mechanical advantage $= \dfrac{\text{load}}{\text{effort}} = \dfrac{20 \text{ lb.}}{\frac{1}{2} \text{ lb.}} = 40$.

(b) When the applied force moves once round the circular head, it has moved a distance, $x$, of 15 in. The point of the screw has then travelled a distance, $y$, of $\frac{1}{6}$ in., which is the horizontal distance between successive threads (see p. 98).

$$\therefore \quad \text{velocity ratio} = \frac{x}{y} = \frac{15 \text{ in.}}{\frac{1}{6} \text{ in.}} = 90.$$

**2.** *Draw a labelled diagram of a hand-operated machine suitable for lifting a load of about 3 cwt. through a distance of several feet; explain the relatively small effort required. Describe how you would determine experimentally the mechanical advantage of the machine. A machine with the velocity ratio of 5 requires 200 ft. lb. of energy to raise a weight of 50 lb. through a vertical distance of 3 ft. Calculate the efficiency and the mechanical advantage of the machine. (N.)*

A pulley system can be used; the relatively small effort $P$ required is explained from the relation between $P$ and the load $W$, as this shows that $P$ is a fraction of $W$ (see p. 91).

$$\text{Efficiency} = \frac{\text{energy obtained from machine}}{\text{energy supplied to machine}} \times 100\%.$$

But energy obtained = force × distance = 50 × 3 = 150 ft. lb.

$$\therefore \quad \text{efficiency} = \tfrac{150}{200} \times 100\% = 75\%.$$

To obtain the mechanical advantage, M.A., we utilize the relation

$$\text{efficiency} = \frac{\text{M.A.}}{\text{velocity ratio}} \times 100\% \text{ (p. 96).}$$

Substituting in this formula,

$$\therefore \quad 75 = \frac{\text{M.A.}}{5} \times 100$$

$$\therefore \quad \text{M.A.} = \frac{75 \times 5}{100} = 3\tfrac{3}{4}.$$

**3.** *Draw and explain diagrams of two different types of machine possessing a velocity ratio 6. A system of pulleys in two blocks, one fixed and the other movable, has a velocity ratio 6 and an efficiency of 75 per cent. An effort of 16 lb. wt. is able to support a useful load of 50 lb. wt. Find (a) the mass of the lower block, (b) the useful work expended when the effort moves through 20 ft. (Neglect friction.) (L.)*

A velocity ratio of 6 can be obtained with 6 pulleys of the system shown in Fig. 67. A lever whose "arms" are in the ratio 6 : 1 has a velocity ratio of 6; a wheel and axle whose radii are in the ratio 6 : 1 has a velocity ratio of 6.

In the system of pulleys, the lower block is movable, and the load $W$ of 50 lb. wt. is attached to it.

(a)    Since efficiency = $\dfrac{\text{mechanical advantage (M.A.)}}{\text{velocity ratio}} \times 100\%$,

$$\therefore \quad 75 = \frac{\text{M.A.}}{6} \times 100$$

$$\therefore \quad \text{M.A.} = 4\tfrac{1}{2}$$

$$\therefore \quad \frac{\text{Load}}{\text{Effort}} = 4\tfrac{1}{2}$$

$$\therefore \quad \frac{\text{Load}}{16 \text{ lb. wt.}} = 4\tfrac{1}{2}$$

$$\therefore \quad \text{Load} = 4\tfrac{1}{2} \times 16 \text{ lb. wt.} = 72 \text{ lb. wt.}$$

But useful load = 50 lb. wt.

$$\therefore \quad \text{mass of lower block} = 72 - 50 = 22 \text{ lb.}$$

Work done by effort = force × distance = $16 \times 20 = 320$ ft. lb. wt.

Since efficiency = 75% = $\tfrac{3}{4}$,

Work obtained from machine = $\tfrac{3}{4} \times 320 = 240$ ft. lb. wt.

$$\therefore \quad \text{useful work expended (i.e. lost)} = 320 - 240 = 80 \text{ ft. lb. wt.}$$

## EXERCISES ON CHAPTER VI

**1.** Name four practical applications of pulleys in industry.

**2.** By means of a machine a load of 100 lb. wt. is raised $3\tfrac{1}{2}$ ft. by a force of 50 lb. wt. moving through a distance of 10 ft. What work is done by the effort and what work is obtained from the machine? Calculate the efficiency of the machine, and explain why it is not 100%.

**3.** In question 2, find the mechanical advantage and the velocity ratio of the machine. How can the efficiency be calculated from these two values?

**4.** A pulley system has 6 pulleys, with a string passing continuously round them (see Fig. 67). The load attached to the lowest pulley is 300 lb. wt. Assuming there is no friction, and neglecting the weight of the pulleys, find the force just required to raise the load. What is the mechanical advantage of the system, its velocity ratio, and its efficiency?

**5.** A wheel and axle system is used for raising 60 lb. of water from the bottom of a well. The wheel has a radius of 2 ft. and the axle has a radius of 2 in. Find the effort required and the mechanical advantage of the system, neglecting friction.

**6.** A screw-driver has a handle of average radius $\frac{1}{2}$ in., and is turned through one revolution by an average force of 10 lb. wt. when driving a screw into wood. What work is done by the force ? If the pitch of the screw is $\frac{1}{24}$ in., calculate the resistance of the wood.

**7.** Define *velocity ratio* for a machine and give a diagram of a pulley system with a velocity ratio of 3. (*L.*)

**8.** Define *mechanical advantage*, *velocity ratio*, and *efficiency* of a machine. Describe, by the aid of a diagram, a machine, other than a simple lever, suitable for lifting a load of 200 lb. wt. by exerting a force of about 50 lb. wt. Explain briefly how you would determine experimentally the efficiency of your machine for a given load. (*L.*)

**9.** Name examples, *one* in each case, of machines which have (*a*) a velocity ratio greater than unity, (*b*) a velocity ratio less than unity. A machine is operated by a 200 lb. weight which falls 30 ft. and lifts a load of 1000 lb. Calculate (*c*) the loss of potential energy of the driving weight, (*d*) the greatest vertical height through which the load can be raised. (*N.*)

**10.** Explain the meaning of *mechanical advantage*, *velocity-ratio*, and *efficiency* of a machine. Give a diagram of any machine, other than a simple lever, which has a practical mechanical advantage of about 5. Describe briefly how you would measure the mechanical advantage and velocity-ratio of your machine for a given load, and explain why you would not expect its efficiency to be 100 per cent. A man lifts a load of 300 lb. by means of a machine, whose efficiency is 75 per cent. and velocity ratio $6\frac{2}{3}$. Find the effort required and the work done by the man in lifting the load 10 ft. (*C.W.B.*)

**11.** Explain what is meant by the term *machine*. Describe the hydraulic (*or* Bramah) press and explain its action, stating the principle on which this is based. Give a diagram and use it to illustrate the meaning of the terms *velocity ratio* and *mechanical advantage*. (*N.*) (*See* p. 124.)

**12.** What do you understand by a machine ? Give diagrams of two different types of machine which, if frictionless, would enable you to lift a useful load of 100 lb. wt. by applying an effort of 20 lb. wt. Describe an experiment to find how the efficiency of a large pulley system consisting of a fixed block and a movable block each containing two pulleys varies with the load. What result would you expect to find ? (*L.*)

**13.** Describe how you would determine the mechanical advantage of a wheel-and-axle machine. Prove that the velocity ratio of such a machine is equal to the ratio of the wheel radius to the axle radius. A wheel and axle whose radii are in the ratio of 6 to 1 has an efficiency of 80 per cent when the load is 420 lb. wt. Calculate (*a*) the least force necessary to lift this load, (*b*) the work done by the effort when this load has been raised a vertical distance of 2 ft. (*N.*)

**14.** Explain the meaning of *velocity ratio, mechanical advantage*, and *efficiency* of a machine, and find the relation between them. A wheel and axle have diameters of 12 in. and 3 in. respectively, and it is necessary to apply a force of 20 lb. wt. at the rim of the wheel in order to lift a mass of 70 lb. hanging from the axle. How much work must be done to raise this mass through 10 ft. ? What fraction of this work is used in overcoming friction in the bearings ? (*L.*)

**15.** Explain how it is possible by the use of a machine to raise a heavy weight by the application of a small force, and illustrate your answer by reference to (*a*) a lever, and (*b*) a block and tackle system of pulleys. A block and tackle has three pulleys in each block, and the weight of the lower block is negligible. If it is required to raise a mass of 120 lb. and the efficiency is 80 per cent, calculate the velocity ratio, the effort which must be applied, and the mechanical advantage.

(*O. & C.*)

**16.** Define *mechanical advantage, velocity ratio*, and *efficiency* as applied to machines. Why do the numerical values of mechanical advantage and velocity ratio differ from each other ? Draw a diagram of a block and tackle system of pulleys for which the velocity ratio is 5.

(*O. & C.*)

**17.** Draw a diagram of a pulley system consisting of four pulleys, pointing out how the diagrammatic representation differs from the practical system. State the velocity ratio of the system you draw and calculate the load which could be raised with an effort of 20 lb. wt. if the efficiency is 65 per cent. (*L.*)

# CHAPTER VII

## DENSITY AND SPECIFIC GRAVITY
## ARCHIMEDES' PRINCIPLE

IF an iron and an aluminium kettle are picked up, one notices immediately the difference in their weight. Cork is a light substance compared with glass or water. If, however, we wish to make a fair comparison between the relative heaviness or lightness of substances, we must take *equal volumes* of them.

The *mass per unit volume* of a substance is known as its *density*;

thus $$\text{density} = \frac{\text{mass of substance}}{\text{volume of substance}} \qquad . \qquad . \qquad (1)$$

The density of iron is 8·5 gm. per c.c., of aluminium 2·7 gm. per c.c., and of cork 0·24 gm. per c.c. Water has a density of 1 gm. per c.c. at a temperature of 4° C., while air has a density of about 0·0013 gm. per c.c. at standard temperature and pressure (p. 225). The densest liquid is mercury, which has a density of about 13·6 gm. per c.c.

The density of a substance may also be expressed in "lb. per cu. ft."; e.g. water has a density of about 62½ lb. per cu. ft. As each of the known elements has a different density, an element can be identified by measuring its density.

From the definition of density in equation (1) above, the mass of a substance can be calculated by

$$\text{mass} = \text{volume} \times \text{density} \qquad . \qquad . \qquad (2)$$

In this relation the mass is in gm. if the density is in gm. per c.c. and the volume is in c.c.; the mass is in lb. if the density is in lb. per cu. ft. and the volume is in cu. ft.

**Specific gravity.** Water is one of the most common substances in the world, and scientists have found it convenient to express the weight of a substance *relative to the weight of water which has* *the same volume.* As an illustration, 10 c.c. of iron has a weight of 85 gm. wt., and 10 c.c. of water has a weight of 10 gm. wt.; consequently the ratio between the weights of equal volumes of iron and water is 8·5. This number is known as the *specific gravity* of the iron, and, in general, the specific gravity of a substance is given by

$$\text{specific gravity } (s) = \frac{W}{w} \quad . \quad . \quad . \quad (3)$$

where $W$, $w$ are the respective weights of *equal volumes* of the substance and water. If we consider a volume of 1 c.c. (or 1 cu. ft.) of iron, the value of $W$ is the *density* of iron, and the value of $w$ is then the density of water. Thus the specific gravity of a substance is the magnitude of its density relative to that of water, and this may also be used as a definition of specific gravity.

The difference between the "density" and "specific gravity" of a substance should be carefully noted. "Density" has units of "gm. per c.c." or "lb. per cu. ft."; "specific gravity" has no units, since it is the ratio of two similar quantities, each of them weights, and has always the same magnitude for any given substance whatever the units in which the weights are measured. Further, "specific gravity" always relates the weight of the substance to that of water, whereas "density" makes no reference to water.

**Measurement of specific gravity.** The *specific gravity bottle* is

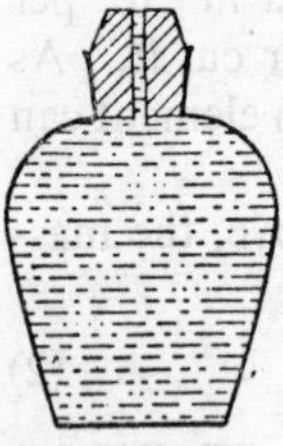

FIG. 76.
Specific gravity bottle

one designed to measure the specific gravity of liquids, and has a ground stopper with a fine hole through it so that the volume of a liquid filling it is always constant (Fig. 76). When the bottle is filled with a liquid and the cork is fitted firmly, the excess of liquid passes through the fine hole and can be wiped off.

(*a*) In measuring the *specific gravity of a liquid*, the weight ($W$) of it filling the bottle is measured. The liquid is then poured out, the bottle is cleaned, and the weight ($w$) of water filling it is then measured. The specific gravity of the liquid is calculated from the ratio $\frac{W}{w}$.

(b) The specific gravity bottle can also be used to find the *specific gravity of lead shot or sand*, substances which are insoluble in water.

As an illustration, suppose that the specific gravity of lead shot is required and that the following measurements are made—

| 1. Weight of bottle empty = 15·0 gm. | 3. Weight of bottle *partly* filled with lead shot = 95·2 gm. |
|---|---|
| 2. Weight of bottle filled with water = 70·4 gm. | 4. Weight of filled bottle when water is added to the lead shot = 143·1 gm. |

The weight of lead shot used is thus 95·2 − 15·0, or 80·2 gm. The weight of water filling the bottle is 70·4 − 15·0, or 55·4 gm.; and the weight of water filling the remainder of bottle when lead shot is present is 143·1 − 95·2, or 47·9 gm. It therefore follows that the weight of water having the same volume as the lead shot = 55·4 − 47·9 = 7·5 gm.

$$\therefore \text{ specific gravity of lead} = \frac{\text{weight of lead}}{\text{weight of water having the same volume}}$$

$$= \frac{80 \cdot 2}{7 \cdot 5} = 10 \cdot 7.$$

The reader should draw sketches of the bottle and its contents corresponding to the four weighings in the table above.

## ARCHIMEDES' PRINCIPLE

**The upthrust of a liquid.** An object experiences an *upward* force when it is immersed in a liquid, owing to the pressure exerted on it by the liquid. The upward direction of the force is evident when a piece of cork is held below the surface of water and then released; for the cork immediately rises to the surface. Although it is not so evident in other cases, there is an upward force on *any* object immersed in a liquid, such as a ship or a submerged submarine. If it were not for the upward force of the water we would not be able to swim, and ships would sink.

The upward force due to a liquid is known as its *upthrust* on the object concerned. ARCHIMEDES is reputed to have considered the problem of the upthrust on objects in a liquid while having his bath. He arrived at a correct solution, known as *Archimedes' Principle*, which states—

**The upthrust on an object immersed in a liquid is equal to the weight of liquid which it displaces.**

Thus if a piece of iron of volume 500 c.c. is completely immersed in water, the upthrust on it is equal to the weight of 500 c.c. of water by Archimedes' Principle, i.e. to 500 gm. wt., as the density of water is 1 gm. per c.c. If the same piece of iron is completely immersed in copper sulphate solution of density 1·1 gm. per c.c., the upthrust = the weight of 500 c.c. of copper sulphate solution = $500 \times 1·1 = 550$ gm. wt. Archimedes' Principle, it should be noted, applies to objects in gases as well as in liquids. An aeroplane in flight displaces its own volume of air, and the upthrust on it is equal to the weight of air displaced (see p. 115).

Archimedes' Principle is used by engineers designing ships and submarines; it is concerned, for example, in calculations on their stability, and on the safe load which they can carry (p. 113).

**Measurement of upthrust. Verifying Archimedes' Principle.** The upthrust of a liquid on an object can easily be measured on a

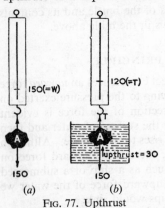

small scale. Suppose, for example, that a solid brass object $A$ is attached to a spring-balance which then registers 150 gm. wt., the natural weight of $A$ (Fig. 77 (*a*)). If the latter is now completely immersed in a liquid, the tension $T$ in the string joining $A$ to the spring-balance is less than 150 gm. wt. by the magnitude of the upthrust of the liquid (Fig. 77 (*b*)). If the spring-balance registers 120 gm. wt., this is the magnitude of the tension $T$. The upthrust of the liquid must hence be 30 gm. wt., the difference between the weight and the tension.

FIG. 77. Upthrust

The reading of the spring-balance when $A$ is immersed in the liquid is sometimes called the "apparent weight" of $A$ in the liquid; the difference between the natural and "apparent" weight of $A$ is known as the "apparent loss in weight." The "apparent weight," it should be noted, is the tension $T$ in the string in Fig. 77 $(b)$, and the "apparent loss in weight" is the upthrust of the liquid.

**Archimedes' Principle can be verified experimentally** by the "bucket" arrangement shown in Fig. 78$(a)$. A solid brass cylinder $B$ is attached to a

hollow cylinder $A$ *of exactly the same volume*, and the arrangement is suspended from a spring-balance which registers the total weight of $A$ and $B$, say 180 gm. wt. When $B$ is totally immersed in water, the spring-balance reading decreases to 140 gm. wt. on account of the upthrust of the water on $B$. But experiment shows that the reading increases to 180 gm. wt. again when the hollow vessel $A$ is filled to the brim with water (Fig. 78$(b)$ ). The upthrust on $B$ must hence be equal to the weight of the water filling $A$, and this is the same as the weight of the water displaced by $B$, as $A$ and $B$ have the same volume. This verifies Archimedes' Principle for the case of $B$ immersed in water.

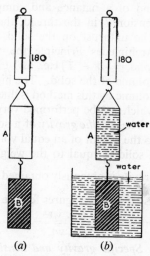

The experiment can be repeated with $B$ immersed in copper sulphate solution. It is then found that the spring-balance registers 180 gm. wt. again when $A$ is completely filled with the same solution. If, however, $A$ is completely filled with *water* while $B$ is completely immersed in

$(a)$          $(b)$

Fig. 78. Upthrust on $B$ in water is balanced by weight of water in $A$

the copper sulphate solution, the reading on the spring-balance is *less* than 180 gm. wt.; showing that the upthrust is greater than the weight of water having the same volume as $B$.

## Measuring specific gravity and density by using Archimedes' Principle.

*Specific gravity and density of a solid.* Suppose that a brass object $A$ weighs 150 gm. wt. in air, and that 132 gm. wt. is its "apparent weight" when $A$ is completely immersed in water (see Fig. 77). The upthrust of the water is then 18 gm. wt., and, by Archimedes' Principle, this is the weight of water *displaced by A*. Since the density of water is 1 gm. per c.c., the volume of $A$ is 18 c.c. Consequently

$$\text{density of brass} = \frac{\text{mass of } A}{\text{volume of } A} = \frac{150 \text{ gm.}}{18 \text{ c.c.}} = 8.3 \text{ gm. per c.c.}$$

It should be noted that Archimedes' Principle enables the volume of a solid of *any* shape to be accurately determined. The solid is weighed in air $(W)$, then suspended by a thread attached to one end of a balance and completely immersed in water, and the tension $T$ in the thread, the "apparent weight", is determined. The upthrust on the solid $= (W - T)$ gm. wt., and hence, by Archimedes' Principle, the volume of the water displaced by the solid $= (W - T)$ c.c., as 1 c.c. of water weighs 1 gm. This is the volume of the solid. The high degree of accuracy in determining volume by this method is due to the fact that weighing is a process which can be performed very accurately.

The *specific gravity* of a solid is the weight of the solid relative to the weight of an equal volume of water. Since the upthrust on a solid is equal to the weight of water it displaces, the specific gravity is obviously obtained from the ratio $\dfrac{\text{weight of solid}}{\text{upthrust}}$. If we use the above figures for the brass object $A$, the specific gravity of brass $= \dfrac{150 \text{ gm. wt.}}{18 \text{ gm. wt.}} = 8.3$.

***Specific gravity and density of a liquid.*** The specific gravity (or density) of a liquid can be determined by (a) weighing a solid, $B$, of any shape in air, (b) then determining the tension, $T_1$, in the thread supporting it when $B$ is completely immersed in the liquid, (c) then determining the tension, $T_2$, in the thread supporting it in water. The weight of liquid displaced by $B = W - T_1$, by Archimedes' Principle; and the weight of water displaced by $B = W - T_2$. As the volumes of the liquid and the water are the same, each corresponding to the volume of $B$, the specific gravity of the liquid $= \dfrac{\text{weight of liquid}}{\text{weight of equal volume of water}} = \dfrac{W - T_1}{W - T_2}$. Thus the specific gravity can be calculated.

**Floating objects and application of Archimedes' Principle.** So far we have discussed the upthrust on objects totally immersed in liquids; *floating* objects, such as ships, are also subject to an upthrust. From Archimedes' Principle, the upthrust on a submarine floating on the surface is less than when it is totally submerged, since the volume of liquid displaced is then less.

When an object, such as a ship, is floating on water, it is subject to two forces: (*a*) its weight; (*b*) the upthrust of the liquid. Now a floating object is in vertical equilibrium. *Hence the weight of liquid displaced (the upthrust) is equal to the weight of the object.* This is called the Principle of Flotation. See Experiment, p. 158c.

Suppose the volume of a block of wood floating in water is 50 c.c. and the density of the wood is 0·8 gm. per c.c.; the weight is then 40 gm. wt. The upthrust, which is equal to the weight, is hence 40 gm. wt. But the upthrust is equal to the

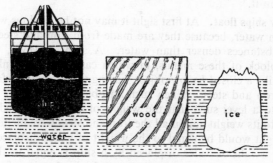

FIG. 79. Floating objects

weight of water displaced, by Archimedes' Principle, and consequently the block displaces 40 c.c. of water (Fig. 79).

It should be noted that the fraction of the volume of the block immersed is $\dfrac{40 \text{ c.c.}}{50 \text{ c.c.}}$, or $\frac{4}{5}$, which is the specific gravity of the wood. *Ice* has a density of about 0·9 gm. per c.c., and, by similar reasoning, it can be seen that ice floats in water with about $\frac{9}{10}$ths of its volume submerged.

Consider a block of iron of volume 30 c.c. and density 6 gm. per c.c.; its weight is 180 gm. wt. If the block is placed in mercury of density 13·6 gm.

per c.c. and floats with a volume $V$ c.c. immersed, the upthrust is $13 \cdot 6V$ gm. wt. by Archimedes' Principle.

$$\therefore \quad 13 \cdot 6V = \text{weight of iron} = 180$$

$$\therefore \quad V = \frac{180}{13 \cdot 6} = 13 \cdot 2 \text{ c.c.}$$

Suppose now that the iron block is held below water so that it is completely submerged; the upthrust of the water is then 30 gm. wt., since the volume of water displaced is 30 c.c. Now this is less than the weight, 180 gm. wt., of the iron; consequently the iron *sinks* when it is not supported.

It can thus be seen that, generally, *a solid floats in a liquid if its density is less than that of the liquid.* Mercury is a dense liquid and hence, as shown above in the case of iron, dense objects float in mercury. The density of the Dead Sea is very high owing to the concentration of salts in it, and swimmers float very easily in it.

**Why ships float.** At first sight it may not be obvious why ships float in water, because they are made from iron and steel, which are substances denser than water. A ship, however, is not a *solid* block of these metals, in which case it would sink, but is more like a closed *hollow* object, having the exterior surface lined with iron and steel. When such an object is placed in water, it sinks to a level such that the weight of water displaced by it is equal to its weight and, if all the iron and steel were melted down, its weight would be equal to the weight of water displaced.

The weight of water displaced by a ship is equal to the ship's weight. It is therefore common to refer to the size of a ship as "8000 tons wt. displacement," for example, which means that this is the weight of water displaced. Without a cargo, a ship stands more out of the water, since the weight of water displaced is less.

Submarines are vessels having separate large water-tight tanks in the bow and stern which can be filled with water or emptied; there are smaller tanks in the middle of the submarine. When the submarine is to be submerged, water is admitted to the tanks through valves. This increases the total weight, and horizontal rudders on the sides of the vessel are tilted slightly with the engines running so that the submarine gradually moves downwards. When the submarine wishes to surface again, the water is pumped out of the tanks and discharged into the sea, thus making the submarine lighter.

**The Plimsoll line.** A ship sinks in water until its weight is equal to the upthrust on it, which is the weight of water displaced. Now the density (or specific gravity) of sea-water varies in different parts of the world, according to whether the climate is hot or cold, and hence a ship sinks to different levels if its journey takes it to different regions. When a ship sinks too low in water, it is unsafe in heavy seas, and the danger increases if the cargo carried is very heavy. Accordingly, in the nineteenth

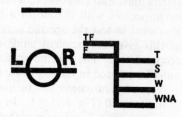

FIG. 80. Plimsoll line

century, PLIMSOLL agitated for a safety line to be drawn clearly on the side of a ship, and it is now illegal for a ship to be loaded so that the Plimsoll line, as it is called, is below the water level. Fig. 80 illustrates markings of the Plimsoll line; the letters $LR$ stand for Lloyd's Register of Shipping, $TF$ for tropical fresh water, $F$ for fresh water, $T$ for tropics, $S$ for summer, $W$ for winter, $WNA$ for winter in the North Atlantic.

**The spirit-level,** used by surveyors and builders to determine whether a plane is horizontal, is an application of the fact that a gas-bubble rises to the top of a liquid because it is less dense. It consists of a small closed glass vessel, $V$, containing a light oil,

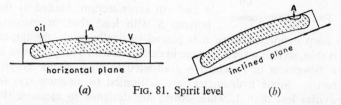

(a)      FIG. 81. Spirit level      (b)

and a small air-bubble $A$. When the spirit-level is placed on a *horizontal* plane, the bubble is exactly in the middle of $V$, Fig. 81 (a). When the spirit-level is on an inclined plane, the bubble $A$ rises to the highest point in $V$, Fig. 81 (b).

**The hydrometer.** Hydrometers are instruments which measure directly the density or specific gravity of a liquid. They are used, for instance, by inspectors to test the richness, or otherwise, of milk, which is intimately related to its specific gravity; the instrument is then given the name of *lactometer*. The strength of spirits must also conform to a certain standard, which is checked by inspectors with the aid of a hydrometer. The instrument is also used to test the specific gravity of the acid in accumulators, as this gives an excellent guide to the general condition of the accumulator. The specific gravity of the water in a ship's dock must be entered in the log book by the captain.

A *simple hydrometer* can be made from a test tube $A$ having its bulb flattened (Fig. 82 (*a*) ). In order to keep it upright when it

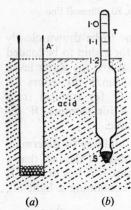

is placed in a liquid, some sand or lead shot is placed inside $A$. As we have learned, the test tube sinks until it displaces a volume of liquid equal to its weight; hence, the denser the liquid is, the *less* the test tube is immersed in it. The tube can be calibrated by pasting paper round it, then placing it in various liquids whose densities are known, and marking the paper in line with liquid surfaces.

Fig. 82 (*b*) illustrates *Twaddell's hydrometer*, named after the designer. It consists of a hollow glass tube $T$ with a uniform cross-section, loaded at the bottom $S$ with lead shot or mercury

(*a*)                  (*b*)

FIG. 82. Hydrometers

to keep it upright when it is placed vertically inside the liquid. $T$ is thin, so that a small change in density causes a large change in the movement of $T$, which is graduated to read specific gravity directly. Some hydrometers are calibrated to measure specific gravities less than 1, while others are designed to measure the specific gravities of liquids denser than water.

**A diver's equipment** includes two heavy lead weights, each above 40 lb., which are fastened to the front and back of the diving suit. These weights enable him to sink to the bottom of the ocean, by overcoming the upthrust of the water on his body. The diver also wears heavy boots, with lead soles and metal toe-caps, which assist him in maintaining stability (see Fig. 83).

**The density of cork.** A piece of cork, which floats on water, can be submerged by tying a weight to it, in which case the weight is known as a *sinker*. (Compare the case of the diver.)

Suppose the following weighings are determined—

Weight of cork in air = 4·4 gm. wt.

Weight of sinker in air = 50 gm. wt.

Weight of cork plus attached sinker in water = 30·5 gm. wt.

Weight of sinker in water = 43·7 gm. wt.

It follows that (*a*) the upthrust on the cork plus attached sinker in water = 4·4 + 50 − 30·5 = 23·9 gm. wt., (*b*) the upthrust on the sinker in water = 50 − 43·7 = 6·3 gm. wt. Thus the upthrust on the cork in water = 23·9 − 6·3 = 17·6 gm. wt. and hence the volume of the cork is 17·6 c.c.

$$\therefore \quad \text{density of cork} = \frac{\text{mass}}{\text{volume}} = \frac{4·4}{17·6}$$

$$= 0·25 \text{ gm. per c.c.}$$

**Archimedes' Principle applied to gases.** Although Archimedes' Principle was originally applied to solids in liquids, it also holds for solids in

FIG. 83. DIVER, FULLY EQUIPPED

*gases*, such as air. For example, an aeroplane or airship is subject to an upthrust equal to the weight of air displaced in the region where it is travelling, and in this connection we speak of the "buoyancy" of the air.

Early flights were made with the aid of large balloons filled with a light gas, such as hydrogen or hot air. The navigators travelled in a basket attached to the balloon, which rose when the upthrust due to the air displaced by the whole arrangement was greater than its total weight, including the weight of the light gas used.

To rise higher, ballast was pushed out of the basket; if a descent was required, some gas was let out of the balloon to deflate it. Enormous air-ships containing hydrogen or helium gas were designed up to 1931, but as the result of a series of disasters their development was largely abandoned.

Nowadays, meteorological stations use small balloons which contain instruments for determining the condition of the air. These balloons, owing to the upthrust on them, ascend to heights which an aeroplane is unable to reach, and return with valuable data which assists in weather-forecasting (see p. 241).

When an object is weighed in air, the result is slightly less than the true weight, since an upthrust acts on the object equal to the weight of the air displaced. In scientific experiments where highly accurate weighing is required, a correction must be made for the "buoyancy" effect of the air.

─────────── Summary ───────────

1. **Density** = **mass/volume** (the units are "gm. per c.c." or "lb. per cu. ft.").
**Specific gravity** = **weight of substance/weight of equal volume of water** = density of substance/density of water (specific gravity has no units).

2. **Archimedes' Principle states : The upthrust on an object immersed in a fluid = the weight of fluid displaced.** (When objects float, their weight = the weight of liquid displaced; principle of flotation.)

3. The specific gravity of a solid = wt. of solid/apparent loss in wt. in water. The specific gravity of a liquid = apparent loss in wt. of solid in liquid/apparent loss in wt. in water.

### WORKED EXAMPLES

1. *Define specific gravity. Describe a method of finding the specific gravity of a piece of cork. A barge 120 ft. long and 20 ft. broad, whose sides are vertical, floats partially loaded in water. How many inches will it sink when 12·5 tons of cargo are added? (1 cu. ft. of water weighs 62·5 lb.) (L.)*

The increased upthrust of the water = 12·5 tons = 12·5 × 2240 lb. wt.

$$\therefore \text{ increase in volume of water displaced} = \frac{12 \cdot 5 \times 2240}{62 \cdot 5} = 448 \text{ cu. ft.}$$

$$\therefore \text{ distance barge sinks} = \frac{\text{volume of water}}{\text{area of water}}$$

$$= \frac{448}{120 \times 20} \text{ ft.}$$

$$= \frac{448 \times 12}{120 \times 20} \text{ ft.}$$

$$= 2 \cdot 24 \text{ in.}$$

**2.** *State the law of flotation and describe an experiment to verify it. A cylindrical glass beaker, of uniform cross-sectional area 50 sq. cm. and of height 12 cm., floats in water with one-third of its volume immersed. What volume of glycerine, of specific gravity 1·25, can be gently poured into the beaker before it completely sinks into the water?* (*C.W.B.*)

The weight of the beaker = weight of water displaced, since the beaker floats.

$$\therefore \text{ weight of beaker} = \tfrac{1}{3} \times 50 \times 12 = 200 \text{ gm. wt.}$$

When the beaker is completely immersed in the water, the upthrust of the water = wt. of water displaced = $50 \times 12 = 600$ gm. wt. But the weight of the beaker is 200 gm. wt., and acts downwards.

$$\therefore \text{ weight of glycerine required to sink beaker} = 600 - 200 = 400 \text{ gm. wt.}$$

$$\therefore \text{ volume of glycerine} = \frac{400}{1 \cdot 25} = 320 \text{ c.c.}$$

**3.** *State Archimedes' Principle. Describe an experiment in which this principle is used to determine the specific gravity of a lump of aluminium. Show clearly how the specific gravity is calculated. What volume of wood, of density 0·9 gm. per c.c, must be tied to a lump of aluminium of mass 10 gm. and density 2·7 gm. per c.c., in order that the whole mass may have a mean density of 1 gm. per c.c., i.e. that it will just float?* (*L.*)

Let $V$ = vol. of wood in c.c.

Then mass = volume × density = $0.9 V$ gm.

$$\therefore \text{ total of mass of wood + aluminium} = (0 \cdot 9V + 10) \text{ gm.} \qquad . \qquad . \quad (1)$$

$$\text{Also, total volume of wood + aluminium} = \left(V + \frac{10}{2 \cdot 7}\right) \text{ c.c.} \qquad . \qquad . \quad (2)$$

$$\text{since the vol. of the aluminium} = \frac{10}{2 \cdot 7} \text{ c.c.}$$

$$\therefore \text{ the mean density of the whole mass} = \frac{\text{total mass (1)}}{\text{total volume (2)}}$$

$$= \left(\frac{0 \cdot 9V + 10}{V + \dfrac{10}{2 \cdot 7}}\right) \text{ gm. per c.c.}$$

5

But mean density = 1 gm. per c.c.

$$\therefore \quad \frac{0 \cdot 9V + 10}{V + \dfrac{10}{2 \cdot 7}} = 1$$

$$\therefore \quad 0 \cdot 9V + 10 = V + \frac{10}{2 \cdot 7}$$

$$\therefore \quad 10 - \frac{10}{2 \cdot 7} = V - 0 \cdot 9V = 0 \cdot 1V$$

$$\therefore \quad V = 10 \left( 10 - \frac{10}{2 \cdot 7} \right) = 63 \text{ c.c. (approx.)}$$

**4.** *A balloon is inflated, at constant temperature and atmospheric pressure, from 20 cylinders each containing 4 cu. ft. of hydrogen at a pressure of 51 atmospheres. Find (a) the volume of the balloon when inflated, (b) the maximum load, including envelope and fittings, that can be lifted. The densities of air and hydrogen at atmospheric pressure are 0·08 and 0·0056 lb. per cu. ft. respectively. Explain each calculation as fully as you can. (N.)*

(a) The total volume of the gas is 80 cu. ft., at a pressure of 51 atmospheres. Since its temperature remains constant, the volume, $V$, of the gas at a pressure of 1 atmosphere is given by Boyle's law (see p. 133). Hence, as $pV$ is a constant,

$$1 \times V = 51 \times 80, \text{ or } V = 4080 \text{ cu. ft.}$$

As $20 \times 4$ or 80 cu. ft. remains in cylinders, 4000 cu. ft. is released.

(b) The balloon displaces a volume of air equal to 4000 cu. ft. Hence from Archimedes' Principle, the upthrust on the balloon = weight of air displaced = $4000 \times 0 \cdot 08 = 320$ lb. wt. The weight of hydrogen, which acts downwards, = $4000 \times 0 \cdot 0056 = 22 \cdot 4$ lb. wt.

$$\therefore \quad \text{net upward force on balloon} = 320 - 22 \cdot 4 = 297 \cdot 6 \text{ lb. wt.}$$

$$\therefore \quad \text{maximum load that can be lifted} = 297 \cdot 6 \text{ lb. wt.}$$

## EXERCISES ON CHAPTER VII

**1.** The side of a solid cube of iron is 10 cm. Calculate its mass if the density of iron is 6·5 gm per c.c. What is the volume of a piece of iron which weighs 130 gm. wt. ?

**2.** The specific gravity of brass is 8·5. What does this statement mean ? Find the weight of 2 cu. ft. of brass if 1 cu. ft. of water weighs $62\frac{1}{2}$ lb. wt.

**3.** A fully-equipped diver has a total volume of $2\frac{1}{2}$ cu. ft. Find the upthrust on him when he is totally immersed in (*i*) water of density $62\frac{1}{2}$ lb. per cu. ft., (ii) salt water of density 66 lb. per cu. ft.

**4.** A solid of 800 gm. wt. is suspended by a string in water. The tension in the string is 600 gm. wt. Calculate (i) the upthrust on the solid, (ii) the volume of the solid, (iii) its density.

**5.** A piece of marble of 200 gm. wt. is suspended in turn in water and in copper sulphate solution, being totally immersed each time. If the respective apparent weights of the marble are 180 and 176 gm. wt., calculate the density of the copper sulphate solution. *Explain* your calculation.

**6.** A cube of cork of mass 8 gm. floats in water. What is (i) the upthrust on the cork, (ii) the volume immersed?

**7.** A hydrometer has a weight of 40 gm. wt. What is the volume immersed when it floats in water? What is the volume immersed if it floats in (i) acid of density 1·25 gm. per c.c., (ii) oil of density of 0·8 gm. per c.c.? Draw a sketch of the hydrometer when it floats in the oil, in water, and in the acid.

**8.** A ship has a volume of 200,000 cu. ft. up to the Plimsoll line, and is immersed in sea-water of specific gravity 1·1. The ship weighs 5000 tons wt., and a cargo of 1060 tons wt. is taken on board. Find if the ship will be allowed to leave port. (Density of water = 62$\frac{1}{2}$ lb. per cu. ft.)

**9.** State the principle of Archimedes and describe how it may be verified experimentally. A wooden block, of specific gravity 0·6 and of mass 100 gm., floats in water with a block of metal, of specific gravity 8·0 and of mass 30 gm., tied underneath it. What fraction of the volume of the wood is submerged? (*L.*)

**10.** Distinguish between *density* and *specific gravity*. How would you use a specific gravity bottle to determine the density of (*a*) turpentine, (*b*) sand? (*L.*)

**11.** A mine is anchored to the sea bed so as to lie a little way below the surface of the water. Show on a diagram the forces acting on the mine and explain why it rises to the surface if the cable breaks. (*N.*)

**12.** Explain why a ship made of iron will float whereas a piece of iron sinks. (*N.*)

**13.** What is the relation between the upthrust on a body and the liquid displaced when the body is immersed in the liquid? How would you demonstrate that, in the case of a solid suspended by a thread and immersed in a liquid which is less dense than the solid, there is a downthrust on the liquid equal to the upthrust on the solid? A mine, of volume 25 litres and mean specific gravity 0·9, is held under sea-water by a light chain fastened to the sea-bed. Calculate the upthrust on the mine if the sea-water is of specific gravity 1·02. Hence determine the tension of the chain. (*C.W.B.*)

**14.** A uniform metre stick is suspended at its centre and carries a block of metal suspended by a thread at one end. It is balanced by hanging a mass of 100 gm. at a distance of 30 cm. from the centre. On immersing the block in water, the 100 gm. mass has to be moved 3 cm. towards the centre to restore the balance. Calculate the specific gravity of the metal. (*L.*)

**15.** (*a*) Describe how you would determine the specific gravity of a specimen of cork. (*b*) At the bottom of a well which is half full of water, a bucket rests on its side with a rope attached to the handle. Describe (or sketch a graph to show) how the tension in the rope varies as the bucket is slowly hauled to the top of the well. Explain the changes in tension which occur. (*N.*)

**16.** State Archimedes' Principle and briefly describe an experiment to verify it.

If you were provided with a spring balance, some cotton, a lump of quartz, a beaker of distilled water and one of methylated spirit, how would you determine the specific gravity of (*a*) the quartz, (*b*) the methylated spirit ? Explain your calculations. (*C.W.B.*)

**17.** Describe how you would make a simple form of hydrometer with the help of a uniform test tube and some lead shot. What is the principle underlying its action ? The stem of a hydrometer has a diameter of 4 mm., and it floats in water with the bulb and part of the stem immersed, the volume immersed being 5 c.c. How much more of the stem will be immersed in a liquid of specific gravity 0.80 ? (*L.*)

**18.** Draw a labelled diagram of a common (constant-weight) hydrometer and explain the special features of its design. A rod, 20 cm. long and of uniform cross-section, floats upright and is to be used as a hydrometer. If its mean density is 0·8 gm. per c.c., how far will the 1·0 and 1·1 scale marks be from the bottom ? (*C.*)

**19.** State Archimedes' Principle and explain how the principle is applied to (*a*) the hydrometer, (*b*) the balloon. An ice-berg has 10,000 cu. ft. of its volume above the surface of the sea. Calculate the volume hidden beneath the water surface, given that the specific gravity of ice is 0·92 and of sea-water 1·03. (*O. & C.*)

**20.** Define *specific gravity*. Describe how you would measure the specific gravity of a liquid such as alcohol. A cube of cork, with sides of length 5 in., floats in water with the top face horizontal and $1\frac{1}{2}$ in. above the surface of the water. If it floats in a similar manner in a liquid of density 0·8 gm. per c.c., what distance will the top face be above the surface of the liquid ? (*O. & C.*)

# CHAPTER VIII

## PRESSURE OF LIQUIDS
## ATMOSPHERIC PRESSURE

IF a heavy parcel is wrapped round with thin string and carried by the string for some distance, a considerable pain in the fingers is experienced. We say that the cause of the pain is the *pressure* exerted by the string on the flesh. If a wide handle is provided for the same parcel, however, little pain is then experienced, despite the fact that the force on the finger (the weight of the parcel) is the same as before. This is due to the larger area of contact between the handle and the flesh; there is less force per unit area on the flesh than when thin string is used.

Scientists define **pressure** as *force per unit area.* Thus the pressure on the head of a diver at work on a sea-bed is the weight of water affecting each square inch of the surface of his head. The units of pressure are "lb. wt. per sq. in." or "dynes (or gm. wt.) per sq. cm." If the average pressure on a diver's head-gear of surface area 100 sq. in. is 50 lb. wt. per sq. in., the *force* on the head-gear is 5000 lb. wt., the product of the area and the pressure. If the total force due to the pressure of water on a plate in the side of a ship is 200,000 lb. wt. and the area of the plate is 5000 sq. in., the average pressure on the latter is 40 lb. wt. per sq. in. Thus the average pressure (*p*) over a surface is given by

$$p = \frac{\text{total force on surface } (P)}{\text{area of surface } (A)} \qquad . \qquad . \quad (1)$$

*Hence "pressure" is not the same as "force."* The force (*P*) on a surface of area *A* can be calculated, if the average pressure *p* is known, by the relation

$$P = p \times A \qquad . \qquad . \qquad . \quad (2)$$

**The magnitude of the pressure round a point in a liquid.** We shall now obtain a formula for the pressure at a place below the surface of a liquid of density $d$ gm. per c.c. Suppose a horizontal area $B$ of 1 sq. cm. is placed $h$ cm. below the surface. The pressure on $B$ is due to the weight of liquid acting on it, and the volume of this liquid $= 1 \times h = h$ c.c. (Fig. 84).

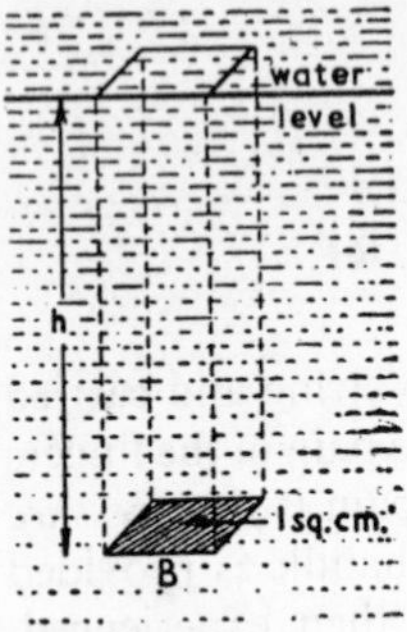

Hence the mass of the liquid $=$ volume $\times$ density $= hd$ gm., and its *weight* (which is a force, p. 33) $= hd$ gm. wt.

$$\therefore \text{ pressure on } B = \frac{\text{force on } B}{\text{area of } B} = \frac{hd}{1}$$

$$= hd \text{ gm. wt. per sq. cm.}$$

In general, **pressure** $= hd$ . . . (3)

**FIG. 84.**
Pressure formula

If $h$ is in ft. and $d$ is in lb. per cu. ft., the pressure is in lb. wt. per sq. ft. If $h$ is in cm. and $d$ is in gm. per c.c., the pressure $p$ is in gm. wt. per sq. cm., or, since $g$ (980) dynes $= 1$ gm. wt.

**pressure** $= hdg$ **dynes per sq. cm.** . . (4)

**A liquid finds its own level.** From the formulae for the pressure in (3) or (4), it follows that the pressure in a liquid increases proportionately with the depth $h$ of the place below the surface, and with the density $d$ of the liquid. The pressure at the bottom of a narrow vessel filled to a height of 10 in. with mercury is thus the same as the pressure at the bottom of a wide vessel filled to the

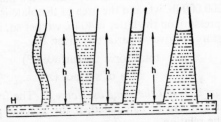

FIG. 85. Pressure and depth

same height with mercury. Of course the weight of the mercury in the latter case is much more than in the former case, but as it is distributed over a much bigger area at the bottom the pressure (force per unit area) is the same in each case.

When a liquid is poured into a vessel having tubes of different shapes and sizes connected together, the liquid is observed to rise to the same height, *h*, in each of the tubes (Fig. 85). Since the pressure along the same horizontal line *HH* in the stationary liquid must be the same (if it were not, the liquid would move until the pressure at different points on *HH* were equalized), this is another demonstration of the fact that the pressure at a point in a given liquid depends only on the depth of that point below the surface. In popular language, it is said that "water finds its own level" in different parts of the same vessel. A water gauge in a boiler utilizes the fact that "water finds its own level", and is used to determine the level of the water in the boiler, which cannot be seen directly (Fig. 86).

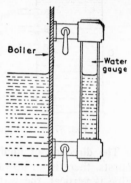

FIG. 86. Water gauge

Water flows naturally from one place to another at a lower level; without some pumping device it is impossible for water to flow from one place to another at a higher level. The water supply in many parts of the country is obtained from lakes situated high above the area concerned, and the water flows naturally downwards. Water-dams and lochs are places where water is stored up to increase the potential energy of the water, and it flows naturally downwards through pipes in hydro-electric installations.

**A simple demonstration** which shows roughly that the pressure increases with the depth is illustrated in Fig. 87. A tall can *D* has short corked tubes of similar size at depths *A*, *B*, *C*, and is filled with water to a level above *A*. When the corks are taken out, the water flowing from *C* is observed to travel the farthest distance from the can, while the water from *A* travels the shortest distance. Consequently the pressure of the water at *C* is greater than at *B*, which in turn is greater than that at *A*. See also Experiment, p. 158b.

The hot water from a tap on the ground floor of a building issues with a greater speed than the water from a tap on the third floor, for the same reason, and the bottom of a dam is made much thicker than the top because the pressure of the water increases with the depth.

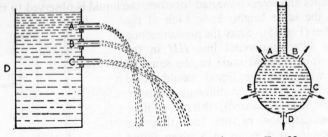

FIG. 87.          Liquid pressure and direction          FIG. 88.

**The direction of liquid pressure.** If a horizontal water-pipe bursts, there is an upward rush of water from the pipe. If the pipe is vertical, the water rushes out in a horizontal direction. These observations show that the pressure of the water is *perpendicular* to the surface on which it acts.

The direction of the liquid pressure at a point on a surface can be demonstrated by means of a glass vessel with small openings such as *A, B, C, D, E,* all round the spherical-shaped bottom. If

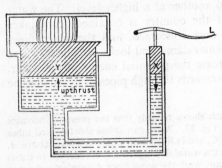

FIG. 89. Hydraulic press

the vessel is filled with water it immediately pours out in directions perpendicular to the surface at the respective points, as illustrated by the arrows in Fig. 88, which shows that the pressure at any point of a surface acts *normally* to the surface at that point.

**The hydraulic press.**

In 1795 JOSEPH BRAMAH invented a *hydraulic press*, a machine enabling a large force to be obtained from an applied small force by the aid of a liquid.

When a pressure is exerted on a liquid, it is transmitted equally in all directions. This is known as the *Law of Transmission of Pressure*. In the last war depth charges exploding at some point below the water gave rise to enormous forces at least 50 yds. from the place of the actual explosion. This transmission of pressure through a liquid is utilized in the hydraulic press, whose principle of action is illustrated in Fig. 89.

By means of a lever $L$ a small force, 20 lb. wt. for example, is applied directly to a piston $X$ of cross-sectional area 30 sq. in., say. The *pressure* on the liquid due to the movement of $X$ is then $\frac{20}{30}$, or $\frac{2}{3}$ lb. wt. per sq. in., and this is transmitted through the water filling the vessel to a tight-fitting piston $Y$ of much greater area of cross-section than $X$. If $Y$ has an area of 9 sq. ft., or 1296 sq. in., the *upward force $F$* exerted on it is given by $F =$ pressure $\times$ area $= \frac{2}{3} \times 1296 = 864$ lb. wt. Thus a large force is obtained by the application of the comparatively small force of 20 lb. wt. It can be seen that the theoretical mechanical advantage (p. 90) of the machine is the ratio $\dfrac{\text{area of } Y}{\text{area of } X}$. Further, from the principle of conservation of energy, the work done by the small piston $X =$ the work done by

FIG. 90. HYDRAULIC PRESS FORGING HOT STEEL

the large piston $Y$, neglecting friction. Consequently the distance moved by $Y$ is much smaller than the corresponding distance moved by $X$.

5*

The hydraulic press is used extensively in industry. One form of it, for example, is used in steel works to operate a large piston which descends on, and forges, white-hot steel held beneath it (Fig. 90). The press is also used to compress bales of wool. Another machine, which exerts a large force by the transmission of pressure through oil, is the *hydraulic jack*, frequently used for raising cars and other vehicles.

## ATMOSPHERIC PRESSURE

**Torricelli's vacuum.** If a long tube $A$, about a yard long, is filled with water and inverted in a vessel $V$, also containing water, experiment shows that the water remains in $A$ (Fig. 91 ($a$) ). It is kept up by the *atmospheric pressure*, which acts on the water surface $S$, and is transmitted through the liquid to the water in $A$; just as the pressure on the smaller piston of the hydraulic press is transmitted through the liquid to the larger piston (p. 125). If the

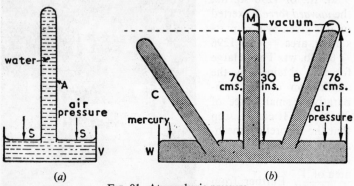

FIG. 91. Atmospheric pressure

height of the tube inverted in $V$ is increased by using a longer tube, experiment shows that the atmospheric pressure is easily able to support this height of water. In 1662, however, TORRICELLI, an Italian scientist, completely filled a tube about a yard long with mercury, the densest liquid, and inverted it in a vessel of mercury. He found, that a space $M$ *was left above the mercury* in the tube, and that the mercury occupied a length of about 76 cm., or 30 in. (Fig. 91 (*b*) ). Torricelli suspected that a *vacuum* had been obtained in *M*, since the air had been excluded from the tube in the first place, and that consequently the pressure of the air was equal to the pressure at the bottom of a height of 76 cm., or 30 in., of mercury. If the tube is inclined slightly from the vertical, more mercury goes into it (see *B*, Fig. 91 (*b*) ), but the height of the mercury level in the tube remains the same. We can understand this result from the fact that the pressure at the bottom $= hd$, where $h$ is the vertical height of the mercury tube above the surface in the vessel. If the tube is inclined at a large angle to the vertical, as shown by *C* in Fig. 91 (*b*), the whole of the tube is filled with mercury, as the top of the tube is now less than 76 cm. above the mercury surface in the vessel *W*.

In order to test Torricelli's theory that the mercury in the tube was supported by air pressure, PASCAL (1623-1662) wrote to a relative who lived in a mountainous district about 5000 ft. above sea-level and asked him to repeat the experiment. His idea was that the air-pressure at a place, which is due to the weight of the air above that place, decreases as we go higher, and the height of mercury supported in the tube should be less than 76 cm. in this case. This was found to be true. Further, if the lower part of the vertical tube and the vessel *W* in Fig. 91 (*b*) is contained, in a sealed glass vessel from which the air is gradually pumped out, the mercury in the tube is observed to diminish slowly in height until it stands only slightly higher than the level of the mercury surface in *W*. The sealed glass vessel is then nearly evacuated.

**The magnitude of atmospheric pressure. The Magdeburg hemispheres experiment.** The magnitude of the air-pressure is frequently represented by the height of mercury which it will support. A unit of pressure known as *one atmosphere* is defined as that pressure which supports a column of mercury 76 cm. high.

It is inaccurate, however, to state that "pressure" (force per unit area) is measured in the same units as "length". The pressure $p$ when the column of mercury supported is 76 cm. long is actually given by $p = hd$, where $h$ is 76 cm., $d =$ density of mercury = 13·6 gm. per c.c., and $p$ is in gm. wt. per sq. cm. (see p. 112). Thus

$$p = 76 \times 13·6 = 1034 \text{ gm. wt. per sq. cm.}$$

$$= 14·5 \text{ lb. wt. per sq. in. (approx.)}$$

If $h$ is the height of a column of *water* supported by one atmosphere of pressure, then, as $p = hd$,

$$h \times 1 = 76 \times 13·6,$$

since the density of water is 1 gm. per c.c. Thus $h = 76 \times 13·6$ = 1034 cm. = 34 ft. (approx.). One atmosphere of pressure therefore supports a height of about 34 ft. of water, and accordingly it is not surprising that water filling a long inverted glass tube placed in a vessel of water is easily supported by the atmospheric pressure (see Fig. 91 (*a*) ).

From the above, it follows that our head and body support a pressure of about 14½ lb. wt. per sq. in. due to the atmosphere; for an area of 5 sq. ft. the total force is hence about 10,000 lb. wt. We suffer no discomfort, however, as the pressure of the blood and its dissolved gases counteracts the atmospheric pressure; but airmen ascending to great heights need special apparatus on account of the reduced air-pressure, which results in bleeding through the nose and other parts where the tissues are thin, owing to the greater pressure of the blood. Divers working at great depths on salvage work, and men engaged in tunnelling operations in compressed air (situations where the pressure is much greater than atmospheric pressure) have to ascend by degrees to the surface, as a sudden decrease in pressure is harmful to the body.

OTTO VON GUERICKE of Magdeburg was keenly interested in air pressure, and is said to have invented the first pump. In 1651 he carried out a demonstration of air-pressure which aroused great curiosity and wonder. He used two large strong copper hemispheres which were made air-tight when they were joined together. The air was pumped out of the hemispheres, and a horse was harnessed to each of them; the hemispheres could not be pulled apart. A team of eight horses on each side was finally required to separate them ! As the atmospheric pressure is about 14½ lb. wt. per sq. in., the total force keeping the large hemispheres together was enormous, perhaps many tons weight in magnitude, and only the pull of sixteen horses could overcome the force.

**The measurement of atmospheric pressure.**    When the weather improves, the atmospheric pressure increases; as the weather worsens, the pressure decreases.    Throughout the world, meteorological stations are engaged in taking readings of the air pressure in order to forecast the weather, a matter of vital importance to sailors, airmen, and farmers, among others.

**Fortin's barometer.**    The *barometer* is an instrument for measuring air pressure.    The most accurate form of barometer was designed by FORTIN, and is now commonly used in laboratories. (Fig. 92).    As in the days of Torricelli, a vertical glass tube containing mercury is placed in a leather bag $W$ containing this liquid, and a vacuum exists above the mercury in the tube.    The

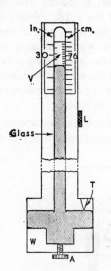

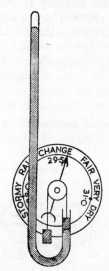

FIG. 92. Fortin barometer    Fig. 93. Siphon barometer

height of the top of the mercury can be read by means of a fixed metal scale graduated in centimetres on one side and inches on the other side, and a vernier scale $V$, which is adjusted by a screw $L$, is used to measure the height to a greater degree of accuracy. A fixed ivory tooth $T$ is an important feature of the Fortin barometer. The metal scale is graduated from the bottom of $T$. Before the reading of the mercury height is taken, the basic level of the surface in the leather bag $W$ is adjusted by means of a screw $A$ until it just touches the bottom of the tooth. In this way Fortin overcame the difficulty of allowing for the rise and fall of the mercury in the tube; this affects the mercury level in the containing vessel, from which the height of the mercury in the tube must always be measured. (Fig. 91 (*b*).)

**The siphon barometer.** Fig. 93 illustrates a *siphon barometer*, used domestically. It consists of a bent tube containing mercury and having a vacuum at the closed end. An iron weight is attached to the ends of a light cord passing round a grooved wheel, and rests on the mercury at the open end; a second and smaller weight is attached to the other end of the cord. When the air pressure decreases, the mercury rises in the open part of the tube, the iron weight is pushed up, and the wheel turns; a pointer attached to the axle of the wheel then rotates over a scale previously calibrated in inches of mercury. When the air pressure increases, the mercury and the iron weight in the open part move downwards, and the pointer rotates in the opposite direction.

**The aneroid barometer** is a form of domestic barometer which contains no liquid. This barometer consists essentially of a steel

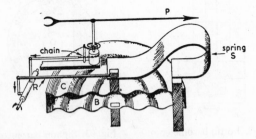

FIG. 94. Aneroid barometer

corrugated cylinder $B$ partially evacuated of air, and the top face $C$ is supported by a strong spring $S$ to prevent $B$ from collapsing under the atmospheric pressure (Fig. 94). When the pressure increases, the upper face of *B* is slightly depressed, and the reverse occurs when the atmospheric pressure decreases. The slight movements of *B* are magnified by means of an attached rod *R* with a system of small jointed levers, which operate a pointer *P* controlled by a spring. *P* moves over a circular dial (not shown) which is graduated in inches of mercury by comparison with a Fortin barometer. The words *stormy, rain, change, fair, very fair,* are printed on the dial (not shown) but are only a very approximate guide to the weather.

**The altimeter.** As Pascal had discovered (p. 127), the pressure of the air decreases as we go higher up, because there is then less weight of air above us. Experiment shows that, approximately, the pressure falls by 1 in. of mercury for every 1000 ft. above the earth. The dial of an aneroid barometer can thus be calibrated to read heights in feet, in place of the pressure in inches of mercury, and the instrument is used in an aeroplane as an *altimeter.* Mountaineers, and cars in hilly countries, carry altimeters to indicate the height they have reached.

**The siphon.** A *siphon* is a simple arrangement for emptying a liquid from a fixed vessel such as a petrol tank *C*, which is difficult to empty directly (Fig. 95). It consists of an open tube *G* filled with the liquid, which is then placed with one end below the

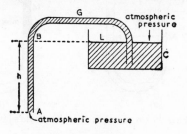

FIG. 95. Siphon

liquid in the vessel, as shown. The liquid in *C* is then observed to run out of the vessel through *A*, as long as the other end of *G* is dipping below the surface of the liquid, and *A* is kept below the level *BL*.

The explanation of the siphon action is as follows. Since the pressure at points in the same horizontal level of a liquid is the same (p. 123), the pressure inside the liquid at B is equal to the pressure at L, which is the atmospheric pressure. The pressure of the liquid at A is therefore greater than the atmospheric pressure by the amount hd, where d is the density of the liquid and h is the height of AB. But the pressure of the air at A is atmospheric, and is therefore unable to support the liquid. It therefore runs out at A. If G is *empty* and is placed with one end below the liquid in C the siphon does not work; there is now no net force on the liquid, as the pressure on it at L, which is atmospheric pressure, is also equal to the air pressure inside the tube G.

**Measuring gas pressure. The manometer.** The pressure of the domestic gas supply, which is not much greater than atmospheric pressure, can easily be measured by joining the gas-tap with rubber tubing to one side of a U-tube containing water (Fig. 96 (a) ). When the pressure is steady, the water is higher on the open side

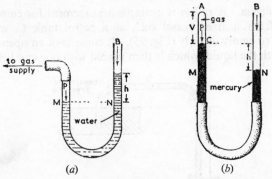

FIG. 96 (a). Gas pressure (b) Boyle's law

of the tube than on the left side, and the difference in levels, h, is noted. Since the pressure is the same at M and N, points in the same horizontal level of the water, the gas pressure, p, which is the pressure at M, is equal to the pressure at N. As the right side of the tube is open to the air, the pressure at N is equal to the atmospheric pressure ($B$) plus that due to a column of water of height $h$,

$$\therefore \quad p = B + \text{pressure due to height } h \text{ of water.}$$

If $h = 20$ cm., the height of mercury giving the same pressure $= \dfrac{20}{13\cdot6} = 1\cdot47$ cm., as the density of mercury $= 13\cdot6$ gm. per c.c. Thus if $B = 76$ cm. of mercury, the gas pressure $p = 76 + 1\cdot47 = 77\cdot47$ cm. of mercury. An arrangement for measuring pressure is known as a *manometer*; Fig. 96 (*a*) represents a water manometer.

Fig. 96 (*b*) illustrates a gas trapped by mercury in a closed tube, with the mercury level $N$ on the open side $h$ cm. *lower* than the mercury level $M$ containing the gas (compare Fig. 96 (*a*) ). The pressure at $N = B$ (atmospheric pressure) = the pressure at $M$, a point on the same horizontal level of the mercury. But the pressure at $M = (p + h)$ cm. of mercury, where $p$ is the gas pressure in cm. of mercury.

$$\therefore \quad p + h = B,$$

if $B$ is in cm. of mercury, and hence $p = B - h$. The gas pressure is thus *less* than the atmospheric pressure by the difference in levels of the mercury.

**Boyle's law.** If a given mass of gas is confined in a cylinder by an airtight piston and the pressure on the gas is increased by the movement of the piston, the volume of the gas decreases. When the pressure on the gas is decreased, the volume of the gas increases. A study of the relation between the pressure and volume of a gas is important for the design of "heat engines" such as those used on the railway and in the motor-car. In the 17th century ROBERT BOYLE discovered a simple law connecting the pressure ($p$) and volume ($V$) of a fixed mass of gas. *Boyle's Law* states—

**For a fixed mass of gas at a given temperature, the product of the pressure and volume is constant; i.e. $pV$ is constant**

Boyle's law can be verified by means of the apparatus shown in Fig. 96 (*b*). A fixed mass of air is trapped in a tube $AC$ of uniform cross-sectional area by mercury, which also fills the right-hand side open tube to a level $N$. The volume ($V$) of the fixed mass of air is proportional to the length $AC$, since the tube here is of uniform cross-section, and the pressure ($p$) of the air is $(B - h)$ cm. of mercury, where $B$ is the barometric pressure. (See p. 133.) By moving the open tube up and down and waiting until the mercury levels are steady each time, many measurements of $p$ and $V$ can be obtained. Of course, the pressure is $(B + h)$ if the mercury on the open side is $h$ cm. *above* the level on the other side. If the different pairs of values of $p$ and $V$ are multiplied it will be found that the product is fairly constant, thus verifying Boyle's law within the limits of experimental error. The following is a list of measurements made by a boy—

Atmospheric pressure = 77·0 cm. ($B$)

| Volume ($V$) | Diff. in Levels ($h$) | Gas Pressure ($p$) | | $pV$ |
|---|---|---|---|---|
| 7·6 units | 0 cm. | 77 cm. | | 585 |
| 7·1 units | 5·8 cm. | 82·8 cm. | $B + h$ | 588 |
| 6·4 units | 15·1 cm. | 92·1 cm. | | 589 |
| 5·8 units | 24·5 cm. | 101·5 cm. | | 589 |
| 9·1 units | 12·2 cm. | 64·8 cm. | $B - h$ | 589 |
| 8·0 units | 3·3 cm. | 73·7 cm. | | 590 |

Another way of expressing Boyle's law is to say that the volume of a given mass of gas is *inversely proportional* to the pressure when the temperature is constant; or, using symbols, that $V \propto \dfrac{1}{p}$ in these circumstances. Thus if the pressure on the gas is halved, its volume becomes doubled; if the pressure is doubled, the volume becomes halved. This type of relation (an inversely-proportional relation) between $p$ and $V$ is discussed more fully on p. 6, to which the reader should now refer. If the values of $\dfrac{1}{V}$ are calculated from the table in decimals and plotted against $p$, a straight line graph is obtained; this is left as an exercise for the reader.

**A diving bell** is used in laying the foundations of lighthouses, and also when repairs below the sea have to be made to the walls of docks or to bridges. As the bell, $D$, is lowered, the air inside it becomes compressed; the pressure at $A$, for example, is equal to the atmospheric pressure plus that due to the height $h$ of water above $A$. The volume of the fixed mass of air in $D$ thus diminishes, according to Boyle's law (Fig. 97). In order to keep the water out of the bell, and so enable workmen on platforms inside the bell to work, compressed air is pumped in continuously through a pipe, $P$. This forces down the water-level.

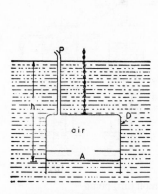

FIG. 97.  Diving bell

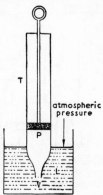

FIG. 98.  Syringe

**The garden syringe.** A garden syringe consists essentially of a tube $T$ containing a movable piston $P$, with an opening $V$ at the bottom. Fig. 98 illustrates the syringe placed in some liquid $L$. The pressure of the air below $P$ decreases as its volume increases, by Boyle's law, so that when the piston is raised the air pressure below it becomes less than the atmospheric pressure. Consequently some of the liquid is forced into the syringe through $V$, and can be expelled by depressing the piston after the syringe is taken out of the liquid $L$.

**The lift pump.** The common or "lift" pump, used for obtaining water from wells in some parts of the country, consists of a pipe $X$ and a cylinder $C$ containing a piston $P$. The cylinder and piston have each a valve, represented by $V_1$ and $V_2$, and we shall suppose that first the piston is at the bottom of its stroke (Fig. 99 $(a)$). As the piston is raised, the pressure of the air enclosed between it and the lower part of the cylinder decreases; and at some stage the pressure becomes lower than the pressure in the pipe. The valve

$V_1$ then opens, $V_2$ being closed, and the atmospheric pressure forces water into the pipe and the cylinder (Fig. 99 (b) ). When the piston is lowered the valve $V_2$ opens, $V_1$ becomes closed, and some water is forced above the piston (Fig. 99 (c) ). As the piston

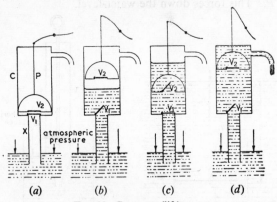

(a)          (b)          (c)          (d)

FIG. 99. Common (lift) pump

is raised, $V_2$ becomes closed, $V_1$ opens, and the water is forced out of the spout (Fig. 99 (d) ). On the downstroke the sequence begins again. Generally, several strokes of the plunger are first necessary to set the pump in full action, because of the residual air in the pipe. It should be noted that the atmospheric pressure does the work of pushing the water from the well into the pipe, and since the atmospheric pressure supports a column of water about 34 ft. high, the distance from the surface of the water in the well to the valve $V_1$ should not exceed this height theoretically. In practice, however, this distance is less than 34 ft., e.g. 24 ft., because the valves and piston leak somewhat.

**The force pump.** The force pump is able to deliver water at a much higher level than the lift pump (Fig. 100). It has two valves $V_1$, $V_2$, which open alternately as in the lift pump, but $V_2$ allows water to flow to an outlet pipe $A$ which leads to the spout. When the piston $P$ is raised, water is forced through the open valve $V_1$ by the atmospheric pressure, the valve $V_2$ being closed. On the downstroke of the piston, water is forced through the valve $V_2$ to the outlet pipe $A$, and the valve $V_1$ becomes closed. The action is repeated, and (as in the lift pump) the valves open and close alternately. In distinction from the lift pump, however, the valve $V_2$ is not in the piston. The force pump is so called because water is forced into the pipe $A$ by the piston, which is solid. This type of pump also should not have $V_1$ more than 24 ft. above the water level.

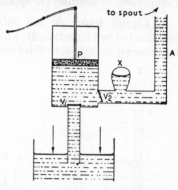

In order to make the pump function on the upstroke of the piston as well as its downstroke, a vessel, $X$, containing air is introduced between $V_2$ and the pipe $A$. On the downstroke the air in $X$ is compressed; on the upstroke, the pressure on the gas is released, which therefore expands and

FIG. 100. Force pump

pushes the water in $A$ through the spout. Water is thus obtained on both strokes of the piston.

**The bicycle pump.** In the common pump, air is withdrawn from the part of the barrel below the piston when the pump is working. In the bicycle pump, however, air is *compressed* below the piston, so that air enters the tyre inner tube. Fig. 101 illustrates the interior of the pump, which contains a piston $P$ with a cup-shaped leather washer, $C$, at the end of it. The "cup" points towards the

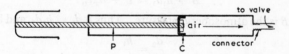

FIG. 101. Bicycle pump

end of the pump to which the pump-connector is attached, and when the piston is raised, air flows round $C$ below the piston. When the piston is pushed down, the air below the piston is compressed, and presses $C$ against the sides of the barrel; thus no air flows back above the piston. When the pressure of the air exceeds the resistance of the valve in the inner tube, the compressed air flows through the valve into the tube.

**Comparison of densities (or specific gravities) of liquids.**

1. *Liquids which do not mix.* The densities (or specific gravities) of two liquids $L$, $M$, which do not mix, may be compared by using an open U-tube filled with the two liquids (Fig. 102 (a) ).

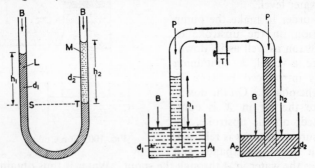

FIG. 102 (a). U-tube—liquids which do not mix

FIG. 102 (b). Hare's apparatus—liquids which mix

Suppose $T$ is the surface of separation of $L$ and $M$, and $TS$ is the horizontal through $T$. Then, since the pressure is the same on the same horizontal level of a liquid (p. 123),

the pressure at $S$ = the pressure at $T$

$$\therefore \quad B + h_1 d_1 = B + h_2 d_2,$$

where $B$ is the atmospheric pressure. Thus $h_1 d_1 = h_2 d_2$, and hence

$$\frac{d_2}{d_1} = \frac{h_1}{h_2}.$$

Consequently the densities of the two liquids are *inversely proportional* to the heights of the liquids *above the surface of separation*. If the latter is in the bend of the U-tube, more of one liquid must be added to move the surface of separation to a vertical limb, as shown in Fig. 102 (a).

**2. *Liquids which mix. Hare's apparatus.*** HARE used a simple apparatus to compare the densities of two liquids which would mix if brought together, for which the U-tube method (p. 138) is useless. The apparatus consists of two vertical glass tubes joined by a rubber T-piece, the ends of the tubes being placed in the liquids $A_1$, $A_2$, respectively (Fig. 102 (*b*) ). By suction at $T$ the two liquids are drawn up the respective tubes. $T$ is then closed to the air by a clip, and the two heights $h_1$, $h_2$, above the levels of the liquids in the respective beakers are measured.

Suppose $d_1$, $d_2$ are the respective densities of the liquids and $p$ is the pressure of the air above the liquids in the apparatus. Then

$$p + h_1 d_1 = p + h_2 d_2 = B \qquad . \qquad . \qquad . \quad (5)$$

where $B$ is the atmospheric pressure, which acts on each of the surfaces of the liquids $A_1$, $A_2$ (see Fig. 102 (*b*) ). Thus $h_1 d_1 = h_2 d_2$, from (5), and hence

$$\frac{d_2}{d_1} = \frac{h_1}{h_2}.$$

The densities are therefore *inversely proportional* to the heights in the two tubes. Hare's apparatus is very suitable for comparing the densities of different concentrations of sulphuric acid, or of copper sulphate solution, to quote two examples.

─────────────────Summary─────────────────

1. **Pressure = force per unit area** (the units are "lb. wt. per sq. in.," or "dynes per sq. cm."). **Pressure = *hd*,** where $h$ is the depth and $d$ is the liquid density.

2. The mechanical advantage of a hydraulic press = cross-sectional area of large piston/cross-sectional area of small piston, neglecting friction.

3. Atmospheric pressure supports a column of mercury about 76 cm., or 30 in., long. "One atmosphere" = $14\frac{1}{2}$ lb. wt. per sq. in. (approx.).

4. **Boyle's law states: For a given mass of gas at a constant temperature, pressure × volume is a constant.**

## WORKED EXAMPLES

1. *Explain the terms force and pressure. Describe the hydraulic press and explain its mode of action. The larger piston of a hydraulic press is 12 in. in diameter and the smaller piston ½ in. in diameter. What force will the larger piston exert when the smaller has an effort of 30 lb. wt. applied to it?* (L.)

$$\text{Pressure} = \frac{\text{force}}{\text{area}},$$

$\therefore$ pressure on smaller piston, $p = \dfrac{30 \text{ lb. wt.}}{\pi(\frac{1}{4})^2 \text{ sq. in.}} = \dfrac{30 \times 16}{\pi}$ lb. wt. per sq. in.

This pressure is transmitted to the larger piston, which has an area of $\pi(6)^2$ sq. in.

$\therefore$ force on large piston $= p \times \text{area} = \dfrac{30 \times 16}{\pi} \times \pi(6)^2$

$$= 30 \times 16 \times 36 = 17,280 \text{ lb. wt.}$$

This is the force the large piston exerts.

2. *Describe how you would determine the specific gravity of alcohol. Explain the theory of the method you describe. The two limbs of a U-tube are uniform cylinders of diameter 1·4 cm. and 0·8 cm. respectively. The lower portion of the*

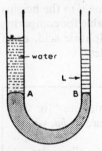

FIG. 103.
Pressure calculation

*U-tube contains mercury. The wider limb contains 12·25 gm. of water standing above the mercury. What mass of liquid must be poured into the narrower limb in order that the mercury surfaces in the two limbs shall be at the same horizontal level?* (N.)

In Fig. 103, suppose $L$ is the liquid in the narrow limb which causes the mercury surfaces $A$, $B$ to be at the same horizontal level. The pressure is the same at the same horizontal level of a liquid; hence the pressure at $A$ = the pressure at $B$.

But pressure at $A = \dfrac{\text{weight of water above } A}{\text{area of tube above } A}$

$+$ atmospheric pressure.

and pressure at $B = \dfrac{\text{weight of liquid } L \text{ above } B}{\text{area of tube above } B} +$ atmospheric pressure.

$\therefore \dfrac{\text{weight of liquid above } B \, (w)}{\text{area of tube above } B} = \dfrac{\text{weight of water above } A}{\text{area of tube above } B}$

$\therefore \dfrac{w}{\pi \times 0\cdot4^2} = \dfrac{12\cdot25 \text{ gm. wt.}}{\pi \times 0\cdot7^2}$

as the area of cross-section of a cylinder $= \pi r^2$, and the radii are $\dfrac{1\cdot4}{2}$ cm. and $\dfrac{0\cdot8}{2}$ cm. respectively.

$$\therefore \quad w = \frac{\pi \times 0\cdot4^2 \times 12\cdot25}{\pi \times 0\cdot7^2} = 4 \text{ gm. wt.}$$

3. *Describe an aneroid barometer, giving a diagram and explaining its action. Explain its use for measuring heights. Calculate the pressure in bars if the barometric height is 740 mm. of mercury. (1 bar = 1 million dynes per sq. cm.; density of mercury = 13·6 gm. per c.c.; g = 981 cm. per sec.²)* (N.)

The pressure, $p = hdg$ dynes per sq. cm., where $h$ is in cm., $d$ is the density in gm. per c.c., $g = 981$ numerically

$$\therefore \quad p = 74 \times 13\cdot6 \times 981 \text{ dynes per sq. cm.}$$

$$= \frac{74 \times 13\cdot6 \times 981}{1,000,000} \text{ bars} = 0\cdot987 \text{ bars}$$

4. *Describe how you would construct a simple type of mercury barometer. The tube of a faulty barometer containing a little air above the mercury projects 80 cm. above the level in the reservoir. When the atmospheric pressure is 76 cm. of mercury the baorometer reading is 72 cm. What will be the atmospheric pressure when the barometer reads 74 cm.?* (L.)

Let Fig. 104 (a) represent the barometer tube when the atmospheric pressure is 76 cm. of mercury. Since the pressure inside the tube at the bottom, $A$, is $(p_1 + 72)$ cm. of mercury, where $p_1 =$ the pressure of the air at the top of the tube,

$$p_1 + 72 = 76, \text{ i.e. } p_1 = 4 \text{ cm.}$$

The volume of air $= 80 - 72 = 8$ units.

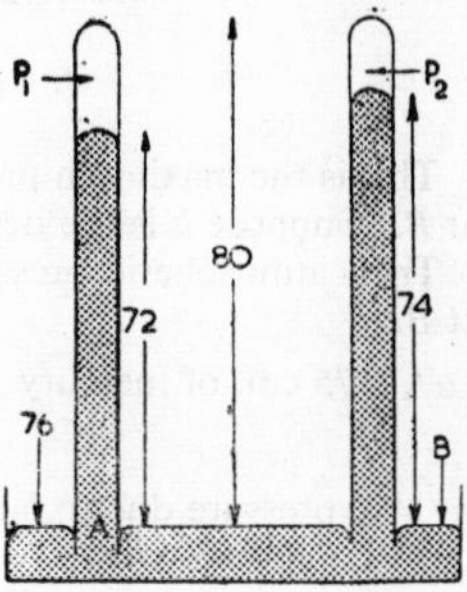

Let Fig. 104 (b) represent the second case, in which the pressure of the air in the tube is $p_2$ and the atmosphere pressure is $B$, say. Then, as above,

FIG. 104. Boyle's law

$$p_2 + 74 = B, \text{ i.e. } p_2 = B - 74$$

The volume is now $80 - 74 = 6$ units.

Applying *Boyle's law* for the air, we know that $pV$ is a constant.

$$\therefore \quad 4 \times 8 = (B - 74) \times 6, \text{ from the above figures}$$

$$\therefore \quad 32 = 6B - 444$$

$$\therefore \quad B = \tfrac{476}{6} = 79\tfrac{1}{3} \text{ cm. of mercury}$$

5. *State Boyle's law and describe an experiment to verify it. A cylindrical iron tube, closed at one end and 30 cm. long, is coated on the inside with a*

*soluble pigment. It is carefully lowered into water, with the open end downwards
and when withdrawn it is found that the pigment has been disolved over a length
of 5 cm. from the bottom. What was the maximum pressure of the air in the*

FIG. 105. Calculation

*cylinder and to what depth was the bottom
of the cylinder sunk? (Barometric height
= 75 cm. of mercury.) (L.)*

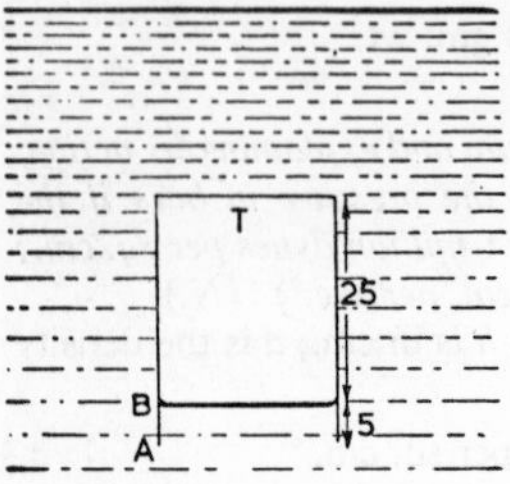

Suppose $T$ is the iron tube (Fig. 105).
Then, since the pigment has been dissolved
over a length of 5 cm. from the bottom, the
level of water inside the tube has risen 5 cm.
from $A$ to $B$.

Now, originally, the air filling the whole
of the cylinder when it was empty was at the
atmospheric pressure of 75 cm. of mercury;
and its volume was 30 cm. $\times a$ sq. cm.,
where $a$ is the cross-sectional area.

In Fig. 105, let the pressure of the air be
$p$ cm. of mercury; its volume is 25 cm. $\times a$ sq. cm. Then, from Boyle's
law, as $pV$ is constant,

$$p \times (25 \times a) = 75 \times (30 \times a)$$

$$\therefore \quad p = \frac{75 \times 30}{25} = 90 \text{ cm. of mercury.}$$

This is the maximum pressure of the air in the cylinder, and is the pressure
at $B$. Suppose $h$ is the depth of $B$ below the surface.

Then atmospheric pressure + pressure due to $h$ cm. of water = pressure
at $B$.

$\therefore$   75 cm. of mercury + pressure due to $h$ cm. of water

= 90 cm. of mercury

$\therefore$   pressure due to $h$ cm. of water = 90 − 75 = 15 cm. of mercury

$$\therefore \quad h = 15 \times 13 \cdot 6 \text{ cm.}$$

since the density of mercury is 13·6 gm. per c.c., i.e. 13·6 times the density
of water.

$$\therefore \quad h = 204 \text{ cm.}$$

$$\therefore \quad \text{depth of } A = 204 + 5 = 209 \text{ cm.}$$

## EXERCISES ON CHAPTER VIII

**1.** Name two different units of pressure. Calculate the pressure at a
point 50 ft. below water if its density is $62\frac{1}{2}$ lb. per cu. ft.

**2.** The weight of a book resting on a table is 270 gm. wt., and the
area of the book in contact with the table is 120 sq. cm. What is the
pressure on the table (i) in gm. wt. per sq. cm., (ii) in dynes per sq. cm. ?

**3.** The height of a water barometer is 32 ft. Calculate the pressure on the earth due to the atmosphere. Taking atmospheric pressure into account, what is the pressure 38 ft. below the surface of a lake? (Density of water = 62½ lb. per cu. ft.)

**4.** The height of a mercury barometer is 75 cm. Calculate the air pressure in dynes per sq. cm. and the force due to the atmosphere on an area of 2000 sq. cm. (Density of mercury = 13·6 gm. per c.c.)

**5.** Describe the action of (i) a bicycle pump, (ii) a force pump, drawing labelled diagrams.

**6.** Write down two everyday observations which show that the pressure in a liquid increases with its depth. How can this be shown experimentally?

**7.** Why is mercury the common liquid in a barometer? Draw sketches of a simple barometer and Fortin barometer, and name the important features of the latter.

**8.** A U-tube contains oil and water. On one side the height of oil above the surface of separation with the water is 25 cm.; on the other side the height of water above the surface of separation is 15 cm. Find the density of the oil, and explain your calculations.

**9.** Describe with sketches the action of (i) a garden syringe, (ii) a common pump. Name one limitation of the common pump.

**10.** Define *pressure* and use the definition to obtain an expression for the pressure at the base of a column of liquid of height $h$ and density $d$. Describe and explain an experiment to compare the densities of two liquids by using this result. (*L.*)

**11.** Describe how you would set up a simple barometer, pointing out the precautions necessary to obtain a reasonably accurate form of instrument. How would you test whether you had a vacuum above the mercury? What would be the effect on the barometer reading of (*a*) using a slightly tapered barometer tube, (*b*) an increase in temperature while the atmospheric pressure remains constant? (*L.*)

**12.** Calculate the pressure at a point 12 metres below the surface of a lake on a day when the atmospheric pressure is 75 cm. of mercury. (S.G. of mercury = 13·6.) (*N.*)

**13.** State *two* structural features which are characteristic of *either* (*a*) a Fortin barometer *or* (*b*) an aneroid barometer. (*N.*)

**14.** Describe a good barometer and explain its mode of action. A perfectly compressible balloon of volume 2 litres is filled with air at atmospheric pressure, which is 33 ft. of water. The balloon is then held under water at a depth of 16·5 ft. below the surface. Calculate (a) the pressure inside the balloon, (b) the new volume of the balloon, assuming its temperature to be unchanged. (*C.W.B.*)

**15.** How would you compare the specific gravity of saturated copper sulphate solution with that of dilute copper sulphate solution? (*L.*)

**16.** Describe and explain the mode of action of a pump suitable for forcing water to a height. (*L.*)

**17.** Describe an experiment to show how the pressure due to a liquid at a point below the surface depends on the depth below the surface. Describe and explain the action of the common pump for raising water from a well. (*L.*)

**18.** How would you show that a column of liquid exerts a pressure at any point below its surface? Describe a machine which uses liquid pressure and explain how you would determine its velocity-ratio and efficiency. (*L.*)

**19.** State *Boyle's law*. A faulty barometer has a little air at the top and reads 74 cm. when the true barometric height is 76 cm. The space at the top of the tube is then 10 c.c. Calculate the true barometric height when the space at the top is 8 c.c. and the instrument reads 74·5 cm.

**20.** How would you set up a mercury barometer? Why must the scale measuring the height of the mercury column be vertical? How does moisture in the tube affect the reading of the barometer? If the density of mercury is 13·6 gm. per c.c. and the reading of your barometer is 760 mm., what is the pressure of the atmosphere in grams weight per square centimetre? (*C.*)

**21.** Describe how you would verify Boyle's law for air. An exhaust pump with a cylinder of cross-section 3 sq. in. and a length of stroke 1 ft. is connected to a vessel containing 144 cu. in. of air at atmospheric pressure. Calculate the pressures of the air in the vessel after the first and second exhaust strokes and the force which must be applied to the piston to hold it when it is at the end of the first exhaust stroke.
(Assume no temperature change occurs. Atmospheric pressure = 15 lb. per sq. in.) (*N.*)

**22.** Distinguish between *pressure* and *atmospheric pressure*. How is a simple barometer set up? A simple mercury barometer which contains a little air reads 76 cm. when the true barometric height is 77 cm., and the internal height of the tube above the mercury in the trough is

85 cm. Calculate the true barometric height when the defective barometer reads 75 cm., temperature remaining constant. (*O. & C.*)

**23.** Describe an accurate form of mercury barometer and explain how it measures the pressures of the atmosphere. What are the advantages and disadvantages of the aneroid barometer as compared with the mercury barometer? (*O. & C.*)

**24.** Define *specific gravity* and hence obtain an expression for the specific gravity of a liquid in terms of its density.

Describe with a sketch Hare's inverted U-tube apparatus for finding the specific gravity of a liquid. Indicate clearly the measurements taken and show how the result is obtained from them.

In such an experiment to determine the specific gravity of methylated spirit, the length of the column of water was 11·6 cm. and that of the methylated spirit 14·5 cm. Calculate the specific gravity of the liquid. If the length of the water was then altered to 15 cm., what would be the height of the column of methylated spirit? (*L.*)

**25.** State Boyle's law and describe an experiment to verify it for air.

A fairly wide, strong walled glass tube, closed at its upper end, cylindrical in shape, 2 ft. long and open at its lower end, was weighted and lowered vertically into the bed of a river until it touched the bottom. When withdrawn, it was found that the water had risen $7\frac{1}{2}$ in. inside the tube. Find the depth of the river. (Atmospheric pressure at the surface of the river may be taken as that due to 33 ft. of river water.) (*L.*)

**26.** State *Boyle's* law and describe an experiment to verify it. Indicate the graphs that should be obtained if (*a*) $p$ were plotted against $1/v$, (*b*) $pv$ were plotted against $p$, where $p$ and $v$ refer to the pressures and volumes obtained from the experiment.

A uniform tube, 96 cm. long, sealed at one end, is lowered vertically with its open end downwards into mercury until the length of the enclosed air column is 84 cm. Find the depth of immersion of the tube in the mercury, given that the atmospheric pressure at the time of the experiment is 77 cm. of mercury. (*L.*)

# CHAPTER IX

## SURFACE TENSION. OSMOSIS. ELASTICITY

WHEN a football bladder is blown up, the surface of the bladder is under tension or strain. The forces in the surface may be termed **surface tension** forces. If a cut is made in the bladder, the material draws away from the air-gap under the action of the forces, which had been in equilibrium before the cut was made.

Surface tension forces also exist in the surface of a liquid, such as water. Thus, a force $T$ acts on both sides of a line $AB$ imagined drawn in the middle of the water surface (Fig. 106 (*c*) ). In the

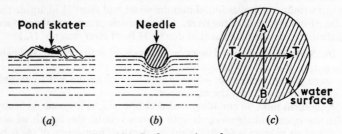

FIG. 106. Surface tension of water

middle $M$ of water in a beaker, the surface is plane; but at the edge $E$ the surface tension force $F$, which is balanced here by an opposite surface tension as at $M$, pulls the water some distance up the glass (see Fig. 107 (*a*) ).

**Effects of surface tension.** An insect known as a pond skater is able to walk across the surface of water, and a greased needle can be supported by water if it is placed gently on it (Fig. 106 (*a*) (*b*) ). These facts suggest that *the surface of water acts like a* *stretched elastic "skin" covering the water.* A water-drop forming slowly at a tap and then breaking away is spherical in shape, almost as if the water was surrounded by a "skin," and small

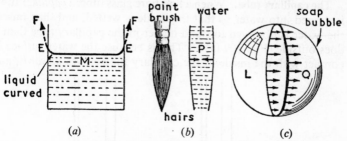

FIG. 107. Surface tension phenomena

particles of mercury on a table gather together in the form of a sphere. The hairs of a paint-brush cling together when it is wet; this is the effect of surface tension forces P acting in the water on the hairs (Fig. 107 (b) ). A soap bubble is kept together in a spherical shape by surface tension forces Q acting over the surface of the bubble. Fig. 107 (c) illustrates one set of such forces, acting

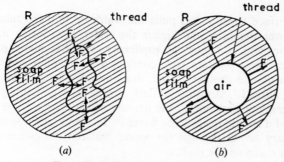

FIG. 108. Surface tension of soap film

on the left half, L. Surface-tension forces can also be revealed by placing a piece of tied thread carefully on a soap film formed on a metal ring R (Fig. 108 (a) ). If the film in the thread is pierced by a rod, the thread is pulled in the shape of a circle. Surface tension forces *F* then act on *one* side of the thread, as illustrated in Fig. 108 (*b*), whereas in Fig. 108 (*a*) the surface tension forces act on both sides of the thread and are in equilibrium.

**The capillary tube.** When a fine-bore glass tube, a *capillary* tube, is dipped into water so that the inside is wetted, and then raised, the water is observed to stand higher in the capillary tube than it does outside (Fig. 109 (*a*) ). This is because the water surface in contact with the inside of the capillary tube curves *upwards*, and

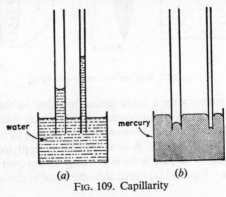

Fig. 109. Capillarity

the surface tension thus pulls the water up into the tube. The narrower the tube, the higher the water rises inside it. This phenomenon is known as **capillarity**, and has many practical applications. Blotting paper contains many very fine pores, which act as minute capillary tubes, drawing up ink from the writing being blotted. Moisture similarly moves through the tiny spaces between soil particles by the process of capillarity. The melted wax of a candle reaches the flame by the same process, rising by capillary action through the spaces between the cotton threads which make up the wick.

If a capillary tube is dipped into mercury, however, the level inside is observed to be *lower* than the level outside (Fig. 109 (*b*) ). This is because the mercury surface in contact with the glass curves *downwards*, and the surface tension thus pulls the mercury down into the tube.

**Contamination and surface tension forces.** The magnitude of the surface tension force is usually reduced if the liquid concerned is contaminated.  An amusing toy, based on this fact, is a celluloid swan which moves slowly across the surface of water.  At the base of the swan is a small piece of camphor $C$, which contaminates the water (Fig. 110).  This decreases the surface tension of the water in the area affected, and unequal forces $P$ and $Q$ act in the surface at the border between the clear and contaminated water.  Since $P$ is greater than $Q$, the swan moves forward.

FIG. 110.  Surface tension force

Another demonstration of the effect of contamination is shown by sprinkling lycopodium powder over the surface of water and touching the surface with a rod previously dipped in soap solution.  The particles of powder are observed to draw away from the place touched by the rod, as the surface tension of the contaminated part of the water is less than that of pure water.

Advantage is taken of this property in the fight against mosquitoes and other insect pests.  Pools containing mosquito larvae are sprayed with an oil which reduces the surface tension of the water.  The larvae, which cling to the underside of the surface to breathe, are then unable to support themselves, and drown.

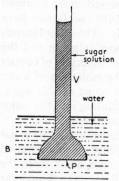

FIG. 111.  Osmosis

**Osmosis.**  If a sugar solution is contained in a long open vessel $V$, having a sheet of parchment covering the lower end, and $V$ is placed inside a beaker of water $B$, observation shows that the level of the liquid inside $V$ begins to rise (Fig. 111).  Water has entered the sugar solution through the parchment $P$.  The flow of water ceases after a time, and the level of the solution in $V$ then remains constant.

The experiment, which was first performed about 1770, shows that the water passes easily through parchment.  However, on testing the water in the beaker $B$, no sugar can be detected;

6

thus sugar does not pass through the parchment. This spontaneous flow of water through the parchment is an example of a phenomenon called *osmosis*.

Parchment is a covering known as a *semi-permeable membrane*. This is the name given to any substance which permits the flow of a solvent (water in Fig. 111) through it, but not the dissolved substance (sugar in Fig. 111). Many plants obtain moisture by the process of osmosis across the semi-permeable membrane of the root cells, while the human body also contains semi-permeable membranes which play an important part in the separation and transfer of liquids from one part of the human system to another.

**Diffusion.** If a beaker *B* is partly filled with a concentrated solution of copper sulphate, and water is then poured carefully on top of it, a surface of separation *A* between the two liquids can

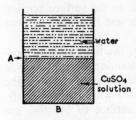

FIG. 112. Diffusion

be observed (Fig. 112). In a few days' time, however, the blue colour will be seen to have moved higher up, so that the copper sulphate solution, which is denser than water, has moved against gravity. As time proceeds, the whole of the solution is observed to be coloured blue, showing that the two liquids have completely mixed. The spontaneous mixing of the copper sulphate and the water is known as *diffusion*. It shows that the molecules of copper sulphate are in motion in the solution, as the molecules of water are (see p. 233). Experiments, beyond the scope of this book, show that the amount of copper sulphate diffusing into the water is proportional to the time and to its concentration.

Gases also diffuse into each other, as the molecules of a gas are in motion (p. 232). Thus, the evil odour of sulphuretted hydrogen ($H_2S$) can be detected some distance away from the fume chamber; and oxygen and hydrogen will mix, even though the denser gas, oxygen, is contained in a jar below a jar of hydrogen. Solids such as lead and gold also diffuse into each other, especially when they are heated, but the process is much slower than in the case of gases and liquids.

**Stretching of wires. Hooke's law.** Before steel is used in the construction of bridges, samples of the metal are subjected to tests. One of the more important tests concerns the behaviour of the metal under forces which tend to *stretch* it, because the material of the bridge is subject to sudden stresses which it must be able to withstand. Similar tests are carried out on the stretching of samples of rubber to be used for conveyor belts (Fig. 113).

In such tests, increasing loads are added to one end of the sample under investigation; this is in the form of a thin wire (or strand, if it is rubber). The other end of the wire is fixed. Up to a point, the wire increases in length by equal small amounts when equal weights are added to it, e.g. a weight of 1 kg. wt. may stretch the wire $\frac{1}{2}$ mm., and weights of 2 and 3 kg. wt. may stretch it 1 mm.

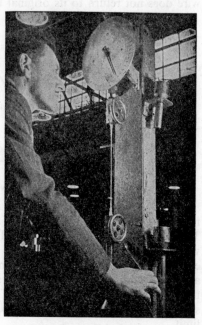

Fig. 113. TESTING RUBBER FOR TENSILE STRENGTH

and $1\frac{1}{2}$ mm. respectively. The graph of the extension plotted against the load is thus a *straight line*, as shown by *OA* in Fig. 114. Further, the wire returns to its original length when the load attached to it is removed. Along *OA*, therefore, we say that the wire appears to be *elastic*, a property possessed by some rubber compounds. It was HOOKE who discovered in 1650 that the extension of a wire is proportional to the tension or force applied, provided the elastic limit (see p. 152) is not exceeded.

As more and more weights are added to the wire under test, the extension at one point, corresponding to a load $OD$ (Fig. 114), becomes relatively much greater than before; the curve $AB$, showing a fairly sharp upward trend, is then obtained. Further, the wire does not return to its original length when the weights are

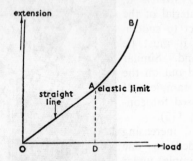

FIG. 114. Hooke's law of elasticity

removed, showing that the wire has suffered a permanent strain. The point $A$ is said to correspond to the *elastic limit* of the wire, since beyond this point the wire ceases to exhibit the elasticity shown from $O$ to $A$. As more weights beyond $OD$ are added to the wire the extension increases rapidly, until finally at the *yield point* the metal "gives". The wire is not safe for withstanding stresses greater than that corresponding to the elastic limit.

**Solid Friction.** It is well known that we are able to walk on surfaces because the force of friction prevents us from slipping on surfaces. Nails are held firmly in wood on account of large frictional forces. Thus friction can be useful. In machines, however, friction is a disadvantage. Some energy is required to overcome friction, and hence less energy is available for the machine itself.

If you move your finger over a surface, you recognize immediately that *friction opposes the motion*. We know now that this is due to very tiny irregularities, humps and depressions, in the surface, however smooth this may appear to the eye.

**Coefficient of friction.** The frictional force $F$ between two wooden surfaces can be studied by placing a block of wood $A$ on a table, and connecting it by means of a light string passing over a pulley to a scale-pan $S$ hanging vertically. Fig. 114A (i). The frictional force is always equal to the force pulling $A$ when the latter does not move. At first, as increasing weights are added to $S$, $A$ remains still; thus the frictional force continues to increase. At one point, however, $A$ begins to slip, and the maximum or *limiting frictional force*, $F$, is now reached. The weight in $S$, together with the weight of the scale-pan, is recorded. The force of the surface on $A$ which acts at right angles to the surface is called the *normal reaction*, $R$, and this is equal to the weight $W$ of $A$.

When $R$ is increased by placing weights on $A$, a new and larger limiting frictional force $F$ is observed. By measurements, it is found that, approximately,

$$\frac{F}{R} = \text{a constant, } \mu.$$

This constant is called the *coefficient of static friction* between the two surfaces.

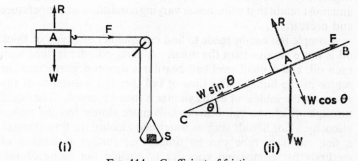

FIG. 114A. Coefficient of friction

The coefficient of friction $\mu$, can also be found by placing the object $A$ on an inclined plane $CB$, and tilting the plane until $A$ is just on the point of slipping down. Fig. 114A (ii). The angle $\theta$ made by $CB$ with the horizontal is then measured, and $\mu$ is found from the relation $\mu = \tan \theta$. To explain this relation, we note that for equilibrium along the plane, $W \sin \theta = F$; and that for equilibrium perpendicular to the plane, $W \cos \theta = R$.

$$\therefore \quad \frac{F}{R} = \mu = \frac{W \sin \theta}{W \cos \theta} = \tan \theta.$$

**Laws of Solid Friction.** The laws of solid friction were known hundreds of years ago, and may be stated as follows:

I. Friction opposes the motion.

II. For two given surfaces, the ratio $F/R$ is a constant approximately, where $F$ is the limiting frictional force and $R$ is the normal reaction.

III. For a given body, the limiting frictional force is independent of the area in contact with the surface.

**Liquid friction or viscosity.** The frictional forces in moving liquids are studied under the heading of *viscosity*, and the subject plays an important part in lubrication, without which metal surfaces would rub against each other. The viscosity of oils in engines, for example, must reach a desired value to be effective, and must retain that value under varying conditions of temperature and pressure.

A simple test can be made to find out whether one type of thick oil is more viscous than the other. A glass cylinder is filled with each oil, and a small steel ball-bearing is dropped gently into the centre of the liquid. The time of fall through a distance of say 10 cm. in the middle of each cylinder is then observed. The liquid through which the ball-bearing falls more slowly has the higher viscosity. For liquids such as water and alcohol the flow through a fine capillary tube can be timed, but further discussion of methods for comparing viscosity is outside the scope of this book.

Some of the differences between solid and liquid friction should be noted. The frictional force between two moving solid surfaces is approximately independent of velocity and the surface area in contact. The frictional force between two moving liquid surfaces, however, depends considerably on the relative velocity and is proportional to the surface area.

---

**Summary**

1. Surface tension phenomena are due to the fact that the surface of liquids act like a stretched elastic skin. Water rises in a capillary tube owing to surface tension.

2. "Osmosis" is the name given to the spontaneous flow of liquid through a membrane separating two solutions.

3. "Diffusion" is the name given to the passage of liquid (or gas) from one region to another.

4. **Hooke's law states: The extension of a spring or wire is proportional to the force or tension acting on it, so long as the elastic limit is not exceeded.**

5. Limiting frictional force is proportional to the normal reaction. Viscosity is the frictional force in liquids.

---

## EXERCISES ON CHAPTER IX

**1.** Name three everyday observations which show that the surface of water acts like a covering "skin".

**2.** Three glass capillary tubes of different diameter are placed inside water in a beaker. Draw sketches showing the level of the water in each of three tubes. What is the name given to the phenomenon? Draw a sketch of the liquid levels when a capillary tube is placed inside a beaker of mercury.

**3.** A little water and a little mercury are each placed on a clean glass horizontal plate. Draw a sketch showing the appearance of the two liquids. How can their appearance be explained?

**4.** Explain the action of blotting-paper in drying ink.

**5.** The mouth of a gas-jar of chlorine, a yellowish gas, is placed below an inverted gas-jar of hydrogen, which is colourless. Chlorine is a much heavier gas than hydrogen, but after a time the upper jar contains a yellowish gas. Explain this observation.

**6.** Describe an experiment which illustrates the diffusion of liquids.

**7.** Samples of steel for bridges are often tested by loading them at one end and observing the extension. What result would you expect for

(i) a good sample, (ii) a bad sample ? Draw graphs to illustrate your answer.

**8.** Red blood corpuscles expand rapidly when they are placed in water, but shrivel when placed in salt solution. What phenomenon does this exhibit, and what is the explanation ?

**9.** Show graphically the relation between the extension of a wire and the load attached to it. Label the axes of your graph and point out the main features shown by the curve. (*L.*)

**10.** What is meant by osmosis ? Illustrate by an example. (*L.*)

**11.** Describe two simple experiments to illustrate the statement that the surface of a liquid behaves like an elastic skin. (*L.*)

**12.** State *two* differences between friction and viscosity.

Describe how you would proceed in order to measure the force of friction between a horizontal wooden surface and a metal block sliding on it. How would the results be affected by placing a second metal block on top of the first ?

A body of mass 5 lb. is placed on a horizontal surface, a horizontal force of 1 lb. wt. is applied to it and the body just moves. Calculate (*a*) the work done against friction when the body is moved a distance of 4 ft., (*b*) the velocity acquired by the body in 2 sec. if a horizontal force of 2 lb. wt. is applied. (Assume that friction remains constant.) (*L.*)

**13.** (*a*) Explain what is meant by the *surface tension* of a liquid. Describe experiments to demonstrate its existence in (i) water, (ii) mercury.

(*b*) Describe briefly an experiment to investigate how the stretching force affects the length of a spiral spring. What result would you expect ?

When a piece of aluminium is suspended in air from the lower end of a vertical spiral spring, of unstretched length 30·6 cm., its length is 35·0 cm. On surrounding the aluminium with water the length of the spring becomes 33·0 cm. Find the specific gravity of the aluminium. (*L.*)

# FURTHER EXPERIMENTS IN MECHANICS AND HYDROSTATICS

**Experiment to demonstrate (1) uniform acceleration, (2) the relation between force and acceleration**

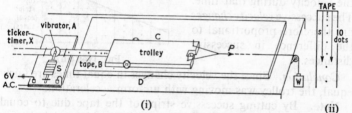

FIG. 114a. Acceleration measurement and relation to force.

*Apparatus.* Ticker-timer $X$, tape $B$, trolley $C$, smooth board $D$, ruler, weights.

*Account.* The ticker-timer consists basically of a solenoid, $S$, to which is connected a low alternating voltage, for example 6 volts, of mains frequency (50 cycles per second). A strip of iron $A$, clamped at one end, passes through the solenoid and between the poles of a permanent magnet (not shown). $A$ vibrates at the mains frequency, and with the aid of carbon paper it marks dots at intervals of 1/50th second on a tape running beneath it.

(1) *Uniform acceleration.* To demonstrate uniform acceleration, one end of a tape was attached to a trolley $C$ placed on a smooth board $D$ at an angle to the horizontal. Fig. 114a (i). The low-voltage mains supply was switched on and the trolley was released from rest. The running tape increases in velocity, and the vibrator $A$ marks dots on it at equal time-intervals. The appearance of the tape after a run is similar to that shown in Fig. 114a (ii).

6*

*Measurements.* The successive distances *s* travelled by the trolley in equal times *t* of 5 dot-intervals, for example, is measured with a ruler from a clear dot near the beginning of the tape run. One result is shown in the table below:

Successive distances (*s*) in equal times (5 dots or 1/10th second)

| *s* (cm.) | 2·5 | 3·2 | 4·0 | 4·7 | 5·4 | 6·1 | 6·9 |
|-----------|-----|-----|-----|-----|-----|-----|-----|
| Vel. change | 0·7 | 0·8 | 0·7 | 0·7 | 0·7 | 0·8 | |

The distance travelled in equal times is proportional to the velocity during that time. The successive *velocity changes* are therefore proportional to the difference in successive distances.

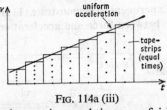

Fig. 114a (iii)

*Conclusion.* Since the velocity changes in equal times are fairly equal, the trolley was moving with uniform acceleration.

**Note.** By cutting successive strips of the tape due to equal time-intervals and pasting them vertically beside each other to represent velocity *v* in equal times, the acceleration is seen to be uniform by joining the tops of the strips. Fig. 114a (iii). (A similar procedure shows variable acceleration.)

(2) *Force and acceleration.* To investigate the relation between acceleration *f* and force *P*, the trolley *C* was placed on the smooth board *D* and the board was tilted slightly from the horizontal until the trolley ran down it with a uniform velocity. In this case the frictional force is balanced by the component of the weight of the trolley down the plane. A string was attached to the end of the trolley and passed over a grooved wheel. Various weights, *W*, were attached to the other end of the string to provide different forces *P*, and each time several runs of the trolley and tape were made to find the acceleration in the way just described.

*Measurements.* The results on the opposite page were obtained for three weights.

*Conclusion.* For a given mass, the acceleration *f* is proportional to the force *P* applied.

10 dot-intervals

| P (force) | s or v | vel. change(f) | average f | P/f |
|-----------|--------|----------------|-----------|-----|
| 50 gm. wt. | 8·5 | | | |
| | 11·1 | > 2·6 | | |
| | 13·6 | > 2·5 | 2·5 | 20 |
| | 16·0 | > 2·4 | | |
| 100 gm. wt. | 4·1 | | | |
| | 8·8 | > 4·7 | | |
| | 13·7 | > 4·9 | 4·7 | 21 |
| | 18·3 | > 4·6 | | |
| 150 gm. wt. | 6·7 | | | |
| | 13·5 | > 6·8 | | |
| | 20·3 | > 6·8 | 6·8 | 22 |
| | 27·2 | > 6·9 | | |

## Experiment to verify conditions of equilibrium for parallel forces

*Apparatus.* Light rod, weights, spring-balances.

*Account.* A light rod *A* was suspended horizontally from spring-balances at either end; and weights *P*, *Q* were hung from two points on the rod. The readings *R*, *S* on the spring-balances were

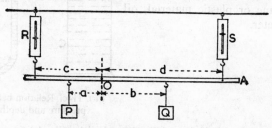

Fig. 114b. Equilibrium of parallel forces

observed, and the distances *a*, *b*, *c*, *d* of the four parallel forces, *P*, *Q*, *R*, *S*, from a point *O* on the rod were measured.

*Measurements.*

| P (gm.) | Q (gm.) | R (gm.) | S (gm.) | a (cm.) | b (cm.) | c(cm.) | d (cm.) |
|---------|---------|---------|---------|---------|---------|--------|---------|
| 500 | 500 | 626 | 380 | 10·3 | 14·8 | 34·2 | 62·6 |

*Calculations.*

(1) Total upward force = $R + S$ = 626 + 380 = 1006 gm. wt.
Total downward force = $P + Q$ = 500 + 500 = 1000 gm. wt.

(2) Total anti-clockwise moment about O = $P \cdot a + S \cdot d$
        = 500 × 10·3 + 380 × 62·6 = 28,940

Total clockwise moment about O = $Q \cdot b + R \cdot c$
        = 500 × 14·8 + 626 × 34·2 = 28,810

*Conclusions.* Allowing for experimental error, (1) the total forces in one direction = the total forces in the opposite direction, (2) the total anti-clockwise moment about any point = the total clockwise moment about the same point.

**Experiment to show the relation between pressure and depth in a liquid**

*Apparatus.* Large measuring cylinder, thistle-funnel covered with cellophane or plastic material, oil manometer.

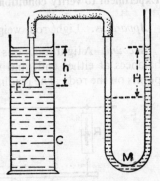

FIG. 114c. Relation between pressure and depth

*Account.* A thistle-funnel $F$, covered with cellophane, was connected by rubber tubing to an oil manometer $M$, Fig. 114c. Initially, there is no difference in levels in $M$. The funnel was then placed at varying depths $h$ below the surface of water contained in a large measuring cylinder $C$, and the corresponding difference in levels $H$ in the manometer $M$ was measured.

*Measurements.*

| Depth ($h$) | $H$ | $h/H$ |
|---|---|---|
| 19·5 units | 8·3 cm. | 2·3 |
| 28 | 12·0 | 2·3 |
| 32 | 13·7 | 2·3 |
| 34 | 14·7 | 2·3 |
| 40 | 16·8 | 2·4 |
| 49 | 20·0 | 2·5 |

*Conclusion.* Since $h/H$ is a constant, allowing for experimental error, the pressure below the surface of a liquid is directly proportional to the depth.

## Experiment to verify the principle of flotation

*Apparatus.* Burette partly filled with water, wooden rod, clamp and stand.

*Account.* A wooden rod $W$ was weighed. The reading on a burette $B$ partly filled with water was then taken, and the rod was then placed in the water so that it floated. The new reading on $B$ was now observed.

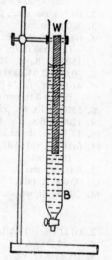

*Measurements.*

| | |
|---|---|
| Weight of wood | = 9·87 gm. wt. |
| Initial reading in burette | = 25·0 c.c. |
| Final reading in burette | = 34·9 c.c. |
| ∴ volume of water displaced by wood | = 9·9 c.c. |
| ∴ weight of water displaced | = 9·9 gm. wt. |

Fig. 114d. Principle of flotation

*Conclusion.* When an object floats in a liquid, the weight of liquid displaced is equal to the weight of the body.

# ANSWERS TO NUMERICAL EXERCISES

## CHAPTER II (p. 28)

2. 440 ft.
3. 75 ft. per sec., $787\frac{1}{2}$ ft.
4. 32 ft. per sec.², 64 ft.
5. 2, 4 sec.
6. 490,849 cm. per sec.²
7. 55,125 cm.
8. 2550 ft., 6 ft. per sec.²
10. 215 m.p.h., N. 22° W.

11. 32·4 m.p.h.
13. 22·5 m.p.h.
14. $3\frac{3}{4}$ miles.
15. 40 m.p.h.
16. (a) 6·93 m.p.h. (b) 4·3 min.
17. 48 ft. per sec., 3 sec.
19. $34\frac{3}{8}$ ft. per sec.
20. (a) 144 ft. (b) 96 ft. per sec.

## CHAPTER III (p. 49)

1. (i) 24,000 pdls. (ii) 750 lb. wt.
   (iii) 75/224 ton wt.
2. 60 cm. per sec.², 6750 cm.
3. 300 ft. lb. wt.
4. (i) 33,600 ft. pdls.
   (ii) 1050 ft. lb. wt.
5. 1500 ft. lb. wt., 1500 ft. lb. wt.
6. 60,000, 2,940,000 ergs.
7. 41 lb. wt. (i) 39 lb. wt.
   (ii) 10·4 ft. per sec.²
8. 5500 ft. lb. wt., 125 lb. wt.
9. 1200 joules, $6 \times 10^9$ dynes.
10. 17,320 ft. lb. wt., 100 lb. wt.
11. (a) 64 ft. (b) (i) 32, 0, (ii) 16, 16
    ft. lb. wt.
12. 8 ft. per sec.²
13. 90,000 ft. pdls.
14. 7·1, $3\frac{1}{8}$ lb. wt.

15. 4 ft. per sec., 150 ft. lb. wt,
16. (a) 32, (b) 27·7 ft. per sec.
    (c) (i) 0, 16, (ii) 12, 0 ft. lb. wt.
17. 343,750 ft. lb. wt. per sec.
18. (a) $10^7$ ft. lb. wt. (b) 800 ft. per
    sec.
19. (a) $451\frac{1}{11}$ lb. wt. (b) $6\frac{9}{16}$ ft. per
    sec.²
20. (a) 16 ft. per sec. (b) 9 ft. lb. wt.
21. (a) 480 h.p. (b) 3388 ft.
22. (a) 2000. (b) 1812·5.
    (c) 1812·5 lb. wt.
23. (a) 8 ft. per sec.² (b) 420 lb. wt.
    (c) $14\frac{2}{3}$ sec. (d) $860\frac{1}{6}$ ft.
24. (a) 4·9 cm. per sec.² (b) 14·7 cm.
    per sec. (c) 22·1 cm., 648,000
    ergs (approx.).
25. 5 lb. wt., 120 ft. lb. wt.

## CHAPTER IV (p. 67)

2. (i) 113·6. (ii) 70. (iii) 94·3 lb. wt.
3. 49·64 lb. wt.
4. 10, 17·3 lb. wt.
5. 330,351 gm. wt.
6. 289,577 lb. wt.
7. 1000 gm. wt.
8. 342; 342,939·7 lb. wt.
9. 1·15, 2·31 lb. wt.
10. 173 gm. wt.

13. 2·9, 5·8 lb. wt.
14. (a) 0·54. (b) 2·07 tons wt.
15. $2\frac{1}{2}$ cwt.
16. (a) 476 lb. wt. (b) 6155 lb. wt.
17. 0·73.
18. (a) 2·31. (b) 1·15 lb. wt.
19. (a) 10·4. (b) 6 lb. wt.
20. (a) 2·1. (b) 0·6 lb. wt.
21. 173,100 gm. wt.

## CHAPTER V (p. 87)

1. 8 ft. lb. wt.
2. (i) 17·3.  (ii) 10 ft. lb. wt.
3. 10·6 ft. lb. wt.
4. 120 ft. lb. wt., 20 lb. wt.
5. 8¾, 33¾ lb. wt.
6. 72, 68 lb. wt.
7. 13, 6, 10·48 in.

10. 10 lb. wt.
11. 3 ft. from $A$.
13. 2 lb.
15. 43 cm. from $A$; 3000 gm. wt.
16. (a) 43·3 lb.  (b) 400 ft. lb.
17. 9·1 in. from edge.
18. 2 in. from middle of square.

## CHAPTER VI (p. 102)

2. 500, 350 ft. lb. wt., 70%.
3. 2, 2$\frac{9}{7}$.
4. 50 lb. wt.; 6, 6, 100%.
5. 5 lb. wt., 12.
6. 2·6 ft. lb. wt., 754·3 lb. wt.
9. (c) 6000 ft. lb. wt.  (d) 6. ft.

10. 60 lb. wt., 4000 ft. lb. wt.
13. (a) 87·5 lb. wt.
    (b) 1050 ft. lb. wt.
14. 800 ft. lb. wt., $\frac{1}{8}$.
15. 6; 25 lb. wt.;  4·8.
17. 4, 52 lb. wt. (continuous string)

## CHAPTER VII (p. 118)

1. 6500 gm., 20 c.c.
2. 1062·5 lb. wt.
3. (i) 156¼, (ii) 165 lb. wt.
4. (i) 200 gm. wt., (ii) 200 c.c.
   (iii) 4 gm. per c.c.
5. 1·2 gm. per c.c.
6. (i) 8 gm. wt. (ii) 8 c.c.
7. 40 c.c. (i) 32,  (ii) 50 c.c.

8. Yes.
9. 0·76.
13. 25,500, 3000 gm. wt.
14. 10.
17. 9·9 cm.
18. 16·0, 14·55 cm.
19. 83,636 cu. ft.
10. $\frac{5}{8}$ in.

## CHAPTER VIII (p. 142)

1. 3125 lb. wt. per sq. ft.
2. 2·25 gm. wt. per sq. cm.,
   2205 dynes per sq. cm.
3. 2000, 4375 lb. wt. per sq. ft.
4. 999,500 dynes per sq. cm.,
   1999 × 10⁶ dynes.
8. 0·6 gm. per c.c.
12. 163·2 cm. mercury.

14. (a) 1½ atmos.  (b) 1⅓ litre.
19. 77 cm.
20. 1033·6 gm. wt. per sq. cm.
21. 12, 9·6 lb. per sq. in.; 9 lb. wt.
22. 75·9 cm.
24. 0·8, 18·75 cm.
25. 15 ft. 7½ in.
26. 23 cm.

## CHAPTER IX (p. 155)

12. (a) 4 ft. lb. wt. (b) 12·8 ft. per sec.   13. (b) 2·2.

# HEAT ENERGY

# CHAPTER X

## MEASUREMENT OF TEMPERATURE

A STEAM engine or locomotive derives its energy from the burning of fuel. The greater the amount of heat supplied, the more work the engine can do. *Heat is thus a form of energy* (p. 210).

When a quantity of hot tea is poured into a cup held in the hand, some heat passes from the liquid to the cup. The cup feels hotter to the touch: that is, it has a higher temperature. *Temperature is a measure of "degree of hotness".*

One of the first actions of the doctor when he is called to a sick patient is to take his temperature with a thermometer. The temperature of a factory or a mill must conform to a value specified by law, otherwise the health of the workers may be impaired. The temperature of the refrigerating plant for storing meat and other food on cargo boats must be a certain value if the food is to remain in good condition over a long period. At the other extreme, furnaces for making glass, cement, and different types of steel must be operated at definite high temperatures. These few examples show the everyday importance of temperature-measurement.

**Types of thermometer. Scales of temperature.** The accurate measurement of temperature requires the careful use of a *thermometer*. This is an instrument which utilizes some physical property of a substance which changes as the temperature is altered. The change in volume of mercury and of alcohol with temperature was first used as a means of temperature-measurement towards the end of the 17th century. To-day, thermometers for everyday use mainly consist of the *mercury-in-glass* type. For very accurate measurements of temperature, however, the change in the electrical resistance of platinum with temperature is utilized in a *platinum resistance thermometer*. A more accurate instrument than the

(a) CHECKING THERMOMETERS by microscopic comparison with
National Physical Laboratory certified instruments

(b) THERMO-ELECTRIC PYROMETER in use for measuring the temperature
of a furnace containing molten metal

FIG. 115. THERMOMETERS

latter is the *gas thermometer*, which utilizes the relation between the volume of a gas at constant pressure and its temperature (p. 223). This type of thermometer is used at the National Physical Laboratory to check the reading on thermometers sent to them by various manufacturing firms, and consequently the gas thermometer acts as a "standard."

The scales of the usual thermometers were chosen many years ago. First, two temperatures which remain *constant* had to be obtained. As NEWTON suggested in 1701, those chosen were (i) the melting point of ice under normal conditions; (ii) the temperature of steam under standard atmospheric pressure of 76 cm. of mercury.

On the **Centigrade system** the first temperature, known as the "ice point" or "lower fixed point" of the thermometer, is given the value 0, while the second temperature, known as the "steam point" or "upper fixed point" of the thermometer, is given the value 100. The scale is divided into 100 equal parts or "degrees."

On the **Fahrenheit system**, the numbers 32 and 212 are used for the ice point and steam point respectively, so that there are 180° between these two temperatures. The Fahrenheit system is adopted for everyday use in this country, but the Centigrade system is used in scientific work owing to the arithmetical advantage of having 100 degrees between the ice and steam points, and zero for the ice point.

The **Réaumur system**, which is still in use in certain parts of the continent, has 0° as the ice point, 80° as the steam point, and divides the interval between them into 80 equal parts.

The Centigrade system was devised by CELSIUS in 1742, while the FAHRENHEIT and RÉAUMUR systems, introduced about 1730, are named after their devisers.

**Changing from one temperature scale to another.** Fig. 116 illustrates the three common scales of temperature, assuming that the same length of glass tubing is used for marking the numbers on the Centigrade (C.), Fahrenheit (F.), and Réaumur (R.) scales. Since a change in temperature of 180° F. is equal to a change in temperature of 100° C., a change of 1° F. is equal to a change in temperature of $\frac{100}{180}$ or $\frac{5}{9}$° C.

To convert a temperature from one scale to the other, suppose that $C$ degrees on the Centigrade scale is the same as $F$ degrees on the Fahrenheit scale, where $C$, $F$ are numbers (Fig. 116). On the

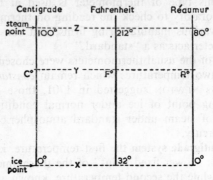

FIG. 116. Scales of temperature

Centigrade thermometer, $XY/XZ = C/100$; on the Fahrenheit thermometer, $XY/XZ = (F - 32)/180$. Hence

$$\frac{C}{100} = \frac{F - 32}{180}.$$

This relation should be memorized. It can be seen that the same temperature, $R$, on the Réaumur scale is given by $\dfrac{R}{80} = \dfrac{C}{100}$, or by $\dfrac{R}{80} = \dfrac{F - 32}{180}$.

**Making a mercury thermometer.** The first step in making a mercury thermometer is to choose a capillary glass tube whose bore has the same cross-sectional area all the way along it. One end of the tube is then sealed, and a bulb is blown at this end. The tube is cooled, a funnel is fitted closely over the open end of the tube top, and mercury is poured into the funnel, the tube being vertical. When the bulb is warmed gently, some air is driven out of the tube through the mercury at the top, and the pressure due to the weight of mercury forces some of the latter inside the bulb when the flame is removed. More mercury is obtained in the bulb by alternate warming and cooling. Finally, the air is driven out of the glass tubing by boiling the mercury;

the glass is then maintained at a temperature a little above the highest at which it will be used, and the top is sealed.

The **steam point** has now to be marked on the glass of the thermometer. For this purpose the thermometer $T$ is placed in an apparatus $H$ known as a **hypsometer**, which contains water, a manometer, $M$, for measuring the pressure of the steam obtained

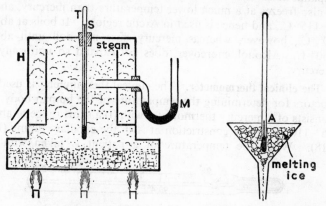

FIG. 117 (*a*). Determination of steam point

FIG. 117 (*b*). Determination of ice point

when the water is boiled, and an outlet tube to the air. Now the temperature of boiling water depends on the dissolved impurities, whereas the temperature of the *steam* obtained is independent of any such impurities. The thermometer $T$ must therefore be placed with its bulb *well above* the water so that the mercury in the tube is all at the temperature of the steam. When the mercury level is steady, its position $S$ is marked on the glass tube, and the pressure of the steam, which is read on the manometer $M$, is noted. The steam point of a thermometer is the temperature of steam at a pressure of 76 cm. of mercury; the marking $S$ would thus require a minor correction if the manometer showed that the pressure was not exactly 76 cm. (Fig. 117 (*a*) ).

To determine the **ice point**, the thermometer is placed with its bulb in *melting ice*, and the level $A$ of the mercury is marked; this is the zero on the centigrade scale (Fig. 117 (*b*) ). Finally, the distance between the ice and steam points is divided into 100 or 180 equal divisions, according to the scale being used.

**The alcohol thermometer.** Mercury has the disadvantage of being expensive, and of freezing at about −40° C. Alcohol is sometimes used as the liquid in a domestic thermometer, since it expands about ten times as much for the same rise in temperature. It also freezes at a much lower temperature than mercury, about −115° C., and hence is used in Arctic regions. It boils at about 80° C., however, whereas mercury does not boil until about 360° C. Alcohol, moreover, does not expand so uniformly as mercury.

**The clinical thermometer.** The clinical thermometer is used by doctors for determining the temperature of the human body. It consists of a mercury thermometer with (i) a short range of about 95°–110° F.; (ii) a constriction at A in the capillary tubing (Fig. 118). When the temperature of a patient is taken with the

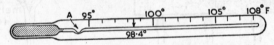

Fig. 118. Clinical thermometer

thermometer, the mercury thread passes A. When the thermometer is removed from the patient the mercury thread to the left of A recedes, but the column to the right remains in position owing to the constriction; by this means, the thermometer can still be read. It is afterwards reset by jerking it sharply, when the thread beyond A rejoins the mercury in the bulb.

**Maximum and minimum thermometer.** In the year 1782, the physicist SIX designed a thermometer, illustrated in Fig. 119, which registers the maximum and minimum temperatures over a period of time, as required in weather forecasting. A large bulb is filled with alcohol, A, and is connected to a bent capillary tube which contains mercury along the part BC. Above C is alcohol, and D contains a little air. A steel index, X, having a light spring attached to keep it in position, is above C; a similar index, N, is above B.

The alcohol in A has the largest volume of the liquids in the thermometer, and alcohol expands relatively much more than mercury for the same temperature change (p. 164). Thus, when the temperature rises, the alcohol $A$ expands and flows past $N$, and the mercury at $C$ rises and pushes $X$ in front of it. When the

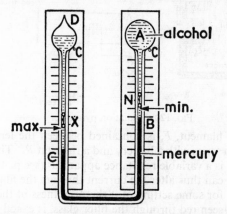

FIG. 119. Combined max. and min. thermometer

temperature falls, the alcohol in $A$ contracts, and the mercury column at $B$ rises and pushes $N$ in front of it. Hence the lower end of $X$ indicates the maximum temperature on the scale shown, and the lower end of $N$ indicates the minimum temperature. After the temperatures have been read, a magnet is used to reset each index above the mercury columns, and the thermometer is then ready for the next observation.

**Pyrometers.** In industry, certain types of steel and glass require heat treatment at definite high temperatures in their manufacture. The usual glass thermometer would melt if placed in contact with the furnace, and accordingly another type of thermometer, known as a *pyrometer*, is used.

Most pyrometers utilize the laws governing *radiation* (see p. 215) from a hot object. The principle of one type, known as an optical pyrometer, is illustrated in Fig. 120. It consists of a lens $A$, which is pointed at the radiation coming through the open furnace door, and a lens $B$, constituting an eyepiece. The light is observed at $E$

through a red filter glass, and a red field of view is obtained *whose brightness depends upon the temperature of the furnace.* An

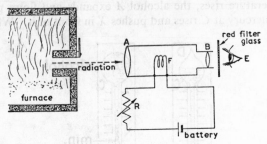

FIG. 120. Radiation pyrometer

electric lamp filament, *F*, is contained between the lenses in the tube, and is in series with a battery and a rheostat *R*. The rheostat

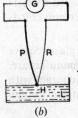

is a variable resistance apparatus (see p. 455), which can thus alter the current through the filament, and for some setting of it the brightness of the filament, seen red through the filter glass, is exactly the same as the brightness of the red field of view. In this case the filament seems to "disappear" into the background. When the rheostat is altered, a pointer rotates round a circular dial graduated in degrees centigrade or fahrenheit, and the temperature of the furnace corresponds to the pointer reading at which the filament "disappears" into the background.

**The thermocouple thermometer.** If two *unlike* metals *A*, *B*, are connected together and one junction *H* is warmed while the temperature of the junction *C* is kept constant (Fig. 121 (*a*) ), an electric current flows along the metals. The two joined metals are known as a *thermocouple*, and provide an example of the direct transformation of heat energy to electrical energy. When the temperature of *C* is kept constant and the temperature of *H* is increased, the magnitude of the electric current increases, and the temperature of *H* can be deduced from the magnitude of the electric current.

FIG. 121. Thermocouple

This is the principle of the *thermoelectric thermometer*, which is used in industry for finding the temperature of molten metals and certain furnaces (see Fig. 115 (*b*) ). Fig. 121 (*b*) illustrates the simple arrangement required. The thermocouple may consist of platinum (*P*) and platinum-rhodium (*R*) wires, with a current-reading instrument, *G*, connected in the circuit. *G* is calibrated to read in °F or °C. The junction *H* of the metals is placed in the furnace or liquid *L*, whose temperature can be read directly from *G*.

───Summary───

1. To change a temperature from °C. to °F., or vice versa, the relation $C/100 = (F - 32)/180$ can be used $\left(\dfrac{F - 32}{C} = \dfrac{9}{5}\right)$.

2. The lower fixed point of a thermometer is the temperature of melting ice, and the upper fixed point is the temperature of steam at 76 cm. mercury pressure.

3. The clinical thermometer has (i) a constriction near the bulb, (ii) a short temperature range (e.g. 94–110° F.).

4. Six's maximum and minimum thermometer utilizes alcohol and mercury, with steel pointers to indicate the maximum and minimum temperatures.

## WORKED EXAMPLES

**1.** *What are the fixed points on a thermometer scale ? How would you determine their positions on a thermometer? The readings of a mercury thermometer which has a scale marked in millimetres are* 10·6 *mm. and* 208·6 *mm. at the standard fixed points. What will be the reading when the thermometer is in a liquid at* 72°*F.?* (*L.*)

Let $x$ = the reading in mm. corresponding to 72°F. (See Fig. 122.)

Then, since the ratios of lengths are equal,

$$\frac{x - 10\cdot6}{208\cdot6 - 10\cdot6} = \frac{72 - 32}{212 - 32} \text{ (Fig. 122)}$$

$$\therefore \frac{x - 10\cdot6}{198} = \frac{40}{180} = \frac{2}{9}$$

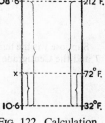

FIG. 122. Calculation

$$\therefore \quad 9(x - 10\cdot6) = 2 \times 198$$
$$\therefore \quad x - 10\cdot6 = 2 \times 22 = 44$$
$$\therefore \quad x = 44 + 10\cdot6 = 54\cdot6 \text{ mm.}$$

2. *The lengths of a degree interval on two mercury thermometers, one having a Centigrade scale and the other a Fahrenheit scale, are $\frac{1}{10}$ in. and $\frac{1}{20}$ in. respectively. Calculate (a) the distance between the upper and lower fixed points of each thermometer; (b) the distance between the lower fixed point and the $-10°$ mark on the Fahrenheit thermometer. (N.J.M.B.)*

(*a*) The interval between the upper and lower fixed points of a Centigrade scale = 100°.

$$\therefore \quad \text{distance} = 100 \times \tfrac{1}{10} = 10 \text{ in.}$$

For the Fahrenheit scale of 180°, distance $= 180 \times \tfrac{1}{20} = 9$ in.

(*b*) The number of degrees $= 32° - (-10°) = 42°$

$$\therefore \quad \text{distance} = 42 \times \tfrac{1}{20} = 2\cdot1 \text{ in.}$$

3. *Describe in detail how the room temperature may be found with the aid of an ungraduated mercury thermometer. The freezing and boiling points of water are marked 20 and 160 on a certain thermo-*

FIG. 123. Calculation

*meter. Calculate (a) the temperature on the Fahrenheit scale corresponding to a reading of 83 on the scale of this thermometer, (b) the temperature on the centigrade scale which is represented by the same reading on this thermometer and on the centigrade thermometer. (L.)*

Fig. 123 illustrates the thermometer $T$, a Fahrenheit thermometer $F$, and a Centigrade thermometer $C$. Suppose $x$ is the temperature on the Fahrenheit thermometer. Then, by proportion (see p. 162),

$$\frac{x - 32}{180} = \frac{83 - 20}{140}$$
$$\therefore \quad (x - 32)140 = 180(83 - 20) = 180 \times 63$$
$$\therefore \quad x - 32 = \frac{180 \times 63}{140} = 81$$
$$\therefore \quad x = 113° \text{ F.}$$

Suppose $y$ is the temperature which is the same number on the thermometer $T$ and the Centigrade thermometer. Then, by proportion,

$$\frac{y - 20}{140} = \frac{y - 0}{100}$$
$$\therefore \quad 100(y - 20) = 140y$$
$$\therefore \quad y = -50° \text{ C.}$$

## EXERCISES ON CHAPTER X

**1.** Name three applications of thermometers in everyday life.

**2.** Convert (i) 63° C., 120° C., − 18° C. to °F., (ii) 212° F., 142° F., 10° F., − 36° F. to °C.

**3.** Describe briefly the construction of a mercury-in-glass thermometer.
Describe a clinical thermometer.

**4.** In which part of the world would an alcohol thermometer be preferred to a mercury thermometer? Name the advantages and disadvantages of an alcohol thermometer over a mercury thermometer.

**5.** Draw a labelled diagram of a combined maximum and minimum thermometer, numbering the scales. How does the instrument work?

**6.** Draw a diagram of a maximum and minimum thermometer, and explain its mode of action. The boiling point of alcohol is 78° C. What is the temperature on the Fahrenheit scale? (*O. & C.*)

**7.** Define the *fixed points* on a thermometer scale. Describe how you would check the fixed points on a Centigrade thermometer. At what temperature is the reading on a Fahrenheit thermometer three times the reading on a Centigrade thermometer? (*L.*)

**8.** Give *two* reasons why water is less suitable than alcohol for use in thermometers. (*N.*)

**9.** Make a labelled diagram of a clinical thermometer. (*N.*)

**10.** Describe the construction of *either* a maximum *or* a minimum thermometer. Explain how it is "set" and how it works. (*C.W.B.*)

**11.** Describe the construction and explain the action of *two* of the following: (*a*) a combined maximum and minimum thermometer, (*b*) a constant volume air thermometer, (*c*) a dew-point hygrometer. (*L.*)

**12.** Describe a thermometer that will record the minimum temperature during the night. An ungraduated mercury thermometer attached to a millimetre scale reads 22·8 mm. in ice and 242·4 mm. in steam at standard pressure. What will it read on a day when the temperature is 22° F.? (*C.*)

**13.** Describe how the accuracy of the fixed point marks on a mercury thermometer can be tested. Explain what factors may have caused any errors revealed by the tests you describe. (*N.*)

**14.** Describe, with experimental details, how you would calibrate a mercury thermometer for use between the temperatures of − 5° C. and 110° C. Describe the construction of a maximum and minimum thermometer. Give reasons why mercury is regarded as a suitable liquid for use in a thermometer. (*O. & C.*)

**15.** (i) Calculate the temperature on the Centigrade scale which has the same numerical value on the Fahrenheit scale, (ii) The ice- and steam-points on a certain thermometer are marked 50 and 170 respectively. Calculate the number on this thermometer corresponding to a temperature of 92° F.

## CHAPTER XI

## EXPANSION OF SOLIDS AND LIQUIDS

IF telegraph wires are observed in winter and in summer, it will be noted that the wires sag more in summer, showing that the length of the wires has increased as the temperature has risen. This and other observations, as well as experiments mentioned in the previous chapter and below, show that *most materials expand when their temperature increases*, and contract when their temperature decreases.

### THE EXPANSION OF SOLIDS

There are many practical consequences of the expansion of solids. The size of the balance-wheel of a watch, which governs the movement of its hands, and the length of a pendulum of a clock, are affected by the change of temperature which occurs from winter to summer. Allowance must be made for the alteration in

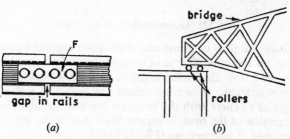

FIG. 124. Allowance for expansion

length of metal bridges and railway lines as the seasons change, since enormous stresses would be set up if the expansion and contraction could not take place freely. Fig. 124 (*a*) illustrates the fish plate, *F*, which joins one section of a railway line to the next section. The holes in the rail are oval, allowing expansion or contraction to take place. The Forth Bridge, the second longest in the British Isles, has a total allowance of about 4 ft. for expansion and contraction, and roller bearings are used at each end of the bridge so that the change in length can take place freely without straining the structure (Fig. 124 (b)). For a similar reason, telegraph wires must have sufficient sag in them, as they would otherwise snap in winter when they contracted.

**Demonstration of expansion of solids.** If a *metal ball*, B, which can just pass through an iron ring, R, is heated, the ball becomes too big to get through the ring, showing that the metal has expanded (Fig. 125 (a)).

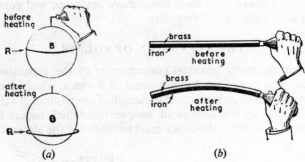

FIG. 125. Demonstrations of expansion

A *bimetallic strip* demonstrates that solids expand by different amounts for the same change of temperature. Two equal lengths of brass and iron are riveted together to form a bimetallic strip, originally straight. When the metals are heated in a flame, the strip begins to *curve* with the brass on the outside, showing that the expansion of the brass is greater than that of the iron for the same change of temperature (Fig. 125 (b)).

A *lever arrangement* can be used to magnify the expansion of a vertical brass rod. One end of the rod is fixed at A in a container C into which hot water is poured. The other end, B, of the rod expands slightly in a vertical direction, and pushes one end of a lever BD. The other end, D, of the lever is thus displaced downward, through a distance DN, and its movement, which may be

50 times as great as the actual expansion of the rod, is sufficient to be observed on a vertical metre ruler, $R$, fixed behind it. (Fig. 126).

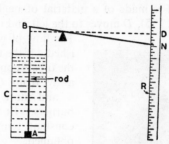

FIG. 126. Linear expansion magnified

**Some uses of expansion of solids.** The expansion of a solid with increase of temperature is put to practical use commercially. Two different metals soldered together, a bimetallic strip, are used in one form of *fire-alarm*. Metal $A$ has a greater expansion than metal $B$ for the same rise in temperature, and hence the strip curves round towards $D$ as the temperature rises (Fig. 127 (a) ). A dangerous rise in temperature closes an electric bell circuit (explained in p. 571) at $D$, and the bell rings.

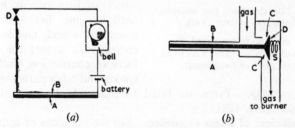

(a)                      (b)

FIG. 127. Applications of expansion

A *thermostat* is a device which maintains a constant temperature. Fig. 127 (b) illustrates the principle of one type, used to control the temperature in gas-cookers and some water-heaters. $B$ is a brass tube with a rod $A$ attached at one end. A valve $D$ is

7

connected to the end of $A$, and allows gas from the mains to pass through the burners, as shown in the diagram. When the temperature of the oven or water-heater increases, the tube $B$ expands, and as the rod $A$ is made of a material of negligible expansion called "invar" (p. 175), $D$ moves to the left and narrows the gaps.

FIG. 128. GAS RING FOR HEATING STEEL LOCOMOTIVE TYRES

The ring is being tested by a workman. When in use it encircles the tyre (*seen in foreground*).

The flow of gas is thus diminished. If the temperature falls, the tube $B$ contracts, $D$ is pulled away from $C$ by the spring $S$, and more gas reaches the burners. In this way a fairly constant temperature can be obtained.

*The bulging walls of a building can be straightened* by passing an iron rod through the two walls, heating the rod, and then tightening nuts on the rod outside the walls. When the rod cools, the large force exerted in contracting causes an appreciable straightening of the walls. In making a cart wheel an iron tyre is heated until it just fits over the assembled wheel, and is then cooled with water; the large force of contraction binds the spokes and other parts together very securely. Tyres are fitted to train wheels by a similar process. (Fig. 128).

**Coefficient of linear expansion.** For the purposes of industry, as well as research, scientists require to know accurately to what extent metals change in length when their temperature alters; and it is obviously most useful to have a table, or list, showing the increase in length of 1 centimetre (or 1 inch) when the temperature of the metal changes by 1° C. (or 1° F.). From a knowledge of this quantity, known as the *coefficient of linear expansion* of the metal, the actual change in length of a metal when its temperature alters can be calculated by simple proportion.

The coefficient of linear expansion ($\alpha$) of a solid is defined as *the increase in length of unit length of the solid when its temperature changes by one degree.* Experiment shows that, over a wide range of temperature, 1 cm. of brass expands by an almost constant length of 0·000018 cm. when its temperature changes by 1° C. Thus $\alpha$ for *brass* is about 0·000018 per °C. Since a change of 1° C. = a change of $\frac{9}{5}$° F. (p. 161), $\alpha$ = 0·000018 per $\frac{9}{5}$° F. change of temperature, or 0·000010 per °F.

Iron has a linear coefficient $\alpha$ = 0·000011 per °C., which is 0·000006 per °F., *glass* has a low value of 0·000008 per °C., and *quartz* has a much smaller linear coefficient than glass. An alloy known as *invar*, which has also an extremely small linear coefficient, was produced in 1904, and is used extensively in industry.

**An experiment to measure the coefficient of linear expansion ($\alpha$)** of a metal is illustrated in Fig. 129. *AB* is a rod of the metal contained in a chamber *C*, and fixed at one end *A* so that no movement of the rod can take place there. *P* is a thermometer,

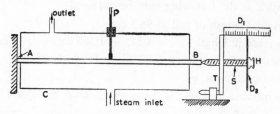

FIG. 129. Linear coefficient measurement

making contact with the rod, to determine its temperature. Since the expansion of a solid is very small, an ordinary ruler by itself is useless for its measurement and a special device known as a **screw gauge**, which measures very small lengths accurately, is therefore employed.

One form of gauge is shown in Fig. 129. It consists of a screw *S* which moves through a fixed support *T* when the knob *H* is turned. *H* is in the centre of a circular dial $D_2$ graduated in 100 equal divisions (Fig. 130), and the movement of $D_2$ can be read from a fixed scale $D_1$ graduated in millimetres. If the point of the screw moves 1 millimetre when it is turned through one revolution, the *pitch* of the screw is said to be 1 mm., or 0·1 cm. A change of 1 division on $D_2$ thus corresponds to a movement of $\frac{1}{100}$ of 0·1 cm., or 0·001 cm., of the point of the screw.

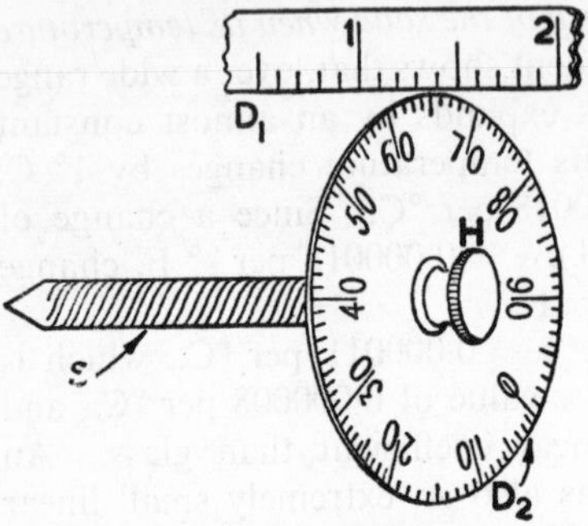

Fig. 130. Screw gauge

First, the screw is adjusted to touch the end $B$ of the metal rod, and the scales $D_1$, $D_2$ are read. Imagine that the reading is 1·3 cm. on $D_1$, and 65 divisions on $D_2$, as in Fig. 130. The reading noted is then 1·365 cm. The screw gauge is now turned back to allow for the expansion of the metal rod, and steam is passed into $C$. When the temperature of $P$ is steady, the screw gauge is adjusted to make contact again with the end $B$ of the rod. Since expansion has occurred, the reading on the scale is altered; suppose the reading is 1·515 cm. Measurements such as the following may be obtained—

original length of rod = 95·3 cm.
increase in length     = 1·515 − 1·365 cm. = 0·15 cm.
change in temperature = 99° C. − 16° C. = 83° C.

∴   mean coefficient, $a$

= increase in length of 1 cm. when temp. changes 1° C.

$$= \frac{0·15}{95·3 \times 83} = 0·000019 \text{ per } °C.$$

**Formulae.** It can be seen that the mean linear coefficient $a$ is the fractional increase of length per degree rise of temperature

i.e.     $$a = \frac{increase \ in \ length}{original \ length \times temp. \ rise} \qquad . \qquad . \quad (1)$$

Further, if the original length of a solid is $l_1$ and its temperature *changes* by $t°$ C.,

*the change in length* $= a l_1 t \quad . \quad . \quad . \quad (2)$

since $\alpha$ is the change in length of unit length for $1°$ C. change in temperature.

The new length, $l_2$, = original length + increase = $l_1 + \alpha l_1 t$

$$\therefore \quad l_2 = l_1(1 + \alpha t) \quad . \quad . \quad . \quad . \quad (3)$$

**Other coefficients of expansion of a solid.** When the temperature of a water-tank or of the metal roof of a shed increases, the surface area increases; and in this connection there is an *area (superficial) coefficient of expansion*, $\beta$. By definition, $\beta$ is the increase in area of 1 sq. cm. of a solid when its temperature increases by $1°$ C., and the new area $A_2$ of a metal is related to the original area $A_1$ when the temperature changes by $t°$ C. by

$$A_2 = A_1(1 + \beta t).$$

This relation has the same mathematical form as $l_2 = l_1(1 + \alpha t)$ for linear expansion, and is derived in a similar way.

Fortunately there is not much trouble in calculating the area coefficient if the linear coefficient of the same substance is known. As shown below,

*area coefficient* ($\beta$) = 2 $\times$ *linear coefficient* ($\alpha$).

For example, if $\alpha = 0.000018$ per $°$C. for brass, the area coefficient $\beta = 0.000036$ per $°$C. Thus 1 sq. cm. of brass increases to $1.000036$ sq. cm. when its temperature is raised $1°$ C.

A glass stopper which has stuck in the neck of a bottle can be removed by warming the neck, which then expands in volume. besides an increase in length and area, then, solids such as a water-tank or a metal sphere expand in volume, and the *volume (cubical) coefficient of expansion*, $\gamma$, of a substance is defined as the increase in volume of 1 c.c. when its temperature increases by $1°$ C. The new volume $V_2$ of a substance, whether hollow or solid, is related to its original volume $V_1$ when its temperature increases by $t°$ C. by

$$V_2 = V_1(1 + \gamma t).$$

As in the case of the area coefficient, the volume coefficient $\gamma$ is easily derived from a knowledge of the linear coefficient, since as proved on p. 178,

*cubical coefficient* ($\gamma$) = 3 $\times$ *linear coefficient* ($\alpha$).

**Proof that $\beta = 2\alpha$ and $\gamma = 3\alpha$.** Suppose a metal square has sides of length 1 cm. and $\alpha$ is its linear coefficient. The increase in length of each side of the square when its temperature changes by $1°$ C. is then $(1 + a)$ cm. and the new area becomes $(1 + a)^2$. The original area is 1 sq. cm., and hence the increase in area $= (1 + a)^2 - 1 = 2a + a^2$. Now in practice $a$ is of the order of $0.00001$ cm.; hence, as the reader can verify for himself, $a^2$ is a negligible quantity compared with $2a$. Consequently the increase in area of 1 sq. cm. when its temperature changes by $1°$ C. is $2a$, which, by definition, is the value of the area coefficient, $\beta$.

To show that the cubical coefficient $\gamma$ is $3a$, consider a cube of side 1 cm. and suppose its temperature changes by $1°$ C. Each side then becomes $(1 + a)$ cm. in length, and hence the volume is $(1 + a)^3$, or $(1 + 3a + 3a^2 + a^3)$. Now in practice $3a^2 + a^3$ are very small numbers compared with $3a$. Hence the increase in volume of 1 c.c. when its temperature increases by $1°$ C. $= (1 + 3a) - 1 = 3a$, which is the value of the cubical coefficient.

**The balance-wheel of a watch** (Fig. 131) oscillates at a rate which depends on its dimensions and the spring, among other factors.

FIG. 131.
Balance-wheel

A change in temperature causes a decrease in the elasticity of the spring and alters the wheel's dimensions, so that the watch gains or loses. Fig. 132 (a) illustrates a method used to compensate for changes of temperature. The rim of the balance-wheel, which consists of segments, is made of a bimetallic strip consisting of an outer strip of brass and an inner strip of invar (p. 175). Invar has a negligible coefficient of expansion, so that the rim curves inwards when the temperature rises, and the size of the wheel, which affects the time of

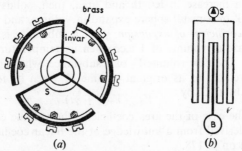

FIG. 132. Compensated balance-wheel and pendulum

oscillation, can be made to compensate for the decrease in elasticity of the spring, $S$.

**Harrison's compensated pendulum.** In the 18th century, HARRISON won a government prize for designing a pendulum clock which kept regular time. He utilized a number of brass and iron rods arranged alternately, as shown in Fig. 132 (b), the bob of the pendulum being B and the point of suspension S. When the temperature changes, the brass rods expand upwards and the iron rods expand downwards; and by suitably adjusting the lengths of the two kinds of rods, the distance of the bob B from S, which affects the period of oscillation, can be made independent of the temperature.

The increase in length of a length $l_b$ of brass when its temperature changes by $t°$ C. is $a_b l_b t$, where $a_b$ is the linear coefficient of brass (p. 176). The increase in length of a length $l$ of iron when its temperature also changes by $t°$ C. is $alt$, where $a$ is the linear coefficient of iron. In order that the distance of the bob B from S should be constant, we must have $alt = a_b l_b t$.

$$\therefore \quad al = a_b l_b$$

$$\therefore \quad \frac{l}{l_b} = \frac{a_b}{a}.$$

Thus the ratio of the lengths of iron and brass must be inversely proportional to their linear coefficients. Since $a_b = 0.000018$ and $a = 0.000011$, $l/l_b = 3/2$ approximately. This is the relation between $l$ and $l_b$ observed in Harrison's compensated pendulum. (Fig. 133.)

FIG. 133.
HARRISON'S
COMPENSATED
PENDULUM

## THE EXPANSION OF LIQUIDS

All liquids change in volume when their temperature alters. This effect is utilized in the construction of thermometers, as we have seen, and enables hot-water systems to function, as explained on p. 250. The expansion of a liquid can be simply demonstrated by the apparatus shown in Fig. 134. $F$ is a flask with a narrow tube $T$ through its cork, and completely filled with coloured water which reaches to a level $A$ in $T$. When the liquid

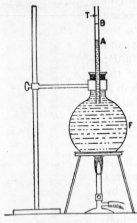

is heated the level is observed to rise gradually to $B$, showing that expansion has taken place.

**Coefficient of cubical (volume) expansion.** A liquid has no definite shape; it always takes the shape of the vessel into which it is poured. We can therefore deal only with *volume* changes of a liquid, unlike the case of a solid.

The *coefficient of cubical (volume) expansion*, $\gamma$, *of a liquid* is in the increase in volume of 1 c.c. when its temperature changes by 1° C. Suppose, for example, that the volume of the liquid at 0° C. is 480 c.c.,

FIG. 134. Liquid expansion

and that the volume increases to 484·8 c.c. when the temperature increases to 20° C. Then 480 c.c. of the liquid at 0° C. expands by 4·8 c.c. for a 20° C. change in temperature, and hence the average or *mean coefficient* of expansion, $\gamma = \dfrac{4 \cdot 8}{480 \times 20} = 0 \cdot 0005$ per °C., for the temperature range 0° C. − 20° C. The magnitude of $\gamma$ varies with the temperature range over which the liquid is heated. The coefficient of cubical expansion of mercury is about 0·00018 per °C.; that of water is about 0·0003 per °C. In general, liquids expand about 5-10 times as much as equal volumes of solids for the same temperature change.

**Formulae.** Suppose 120 c.c. is the original volume of a liquid and 121·44 c.c. its new volume when its temperature changes from 15° C. to 45° C. The increase in volume is hence 1·44 c.c., and the mean cubical coefficient of the liquid from 15° C. to 45° C. $= \dfrac{1 \cdot 44}{120 \times 30} = 0 \cdot 0004$ per °C. Thus, in general, the mean cubical coefficient, $\gamma$, is given by

$$\gamma = \frac{increase\ in\ volume}{original\ volume\ \times\ temp.\ rise} \qquad . \qquad . \quad (4)$$

Suppose $V_1$ is the original volume of a liquid whose temperature *changes* by $t°$ C. and that $\gamma$ is the mean cubical coefficient of expansion. Then, since $\gamma$ is the increase in volume of 1 c.c. when its temperature changes by $1°$ C.,

$$the\ increase\ in\ volume = \gamma V_1 t \qquad . \qquad . \qquad . \quad (5)$$

by simple proportion. Consequently the new volume, $V_2$, of the liquid $= V_1 + \gamma V_1 t$.

$$\therefore \quad V_2 = V_1(1 + \gamma t) \qquad . \qquad . \qquad . \qquad . \quad (6)$$

These formulae are similar to those obtained in connection with the expansion of solids (see pp. 176-7).

**Apparent and true (absolute) coefficients of expansion.** If a liquid is to be heated or cooled, it must be contained in a vessel. Now the vessel, as well as the liquid, alters in volume when the liquid is heated; consequently *direct* observation on expansion is not so easy with liquids as with solids.

Suppose scratches are made at $A$, $B$ on the sides of a glass vessel, at the same level as a liquid contained in the vessel. When the vessel is heated, the level of the liquid is observed to *fall below* the scratches at first, as the heat takes a little time to pass through the glass and reach the liquid. After a time the liquid level reaches $CD$, and because the glass also expands, the scratches on the glass reach the level $A'B'$ (see exaggerated diagram, Fig. 135). The *apparent* expansion of the liquid, estimated from the scratches $A'B'$ to the level $CD$ is thus $A'DCB'$; but the *true* expansion is $ADCB$, which is greater than the apparent expansion by the expansion of the vessel.

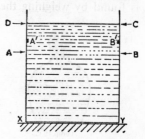

Fig. 135.
True and apparent expansion

We are thus led to speak of, and distinguish between, the *true* (of *absolute*) *coefficient* of expansion of a liquid and the *apparent*

7*

*coefficient*, which concerns the expansion relative to that of the vessel. The former coefficient, $\gamma$, was defined on p. 181; the latter is defined by $\dfrac{apparent\ expansion\ of\ liquid}{original\ volume \times temp.\ rise}$, with the emphasis on the word "apparent." If an experiment measures the apparent coefficient of a liquid, the true (absolute) coefficient is obtained as follows—

*true coefficient*, $\gamma$ = *apparent coefficient* + *c*,

where *c* is the *cubical* coefficient of expansion of the vessel's material. Thus, suppose glass of *linear* coefficient of expansion 0·000008 per °C. is used as the material of the vessel, and the apparent coefficient was found to be 0·0003 per °C.

Then $c$ = 3 × linear coefficient of glass (p. 177) = 0·000024 per °C.

$\therefore$ true coefficient of liquid, $\gamma$

$$= 0\cdot0003 + 0\cdot000024 = 0\cdot000324 \text{ per } °C.$$

**Measurement of apparent coefficient of a liquid. Specific gravity bottle method.** In this method a specific gravity bottle is completely filled with the liquid whose coefficient of cubical expansion is required (Fig. 136 (*a*) ), and the mass of the liquid, 152 gm. (say), is found by weighing the bottle plus liquid and subtracting the

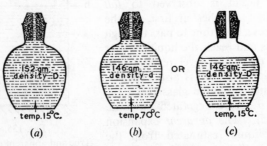

Fig. 136. Apparent coefficient

weight of the bottle alone. The temperature of the surroundings, 15° C., say, is noted.

The bottle is then placed in a beaker containing water so that the bottle is nearly covered (not shown), and the beaker is warmed until the temperature of the water is constant at 70° C., say. Since the liquid expands, some of the liquid is pushed out, or *expelled*, from the bottle through the fine opening in the stopper. Consequently less liquid remains in the bottle at 70° C. The bottle is taken out of the water, excess liquid is wiped off the stopper. On re-weighing, the mass of liquid remaining is found to be 146 gm., say. It will be noticed that, as its temperature falls, the liquid shrinks in volume until it attains the temperature of the surroundings. (See Fig. 136 (*b*), (*c*).)

The volume of a liquid is given by $\dfrac{\text{mass of liquid}}{\text{density of liquid}}$ (p. 105).

Hence the volume of the liquid filling the bottle at 15° C. $= \dfrac{152}{D}$, where $D$ is the density of the liquid at 15° C. (Fig. 136 (*a*)); the volume of the liquid filling the bottle at 70° C. is also given by $\dfrac{146}{d}$ where $d$ is the density of the liquid at 70° C., and the volume of the liquid finally remaining in the bottle at 15° C. is given by $\dfrac{146}{D}$. (See Fig. 136 (*b*), (*c*).)

Now imagine matters reversed, and suppose the liquid *remaining* in the bottle is heated from 15° C. to 70° C. The volume of the liquid then increases from $146/D$ to $146/d$ c.c. If we ignore the expansion of the vessel, we can also say that the volume of the liquid filling the bottle at 70° C. is $152/D$ c.c. Thus

$$\text{apparent expansion of liquid} = \frac{152}{D} - \frac{146}{D} = \frac{6}{D} \text{ c.c.}$$

$$\text{But original volume of liquid} = \frac{146}{D} \text{ c.c.}$$

$$\text{and rise in temperature} = 70° \text{ C.} - 15° \text{ C.} = 55° \text{ C.}$$

$$\therefore \text{ apparent coefficient, } \gamma' = \frac{\text{expansion of liquid}}{\text{original vol.} \times \text{temp. rise}}$$

$$= \frac{6/D}{146/D \times 55}.$$

The density $D$ cancels from the numerator and denominator, and thus

$$\text{apparent coefficient} = \frac{6}{146 \times 55}.$$

Now 6 gm. is the *mass of liquid expelled* from the bottle in the actual experiment, and 146 gm. is the *mass of liquid left* in the experiment. Hence an easily-remembered formula can be given for the apparent coefficient, $\gamma'$ —

$$\gamma' = \frac{\text{mass of liquid expelled}}{\text{mass of liquid } \textbf{left} \times \text{ temp. rise}} \qquad . \qquad (7)$$

A word of warning is necessary here; do not confuse "volume" with "mass" in formula (7). The theory of the experiment just given shows that the denominator concerns the mass left and not the original mass. This should not occasion much surprise, as, from the theory of the experiment given above, the volume expelled is equal to the increase in volume of the mass *left* when the latter is re-heated.

**Flask and pipette experiment.** A direct method of finding

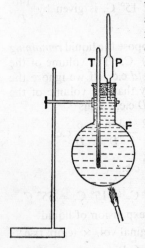

FIG. 137.
Apparent coefficient

the apparent coefficient, which is not as accurate as the specific gravity bottle, is shown in Fig. 137. A flask $F$ is completely filled with the liquid, and an inverted 2 c.c. pipette is pushed through a cork in the neck until the liquid level just reaches the 2 c.c. graduation mark. A thermometer $T$ is placed through another hole in the cork to measure the liquid temperature, and the initial temperature is observed. The liquid is then warmed gently by a low flame, and the liquid begins to expand up the pipette. When it just reaches the top, the new temperature is observed.

In one experiment with water, the volume of the flask was 400 c.c., the initial temperature 15° C., and the final temperature 39° C. The apparent liquid expansion is 2 c.c., since the expansion of the flask is not taken into account. Hence

$$\text{apparent coefficient} = \frac{2 \text{ c.c.}}{400 \text{ c.c.} \times (39 - 15)°C} = 0.0002 \text{ per } °C.$$

**The anomalous expansion of water with temperature.** Ice (a solid) expands slightly when warmed from − 5° C. say to 0° C. ; at 0° C. it forms water and contracts. Fig. 138 (i). When the water is warmed, it *contracts* from 0° C. to 4° C., which is exceptional, but from 4° C. to 100° C. it expands, behaving now like most liquids. When it changes into steam at 100° C. the volume increases 1600 times as much. Water is thus exceptional or "anomalous" in the

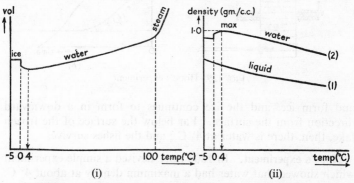

FIG. 138. Volume and density variation (*not to scale*)

range 0° C. to 4° C. Since "density" is "mass/volume", and the mass is unaltered by warming, it follows that *water has a maximum density at about* 4° C. Fig. 138 (ii) shows the density-temperature variation of water compared with most liquids. The latter increase in volume as their temperature rises from 0° C. and hence their density diminishes continuously.

This anomalous expansion of water makes it possible for fish to survive in winter-time in Arctic and other cold regions. Consider the air outside a lake, for example, when the temperature begins to fall. The water at the top increases in density, and falls to the bottom of the lake, where the water is less dense. Other water takes the place of the water which sinks to the bottom, and this in turn becomes cooled and sinks. The downward movement of cold water continues until the temperature at the top reaches 4° C. The water at the bottom is then at this temperature, but as the temperature of the top falls to 3° C. and lower, the *colder water remains at the top* because the density of water below 4° C. is *less* than the density at 4° C. (see Fig. 138). When the temperature of the air reaches 0° C. the water at the top begins to freeze

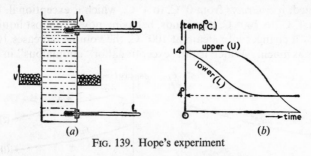

FIG. 139. Hope's experiment

and form ice, and the ice continues to form in a downward direction from the surface. Far below the surface of the frozen lake, then, there is water at 4° C.; and the fishes survive.

**Hope's experiment.** In 1804 HOPE devised a simple experiment which showed that water had a maximum density at about 4° C. A tall vessel *A* containing water is surrounded at the middle by a vessel *V* containing ice and salt (Fig. 139 (*a*) ). At the top and bottom of *A* are thermometers, *U, L,* and readings of their temperatures and the time are taken. The *lower thermometer, L,* drops from its initial value of 14° C., say, as the water cooled at the middle of *A* drops to the bottom; but at a temperature of about 4° C. the reading on *L* remains constant and the *upper thermometer U,* which had been constant at 14° C., now begins to fall. (See Fig. 139 (*b*).) This is due to the fact that the cold water now rises from the middle to the top, as explained above in the case of the lake, and the temperature on *L* continues to fall as more cold water reaches the top and ice is formed there. In this way Hope found the temperature of maximum density of water was about 4° C.

─────────────Summary─────────────

1. The linear coefficient of expansion of a solid ($a$) is the increase in length of unit length for one degree rise in temperature. The area (superficial) coefficient $= 2a$. The volume (cubical) coefficient $= 3a$.

2. **The increase in length of a solid $= alt$, where $l$ is the original length and $t$ is the temperature rise. The new length, $l_t = l(1 + at)$.**

3. The **true (absolute) coefficient** of expansion of a liquid ($\gamma$) is the true increase in volume of 1 c.c. for 1° C. rise in temperature. The **apparent coefficient** ($\gamma'$) is the apparent increase in volume of 1 c.c. for 1° C. temperature rise.

4. The apparent coefficient ($\gamma'$) of a liquid can be measured by a specific gravity bottle: $\gamma' =$ mass of liquid expelled/(mass **left** × temp. rise).

5. Water has a maximum density at about 4° C.

## WORKED EXAMPLES

1. *Describe an experiment to determine the coefficient of linear expansion of copper, provided in the form of either a rod or a hollow tube. A copper cube, of 10 cm. side at 0° C., is raised in temperature to 100° C. Calculate (a) the increase in volume of the cube, (b) the increase in its surface area. The coefficient of linear expansion of copper = 0·000017 per °C. (C.W.B.)*

(*a*) Volume (cubical) coefficient of copper $= 3 \times$ linear coefficient

$$= 3 \times 0\cdot000017 = 0\cdot000051$$

Original volume of copper $= 10^3 = 1000$ c.c.

∴   increase in volume $= 0\cdot000051 \times 1000 \times 100$ c.c.

since 0·000051 c.c. is the increase in volume of 1 c.c. when the temperature is raised 1° C.

∴   increase in volume $= 5\cdot1$ c.c.

(*b*) Area (superficial) coefficient $= 2 \times$ linear coefficient (p. 178)

$$= 2 \times 0\cdot000017 = 0\cdot000034$$

Original surface area $= 6 \times 10^2 = 600$ sq. cm.

∴   increase in area $= 0\cdot000034 \times 600 \times 100 = 2\cdot04$ sq. cm.

2. *Explain the term coefficient of linear expansion, and describe an experiment to find its value for brass. A metal ball has a diameter of 4·02 cm. and a hole in*

*a brass plate has a diameter of 4·00 cm. when both are at 20° C. To what temperature must the plate be heated so that the ball (still at 20° C.) may just pass through the hole? (The coefficient of linear expansion of brass is 0·000018 per °C.) (L.)*

The plate must be heated to a temperature such that the diameter of the hole increases to 4·02 cm. from 4·00 cm. We require the *change in circumference* of the hole, as this affects the diameter.

$$\text{Original circumference} = 2\pi r = 2\pi \times 2\cdot00 = 4\pi \text{ cm.}$$

$$\text{Final circumference} = 2\pi \times 2\cdot01 = 4\cdot02\pi \text{ cm.}$$

$$\therefore \quad \text{change in circumference} = 4\cdot02\pi - 4\pi = 0\cdot02\pi \text{ cm.}$$

Since change in circumference (length) $= \alpha l t$ (p. 176)

where $t$ is the change of temperature, $\alpha = 0\cdot000018$,

$$\text{and } l = \text{original circumference} = 4\pi \text{ cm.,}$$

$$\therefore \quad 0\cdot02\pi = 0\cdot000018 \times 4\pi \times t$$

$$\therefore \quad t = \frac{0\cdot02\pi}{0\cdot000018 \times 4\pi} = 277\cdot8° \text{ C.}$$

$$\therefore \quad \text{new temperature} = 277\cdot8 + 20 = 297\cdot8° \text{ C.}$$

3. *Distinguish between the real and apparent expansion of a liquid. Describe an experiment to find the coefficient of apparent expansion of turpentine.*

*A flask is completely filled with 50·05 gm. of liquid at 15° C. and on heating the flask to 60° C. it is found that 0·73 gm. of liquid overflows. What is the coefficient of apparent expansion of the liquid? (L.)*

$$\text{Apparent expansion } \gamma' = \frac{\text{mass expelled}}{\text{mass LEFT} \times \text{temp. rise}}.$$

Since 0·73 gm. of liquid is expelled,

$$\text{mass left} = 50\cdot05 - 0\cdot73 = 49\cdot32 \text{ gm.}$$

$$\therefore \quad \gamma' = \frac{0\cdot73}{49\cdot32 \times (60 - 15)} = 0\cdot00032 \text{ per °C.}$$

4. *Distinguish between the real and apparent expansion of a liquid. Describe an experiment to determine the coefficient of apparent expansion of a liquid.*

*A mercury thermometer has a bore of 0·028 cm. diameter and the distance between its fixed points is 25 cm. What is the volume of the thermometer bulb up to the lower fixed point mark? The coefficient of real expansion of mercury = 0·00018 and that of the cubical expansion of the glass = 0·000026, both per °C. (C.W.B.)*

The coefficient of apparent expansion of the mercury, $\gamma' = 0\cdot00018 - 0\cdot000026 = 0\cdot000154$ per °C. Suppose that $V_0$ is the volume up to the lower fixed point mark (0° C.). When the thermometer is heated to 100° C., the upper fixed point, the increase in volume of the mercury relative to the glass $= \gamma' V_0 t$ (p. 181) $= 0\cdot000154 \times 100 \times V_0$ c.c.

But this increase in volume $= \pi r^2 h = \pi \times 0.014^2 \times 25$ c.c.

$$\therefore \quad 0.000154 \times 100 \times V_0 = \pi \times 0.014^2 \times 25$$

$$\therefore \quad V_0 = \frac{\pi \times 0.014^2 \times 25}{0.000154 \times 100} = 1 \text{ c.c.}$$

## EXERCISES ON CHAPTER XI

**Expansion of solids**

**1.** A metal rod increases in length from 50·00 cm. at 10° C. to 50·12 cm. at 100° C. Calculate the mean coefficient of linear expansion.

**2.** "Brass has a linear coefficient of expansion of 0·000018 per °C." What does this statement mean ? Find the value of the linear coefficient expressed in °F.

**3.** Using the linear coefficient value in question 2, calculate the new length at 30° C. of a piece of brass which is 100 cm. long at 5° C.

**4.** Brass and iron have linear coefficients respectively of 0·000018, 0·000011 per °C. Two equal lengths of brass and iron, welded together, are heated. Draw a sketch of their subsequent appearance, and explain it. What would have been their appearance if they had been cooled instead of heated ?

**5.** Draw a diagram of a compensated balance-wheel of a watch, and explain how it keeps regular time.

**6.** The linear coefficient of steel is 0·000012 per °C. What are the superficial (area) coefficient and the cubical (volume) coefficient of steel ? The steel roof of a shed has an area of 150 sq. ft. at 12° C. What is the new area (i) in summer when the temperature reaches 18° C., (ii) in winter when the temperature reaches freezing point ?

**7.** Describe two cases in which (i) allowance must be made for the expansion of a metal, (ii) the expansion is utilized.

**8.** An iron rod has a length of 42·64 cm. at 12° C. At what temperature will its length become 42·75 cm. ? (Coefficient of linear expansion of iron = 0·000011 per °C.)

**9.** Explain the statement that the coefficient of linear expansion of brass is 0·0000189 per °C. and describe carefully the experiments you would make to verify it. What is the value of this coefficient if the Fahrenheit scale of temperature is used ? Mention *one* useful application of linear expansion in engineering and give *one* example in which allowance must be made for it. (*L.*)

**10.** Define *coefficient of linear expansion* and describe an experiment to determine its value for an iron rod or tube. A storage shed, 100 ft. long, 50 ft. wide, and 30 ft. high at the normal lowest winter temperature of $-5°$ C., is made of corrugated iron sheets. Assuming free expansion as a whole, determine the change in (*a*) the area of the roof, (*b*) the volume of the shed, when at a summer temperature of 35° C. The coefficient of linear expansion of iron $= 12 \times 10^{-6}$ per °C. (*L.*)

**11.** Explain why temperature variations affect the time-keeping of a pendulum clock and describe *one* method of compensating the pendulum. A copper rod and an aluminium rod have the same length at a temperature of 100° C. Calculate the ratio of their lengths at 0° C. (Coefficient of linear expansion of copper, aluminium $= 17 \times 10^{-6}$, $25 \times 10^{-6}$ per °C.) (*O. & C.*)

**12.** A railway line is laid with 40 ft. lengths of rail on a day when the temperature is 10° C. and gaps of $\frac{1}{4}$ in. are left between successive lengths. At what temperature would the gaps close up ? (Coefficient of linear expansion of steel $= 0.000012$ per °C.) (*C.*)

**13.** Define the coefficient of linear expansion of a substance, and describe an experiment to determine its value for an iron rod. A solid iron hemisphere, having a diameter of 35 cm., is raised in temperature from 0° C. to 100° C. Calculate (*a*) the increase in area of the flat circular face, (*b*) the increase in volume of the hemisphere. Take $\pi$ as $\frac{22}{7}$ and the coefficient of linear expansion of iron as $0.000012$ per °C. [The volume of a sphere $= \frac{4}{3}\pi(\text{radius})^3$.] (*C.W.B.*)

**14.** Define *coefficient of linear expansion* and prove that the coefficient of cubical expansion is approximately three times its linear coefficient. Describe an accurate method for measuring the coefficient of linear expansion of a metal. The tungsten filament of a wireless valve has a length of 2 cm. and a radius of $0.1$ mm. at 20° C. If the temperature of the filament is raised to 3020° C., find (*a*) the increase in its length and (*b*) the increase in its volume. (Coefficient of linear expansion of tungsten $= 0.000007$ per °C. Take $\pi$ as $\frac{22}{7}$.) (*N.*)

### Expansion of liquids

**15.** "The cubical expansion of water is $0.0002$ per °C." What does this statement mean ? Some water has a volume of $80.0$ c.c. at 15° C. Calculate its volume at 45° C.

**16.** Define the terms *apparent coefficient of expansion* and *absolute (real) coefficient of expansion* of a liquid. What is the relation between them ?

**17.** How does the density of a liquid usually vary when its temperature rises? How does the density of water vary as its temperature rises from 0° C.? Draw rough sketches illustrating your answers in both cases.

**18.** Write down the formula for calculating the cubical coefficient of a liquid in the density (or specific gravity) bottle experiment. Describe briefly how the experiment is carried out.

**19.** Describe Hope's apparatus for investigating the variation of the density of water with temperature. Which of the thermometers never reads lower than 4° C.? Why is it possible for fishes to live in Arctic regions?

**20.** A litre bottle full of milk which has just been pasteurized at 60° C. is sealed and cooled to 15° C. Find the volume of the empty space in the bottle at the new temperature, assuming that the coefficients of cubical expansion of milk and glass are 0·00038 and 0·000025 per °C. respectively. (*O. & C.*)

**21.** How would you determine the apparent coefficient of expansion of glycerine? A hollow sealed glass bulb weighted with mercury just floats in water at 4° C. State, giving your reasons, what would be observed if the water were (*a*) cooled to 0° C., (*b*) heated to 10° C. (*N.*)

**22.** Describe an experiment to show that water has a maximum density at 4° C.? What important consequences follow from this peculiar property of water? Will a water (in glass) thermometer show a minimum reading at 4° C. or at a lower or higher temperature than this? Give reasons for your answer. (*L.*)

**23.** Distinguish between the real and apparent coefficients of expansion of a liquid, and describe how you would find the apparent coefficient of expansion of alcohol.

A glass sinker is attached to one arm of a balance and suspended in water at 0° C. Describe and account for the changes in apparent weight that are observed as the water is gradually heated up to 20° C. (*N.*)

**24.** Define the coefficients of apparent and real expansion of a liquid. What is the relation between them? A glass vessel contains when full 816·00 gm. of mercury at 0° C. The mass of mercury which fills it at 100° C. is 803·21 gm. The coefficient of absolute expansion of mercury is 0·000182 per °C. Calculate the coefficient of cubical expansion of glass. (*C.*)

**25.** State the advantages possessed by mercury for use in thermometers. Why is alcohol sometimes used instead of mercury? In an alcohol thermometer the volume of alcohol filling the bulb and stem as far as the 0° C. mark is 3 c.c. at 0° C. Where should the 50° C. mark be made if the coefficient of expansion of alcohol relative to glass is 0·0011 per °C. and the area of cross section of the bore of the stem is 1 sq. mm. ? (L.)

**26.** Explain the meaning of *absolute* and *apparent* coefficients of expansion of a liquid and state how they are related. Describe an experiment to show that water has a maximum density at a temperature of about 4° C. What bearing has this fact on the freezing of ponds in winter ? (O. & C.)

**27.** Define *coefficient of volume expansion*. Distinguish between the coefficients of *real* and *apparent* expansion of a liquid in a glass vessel.

A mercury-in-glass thermometer has a bulb of internal volume 0·30 c.c. and a tube of internal cross-sectional area $2·5 \times 10^{-4}$ sq. cm. If the 0° C. division on the tube is just above the bulb find (a) the apparent increase in volume of the mercury when the temperature of the thermometer rises from 0° C. to 100° C., (b) the distance between the 0° C. and 100° C. divisions. (The coefficient of apparent expansion of mercury in glass may be taken as $16 \times 10^{-5}$ per degree C.) (L.)

**28.** Define *coefficient of expansion*. Distinguish between the coefficients of *real* and *apparent* expansion of a liquid, and state a relation between them.

A glass bottle of volume 10·0 c.c. at 0° C. contains 132·6 gm. of mercury of density 13·60 gm. per c.c. at 0° C. What is the volume of the mercury at 0° C. ? Assuming the coefficients of volume expansion of mercury and glass to be $18 \times 10^{-5}$ per degree C. and $3 \times 10^{-5}$ per degree C. respectively find the temperature at which the mercury just fills the bottle. (L.)

# CHAPTER XII

# SPECIFIC HEAT AND LATENT HEAT

IN the home, heat for cooking is obtained by burning gas or by using electricity. Steam engines burn coal, Diesel engines burn oil; and, in general, the greater the amount of fuel, the greater is the quantity of heat obtainable from it.

**Units of quantity of heat.** The scientist and the engineer have defined *units* of quantity of heat. The unit in scientific work is the **calorie**, which is defined as *the quantity of heat which raises the temperature of 1 gm. of water by 1° C.* In this country, however, commercial gas companies and engineers use a unit of heat, known as the **British Thermal Unit (B.Th.U.),** which is defined as *the quantity of heat which raises the temperature of 1 lb. of water by 1° F.* The unit on which gas companies base their charge, the **therm,** is defined as 100,000 B.Th.U.

## SPECIFIC HEAT

Although 10 calories of heat raise the temperature of 2 gm. of water 5° C., experiment shows that 10 calories of heat raise the temperature of 2 gm. of copper 50° C., and the temperature of 2 gm. of aluminium 20° C. Different substances of equal mass, then, experience different rises of temperature when they absorb equal quantities of heat.

The *specific heat, s,* of a substance is defined as *the amount of heat which raises the temperature of 1 gm. of the substance 1° C.* Thus the specific heat of water is numerically 1, from the definition of the calorie. The specific heat of copper is about 0·1 numerically, and the specific heat of aluminium is about 0·25 numerically. It should be noted that specific heat, s, is expressed in "calories per gm. per °C."; but we shall frequently use the numerical value of s and omit the units.

**Formula for $Q$.** Suppose a quantity of heat, $Q$, raises the temperature of 200 gm. of copper, specific heat ($s$) 0·1 numerically, by 15° C. Since 0·1 calories is the amount of heat which raises the temperature of 1 gm. of copper 1° C., the amount of heat $Q$ is 0·1 × 200 × 15 calories, by simple proportion. It can now be seen that the quantity of heat $Q$ supplied to a mass $m$ of specific heat $s$ when its temperature changes by $t°$ C. is given by the basic formula in heat calculations.

$$Q = mst.$$

**Thermal capacity.** It will be noted that "specific heat" concerns one gram of a material. For a substance of any mass, however, the "thermal capacity" is defined as *the amount of heat which is required to raise its temperature by 1° C.* Consider, for example, a copper vessel of 80 gm. and $s = 0·1$ numerically. Then the quantity of heat required to raise its temperature by 1° C. is given by $Q = 80 × 0·1 × 1$ calories, using the formula $Q = mst$. Thus $Q = 80 × 0·1 = 8$ cal., which is the thermal capacity of the copper vessel. Thus *thermal capacity in calories = mass of substance in grams × specific heat.*

**Transfer of heat.** Suppose a hot aluminium solid of mass 120 gms., $s = 0·25$, and temperature of 90° C., is dropped into a copper vessel of $s = 0·1$ and mass 100 gm., containing 80 gm. of water originally at 20° C. Since the temperature of the solid (90° C.) is greater than the temperature (20° C.) of the water and copper vessel, heat continues to pass from the solid to the water and copper vessel until the temperatures of all three substances concerned are the *same*. The temperature of the mixture will be less than 90° C. and greater than 20° C.

To calculate the final temperature of the mixture, $x°$ C. say, we start with the fact that

*heat lost by solid = heat gained by water and vessel* . (1)

This statement "heat lost = heat gained" occurs in all problems of heat transfer, as the reader will observe in what follows.

Since the change $t$ in temperature of the solid is $(90 − x)°$ C. and of the water and vessel $(x − 20)$ °C.,

heat lost by solid $= mst = 120 × 0·25 × (90 − x)$ cal.
heat gained by water $= mst = 80 × 1 \quad × (x − 20)$ cal.
heat gained by vessel $= mst = 100 × 0·1 \quad × (x − 20)$ cal.

Thus, from (1),

$$120 \times 0.25 \times (90 - x) = 80 \times 1 \times (x - 20) + 100 \times 0.1 \times (x - 20)$$

$$\therefore \quad 30(90 - x) = 80(x - 20) + 10(x - 20) = 90(x - 20)$$

Solving,          $\therefore \quad x = 37.5°$ C.

FIG. 140. HIGH PRESSURE BOILERS

Boiler Room of a large factory, where water is heated to 310° F. at 80 lb. per sq. in pressure (there is a drop of 100° F. round the system). (See p. 205.)

**Water equivalent of calorimeter.** A *calorimeter* is the name given to a vessel concerned in heat transfer, such as the vessel in the above example. Suppose 100 gm. is the mass of a copper calorimeter of specific heat 0.1. Then, if the temperature of the calorimeter increases by 20° C., the heat gained by the calorimeter = $mst$ = $100 \times 0.1 \times 20 = 200$ cal. Now 10 gm. of water whose temperature increased by 20° C., the same increase as the calorimeter, would also gain 200 cal. of heat. We can therefore imagine that, so far as the heat absorbed by the calorimeter is concerned, 10 gm. of *water* are present in place of the 100 gm. of copper. The *water* *equivalent* ($W$) of the calorimeter is thus said to be 10 gm., and, as the reader can check, the numerical value of $W$ in every case is given by

$$W = \text{mass of calorimeter} \times \text{specific heat} \qquad . \qquad (2)$$

On a number of occasions in heat transfer, water is placed inside a calorimeter, and a hot substance is added to the calorimeter. Suppose the mass of water is 150 gm. and the calorimeter has a mass of 100 gm. and $s = 0.1$. Then, since the *water equivalent* of the calorimeter is 10 gm., the "equivalent" mass of water present, from the point of view of quantity of heat, is (150 + 10), or 160 gm.; if the temperature of the water and calorimeter increases by 20° C., the total heat ($mst$) gained is

$$160 \times 1 \times 20 \text{ cal., or } 3200 \text{ cal.}$$

## Determining the specific heat of solids and liquids.

*Method of mixtures.* The specific heat of a metal, such as aluminium or brass, can be found by heating the solid in a beaker containing water until its temperature is steady, and then transferring the solid quickly to a copper calorimeter filled about two-thirds full of water. The latter is stirred, and its final temperature is noted. Knowing the mass of the water, the calorimeter, and the solid, the initial temperatures of the solid and water, and the final temperature of the water, the specific heat of the solid can be calculated by applying the "heat lost = heat gained" equation to the heat transfer which takes place. An example will illustrate the method.

The following measurements were taken in an experiment to find the specific heat of aluminium—

|  |  |
|---|---|
| Mass of calorimeter | = 36·62 gm. |
| Mass of calorimeter + water | = 85·41 gm. |
| Mass of aluminium | = 27·6　gm. |
| Initial temp. of water | = 11·0° C. |
| Initial temp. of aluminium | = 95·5° C. |
| Final temp. of water | = 20·5° C. |

Specific heat of copper = 0·1.

Heat lost by aluminium

$$= \text{Heat gained by calorimeter} + \text{water}$$

$$\therefore \quad 27.6 \times s \times (95.5 - 20.5)$$
$$= 36.62 \times 0.1 \times (20.5 - 11.0) + 48.79 \times 1 \times (20.5 - 11.0)$$

(Note that the final temp. of the aluminium is 20·5° C. and the mass of water used is 48·79 gm.)

Solving this equation, $s = 0.24$

*Note.* The heat gained by the calorimeter + water can also be expressed by $(36.62 \times 0.1 + 48.79)(20.5 - 11.0)$, since the water equivalent of the calorimeter = $ms = 36.62 \times 0.1$.

### Specific heat of a liquid.

The specific heat of a liquid $A$ can be found by the method of mixtures, as described above, provided the liquid does not react chemically, or give rise to heat of solution, with water. If it does, the water in the calorimeter should be replaced by another liquid $B$ of *known* specific heat. The liquid $A$ should be heated to a known temperature, then poured into the calorimeter containing liquid $B$, and the final steady temperature noted. The calculation for the specific heat of $A$ is the same as that given above in the case of a solid.

An alternative method consists of heating a metal of known specific heat to a high temperature, and then transferring it quickly to the liquid whose specific heat is required. Assuming that no chemical action takes place between the metal and liquid, the heat lost by the metal = the heat gained by the liquid and calorimeter; and the unknown specific heat of the liquid can hence be found.

### Determining the temperature of the bunsen burner flame.

The unknown temperature of a bunsen burner flame can be determined by the method of mixtures. A metal solid, such as brass, is suspended in the flame for a brief period of time, and it is then quickly transferred to a calorimeter containing water. The mixture is stirred, and the final constant temperature is noted.

The temperature, $t$, of the bunsen flame is calculated as follows—

| | |
|---|---|
| Mass of brass | = 65.0 gm. |
| Mass of calorimeter | = 42.46 gm. |
| Mass of calorimeter + water | = 93.61 gm. |
| Initial temp. of water | = 14.6° C. |
| Final temp. of water | = 68.4° C. |
| Temperature of burner | = $t$° C. |

Sp. ht. of calorimeter = 0.1, sp. ht. of brass = 0.092.

Heat lost by brass = Heat gained by calorimeter + water

$$\therefore 65 \times 0.092 \times (t - 68.4) = (42.46 \times 0.1 + 51.15)(68.4 - 14.6)$$

$$\therefore t - 68.4 = \frac{55.4 \times 53.8}{65 \times 0.092}$$

$$\therefore t = 569° C.$$

**Determining the water equivalent of a calorimeter.** The water equivalent (p. 195) of a calorimeter can also be determined by the method of mixtures. One such method consists of heating water in a beaker to about 40° C., then pouring the water into the calorimeter until it is about half-full, stirring, and taking the final temperature of the water.

The water equivalent $W$ is then calculated as follows—

Suppose the mass of water is 25·2 gm., its initial temperature is 39·0° C., and its final temperature is 33·5° C. If the initial temperature of the calorimeter is 13·4° C., the heat gained by the calorimeter $= W \times (33·5 - 13·4)$ cal. But the heat lost by the water $= 25·2 \times 1 \times (39·0 - 33·5)$ cal.

$$\therefore \quad W(33·5 - 13·4) = 25·2(39·0 - 33·5)$$

$$\therefore \quad W = \frac{25·2 \times 5·5}{20·1} = 6·9 \text{ gm.}$$

## LATENT HEAT

**Latent heat of fusion.** (*Solid to liquid change*). Ice can be cooled below 0° C. by a freezing mixture. Suppose some ice in a test tube is at $- 8°$ C., and is warmed slowly, its temperature being taken every minute. Observation shows that its temperature rises along a curve such as $AB$ until 0° C. is reached, and at this stage the solid begins to *melt* (Fig. 141). As more heat is supplied more water is formed, *but the temperature remains at* 0° C.; thus a line

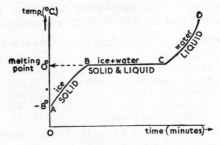

FIG. 141. Solid to liquid change

$BC$ parallel to the time-axis is obtained. When *all* the ice has melted, so that only water is present, the temperature rises as more heat is supplied, and the line $CD$ is now obtained.

Along *BC* then, no temperature increase occurs, although heat is being continuously supplied to the ice. This observation was made many years ago, and in 1761 BLACK gave the name "latent," or "hidden," heat to the heat absorbed during melting.

The *melting point* of a solid such as napthalene, or wax, can be found by warming it until it is a liquid, placing a thermometer in it, and observing the temperature as the naphthalene is allowed to cool. A graph similar to that shown in Fig. 141, but in reverse, is obtained when the temperature is plotted against the time, and the melting point is that temperature which corresponds to the *flat* portion of the graph, such as *BC*.

When the temperature-time curve of an *alloy* of metals is examined, various flat portions are observed on the graph. They correspond to the melting points of the various metals comprising the alloy and, since the melting point of an element is a characteristic property of it, an alloy can be analysed by plotting the temperature-time curve and noting the various melting points obtained.

**Definition of latent heat of fusion.** Black could not explain why the temperature of a substance remained constant as it changed from the solid to the liquid state. We now know, however, that the molecules of a solid are closer together on the average than the molecules of a liquid, and *energy* is therefore required to make the molecules of a solid change to the new position which constitutes a liquid state (p. 233). Thus the latent heat represents energy used in breaking down the forces which hold the atoms of a solid in a regular pattern, and setting them free to slide about, as they do in the liquid state.

The *latent heat of fusion* (*L*) of a substance is defined as the heat required to change 1 gm. of it from a solid at the melting point to a liquid *at the same temperature*. For example, ice melts at 0° C., and the latent heat of fusion (*L*) of ice is the heat required to change 1 gm. of ice at 0° C. to water at 0° C., about 80 calories per gm.

Suppose 5 gm. of ice at 0° C. are heated so that all the ice melts and forms water at a final temperature of 10° C. This change occurs in two stages: (1) the ice melts to water *at 0° C*. (2) the water formed rises in temperature from 0° C. to 10° C. (Fig. 142). The heat required for stage (1) depends on the value of the latent heat of fusion ($L$) for ice, 80 cal. per gm., since it is a change from a solid to a liquid state *at the melting point*. Thus the heat = $5 \times L$ = $5 \times 80 = 400$ cal. (Fig. 142). The heat for stage (2) concerns only the formula $Q = mst$, since 5 gm. of water at 0° C. rises in temperature to 10° C. (no change of state occurs); thus $Q = 5 \times 1 \times 10 = 50$ cal. The *total* amount of heat concerned is thus (400 + 50), or 450 cal.

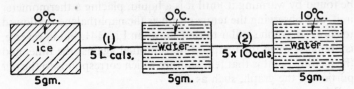

FIG. 142. Calculation of heat gained by ice

**Experiment to find latent heat of fusion of ice.** The latent heat of fusion ($L$) of ice can be found by the method of mixtures. Some ice is dried with blotting-paper and transferred to a calorimeter containing water. The ice is now stirred until it *all* melts, and the final temperature of the water, which is lower than the initial temperature, is noted. The mass of ice melted is found by subtracting the mass of calorimeter plus water from the mass of calorimeter plus water plus melted ice.

Results in an experiment were as follows—

Mass of calorimeter ($s = 0.1$)   = 58·3 gm.
Mass of water                    = 81·35 gm.
Mass of melted ice         = 5·6 gm.
Initial temp. of water      = 13·7° C.
Final temp. of water        = 8·0° C.

The ice has melted and formed *water at 8° C.* Since the heat gained by 5·6 gm. of ice in changing to 5·6 gm. of water at 8° C. is equal to the heat lost by the water and calorimeter,

$$\therefore \quad 5·6L + 5·6 \times 1 \times (8 - 0) = (81·35 + 5·83)(13·7 - 8) \quad . \quad . \quad (3)$$

Solving equation (3),       $L = 81$ cal. per gm.

**Latent heat vaporization.** (*Liquid to vapour change.*) We have now to consider the change from a liquid to a vapour state, when "boiling," or "vaporization," is said to take place.

For convenience, suppose water at 20° C. is heated steadily and its temperature is observed after every minute. The temperature rises until some vapour is observed to come off continuously from the water, and boiling now takes place. At this stage, if the atmospheric pressure is 76 cm. mercury, the temperature is 100° C., and remains steady until *all* the water has changed to vapour, or steam as it is called; further heating increases the temperature of the vapour (Fig. 143). As in the case of melting, when no temperature change occurs, the term "latent heat" is applied to the

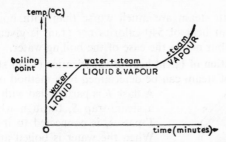

FIG. 143. Liquid to vapour change

heat given to the water as it changes to steam; the heat, however, is used in this case in providing the energy to make the molecules change from the liquid state to the vapour state, in which they are much further apart. (Compare "latent heat of fusion," p. 199.) The *boiling point* of a liquid corresponds to the flat part of the temperature-time graph, such as that in Fig. 143.

The *latent heat of vaporization* (*L*) of a substance is defined as *the amount of heat required to change* 1 *gram of the substance from its liquid to its vapour state at the boiling point.* About 540 calories of heat are needed to change 1 gram of water at the boiling point to vapour at the same temperature, so that $L = 540$ cal. per gm. for steam.

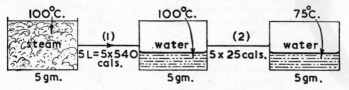

FIG. 144. Calculation of heat given up by steam

If 5 gm. of steam at 100° C., say, condense to water *at* 100° C., $5 \times 540$ calories of heat are given up (Fig. 144). If the steam condenses to water which reaches a temperature of 75° C., however, the heat given up is the sum of two quantities:

(1) The heat given up when 5 gm. of steam at 100° C. condenses to water *at* 100° C., which is $5L = 5 \times 540$ cal. (2) The heat given up when the water decreases in temperature from 100° C. to 75° C.; from $Q = mst$, this is $5 \times 1 \times (100 - 75)$, or 125 cal. Thus the total heat given up = 2700 + 125 = 2825 cal.

Scalds with steam are much worse than with boiling water, because latent heat of 540 calories per gram is given up in the former case but not in the case of the boiling water.

**Determination of latent heat of vaporization ($L$) of steam.** The latent heat of steam can be determined by a method of mixtures.

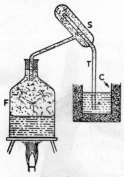

FIG. 145.
Latent heat of steam

A flask $F$ is partly filled with water, and a steam-trap $S$, through which a tube $T$ passes, is connected to it (Fig. 145). When the water is boiled and steam is issuing freely from the end of $T$, a calorimeter $C$ with water is placed just below $T$. The steam condenses in the water, whose temperature then rises, and after a suitable rise in temperature the calorimeter $C$ is taken away from $T$, the water in it is stirred, and the final temperature is noted. The mass of steam which has condensed is measured by subtracting the mass of calorimeter and water before steam is passed into it from the mass of calorimeter and water at the end of the experiment.

The following results were obtained in an experiment—

|  |  |
|---|---|
| Mass of copper calorimeter | = 56·9 gm. |
| Mass of calorimeter + water | = 140·82 gm. |
| Mass of calorimeter + water + steam | = 144·53 gm. |
| ∴ mass of water | = 83·92 gm. |
| ∴ mass of steam | = 3·71 gm. |
| Initial temp. of water | = 17·5° C. |
| Final temp. of water | = 43·0° C. |
| Temperature of steam | = 100° C. |

Heat lost by steam in condensing to water at 43·0° C. = heat gained by calorimeter + water

∴ $3·71L + 3·71 \times 1 \times (100 - 43) = (56·9 \times 0·1 + 83·92)(43 - 17·5)$

Solving the equation, $L = 559$ cal. per gm.

If the steam-trap were omitted from the apparatus, some water, condensed on the side of $T$, would pass into the calorimeter in addition to the steam. This would lead to error in determining the mass of steam passing into the water, with a consequent error in the calculated latent heat value.

**Why pipes burst in cold weather.** Ice-bergs float in water, since the density of ice is less than that of water (p. 111). Now volume $= \dfrac{\text{mass}}{\text{density}}$. Consequently the volume of 1 gram of ice is greater than the volume of 1 gram of water. In very cold weather, when the temperature is below freezing point and the water in household pipes happens to freeze, the ice formed is greater in volume than the mass of water frozen. The pipe is hence subjected to a very great force, and the metal becomes cracked. When the temperature rises above the freezing point, the ice in the pipe melts and the water pours through the cracks.

**The effect of pressure on the melting point of a solid.** When ice is subjected to a large pressure its volume tends to decrease. Now when ice changes to water, its volume becomes smaller, and hence increased pressure on ice helps it to change to water. It follows that ice can be melted if sufficient pressure is applied to it and, in general, *the melting point of ice is lowered when it is subjected to increased pressure.* It has been calculated that the melting point for ice is lowered by about 0·008° C. for an increase in pressure of one atmosphere.

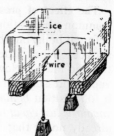

Fig. 146. Regelation

A vivid demonstration of the phenomenon was given by TYNDALL. A large block of ice was suspended, and a thin wire, with heavy weights attached at both ends, was hung round the middle of the block. The ice in contact with the wire was thus subjected to a great pressure and melted; the wire then sank a little into the ice through the water that was formed, and the latter, being freed from the pressure, immediately froze again on the other side of the wire. In this way the wire gradually sank or "cut" into the ice-block, and eventually the wire passed completely through the block, leaving it unaltered. This phenomenon is known as **regulation** (re-freezing). (See Fig. 146.)

Ice-skaters are able to move freely on account of the decrease of the melting point of ice under increased pressure. The knife-edge bottom of the skate has a very small cross-sectional area, and hence the pressure under the knife-edge $\left( \dfrac{\text{weight of skater}}{\text{area of knife-edge}} \right)$ is very large. The ice thus melts, and the skater moves easily through a thin film of water on top of the ice. If it is too cold, the drop of the melting point caused by the increased pressure is insufficient to make the ice melt, and the skate moves with difficulty over the ice.

Snowballs are made by compressing the snow very strongly, when some of it melts under the increased pressure. When the pressure is released, the water that was formed freezes, thus making a hard ball. If it is too cold, the snow does not melt under increased pressure, and it is difficult to make a snowball.

Many substances *expand* when they change from the solid to the liquid state, unlike ice. Tin and paraffin-wax are two examples of these substances. Since increased pressure is unfavourable to a change from the solid to the liquid state in this case, increased pressure *raises* the melting point.

**The effect of pressure on the boiling point of water.** It is a common observation that *bubbles* form in the interior of water before it boils, but that these bubbles collapse before reaching the surface. When water is boiling, however, the bubbles rise to the top and burst. Now the bubbles contain water-vapour, and the vapour pressure increases when the liquid is heated (p. 235). Hence it follows that *boiling takes place when the pressure of the water-vapour is just equal to the external atmospheric pressure.* We have already noted that water boils at 100° C. when the pressure is 1 atmosphere (76 cm. of mercury), and thus water boils at a lower temperature than 100° C. when the external pressure is less than 1 atmosphere. BENJAMIN FRANKLIN, a noted American scientist of the 18th century (p. 411), demonstrated this fact by a simple experiment, which consisted of boiling water in a flask and then sealing the top by a cork. When the burner was removed and the temperature of the water decreased slightly, the bubbling at the water surface died down; boiling had ceased. Franklin then inverted the flask and poured cold water over its outside, whereupon the water inside the flask began to boil again. The cold water caused some of the steam to condense, and the pressure above the water in the flask was then *less* than atmospheric pressure; hence a decrease in the external pressure causes a drop in the boiling point of water. If the pressure is increased, the boiling point increases; water boils at 200° C. at about 15 atmospheres pressure!

As we go higher up, the atmospheric pressure decreases. At about 10,000 ft. above sea-level water boils at 90° C., and at 14,000 ft. it boils at 85° C. Certain foods, such as beans, cannot be properly cooked in the open in very hilly regions, as the boiling point of water in these places is too low. The decrease of the boiling point of water with decreasing pressure is put to practical use at stages in the manufacture of sugar, when the latter is boiled in "vacuum pans." These are containers in which the pressure is low, so that the sugar is prevented from charring. It is of interest to note that the hypsometer, the apparatus for determining the boiling point of a liquid (see p. 163), was formerly used by mountaineers to determine the height they climbed, as the boiling point of water depends on the height. The hypsometer has been superseded by the altimeter (p. 131).

---

## Summary

1. The units of heat are: calorie, British Thermal Unit, therm. **The specific heat of a substance is the heat required to raise the temperature of 1 gram by 1° C.**

2. **Water equivalent** of a calorimeter (in grams of water) = **mass × specific heat.** Thermal capacity of a substance (in calories per °C.) = mass × specific heat.

3. The specific heat of a solid (or liquid) can be found by the method of mixtures, using "heat lost = heat gained."

4. The **latent heat of fusion of a solid** is the heat required to change 1 gram of the solid at its melting point to liquid at the same temperature. The **latent heat of vaporization of a liquid** is the heat required to change 1 gram of the liquid at its boiling point to vapour at the same temperature.

5. Increased pressure lowers the freezing point of ice, and raises the boiling point of water.

8.

## WORKED EXAMPLES

1. *Calculate the quantity of heat given out (a) when 10 gm. of steam at 100° C. is condensed at 100° C. and cooled to 15° C., (b) when 40 lb. of water at 100° F. is cooled to 32° F.  (Latent heat of steam = 540 cal. per gm.) (N.)*

(*a*) Heat given out when steam condensed at 100° C.

$$= mL = 10 \times 540 = 5400 \text{ cal.}$$

Heat given out when water formed is cooled to 15° C.

$$= mst = 10 \times 1 \times (100 - 15)$$
$$= 850 \text{ cal.}$$

∴   total heat given out = 5400 + 850 = 6250 cal.

(*b*) In this case there is no latent heat to consider.  The heat, $Q$, given out $= mst = 40 \times 1 \times (100 - 32) = 2720$ B.Th.U., since we are dealing with lb. and °F.

2. *Define latent heat of fusion of a solid.  Describe how you would measure the latent heat of fusion of ice.  What mass of ice at 0° C. must be mixed with 500 gm. of water at 17° C. to lower its temperature to 10° C.?  The water equivalent of the vessel containing the water is 13 gm.  (The latent heat of fusion of ice is 80 cal. per gm.) (L.)*

Let $m$ = the mass of ice in grams.

Then heat gained by ice in melting to water at 10° C.

$$= mL + m \times 1 \times (10 - 0)$$
$$= 80m + 10m = 90m \text{ cal.}$$

Heat lost by water + calorimeter $= mst = (500 + 13) \times 1 \times (17 - 10)$, since the "effective" total mass of water is $(500 + 13)$ gm.

$$\therefore \quad 90m = 513 \times 7$$
$$\therefore \quad m = \frac{513 \times 7}{90} = 39 \cdot 9 \text{ gm.}$$

3. *Ascertain the temperature of 460 gm. of water, initially at 20° C., contained in a copper kettle weighing 300 gm., if 10 gm. of steam at 100° C. are passed into it.  (Latent heat of steam = 540 cal. per gm. ; specific heat of copper = 0·1.) (N.)*

Let $t$ = final temperature of water in °C.

∴   Heat lost by steam

$$= 10L + 10 \times 1 \times (100 - t) = 5400 + 10(100 - t) \text{ cal.}$$

Heat gained by water + kettle

$$= 460 \times 1 \times (t - 20) + 300 \times 0 \cdot 1 \times (t - 20)$$
$$= 490(t - 20).$$

Since heat gained by water + kettle = heat lost by steam,

$$490(t - 20) = 5400 + 10(100 - t),$$
$$\therefore \quad 490t - 9800 = 5400 + 1000 - 10t$$
$$\therefore \quad 500t = 16,200$$
$$\therefore \quad t = 32 \cdot 4° \text{ C}.$$

4. *Describe how you would determine the latent heat of fusion of ice. A copper can of negligible weight containing 200 gm. of water at 16° C. is placed in a refrigerator which abstracts heat at the rate of 240 cal. per min. Calculate the time taken for water to be converted to ice at 0° C.; explain the fact that this time would not be much affected even if the can were as heavy as the water. (Latent heat of fusion of ice = 80 cal. per gm.) (N.)*

(*a*) The heat given up when the temperature of the water changes from 16° C. to 0° C. is given by $Q = mst = 200 \times 1 \times (16 - 0) = 3200$ cal.

The heat required to change 200 gm. of water at 0° C. to ice at 0° C. is given by $Q = mL = 200 \times 80 = 16,000$ cal.

$\therefore$ total heat required to change water to ice = 19,200 cal.

But heat is abstracted at rate of 240 cal. per min.

$$\therefore \quad \text{time} = \frac{19,200}{240} = 80 \text{ min.}$$

(*b*) Suppose the copper can had a mass of 200 gm., the same as the water. Since copper has a specific heat, *s*, of about 0·1 numerically, the heat given up when its temperature changes from 16° C. to 0° C. is given by $Q = mst = 200 \times 0 \cdot 1 \times (16 - 0) = 320$ cal. Since this is a small amount of heat compared with 19,200 cal., the time is not much affected even if the can were as heavy as the water.

## EXERCISES ON CHAPTER XII

1. "The specific heat of copper is 0·1 numerically." What does this statement mean? What is the specific heat of water?

2. Define the calorie, the British Thermal Unit, and the therm. Which units are used (*a*) in scientific work, (*b*) in commercial practice?

3. 500 gm. of iron is heated from 12° C. to 100° C. What heat is supplied to it? If the hot iron is dropped into 1000 gm. of water at 15° C., calculate the new temperature of the water (specific heat of iron = 0·12 cal. per gm. per °C.).

4. 200 lb. of copper of specific heat 0·1 numerically is heated from 40° F. to 212° F. How much heat is supplied?

5. A copper calorimeter, specific heat 0·1, weighs 150 gm. wt. What is the *water equivalent* of the calorimeter? 85 gm. of water at 15° C. are poured into the calorimeter, and it is then heated to 60° C. Calculate the heat gained by the water and calorimeter.

**6.** A hot aluminium metal of 200 gm. wt. and temperature 90° C. is dropped into a copper calorimeter of 120 gm. wt. containing 108 gm. of water at a temperature of 10° C. Find the final temperature of the mixture (specific heat of copper and aluminium = 0·1 and 0·24 respectively).

**7.** Write down a list of all measurements you would make to measure the specific heat of brass. By using imaginary figures, explain how the result is calculated.

**8.** A hot piece of lead, of specific heat 0·03 and 500 gm. wt., is dropped into 60 gm. of water at 20° C. contained in a copper calorimeter of 150 gm. wt. and specific heat 0·1. If the temperature of the mixture is 42° C., find the initial temperature of the hot lead.

**9.** "The latent heat of fusion of ice is 80 calories per gram." What does this statement mean ? 4 gm. of ice are dropped into 65 gm. of water at 12° C. contained in a copper vessel of 230 gm. The final temperature when the ice has all melted is 8° C. Calculate a value for the latent heat of fusion of ice. (Sp. ht. of copper = 0·1 cal. per gm. per °C.)

**10.** Explain the meaning of the statement that the latent heat of steam is 540 calories per gm. What mass of steam at 100° C. must be condensed in 500 gm. of water, contained in a calorimeter of mass 200 gm. and specific heat 0·2, to raise the water from 20° C. to its boiling point ? State the reasons why the mass of steam condensed in an experiment such as the above would be greater than the calculated value. (L.)

**11.** A block of brass weighing 90 gm. and having a thermal capacity 10 cal. per °C. is observed to cool at the rate of $\frac{1}{10}$° C. per sec. when placed in a refrigerator. Calculate (a) the rate at which the block is losing heat, (b) the specific heat of brass. (N.)

**12.** What is meant by the *latent heat of fusion* of ice ? Describe an experiment to determine the value of the latent heat of fusion of ice, pointing out the necessary precautions to ensure an accurate result, and showing how you would calculate the value from your observations. 20 gm. of ice contained in a vessel of water equivalent 10 gm. are heated by a bunsen burner which supplies 60 calories of heat per sec. Assuming that 50 per cent. of the heat supplied is utilized and that both the ice and vessel are initially at 0° C., calculate the time required to vaporize all the ice. [Latent heat of fusion of ice = 80 cal. per gm.; latent heat of vaporization of water = 540 cal. per gm.] (C.W.B.)

**13.** Explain what is meant by the specific heat and by the water equivalent of a body. An aluminium calorimeter weighs 28·40 gm. empty, and 80·70 gm. after some water has been poured into it. An aluminium cylinder weighing 79·60 gm. is heated to 98·8° C., and dropped into the water. The temperature of the water rises from 15·2° C. to 34·0° C. Calculate the specific heat of the aluminium and the water equivalent of the calorimeter. (*C.*)

**14.** Describe a method of finding the latent heat of fusion of ice. State and explain the effect, if any, on the result if (*a*) the barometric pressure was low, (*b*) the ice was not dry. (*N.*)

**15.** Explain the meaning of *water equivalent of a calorimeter* and describe how you would determine its value experimentally. A copper calorimeter, of water equivalent 12 gm., contains 42 gm. of a liquid of specific heat 0·5 and at a temperature of 16° C. A lump of aluminium of specific heat 0·22 is heated for a time in boiling water at 100° C. and then lowered into the calorimeter. The temperature of the liquid rises to 30° C. What is the mass of the aluminium ? (*L.*)

**16.** Define the terms latent heat, specific heat. 400 gm. of a liquid at 16° C. are contained in a copper calorimeter of mass 100 gm. Heat is supplied at a constant rate by a bunsen burner, and it takes 5 min. to bring the liquid to its boiling point 156° C. How long will it take to evaporate the whole of the liquid ? Neglect radiation losses. (Latent heat of evaporation of liquid = 69 cal. per gm.; sp. ht. of liquid = 0·47; sp. ht. of copper = 0·1.) (*C.*)

**17.** Describe how you would determine the latent heat of steam. A kettle, of negligible thermal capacity, contains 3 lb. of ice-cold water and is heated at a constant rate. After 1½ min. its temperature is found to be 68° F. Calculate, neglecting heat losses, (*a*) the total time for the water to reach boiling point, (*b*) the rate at which water will be boiled away. State the ways in which heat may be lost if the kettle is heated by a gas ring. (Latent heat of steam 970 B.Th.U. per lb.) (*N.*)

**18.** Define *specific heat*. Would the numerical value for the specific heat of copper change if temperatures were measured on the Fahrenheit scale instead of on the Centigrade scale ? Give a reason. A spiral of wire carrying an electric current is immersed in 500 gm. of water in a Thermos flask. The water, initially at a temperature of 20° C., begins to boil in 10 min. Neglecting the heat capacity of the flask, calculate the time taken to boil away half the water in the flask. (Latent heat of steam = 540 calories per gm.) (*O. & C.*)

# CHAPTER XIII

# HEAT AND MECHANICAL ENERGY

EARLY scientists considered that heat was a material substance which they called *caloric*. Thus a body gaining heat increased its caloric content, and one losing heat diminished its caloric content. This conception of heat was generally accepted up to about 1800, when experimental results were obtained which could not be explained by the caloric theory of heat. Similar cases of "experiment *versus* theory" had occurred in the past in science (for example, the experimental results on falling objects were contrary to Aristotle's theory (p. 2), and they will doubtless occur in the future. In accordance with the scientific spirit, which abides by experimental results, the caloric theory was rejected.

One of the first experiments which spelt the doom of the caloric theory was performed by BENJAMIN THOMPSON, or COUNT RUMFORD, as he was later called. Rumford was an American engaged in 1798 in supervising the boring of cannon in a munitions factory in Bavaria. He noted that the borer, the metal bored, and the metal chips all became hot, and that heat continued to be produced as long as the boring took place. It seemed highly improbable to Rumford that such a small amount of metal with which he was dealing could contain such an enormous amount of caloric, and he realized that the heat produced was related to the mechanical energy used in boring the cannon. To confirm his view he surrounded a gun-barrel and a borer in a container of water and, to the astonishment of the onlookers, who knew no flame was being used, the water was made to boil. These demonstrations implied that **heat was a form of energy**, and not a material substance. Later, experimental evidence obtained by JAMES PRESCOTT JOULE showed conclusively that this was the true conception of the nature of heat.

**Joule's experiments.** Joule was born near Manchester in 1818, and spent much of his life in researches on heat. Fig. 147 illustrates the esssential features of Joule's most famous experiment performed in 1842. Two weights, $A$, $B$, were connected to strings passing in opposite directions round a cylinder $D$, so that an axle $H$ passing through $D$ was rotated as the weights fell. The axle in turn caused a paddle $P$ to rotate in a calorimeter $C$ containing water, and vanes from $C$ caused the water to be churned by $P$, and not rotated, so that *all* the mechanical energy of $P$ was converted to heat by friction between the latter and the water. The

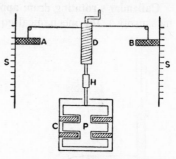

FIG. 147. Joule's experiment

weights $A$, $B$, were released simultaneously from the same height above the ground, and the rise in temperature in the water as they just reached the ground was noted on a sensitive thermometer graduated in °F.

Suppose $W$ is the total weight of $A$, $B$ in lb. wt., and $h$ is the height in feet above the ground when they are released. Then the amount of mechanical energy expended by $P = Wh$ ft. lb. wt. (p. 42). If the mass of water in $C$ is $m$ lb. the water equivalent of the calorimeter $w$ lb., and the rise in temperature $\theta°$ F., the heat produced $= (m + w)\theta$ B.Th.U.

∴ $(m + w)\theta$ B.Th.U. are produced by $Wh$ ft. lb. wt.

∴ 1 B.Th.U. is produced by $\dfrac{Wh}{(m + w)\theta}$ ft. lb. wt.

Joule performed the experiment many times, with different weights and heights. He found that about 778 ft. lb. wt. of mechanical energy were *always* required to produce 1 B.Th.U. of heat, or that about 42 million ergs of mechanical energy were always required to produce 1 calorie of heat. The mechanical energy required to produce 1 unit of heat is known as the **mechanical equivalent of heat,** and the latter is denoted by the symbol J, after Joule. In his honour, too, a unit of mechanical energy is named after him. This is the joule, defined as 10 million ergs (p. 38), so that the mechanical equivalent of heat can be expressed as approximately 4·2 joules per calorie.

**Callendar's rotating drum apparatus.** A laboratory method of measuring J, the mechanical equivalent of heat, was devised by

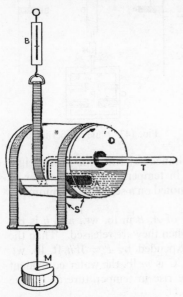

CALLENDAR, the inventor of the platinum resistance thermometer (p. 159). The apparatus consists of a brass drum D containing some water A, whose temperature can be measured on a thermometer T of a special shape (Fig. 148). A silk band S is wound tightly round the drum, and a large mass M, and a spring balance B, are attached to the ends of the silk as shown.

When the drum D is rotated in a clockwise direction at a suitable steady speed, the silk band, the weight M, and the spring-balance reading all remain stationary, and friction is created between the band and drum. The drum and water are thus warmed, and after a definite number or revolutions the rise in temperature is observed on T.

FIG. 148. Callendar's drum

**Measurements.** The following measurements were taken in an experiment—

|  |  |
|---|---|
| Mass of water | = 100 gm. |
| Mass of brass drum | = 42·86 gm. |
| Sp. ht. of brass | = 0·092 |
| No. of revolutions | = 473 |
| Mass of M | = 1000 gm. |
| Tension in B | = 50 gm. |
| Initial temp. of water | = 13·1° C. |
| Final temp. of water | = 19·4° C. |
| Radius of drum | = 10 cm. |

**Calculation for J.** *Work done.* The net tension (force) in the silk band = 1000 − 50 = 950 gm. wt., since the weight of $M$ and the tension in $B$ act oppositely to each other.

In 1 revolution the force acts through a distance equal to the circumference of the drum, which is $2\pi r$, or $2\pi \times 10$ cm.

$$\therefore \text{ work done in 1 revolution} = \text{force} \times \text{distance}$$
$$= (950 \times 980) \times 2\pi \times 10 \text{ ergs}$$

since 950 gm. wt. = 950 × 980 dynes.

$\therefore$ work done in 473 revolutions, $W = 950 \times 980 \times 2\pi \times 10 \times 473$ ergs

*Heat produced,* $H = (100 + 42\cdot86 \times 0\cdot092)(19\cdot4 - 13\cdot1)$ cal.

$$= 103\cdot7 \times 6\cdot3 \text{ cal.}$$

$$\therefore \quad J = \frac{W}{H} = \frac{950 \times 980 \times 2\pi \times 10 \times 473}{103\cdot7 \times 6\cdot3} \text{ ergs per cal.}$$

$$= 4\cdot24 \times 10^7 \text{ ergs per cal.}$$

Since 1 joule is $10^7$ ergs, $J$ can also be expressed as 4·24 joules per calorie.

---

## Summary

1. Joule's experiments showed that heat is a form of energy, and not a material substance as earlier scientists had thought.

2. **The mechanical equivalent of heat ($J$) is the mechanical energy required to produce 1 unit of heat.** ($J = 42$ million ergs per calorie, or 4·2 joules per calorie, or 780 ft.-lb. wt. per B.Th.U.)

3. Callendar's rotating drum apparatus can be used to measure the mechanical equivalent of heat.

---

## WORKED EXAMPLES

1. *Give a reason for the fact that water at the bottom of a waterfall is higher in temperature than at the top. State the rise in temperature for a waterfall 780 ft. in height.* ($J = 780$ *ft. lb. per B.Th.U.*) (*N.*)

Let 1 lb. of water drop from the top to the bottom of the waterfall. Then potential energy lost = 1 × 780 ft. lb. = 780 ft. lb.

$$\therefore \text{ heat energy obtained, } Q = \frac{780}{J} = \frac{780}{780} = 1 \text{ B.Th.U.}$$

Since $Q = mst$, we have $1 = 1 \times 1 \times t$, i.e. $t = 1°$ F. Hence the rise in temperature is $1°$ F.

2. *Describe a non-electrical method of finding the relation between work and heat, and define the units of work and heat employed. A steam plant, burning 0·5 lb. of coal per minute, supplies steam to a steam-engine. If the overall efficiency of the system is 25 per cent, determine the horse-power generated by the steam-engine. The calorific value of the coal = 14,960 B.Th.U. per lb., the mechanical equivalent of heat = 780 ft. lb. wt. per B.Th.U., and 1 h.p. = 33,000 ft. lb. wt. per min. (C.W.B.)*

The heat generated by the burning of 1 lb. of coal = 14,960 B.Th.U. Since the efficiency is 25 per cent, or $\frac{1}{4}$, the energy in B.Th.U. generated by the steam-engine = $0.5 \times \frac{1}{4} \times 14,960 = 1870$ B.Th.U.

Now this is the energy generated per minute,

$\therefore$ power of steam-engine = 1870 B.Th.U. per min.

$= 1870 \times 780$ ft. lb. wt. per min.

But $\qquad$ 1 h.p. = 33,000 ft. lb. wt. per min.

$\therefore$ h.p. of steam-engine $= \dfrac{1870 \times 780}{33,000} = 44 \cdot 2$

3. *What do you understand by the statement that the mechanical equivalent of heat is 4·2 joules per calorie? How would you attempt to verify this statement experimentally? A machine is driven by a 2-kilowatt motor. The efficiency of the machine is 79 per cent. Assuming that all the wasted energy is converted into heat which is absorbed by a mass of 50 Kg. of iron of specific heat 0·12, find the rise in temperature of the iron after the machine has been running for 1 minute. (N.)*

Since the efficiency of the machine is 79 per cent, 21 per cent of the energy is converted into heat. Now energy = power × time.

$\therefore$ heat energy $= \frac{21}{100} \times 2000 \times 60$ joules,

since 2 kilowats = 2000 watts, and 1 min. = 60 sec.

$\therefore$ heat in calories $= \dfrac{21 \times 2000 \times 60}{100 \times 4 \cdot 2} = 6000$ cal.

Since $Q = mst$, where $t$ is the temperature rise

$\therefore$ $6000 = 50,000 \times 0 \cdot 12 \times t$

$\therefore$ $t = \dfrac{6000}{50,000 \times 0 \cdot 12} = 1^\circ$ C.

## EXERCISES ON CHAPTER XIII

1. Name two cases in which heat is converted into mechanical energy, and two cases in which mechanical energy is converted into heat.

2. Define the term *mechanical equivalent of heat*, and state the units in which it is measured. Calculate the mechanical equivalent of heat when (i) 8·4 joules of mechanical energy are obtained from 2 calories of heat (ii) 2340 ft. lb. wt. of energy is derived from 3 B.Th.U.

**3.** 250 B.Th.U. are utilized in an engine. How much mechanical energy is theoretically available? (Mechanical equivalent of heat, $J = 780$ ft. lb. wt. per B.Th.U.)

**4.** An 80 lb. wt. falls from a height of 100 ft. Calculate the heat developed at the ground. ($J = 780$ ft. lb. wt. per B.Th.U.)

**5.** A water-fall has a height of 5000 cm. Considering 1 gm. of water and the energy it loses, calculate the difference in temperature between the top and bottom of the water-fall. ($J = 42$ million ergs per cal.)

**6.** Describe an experiment to measure the mechanical equivalent of heat. Choosing imaginary values, show how the result is calculated.

**7.** A leather belt is wound round a wheel of diameter 60 cm. The wheel rotates while the belt is kept stationary, and the average frictional force generated is 10,000 gm. wt. What heat is generated when the wheel rotates through 20 complete revolutions? ($J = 42$ million ergs per cal.)

**8.** Explain what is meant by the *mechanical equivalent of heat* and describe fully its determination by *either* (*a*) an experiment performed by Joule *or* (*b*) an experiment you have seen performed. Why are the results obtained in ordinary laboratory experiments usually a little above the true value? (*N.*)

**9.** State briefly the meaning of *mechanical equivalent of heat*. Give an example of a process in which heat energy is translated into mechanical energy. In an experiment, a mass of lead-shot contained in a vertical cardboard cylinder, falls 60 cm. when the cylinder is inverted. Calculate the rise of temperature caused by 70 inversions, taking the specific heat of lead as 0·032. What are the probable sources of error in this experiment? (*L.*)

**10.** Explain fully what is meant by the Mechanical Equivalent of Heat. How long should a motor-cycle working at 4 h.p. be able to run on 1 gallon of petrol, if one-fifth of the energy of the fuel is converted into useful work? (1 gallon of petrol in burning yields 120,000 B.Th.U.; 1 h.p. = 33,000 ft. lb. per min.; mech. equiv. of ht. = 770 ft. lb. per B.Th.U.) (*C.*)

**11.** In a rough determination of the mechanical equivalent of heat a quantity of lead shot was placed in a cardboard cylinder of length 1 metre and was allowed to fall backwards and forwards from one end of the tube to the other 50 times in succession. The temperature of the shot was found to have risen 3·7° C. Calculate the mechanical equivalent of heat. (Sp. ht. of lead = 0·03, 1 cm. g. = 981 ergs.) (*O. & C.*)

**12.** How do you explain the rise in temperature of a bicycle pump during the action of pumping up a tyre ? (*L.*)

**13.** Define *erg*, *calorie*. Explain the meaning of the term *mechanical equivalent*. Describe a method of determining the mechanical equivalent of heat and explain how the result is calculated from the readings. (*O. & C.*)

**14.** Describe a simple method of measuring the *mechanical equivalent of heat* and explain how the result is calculated from the readings. Fifty litres of water are heated by a rotating paddle driven by a ¼ h.p. motor. Assuming that all the work done is used to heat the water, calculate the time required to raise the temperature of the water 5° C. neglecting heat losses. [Take 1 h.p. = 746 joules per sec., and 4·2 joules = 1 cal.] (*O. & C.*)

**15.** Describe an experiment to show that when mechanical energy is transformed into heat energy the ratio of the work done, expressed in ergs, to the heat generated, in calories, is a constant.

A 20 watt electric heater immersed in water in an open Dewar flask just keeps the water boiling. If no heat is lost by conduction, convection and radiation, find how many grams of water boil away in 20 min. Explain briefly what has happened to the electrical energy supplied to the heater. (*J* = 4·2 joules per calorie. Latent heat of steam = 540 cal./gm.) (*O. & C.*)

**16.** Explain the statement, "the mechanical equivalent of heat is 4·2 joules cal.⁻¹", and describe an experiment to verify this value.

A piece of lead falls 3 metres from rest, coming to rest again on the ground. Calculate the rise in its temperature, the specific heat of lead being 0·032. State the assumptions you make in the calculation. (*L.*)

**17.** State two distinct ways in which mechanical energy may be transformed into heat energy.

Describe and explain an experiment to determine the relation between a unit of heat and a unit of work, stating the units used.

A steam engine develops 195 h.p. and its boiler uses 400 lb. of coal per hour. If the calorific value of the coal is 13,500 B.Th.U. per lb., find the efficiency of the plant.

(1 B.Th.U. = 780 ft. lb. wt. and 1 h.p. = 550 ft. lb. wt. per sec.) (*L.*)

# CHAPTER XIV

## PROPERTIES OF GASES. THE GAS LAWS

WHEN a steam engine, a motor car engine, or an aeroplane engine is functioning, gases inside cylinders are expanding and contracting in volume, thus operating pistons which make the machine move. The mechanical engineer must therefore know something of the laws governing the behaviour of gases; he must know, for example, how the volume of a gas changes when its pressure and temperature alter.

The condition or *state* of a given mass of gas depends on its volume, temperature, and pressure. Solids and liquids are practically unaffected by small changes of pressure, so that there is no need to take pressure into account in discussing their changes in volume with temperature. A small change in pressure can, however, considerably affect the volume of a gas. Since there are *three* variable quantities for a given mass of gas (volume, pressure, temperature), it is best to keep one of the quantities fixed and discuss the variation of the remaining two quantities.

**Volume coefficient of a gas.** Suppose that the pressure of a given mass of gas is kept constant. Experiment then shows that the volume of the gas increases regularly as the temperature is raised (see p. 219); and, as in the expansion of liquids, scientists define a *volume coefficient* (a) of a gas *at constant pressure*. The volume coefficient is given by *the increase in volume of* 1 *c.c. of the gas* **at** 0° **C.** *for* 1° **C.** *rise in temperature, the pressure remaining constant.*

Suppose the volume of a gas at 0° C. is 200 c.c. and the volume at 50° C. is 236 c.c., the pressure being kept constant. Then the expansion in volume of 200 c.c. at 0° C. is 36 c.c., and hence the volume coefficient (a) at constant pressure is given by

$$a = \frac{36}{200 \times 50} = 0 \cdot 0036 \text{ per } °C.$$

217

**Formulae.** Let $V_o$ be the volume of a gas at 0° C., and $V_t$ the new volume at $t$° C., the pressure remaining constant. The increase in volume is then $(V_t - V_o)$, and hence

$$a = \frac{V_t - V_o}{V_o \times t} \qquad . \qquad . \qquad . \qquad . \quad (1)$$

Thus $V_t - V_o$ = increase in volume = $V_oat$, and hence

$$V_t = V_o(1 + at) \qquad . \qquad . \qquad . \qquad . \quad (2)$$

When the expansion of solids and liquids was discussed, we referred to a "mean coefficient" of expansion over a temperature range, e.g. from 15° C. to 40° C. Now experiments with gases show that the values so obtained depend on the initial temperature of the gas. The coefficient of expansion of a gas is therefore always referred to its volume *at 0° C.*; and if the volume at 0° C. cannot be measured directly, it must be found indirectly, as shown below.

**An experiment to determine** $a$. The magnitude of $a$, the volume coefficient, can be determined by the apparatus shown in Fig. 149. Air is trapped by mercury in the volume *AB* of a tube of *uniform* cross-sectional area. When the tube *ED* is raised or lowered, the mercury level at *A* rises or falls; *and throughout the experiment the mercury level is adjusted to be the same at A and D, so that the pressure is always atmospheric pressure and therefore constant.*

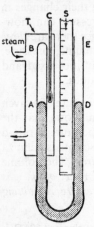

A glass tube *T* surrounds *AB*, and steam is first passed into *T*. The temperature is now observed on a thermometer *C*, the tube *ED* is moved until the mercury levels are the same on both sides, and the volume of the air, which is proportional to the length *AB*, is observed by means of the metre rule *S*. The steam is then cut off, and the temperature commences to fall. When it reaches 80° C., for example, the tube *ED* is raised so that the levels of mercury are the same on both sides, and the new volume of the

FIG. 149.
Volume coefficient

air in *AB* is again measured by *S*. Further readings of the volume at constant pressure are observed as the temperature decreases, and below are readings taken in an experiment to determine *a*—

| Volume (∝ length *AB*) | Temp. (*t*° C.) | Absolute temp. (*T*) (*see p. 220*) |
|---|---|---|
| 9·6 | 95° C. | 368 |
| 9·2 | 80° C. | 353 |
| 8·9 | 70° C. | 343 |
| 8·5 | 55° C. | 328 |
| 8·3 | 45° C. | 318 |
| 7·9 | 30° C. | 303 |
| 7·5 | 15° C. | 288 |

**Deductions from the experiment.** When the readings of volume (*V*) are plotted against the temperature in °C., a straight line graph *XY* is obtained (Fig. 150). Thus the volume of a gas at constant pressure increases regularly with the temperature.

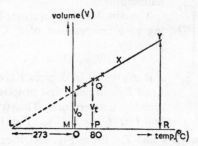

FIG. 150. Graph of volume v. temperature

To find the volume coefficient, *a*, the line *YX* is produced to cut the volume-axis at *N*, thus determining the volume $V_0$ of the gas at 0° C. In the present case, $MN = V_0 = 7·1$. The volume $V_t$ at any other temperature, e.g. 80° C., is then taken, and $PQ = V_t = 9·2$. The volume coefficient, *a*, is calculated from

$$a = \frac{V_t - V_0}{V_0 \times t} = \frac{(9·2 - 7·1)}{7·1 \times 80}$$

$$= 0·0037 \text{ per °C.}$$

Accurate experiment shows that $\alpha$ is 0·00366, or $\frac{1}{273}$, per °C. (approx.) for all gases, a fact which was first discovered by J. A. C. CHARLES in 1780. **Charles' law** states that *a given mass of gas increases in volume by $\frac{1}{273}$ of its volume at 0° C. for every degree Centigrade rise in temperature, the pressure remaining constant.*

**Absolute temperature, T.** When the temperature of the air in the above experiment is reduced, the pressure being kept constant, the volume diminishes along the straight line *YX* (Fig. 150). Below 0° **C.**, if a freezing mixture were available, the air would diminish in volume along the line *YNL*, and at a temperature corresponding to *L*, where the temperature axis is cut, we have the startling deduction that the volume of the air would *theoretically* become *zero*. The temperature *L* (actually − 273° C.) is given the name of **absolute zero.** In practice, a gas liquefies before it approaches this temperature.

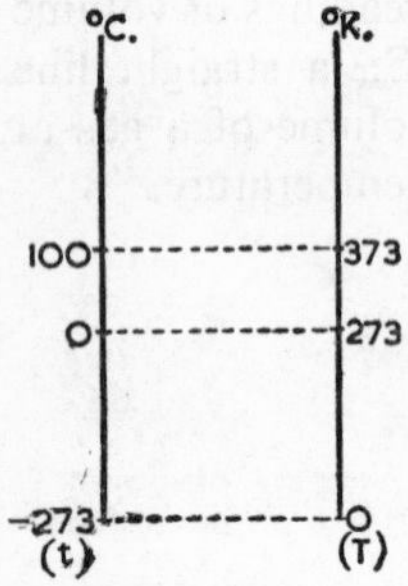

FIG. 151.
$T = 273 + t$

The absolute temperature, *T*, corresponding to a temperature of $t°$ C. is obviously given by

$$T = 273 + t \qquad . \qquad . \qquad . \quad (3)$$

*T* is expressed in degrees Kelvin (°K.), after Lord Kelvin, who first proposed the idea of absolute temperature. Thus 0° C. = 273° K. and 100° C. = 373° K. (Fig. 151).

**Relation between *V* and *T*.** From Fig. 150, it can be seen that $YR/QP = LR/LP$, from similar triangles *LRY,LPQ*. But *QP* and *YR* are the volumes of the gas corresponding to absolute temperatures *LP, LR* respectively. Hence we arrive at the important result that *the volume of a given mass of gas at constant pressure is directly proportional to its absolute temperature,*

i.e. *volume, V* $\propto$ *absolute temperature, T* $\qquad$ . $\quad$ (4)

This can be shown experimentally by dividing the volume readings in the table on p. 219 by the corresponding absolute temperature, when a constant number is obtained. The calculation should be verified by the reader.

Suppose that a gas has a volume of 50 c.c. at 15° C. and that its temperature is altered to 100° C., the pressure remaining constant. The absolute temperature corresponding to 15° C. = 273 + 15 = 288° K., and the absolute temperature of 100° C. = 273 + 100 = 373° K. Then, from (4), the new volume $V$ of the gas is given by

$$\frac{V}{50} = \frac{373}{288}$$

$$\therefore \quad V = 50 \times \frac{373}{288} = 64 \cdot 8 \text{ c.c.}$$

Again, suppose that a gas has a volume of 40 c.c. at 20° C. and its volume becomes 80 c.c. at a higher temperature, the pressure being kept constant. From (4), the absolute temperature $T$ of the gas at the higher temperature is given by

$$\frac{T}{273 + 20} = \frac{80}{40}$$

$$\therefore \quad T = \frac{80}{40} \times 293 = 586°$$

$$\therefore \quad \text{temp. in °C.} = 586 - 273 = 313°$$

**The pressure coefficient of a gas.** We have now to consider the changes of *pressure* of a gas with temperature, the volume being kept constant.

Experiments similar to that described below show that *the pressure of a given mass of gas increases by $\frac{1}{273}$ of its pressure at 0° C. for every °C. rise in temperature, the volume being kept constant.* Thus $\frac{1}{273}$ per °C. is known as the *pressure coefficient* of a gas at constant volume. It should be noted that the definition of pressure coefficient refers to the original pressure of the gas at 0° C. (compare "volume coefficient"), and that the pressure coefficient has the same magnitude as the volume coefficient. On the latter account, we shall use the same symbol, $\alpha$, for the pressure coefficient.

**Experiment to find pressure coefficient of a gas.** Fig. 152 illustrates one form of apparatus for measuring the pressure coefficient of air. *B* is a large bulb of air, with a narrow (capillary) tube *L* connected to a mercury manometer for measuring its pressure. *B* is surrounded by a vessel containing water, which is heated to different temperatures to alter the temperature of the air. The volume of the air will then alter, but by raising or lowering the tube *GD*, which is open to the air, the mercury on the other side can always be brought back to a fixed point $C$. In this way, *the volume of the air above $C$ is kept constant.*

The pressure of the air in the bulb $B$ is given by $(A + h)$ cm., where $A$ is the atmospheric pressure in centimetres of mercury and $h$ is the difference in level, $GD$, of the mercury levels. If the level in the right-hand tube is *less* than the level at $C$, the pressure of the

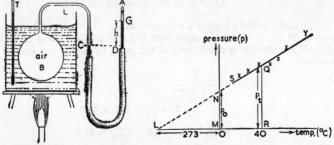

FIG. 152. Pressure coefficient     FIG. 153. Graph of $p$ v. $t°$ C (*not to scale*)

air in $B$ is given by $(A - h')$ cm., where $h'$ is the difference in levels (see p. 132). When the pressure, $p$, is plotted against the temperature in °C., a straight line graph $YS$ is obtained. The pressure $p_o$ which the gas would have at 0° C. is obtained by producing $YS$ to meet the pressure-axis at $N$, and the pressure coefficient, $\alpha$, can now be calculated. (Fig. 153.)

Suppose $p_o = 69 \cdot 8$ cm. of mercury and that the pressure, $QR$, at 40° C. is 80·0 cm. The increase in pressure from 0° C. $= 80 \cdot 0 - 69 \cdot 8 = 10 \cdot 2$ cm.

$$\therefore \text{pressure coefficient} = \frac{\text{increase in pressure from 0° C.}}{\text{original pressure at 0° C.} \times \text{temp. rise}}$$

$$= \frac{10 \cdot 2}{69 \cdot 8 \times 40} = \frac{1}{273} \text{ per °C.}$$

**Formulae.** If $p_t$ is the pressure at $t°$ C. and $p_o$ is the pressure at 0° C., the pressure coefficient $\alpha$ is given by

$$\alpha = \frac{p_t - p_o}{p_o \times t} \qquad . \qquad . \qquad . \qquad (5)$$

Thus $p_t - p_o = $ increase in pressure $= p_o \alpha t$, i.e. $p_t = p_o + p_o \alpha t$.

$$\therefore \quad p_t = p_o(1 + \alpha t) \qquad . \qquad . \qquad . \qquad (6)$$

Since $a = \frac{1}{273}$ per °C., $p_t = p_o(1 + \frac{1}{273}t)$. Thus the pressure of a gas is theoretically zero at $t = -273°$ C., when the bracket term becomes zero, and $LM = 273°$ in Fig. 153.

From Fig. 153, it can be seen that *the pressure of a gas at constant volume*, e.g. *QR, is proportional to its absolute temperature*, i.e. the length from $L$ to $R$. This result is similar to that in the case of the volume of a gas at constant pressure (p. 220).

If the pressure of a gas at $-10°$ C. is 50 cm. of mercury, its pressure $p$ at 80° C. when the volume is kept constant is given by

$$\frac{p}{50} = \frac{273 + 80}{273 - 10} = \frac{353}{263}$$

$$\therefore \quad p = \frac{50 \times 353}{263} = 67 \cdot 1 \text{ cm.}$$

**The constant volume gas thermometer.** For a given alteration of temperature, the pressure change of a gas at constant volume is relatively much greater than the volume change in liquids. For example, the pressure coefficient of a gas is about $0 \cdot 0037$, while the volume coefficient of mercury is only about $0 \cdot 00018$. Scientists therefore rely on a *gas thermometer* for very accurate work. There is an elaborate gas thermometer at the National Physical Laboratory, but the principle is shown by the simple apparatus in Fig. 152.

To find the lower fixed point of a gas thermometer, the bulb $B$ (Fig. 152) is entirely surrounded by melting ice, and the pressure ($p_o$) at constant volume is measured by adjusting the mercury level until it reaches $C$. Suppose $p_o = 50$ cm. of mercury. The upper fixed point is obtained by surrounding $B$ with steam at normal atmospheric pressure, and bringing the level of the mercury back to $C$. Suppose the pressure of the gas is now denoted by $P$ and is 70 cm. of mercury (Fig. 154 (*a*)). Just as for the mercury thermometer, the interval between

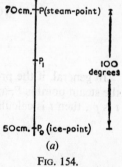

FIG. 154.

Temperature measurement

50 cm. and 70 cm. pressure is divided into 100 equal parts or degrees on the Centigrade scale; so that a pressure of 50 cm.

corresponds to the temperature of 0° C., and a pressure of 70 cm. corresponds to the temperature of 100° C.

To measure the unknown temperature, $t$, of a liquid, for example, the bulb $B$ of the gas thermometer is placed completely inside it, and the volume of the gas is adjusted to be constant. The pressure is then read; suppose it is 55 cm. of mercury. The interval of $t$ degrees from 0° C. to $t°$ C. is 55 cm. − 50 cm., or 5 cm. pressure; and the interval of 100 degrees from 0° C. to 100° C. is 70 cm. − 50 cm., or 20 cm. pressure. Consequently, by proportion,

$$\frac{t}{100} = \frac{5}{20}$$

$$\therefore \quad t = \frac{5}{20} \times 100 = 25° \text{ C.}$$

Suppose that a higher temperature, $x$, was measured by the same gas thermometer, and the pressure indicated was 64 cm. of mercury. Then $x$ degrees corresponds to an interval from 64 cm. ($x°$ C.) to 50 cm. (0° C.); and 100 degrees corresponds to an interval from 70 cm. (100° C.) to 50 cm. (0° C.). Consequently, by proportion,

$$\frac{x}{100} = \frac{64 - 50}{70 - 50} = \frac{14}{20}$$

$$\therefore \quad x = \frac{14}{20} \times 100 = 70° \text{ C.}$$

In general, if the pressure at the ice point is $p_0$, the pressure at the steam point is $P$, and the pressure at an unknown temperature $t$ is $p_1$, then $t$ is calculated in °C. from the relation

$$\frac{t}{100} = \frac{p_1 - p_0}{P - p_0}.$$

**An accurate constant volume gas thermometer.** An accurate form of constant volume gas thermometer is shown in Fig. 154 (*b*). The bulb $A$ contains nitrogen, or hydrogen, and is connected by a capillary tube to a manometer $M$. When $A$ is at a certain temperature and the gas pressure is required, the mercury column $B$ is moved until the mercury in $M$ reaches the tip of a pointer

*P*. The same procedure is adopted at any other temperature, thus keeping the volume of the gas constant. A barometer is also in the apparatus; it corresponds to the tube *RH*, a vacuum existing in the closed tube above *H*. Besides the adjustment at *P*, the mercury at *H* is adjusted so that it just touches the tip of a pointer *Q*; this is done by raising or lowering *RH*. As the tip of *Q* is at the zero of pressure, the pressure of the gas in *A* then corresponds to *h* cm. of mercury.

**The relation between *p*, *V*, *T* for a given mass of gas.** From pp. 223, 220, it can be stated that, for a *given mass* of a gas,

$$p \propto T \text{ (volume kept constant)},$$

and

$$V \propto T \text{ (pressure kept constant)}.$$

If these two relations are combined into one, so that all three quantities, *p*. *V*, *T*, are considered to vary, we have

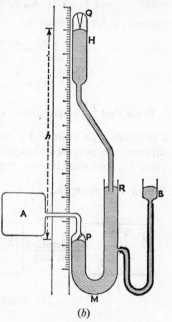

FIG. 154. Gas thermometer

$$pV \propto T,$$

or $\dfrac{pV}{T} = $ a constant value (see proof on p. 226).

The value of the constant depends on the nature and mass of the gas used. But for a *fixed* mass of gas,

$$\frac{p_1 V_1}{T_1} = \frac{p_2 V_2}{T_2} \qquad . \qquad . \qquad . \qquad (7)$$

where the symbols in (7) represent the respective values of pressure and volume at two different *absolute* temperatures $T_1$, $T_2$. It will be noted that $p_1 V_1 = p_2 V_2$ when the temperatures $T_1$, $T_2$ are the same, from (7) which is Boyle's law (p. 133).

On many occasions, the volume of a gas collected in a chemical experiment is "reduced" to its volume at **S.T.P.**, i.e. standard temperature (0° C.) and pressure (76 cm. of mercury).

As an illustration, suppose the volume of a given mass of oxygen at 75 cm. pressure and 12° C. is 85 c.c. Then $p_1 = 75$ cm., $V_1 = 85$ c.c., $T_1 = 273 + 12 = 285°$ K., and $T_2 = 273°$ K., $p_2 = 76$ cm.; thus, since $\dfrac{p_1 V_1}{T_1} = \dfrac{p_2 V_2}{T_2}$, the volume $V_2$ at S.T.P. is given by

$$\frac{75 \times 85}{285} = \frac{76 \times V_2}{273}$$

$$\therefore \quad V_2 = \frac{75 \times 85 \times 273}{285 \times 76} = 80 \cdot 4 \text{ c.c.}$$

**Proof that $\dfrac{pV}{T} = $ constant.** Consider a fixed mass of gas having a pressure $p_1$, a volume $V_1$, and an absolute temperature $T_1$, and suppose the pressure changes to a value $p_2$, the volume to a value $V_2$, and the absolute temperature to a value $T_2$ (Fig. 155). We have to show that $\dfrac{p_1 V_1}{T_1} = \dfrac{p_2 V_2}{T_2}$.

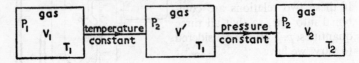

FIG. 155. Proof of $\dfrac{pV}{T} = $ constant

*Stage* 1. *Keeping the absolute temperature constant at $T_1$*, let the pressure be altered to the value $p_2$. The volume of the gas then changes to a value $V'$ such that $p_1 V_1 = p_2 V'$, from Boyle's law.

Thus $$V' = \frac{p_1 V_1}{p_2} \qquad . \qquad . \qquad . \qquad . \quad (8)$$

*Stage* 2. *Keeping the pressure constant at $p_2$*, let the absolute temperature change from $T_1$ to the value $T_2$, when the volume changes to the value $V_2$ (Fig. 135). Since the volumes are proportional to the absolute temperatures, from Charles' law,

$$\frac{V'}{V_2} = \frac{T_1}{T_2}.$$

Thus
$$V' = \frac{V_2 T_1}{T_2} \qquad . \qquad . \qquad . \qquad . \qquad (9)$$

From (8) and (9), it follows that

$$\frac{p_1 V_1}{p_2} = \frac{V_2 T_1}{T_2}$$

$$\therefore \quad \frac{p_1 V_1}{T_1} = \frac{p_2 V_2}{T_2}.$$

Thus $\dfrac{pV}{T} = $ a constant value for a fixed mass of gas.

---

## Summary

1. The **volume coefficient** of a gas is the increase in volume of 1 c.c. **at** $0°$ **C.** for $1°$ C. temperature rise. The **pressure coefficient** is the increase in pressure per unit pressure **at** $0°$ **C.** for $1°$ C. temperature rise.

2. Absolute temperature, $T°$ K. $= 273 + t$, where $t$ is the temperature in $°$C.

3. Charles' law states: For a given mass of gas at constant pressure, the volume increases by $\frac{1}{273}$ of its volume at $0°$ C. for each degree rise in temperature. Thus, **the volume is proportional to the absolute temperature.**

4. **For a given mass of gas, $\dfrac{pV}{T} = $ constant.**

5. In a constant volume gas thermometer, the lower fixed point is the pressure of the gas at $0°$ C. and the upper fixed point is the pressure of the gas at $100°$ C.

---

### WORKED EXAMPLES

1. *Describe a constant-volume air thermometer and explain how you would use it to find how the pressure of a mass of gas maintained at constant volume varies with the temperature. A high-altitude aircraft carries a supply of oxygen in cylinders at a pressure of 180 lb. per sq. in. measured at 27° C. What will be the pressure in the cylinders when the temperature inside the aircraft falls to − 3° C.?* (N.)

Since the volume of the oxygen remains constant, the pressure of the gas is proportional to the *absolute* temperature.

The absolute temperature of $27°$ C. $= 273 + 27 = 300°$ K.

The absolute temperature of $-3°$ C. $= 273 - 3 = 270°$ K.

$$\therefore \quad \frac{p}{180} = \frac{270}{300}$$

where $p$ is the new pressure in the cylinders.

$$\therefore \quad p = \frac{270}{300} \times 180 = 162 \text{ lb. per sq. in.}$$

2. *State Charles's law and describe an experiment to verify it. The gas required to fill a barrage balloon to a pressure of 75 cm. of mercury at 12° C. is stored in 10 cylinders each of volume 0·25 cu. metre. The gas in the cylinder is at a temperature of 15·8° C. and at a pressure of 84 standard atmospheres (one standard atmosphere = 76 cm. of mercury). Find the capacity of the balloon in cu. metres.* (*N.*)

The total volume of the gas is $10 \times 0.25$ cu. metre, at a temperature of $15.8°$ C. and a pressure of $84 \times 76$ cm. of mercury. The mass of the gas is constant; hence $\frac{pV}{T}$ is constant (p. 225). Originally, $p_1 = 84 \times 76$ cm., $V_1 = 2.5$ cu. metres, $T_1 = 273 + 15.8 = 288.8°$ K.; finally, $p_2 = 75$ cm., $T_2 = 273 + 12 = 285°$ K., and $V_2 =$ volume of balloon.

Since

$$\frac{p_2 V_2}{T_2} = \frac{p_1 V_1}{T_1}$$

$$\therefore \quad \frac{75 V_2}{285} = \frac{84 \times 76 \times 2.5}{288.8}$$

$$\therefore \quad V_2 = \frac{285 \times 84 \times 76 \times 2.5}{75 \times 288.8} = 210 \text{ cu. metres.}$$

3. *How would you determine the coefficient of expansion of air at constant pressure? 1 gm. of air, occupying 774 c.c. at 0° C. and at a pressure of 1 atmosphere, is heated to 100° C. at constant pressure. Find (a) the volume of the heated air, (b) the density of the heated air, (c) the pressure required to restore the air to its original volume, the air remaining at 100° C. Explain the method employed in the calculation of (c) . (Coefficient of expansion of air at constant pressure = $\frac{1}{273}$ per °C.) ( N.)*

(*a*) Suppose $V_t$ is the volume of the air at $100°$ C. Then, since

$$V_t = V_0(1 + at),$$

we have

$$V_t = 774(1 + \tfrac{1}{273} \times 100) = 774 \times 1\tfrac{100}{273}$$

$$= 1057 \text{ c.c.}$$

(*b*) The volume of the heated air $= 1057$ c.c., and its mass $= 1$ gm.

$$\therefore \quad \text{density} = \frac{\text{mass}}{\text{volume}} = \frac{1}{1057}$$

$$\therefore \quad \text{density} = 0.000946 \text{ gm. per c.c.}$$

(c) Since the temperature remains constant (at 100° C.), we can apply Boyle's law to calculate the pressure, $p$, required to restore the air to its original volume of 774 c.c. The pressure of the gas is 1 atmosphere, and the volume of the gas is 1057 c.c. Since $pV$ is a constant, we have

$$1 \times 1057 = p \times 774$$

$$\therefore \quad p = \frac{1 \times 1057}{774} = 1 \cdot 366 \text{ atmospheres.}$$

## EXERCISES ON CHAPTER XIV

**1.** Name three machines whose design requires a knowledge of the Gas Laws.

**2.** A gas has a volume of 341 c.c. at 0° C. and is heated while its pressure is kept constant. If its volume is 351 c.c. at 8° C., calculate the volume coefficient of the gas.

**3.** Define in words the *volume coefficient* of a gas. How does the definition differ from that of the volume coefficient of a liquid?

**4.** "The volume of a gas at constant pressure is proportional to its temperature." Why is the statement incorrect? A gas is heated at constant pressure so that it expands from a volume of 80 c.c. at 40° C. to a volume of 160 c.c. Calculate the new temperature of the gas. What is the new volume if the temperature is changed to 7° C. at constant pressure?

**5.** A bicycle tyre is warmed from 13° C. to 20° C. by the sun. If the pressure in it is usually 80 cm. mercury, calculate the new pressure assuming the volume of the tyre is constant.

**6.** Draw a diagram of a simple form of *constant volume gas thermometer*. The pressure of the gas in the thermometer is 80 cm. mercury at 100° C. and 64 cm. mercury at 0° C. When the instrument is used in turn to measure the temperature of three liquids, the pressure is respectively 68, 78, and 90 cm. mercury. Calculate the three temperatures.

**7.** The volume of a mass of hydrogen collected at 15° C. and 75 cm. mercury pressure is 350 c.c. Calculate its volume at S.T.P. (0° C. and 76 cm. mercury pressure). What is the volume at 20° C. and 78 cm. mercury pressure?

**8.** The volume of a gas at 27° C. and 1 atmosphere pressure is 200 c.c. What is the pressure if the same mass of gas has a volume of 180 c.c. at 47° C.?

**9.** A faulty barometer tube contains a little air at the top. When the true atmospheric pressure is 77 cm. mercury, the barometer reading is 75 cm. and the volume of air in the top of the tube is 20 c.c. When the true atmospheric pressure changes to 76 cm., the barometer reading is 74·5 cm. What is the new volume of the air ?

**10.** Calculate the increase pressure of the air in a tyre due to an increase in temperature of 15° C., the tyre having been pumped originally to 28·5 lb. wt. per sq. in. at 12° C. State the law used in your calculation. (*L.*)

**11.** Describe an experiment to show that the volume of a gas at constant pressure is proportional to the absolute temperature. (*L.*)

**12.** A quantity of air, of mass 5·04 gm., occupies a volume of 4560 c.c. at a temperature of 19·5° C. and a pressure of 75 cm. of mercury. Calculate the density of air at N.T.P. (*C.W.B.*)

**13.** Describe some method for determining the relation between the pressure and the temperature of a given mass of gas at constant volume. A gas cylinder contains oxygen at a pressure of 14 atmospheres when the temperature is 7° C. By how much would the pressure increase if the temperature were raised to 37° C. ? (*C.*)

**14.** How would you investigate the relation between the pressure and the volume of a fixed mass of gas ? Show clearly how the observations would be used to verify the law connecting these quantities. A cylinder of hydrogen contains 0·5 cu. ft. of the gas pressure of 120 atmospheres. By how much would the pressure in the cylinder fall on releasing gas occupying 3 cu. ft. at atmospheric pressure, the temperature remaining constant ? (*L.*)

**15.** Describe an experiment to find the coefficient of expansion of air kept at constant pressure. (*L.*)

**16.** Describe *one* experiment to show that the volume occupied by a given mass at constant temperature is inversely proportional to the pressure on it, and *one* other experiment to show that the mass of a constant volume at constant temperature decreases when the pressure decreases. An airman who normally breathes 18 times a minute when the pressure is 30 in. of mercury ascends to a place where the pressure is 22·5 in. Find his new rate of breathing in order that he may inhale the same mass of oxygen per minute, assuming that the composition of the air is unchanged. (*N.*)

**17.** A mass of hydrogen occupies 200 c.c. at 17° C. and 780 mm. of mercury pressure. What is its pressure when the volume is 150 c.c. and the temperature 27° C. ?

A column of air is trapped in a capillary tube sealed at one end, beneath a mercury index. The lengths of the air column at 19·5 C. and 100° C. are 14·7 cm. and 18·6 cm. respectively. Calculate the coefficient of expansion at constant pressure of the air. If this value were correct what would be the temperature of absolute zero ? (C.)

**18.** Describe how you would determine the coefficient of expansion of air at constant pressure. Gas in a container, originally at 0° C., is heated to 540° C. at constant pressure. If the coefficient of expansion of the gas is 0·0037 per °C., show that the volume of the gas has been approximately trebled. If the expansion of the gas is utilized to lift a weight of 30 lb. a vertical distance of 8 in., calculate (a) the work done in lifting the weight, and (b) the heat equivalent of this work. (Mechanical equivalent of heat 780 ft. lb. per B.Th.U.) (N.)

**19.** What is meant by the statement that the coefficient of expansion of air at constant pressure is $\frac{1}{273}$ per °C. ? Describe an experiment to measure the coefficient of expansion of air at *constant pressure* and show how the result is calculated from the observations. Draw a diagram of the apparatus. A capillary tube sealed at one end contains air enclosed by a short thread of mercury. The enclosed air is 20 cm. long at a temperature of 0° C. and 27·4 cm. long at 100° C. Calculate the coefficient of expansion of air at constant pressure. (O. & C.)

**20.** State Charles' law of expansion of gases. Explain the meaning of *absolute temperature* and *absolute zero of temperature*. Calculate the absolute temperatures corresponding to 0° C. and 273° C. Describe a constant-pressure air thermometer, and explain how it is used to measure the thermal expansion of air. (O. & C.)

**21.** What is understood by: (i) S.T.P., (ii) absolute gas scale of temperature ? A fixed mass of air is heated, (a) without altering the pressure, and (b) without altering the volume. State in each case the laws governing the change which takes place.

Three litres of air at 0° C. and a pressure of 1 atmosphere are heated in such a way as to keep its volume unchanged. Calculate the temperature of the air when its pressure becomes 5 atmospheres. The temperature is now kept constant and the pressure reduced again to 1 atmosphere. What will be the resulting volume ? If, finally, this last pressure be maintained, calculate the temperature for which the volume will be reduced to 2 litres. (L.)

# CHAPTER XV

## PROPERTIES OF VAPOURS. HYGROMETRY

IN 1808 JOHN DALTON suggested that elements were made up of tiny particles, **atoms**, which were the same for each element. The **molecule** of an element is the smallest group of its atoms which can exist independently; in many cases (e.g. oxygen, nitrogen), the molecule consists of two atoms. A **chemical compound** consists of molecules made up of atoms of the different constituent elements. All substances, whatever their chemical constitution, can be divided into three large classes or groups—**solids, liquids** and **gases.** Some substances exist in each of these states; for example, ice, water, and steam are three states of the same substance, and there are solid, liquid, and gaseous forms of carbon dioxide.

**The solid state.** Particles of matter have an attraction for each other which increases as their distance apart decreases. These cohesive forces, as they are called, are greatest in the *solid* state, as the molecules then are closest together. The molecules of a solid are not stationary; they vibrate through a very small distance about some mean or "anchor" position, and consequently have an amount of kinetic energy. If an iron is heated so that its temperature rises, the molecules of the metal vibrate faster about their respective mean positions. This is because heat is a form of energy, and causes an increase in the energy of the vibrating molecules.

**The gaseous state.** The molecules of a gas are so far apart from each other on the average that the force of attraction between them is almost negligible compared with that between the molecules of a solid. A gas thus automatically fills any space into which it is led. The motion of the gas molecules is completely random; and one may visualize a particular particle in a gas cylinder striking a wall, rebounding from it like an elastic ball, having momentary collisions with other molecules, and darting about irregularly in the space of the cylinder to make further collisions with the walls. The pressure exerted by the gas is due to the bombardment of the cylinder walls by the molecules. CLERK MAXWELL, one of the greatest of mathematical physicists, showed that most of the observed facts about gases, such as Boyle's and Charles' laws, followed from this *kinetic theory of gases*, which treats a gas as a collection of particles each having kinetic energy.

**The liquid state.** The cohesive forces between the molecules of a *liquid* are smaller than those between the molecules of a solid, but greater than those between the molecules of a gas; in fact, the molecules of a liquid move about in a random manner inside the whole volume of the liquid.

The irregular motion of the molecules of a liquid was first confirmed from an observation made by BROWN, an English botanist, towards the end of the last century. Brown was examining some pollen in water through a powerful microscope, and he recorded that some of the small groups of particles were darting about unceasingly inside the liquid in an irregular manner. The *Brownian movement*, as the motion is called, is caused by the actual impact of the molecules of the liquid on the particles. The zig-zag motion of the pollen shows that the motion of the molecules of a liquid is haphazard.

Evaporation occurs when some molecules with sufficient kinetic energy, escape through the surface of the liquid. Wind, high temperature, large surface area, and low relative humidity assists evaporation. **Boiling** should be distinguished from evaporation. Boiling occurs throughout the whole volume of liquid, whereas evaporation is a surface phenomenon. The boiling point is a definite temperature depending on the external pressure (see p. 204), whereas evaporation occurs at all temperatures and is independent of the atmospheric pressure.

In evaporation into a closed space, the escaped molecules remain above the liquid, and exert a pressure on it. If all the liquid evaporates into the space, and the vapour could still take up more molecules, the vapour is said to be "unsaturated". If more liquid is introduced into the space, more molecules escape from the liquid surface; but eventually, when some liquid remains in the closed space, the number of molecules leaving the liquid per second is equal to the number of molecules entering the liquid per second.

This is called **"dynamic" equilibrium**, and the vapour is said to be "saturated" (see p. 235).

Heat increases the energy of the molecules in a liquid, so that more escape from the surface of the liquid, and the concentration of molecules in the vapour increases. The number returning per second to the liquid thus increases, and a new dynamic equilibrium is then reached at the higher temperature. Since the number of molecules per c.c. in the vapour is greater than before, the saturation vapour pressure is also greater.

## VAPOUR PRESSURE

**Saturated and unsaturated vapours.** At ordinary temperatures the *vapour pressure* of water vapour (or any other liquid) can be measured by means of the apparatus shown in Fig. 156. Two long

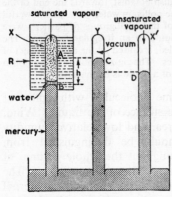

FIG. 156. Saturated and unsaturated vapours

tubes, $X$ and $Y$, closed at one end, are completely filled with mercury and then inverted in a trough of the liquid, a vacuum being formed at the top of each tube (p. 127). By means of a bent pipette a little water is introduced into the bottom of $X$. Since water is less dense than mercury, the water rises to the top of $X$, and evaporates. The pressure of the vapour then forces the mercury from level $C$ to level $D$, as represented in the tube $X'$, and the vapour pressure is equal to the difference in levels of $C$

and $D$. If a little more water is introduced into the space at the top of $X$ it again evaporates, and the mercury level is depressed further. So long as more water evaporates, the space above the mercury is *unsaturated*.

As more water is introduced into $X$, a point is reached when water remains, as at $B$, on the top of the mercury column; the level of the mercury now remains practically constant, however much water is introduced into $X$. The space above the mercury can now take up no more water vapour; it is *saturated* with water vapour.

The saturation vapour pressure (S.V.P.) of water is found to be constant at a given temperature. If, however, the top of $X$ is enclosed in a jacket $R$ containing water which can be warmed (Fig. 156), observation shows that the water at the top of $X$ evaporates as the temperature is increased. If more water is introduced into $X$, a stage is again reached when water is observed at the top of the mercury column, and the space is now saturated at the higher temperature.

The following table shows how the saturation vapour pressure (S.V.P.) of water increases with temperature—

| Temp. in °C. | 0 | 1 | 2 | 3 | 4 | 5 | 6 | 7 | 8 | 9 | 10 | 11 | 12 | 13 | 14 | 15 | 16 |
|---|---|---|---|---|---|---|---|---|---|---|---|---|---|---|---|---|---|
| S.V.P. (mm. mercury) | 4·6 | 4·9 | 5·3 | 5·7 | 6·1 | 6·5 | 7·0 | 7·5 | 8·0 | 8·6 | 9·2 | 9·8 | 10·5 | 11·2 | 11·9 | 12·7 | 13·5 |

The bursting of a boiler is due to the rapid rise of the saturation vapour pressure with temperature. This rise is much more rapid than the rise of pressure with temperature of an unsaturated vapour or gas, which is given by the law on p. 223.

## PRINCIPLES OF HYGROMETRY

On a fine day, with a cool breeze blowing, we are conscious that the air is "dry"; whereas on "muggy" days the air feels "wet" or *humid*. This feeling of comparative wetness or dryness is due to the water vapour in the air. Since, however, the actual mass of water in the air may be more on the "dry" day than on the "muggy" day, the degree of humidity in the air is not measured by the mass of water alone. A *ratio*, known as the **relative humidity**, is chosen as a measure of the degree of "wetness" in the air, and is defined by the relation

$$\text{relative humidity (R.H.)} = \frac{m}{M} \times 100\% \qquad . \qquad . \quad (1)$$

where $m$ is the mass of water vapour actually present in a certain

(a) *Delhi on a July day, 3.15 p.m.* An established cumulus cloud reaches a level where the instability in the atmosphere increases suddenly.

(b) *4 minutes later.* Cloud is growing. Rising warm air has caused a vertical movement inside the cloud.

(c) *10 minutes later.* The cloud billows as it rises, becoming more diffuse.

(d) *15 minutes later.* The rising cloud increases rapidly in size.

volume of the air, and $M$ is the mass of water vapour required to *saturate* the same volume of air at the same temperature.

Suppose that a sealed room had 1·5 gm. of water vapour present inside it. If a hole were made in a wall, and steam (water vapour) passed gently through the opening into the room, the pressure of water vapour would rise at first. Soon, however, it would be noted that the surfaces of polished metal articles in the room suddenly became misty; at this point the water vapour in the room becomes *saturated*. The introduction of more steam causes condensation to take place, but does not lead to any further increase in the amount of water *vapour* in the room.

If the mass of water vapour in the air, $M$, is now 2·5 gm., the original relative humidity, R.H., is given by $\dfrac{1\cdot5}{2\cdot5} \times 100\%$, or 60%, from (1).

**A practical definition of relative humidity.** It is not an easy matter to measure the mass, $m$, of water vapour present in the air and the mass, $M$, required to saturate the same volume at the same temperature. Scientists have therefore developed another method, which assumes Boyle's law applies to vapour.

The mass of water vapour in a given volume is proportional to the density of the water vapour present. But, according to Boyle's law,

$$\text{pressure} \propto 1/\text{volume} \propto \text{density}$$

for a given mass at a given temperature: i.e. the density is proportional to the *pressure* of the water vapour. Consequently, at any given temperature, the mass $m$ of the water vapour present in air is proportional to its vapour pressure $p$; and the mass $M$ of water vapour required to *saturate* the air is proportional to the saturation vapour pressure $P$. Thus,

$$\text{relative humidity} = \frac{m}{M} \times 100\% = \frac{p}{P} \times 100\% \qquad . \quad (2)$$

On a warm day, at say 68° F., the pressure of the water vapour in the air may be 8 mm. mercury, whereas the saturation vapour pressure (S.V.P.) at 68° F. is 17·5 mm. Thus the relative humidity is only 8/17·5 × 100 or 46%, and so sweat evaporates readily and the air feels dry. On a cold day, at say 50° F., the pressure of water vapour may again be 8 mm. mercury but, since the S.V.P. at 50° F. is only 9·2 mm., the relative humidity is as high as 8/9·2 × 100 or 87%. We thus perspire slowly that day and the air feels moist. Our feeling of discomfort increases as the relative humidity of the air approaches 100%.

**The dew point.** If a highly polished metal surface is cooled slowly, the air near it decreases in temperature. If this air contains any water vapour, the air eventually becomes *saturated*, and condensation then takes place on the metal surface, making it misty. The temperature of the air when it becomes saturated with water vapour is known as the *dew point*.

Since the amount of vapour in the air has not altered during the cooling, the pressure of the water vapour in the air at the original temperature is equal to the saturation vapour pressure (S.V.P.) at the dew point. From (2), it follows that

relative humidity

$$= \frac{\text{vapour pressure at original temperature } (p)}{\text{S.V.P. at original temperature } (P)} \times 100\%$$

$$= \frac{\text{S.V.P. at dew point}}{\text{S.V.P. at original temperature}} \times 100\% \qquad . \qquad . \quad (3)$$

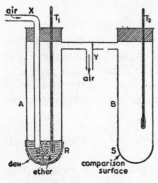

FIG. 158. Dew point hygrometer

This last relation is the basis of accurate methods of measuring relative humidity.

**A dew point hygrometer.** REG-NAULT devised a *hygrometer*, an instrument for measuring relative humidity, whose basic form is still used (Fig. 158). It consists of two test tubes $A$, $B$, connected together, with highly polished metal "thimbles" at the lower ends, $R$, $S$. Liquid ether is contained inside $R$, and by means of an apparatus connected to the tube $Y$ (not shown), air is drawn through the ether *via* $X$ at a steady rate, causing some of the liquid ether to evaporate.

Since heat is required to change a liquid to a vapour (see p. 200), heat is taken from the metal thimble $R$, which thus becomes cooled. As its temperature is lowered, a point is reached when the surface $R$ becomes misty. Comparison with the unaffected surface of $S$ enables this stage to be determined more easily. The temperature of $R$, observed on the thermometer $T_1$, is the *dew point* of the air.

Suppose that it is 4° C. The actual air temperature is observed on the thermometer $T_2$; suppose that it is 11° C. This S.V.P. of water vapour at 4° C. and 11° C. are then obtained from tables, and, as shown above, the relative humidity of the air is given by

$$\frac{\text{S.V.P. at 4° C.}}{\text{S.V.P. at 11° C.}} \times 100\%.$$

The S.V.P. at 4° C. = 6·1 mm., the S.V.P. at 11° C. = 9·8 mm. (p. 235), and hence

$$\text{the relative humidity} = \frac{6·1}{9·8} \times 100\% = 62\%.$$

**Wet-and-dry bulb hygrometer.** The Regnault hygrometer just described requires an experienced observer to give quick and reliable results of relative humidity. The "wet-and-dry bulb" hygrometer, however, can be used by an untrained person. It consists merely of two thermometers $A$, $B$, standing side by side, with a piece of muslin wrapped round the bulb of $B$, and at its lower end dipping into a small jar, $C$, of water. (Fig. 159.)

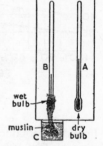

FIG. 159.
Wet-and-dry bulb hygrometer

*Thus the bulb of B is always wet and the bulb of A is dry.*

The water round the bulb of $B$ evaporates into the air, at a rate which depends on the relative humidity of the air, and cools $B$ as it evaporates. To take an extreme case, no water evaporates if the relative humidity is 100 per cent., as the air is then saturated, and the temperature on $B$ is then exactly the same as the temperature on $A$, which records the temperature of the air. If the air were far from being saturated, the water in the muslin surrounding the bulb of *B* would evaporate quickly, and hence the temperature on *B* would be much lower than that on *A*. Thus the *difference in temperature* between *A* and *B* depends on the relative humidity of the air. Tables have been drawn up which relate the difference in temperature, the temperature of *A*, and the corresponding relative

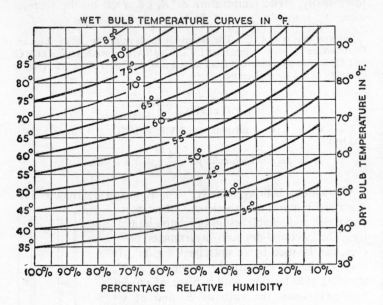

FIG. 160. Relative humidity graph

humidity (obtained accurately with the aid of a Regnault hygrometer). Fig. 160 is a chart from which the relative humidity can be read directly.

**The importance of relative humidity.** One of the factors taken into account in weather-forecasting is the relative humidity of the air, and meteorological stations have special instruments for measuring it. In factories and mills, as a protection for the health of the workers, the relative humidity must not exceed a value prescribed by law. Tobacco in store, and meat in large-scale refrigerating rooms, will not "keep" unless the relative humidity is maintained at a specified level.

**The radio sonde.** Nowadays regular observations are taken of the temperature, pressure and humidity high above the earth. Small balloons, which carry instruments, are released from meteorological stations, and may rise as high as 15 to 20 miles above the earth before they burst (owing to the considerably reduced air pressure at this height). The instruments are brought safely to the ground by an automatic parachute. The whole apparatus is known as a "radio sonde."

The radio sonde contains a small radio transmitter, which automatically sends out signals at short intervals. These signals are derived from (i) a small aneroid box, which enables the pressure (and hence the height) to be determined, (ii) a humidity element, and (iii) a thermometer element.

FIG. 161. AIR CONDITIONING

The installation near the ceiling maintains correct humidity in a tobacco store by means of a fine spray of water.

**Cloud, rain, fog, hail, snow.** Owing to the evaporation which takes place from lakes, rivers, the sea and the earth, the air contains water vapour. If a current of warm air meets cold air, the water vapour in the warm air becomes saturated, and droplets of water are formed. If they are very small, the droplets do not fall to the earth, but form a *cloud*. If the drops coalesce and become heavier, they fall to the ground as *rain*. Moist air above lakes at the bottom of a mountain is forced up into colder regions, so that clouds and rain are obtained; thus "mountains make their own weather." Cloud and rain are also obtained when warm air from the equator meets cold air from the north pole.

*Mist* and *fog* are obtained when cooling of the air takes place after its relative humidity is nearly 100 per cent. Since the air is nearly saturated, droplets of water are formed, usually on dust and dirt particles present in the air. Fog occurs especially in towns, for this reason.

*Hail* is frozen rain, and is probably obtained when raindrops pass through air below 0° C. *Snow* is formed when water vapour at very high altitude changes directly into the solid state without the formation of liquid.

**Refrigerating and ice machines.** We have already seen that heat, known as latent heat, is required to make a liquid vaporize (p. 200). If the vapour pressure of a liquid is high at room temperature, the liquid rapidly evaporates. Heat is thus abstracted from it and the containing vessel, which therefore become colder. This is the basic principle of all refrigerators and ice-making machines. The *volatile liquids* (ammonia, carbon dioxide and sulphur dioxide) all of which have a high vapour-pressure at ordinary temperatures are utilized.

Fig. 162 illustrates the essential features of an *ammonia ice-making machine*. It contains a closed system of two sets of

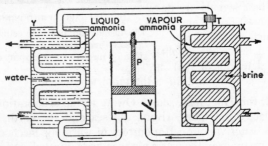

FIG. 162. Refrigerator principle

pipes, connected to a valve *T* which allows a pressure difference to be set up between them. A pump *P* is placed as shown. During the upward stroke, the pump draws ammonia vapour through the valve *V*, and at the same time liquid ammonia passes through $T$ and evaporates in pipes inside $X$. While evaporation occurs, heat is abstracted from a brine bath $X$ surrounding the pipes, and hence the brine cools. The ammonia vapour is compressed by the pump during the downward stroke, and expelled through a second valve to pipes surrounded by a water bath $Y$, which helps to cool the liquid ammonia. As the pump operates and the vaporization and condensation of ammonia take place, the brine bath $X$ becomes colder and colder, and may be cooled far below zero without freezing because it has a lower melting point than pure water.

Ice is made by placing cans of water inside the brine. In cargo ships, refrigerator-holds have pipes along the walls or ceilings, and liquid carbon dioxide is used as the volatile liquid inside the pipes instead of ammonia, as the latter is dangerous if the pipes burst. An electric refrigerator is simply one whose pump, such as $P$ in Fig. 162, is operated by an electric motor.

## Summary

1. A **solid** contains molecules which vibrate about some mean position; a **liquid** contains molecules with varying kinetic energy which move freely and haphazard through the liquid; a **gas** contains molecules with varying kinetic energy whose distance apart is relatively far greater than that of the molecules of a liquid.

2. A saturated vapour is one in equilibrium with its liquid in a closed space.

3. The saturation vapour pressure (S.V.P.) of water increases with temperature (this pressure can be measured by using two barometer tubes, one of which has water at the top of the mercury).

4. **The relative humidity, R.H., of the air** $= (m/M) \times 100\%$, **where** $m$ **is the mass of water-vapour per litre in the air and** $M$ **is the mass required to saturate the volume.**

Also, $\text{R.H.} = \dfrac{\text{S.V.P. at dew point}}{\text{S.V.P. at air temperature}} \times 100\%.$

5. Relative humidity can be measured by means of Regnault's hygrometer or by the wet-and-dry bulb hygrometer.

## EXERCISES ON CHAPTER XV

**1.** Name two everyday observations which show that water-vapour exists in the atmosphere. Do clouds consist of water-vapour ?

**2.** A little liquid ether is poured on a metal plate. What happens to the ether and to the plate ? Explain your answers.

**3.** Pools of water disappear from the road, even in winter time. Explain how this occurs.

**4.** What are the main differences between the *evaporation* and the *boiling* of a liquid ?

**5.** How does the boiling point of water vary with the atmospheric pressure ? Explain how (i) the height of a mountain can be roughly measured by using this phenomenon, (ii) the cooking of vegetables is influenced by living on a mountain-side.

**6.** Draw a diagram of a simple refrigerator, and explain fully how it functions.

**7.** What is the *dew point* ? Describe how it can be found experimentally.

**8.** What is the difference between a saturated and an unsaturated vapour ? Describe how the saturation vapour-pressure of water at 20° C. can be found.

**9.** Give *two* definitions of relative humidity. The dew point of air is found to be 8° C., and the temperature of the air is 14° C. Calculate the relative humidity. Is the relative humidity higher or lower if the dew point is 10° C. instead of 8° C. ? (Saturation vapour-pressure of water at 8° C., 14° C. = 8·0 mm., 12·0 mm. mercury respectively.)

**10.** Describe an experiment by which the saturation pressure of water vapour could be measured at different temperatures and draw a curve to illustrate, in general, the results which would be obtained. (*L.*)

**11.** Describe simple experiments, one in each instance, (*a*) to show the difference between a saturated and an unsaturated vapour, (*b*) to demonstrate the presence of water vapour in the air. A bowl of water is exposed to the atmosphere. What are the conditions which influence the rate of evaporation of the water ? Give reasons. (*L.*)

**12.** What is meant by (*a*) the latent heat of evaporation of a liquid, (*b*) the saturated vapour pressure of a liquid, (*c*) the boiling point of a liquid ? At one time the hypsometer was used by Alpine climbers for estimating height of mountains. Explain briefly the principles upon which this method of estimating altitudes is based. (N.)

**13.** Describe and explain experiments to prove that increase of pressure (a) lowers the melting point of ice, (b) raises the boiling point of water. Describe *two* examples of, or applications of, *one* of these phenomena. (O. & C.)

**14.** State *two* ways in which *saturated* and *unsaturated* vapours differ in their behaviour when subjected to a small decrease in volume, the temperature remaining constant. (N.)

**15.** What is meant by dew point? Describe an experiment to determine the dew point of the air in a room. (L.)

**16.** Define *dew point* and *relative humidity*. Explain how the values of these quantities affect the drying power of the air. Describe a method of measuring the relative humidity of the air. (C.)

**17.** Describe how the following can be demonstrated: (a) a vapour exerts a pressure, (b) this pressure is a maximum when the vapour is saturated, (c) the maximum vapour pressure is independent of the volume occupied by the vapour. The maximum vapour pressure at 40° C. of alcohol is 133 mm. of mercury whilst that of ether is 921 mm. State which has the higher boiling point and give reasons for your answer. (N.)

**18.** Describe the construction and explain the action of *two* of the following: (a) a combined maximum and minimum thermometer, (b) a constant volume air thermometer, (c) a dew point hygrometer. (L.)

**19.** What is meant by the *saturation vapour pressure* of a liquid? How would you measure the saturated vapour pressure of alcohol at room temperature? The vapour pressure of water at 99° C. is 733 mm. of mercury and at 100° C. it is 760 mm. of mercury. Assuming that the graph of vapour pressure against temperature is a straight line between these two temperatures, find the temperature at which the vapour pressure is 750 mm. of mercury. What is the error in the upper fixed point of a mercury thermometer which records 99·2° C. when suspended in steam from water boiling in a hypsometer under a pressure of 750 mm. of mercury? (C.)

**20.** Describe simple experiments, *one* in each case, which you would perform in order to carry out the following instructions. (a) Determine the saturation vapour pressure of alcohol at room temperature; (b) Show that, at constant temperature, the pressure exerted by a saturated

**9\***

vapour is independent of the volume which it occupies; (c) Show that the temperature at which water boils depends on the pressure acting on it. (N.)

**21.** Define the terms *dew point, relative humidity.* Explain the presence of water vapour in the atmosphere and account for the formation of **each** of the following: *dew, fog, hoar-frost, mist.* (O. & C.)

**22.** Explain the meaning of the terms *saturation vapour pressure, dew point, relative humidity.* Describe how you would measure the saturation vapour pressure of alcohol at temperatures between 20° C. and 70° C. (Boiling point of alcohol = 78° C.) (O. & C.)

**23.** Define dew point and relative humidity. Describe an experiment to determine the dew point. How can the relative humidity be determined from a knowledge of it? (L.)

# CHAPTER XVI

## CONDUCTION, CONVECTION, RADIATION OF HEAT

WHEN the water in a kettle or a boiler is heated, the heat from the gas passes through the metal of the container to the water. This is an example of the **conduction** of heat through a metal. Similarly, the end of a spoon dipping into hot tea, and the handle of a poker in a fire, both become hot owing to the heat conducted along the metal from the hot to the cold end.

**How conduction of heat takes place.** The conduction of heat can be explained from our knowledge that heat is a form of energy (p. 210). Consider one end $A$ of a metal bar $AB$ placed in a fire (Fig. 163). We can imagine the whole length of $AB$ to be divided up into many sections, such as $a$, $b$, which are perpendicular to $AB$. As a result of the heat $A$ receives, the molecules at the end section *vibrate faster* than before the heat was applied.

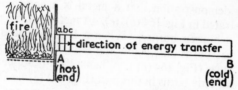

Fig. 163. Heat transfer by conduction

In other words, $A$ gains some energy. The molecules in $A$ then jostle the molecules in the next section $b$, and cause an increase in the energy of these molecules. These, in turn, cause the energy of the molecules in the next section $c$ to increase. Thus the sections pass *energy* along the metal in the form of heat, without leaving their average position.

We may picture this by imagining some boys lined up in a row.

A boy at one end gives a hard push, which is passed on until it reaches the boy at the other end. Energy has been transferred along the line from boy to boy, although the average position of each boy remains unaltered.

**Good conductors of heat.** All metals are good conductors. This can be demonstrated by placing a sheet of paper *P* round a copper rod, for example (Fig. 164 (*a*) ), and holding the paper and rod near a bunsen burner flame. The paper is observed to be unaffected by the heat. The explanation is simply that the heat is conducted away quickly from the paper by the copper, a good conductor, and the temperature of the paper thus remains too low for it to burn. If the experiment is repeated with paper wound round wood, the paper chars, showing that the wood, a poor conductor, conducts the heat away slowly from the paper.

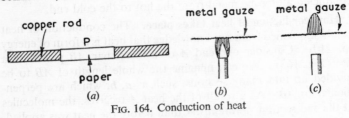

FIG. 164. Conduction of heat

Another demonstration that a metal is a good conductor of heat is illustrated in Fig. 164 (*b*), (*c*). A metal gauze is placed over the gas of a bunsen burner. When a light is placed below the gauze, the gas burns only below the latter; no flame is obtained *above* the gauze (Fig. 164 (*b*) ). When a light is applied *above* the gauze, the gas only burns in this part (Fig. 164 (*c*) ). These results are explained by the fact that the metal gauze is a good conductor, and conducts the heat away rapidly from the gas on the unlighted side of the gauze. Since the temperature on the latter side is too low, the gas does not burn.

Huts built for arctic and antarctic expeditions have double walls and roof as a protection from the cold; but no bolts pass right through the walls, because metal is a good conductor of heat, and icicles would form on the bolts inside the hut. In the area of the Red Sea a person touching the hot metal-work of a ship would get a blister, as heat would pass from the metal to the flesh. If the metal-work of a ship in the antarctic is touched a sensation of extreme cold is experienced, as heat passes from the flesh to the metal, and this can be equally dangerous.

**The miner's safety lamp.** At the beginning of the 19th century SIR HUMPHREY DAVY an eminent scientist of the day, was urged to invent a device to protect miners from gas explosions. The result was the miner's safety lamp (Fig. 165), in which a gauze metal cylinder surrounds a burning wick "fed" by oil from the base of the lamp. There are holes at the top of the lamp to let the air in and the combustion products out. As with the metal gauze (Fig. 164 (b) ), the heat of the flame is conducted by the metal cage away from the gas outside, and the latter is thus kept below its temperature of ignition. When there is inflammable gas in the mine, it enters through the gauze and burns with a blue flame outside the white flame of the oil, and gives warning of danger.

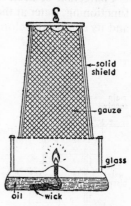

FIG. 165. Miner's lamp

**Bad conductors of heat.** Air, glass, paper, wood and water are examples of bad conductors of heat; i.e. heat will not pass easily along them. The fact that water is a bad conductor of heat is demonstrated by placing a lump of ice weighted with copper wire at the bottom of a test-tube filled with water, and heating the *top* of the test-tube. Here the water can be made to boil, but the ice at the bottom remains unaffected, showing that little heat passes from the top to the bottom of the water. Everyday examples show that air, too, is a bad conductor of heat. In winter time a robin appears bulkier than in summer. It raises its feathers in the winter, and creates pockets of air which protect it from the cold. The air pockets in wool provide protection against the cold. A rug is a worse conductor of heat than linoleum; it is more comfortable to jump out of bed in the morning on to a rug, since the rug conducts heat away from the feet more slowly than linoleum.

The *hay box* is a simple, but efficient, method of keeping food containers warm in a nest of hay. The air in the hay acts as an insulator, and prevents loss of heat from the box. Experiment shows, in fact, that all gases are bad conductors of heat.

**Convection of heat.** When a *hot water circulation system* is functioning, water at the very bottom near the fire becomes warm and its density therefore decreases (see p. 185). The warm water then rises, and carries heat *with it* up the pipe. The colder water, which is denser, falls to the bottom, where it in turn becomes hot and rises. In time, the water in the hot water tank becomes nearly as hot as that in the boiler.

Fig. 166 illustrates a simple hot water circulation system, the arrows denoting the movement of the water.

The transfer of heat from one part of a liquid to another by the movement of the liquid itself, is called *convection*. Heat passes far more easily through liquids by convection than by conduction; in the case of gases, the difference is even greater. The hot air above an open fire moves upwards through the chimney, its place being taken by other

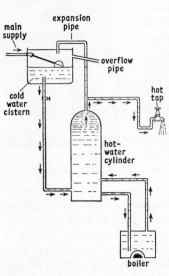

FIG. 166. Convection boiler system

colder air; a convection current of air, which assists materially in keeping the room fresh, is thus caused. On the same principle, years ago mines had several shafts open to the air, with a protected fire at the bottom of one shaft; convection currents circulated down one shaft through the mine and up through another shaft, keeping the air fresh. The so-called "radiators" of a central heating system heat the rooms more by convection than by radiation (see below). The air near a radiator becomes warm and rises, its place being taken by cooler air, and in this way the room becomes warmed.

**Radiation of heat.** Besides the processes of conduction and convection, heat can be transferred by *radiation*. The heat of the sun reaches us across space in which very little gas is present, and the medium through which it passes plays no part in the transfer of the heat. Most of the heat from a fire reaches us by radiation; very little heat is conducted through the air, which is a bad conductor (p. 249). Radiation passes most easily through a vacuum.

The radiating powers of different surfaces at the same temperature can be investigated by means of a metal vessel known as LESLIE's cube. This cube has one vertical side *A* blackened, another side *B* with highly-polished silvering, and the remaining two vertical sides painted with different colours (Fig. 167). Hot water is poured into the cube, and a **thermopile** *T* (see p. 166), is

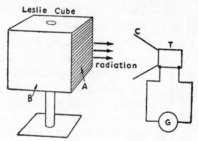

FIG. 167. Radiation of heat

placed at some distance from the blackened side *A*. The thermopile has a cone *C* to collect the radiation from a required direction, which is measured by the deflection on a galvanometer, *G* (see p. 588). When *T* is placed at the same distance from each side of the cube in turn, the heat radiated from the black surface is observed to be greater than the heat radiated from any other surface, and the least of heat is radiated from the highly polished silvered surface. In general, the amount of radiation emitted by an object also depends greatly on its temperature, as can be shown by pouring hotter water into the cube. Experiment also shows that the heat *absorbed* by a surface is large when it is blackened, and small when it is highly polished silver.

**Applications.** A radiator in a building is painted black, so as to radiate as much heat as possible from its surface. White clothes are worn in hot climates to reduce the heat absorbed. Factory roofs are sometimes coated with an aluminium (light) paint, which reduces absorption of heat during the day and reduces radiation during the night, so that a fairly steady temperature is obtained in the factory. The cooling of the earth at night is due to radiation; on a cloudy night much heat is reflected back, so that it is not so cold as on a clear night.

Glass appears to be transparent to radiation from bodies, such as the sun, which are hotter than about 500° C., and opaque to radiation from cooler bodies. A greenhouse thus acts as a "trap" for the heat from the sun. Good absorption of radiation by dark bodies causes ice or snow on a mountain to melt where a stone "hides" it from the sun, while it remains frozen in full sunlight.

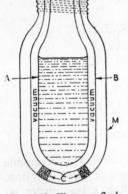

FIG. 168. Thermos flask

**The thermos flask.** The experimental facts on radiation just outlined were put to a useful purpose by SIR JAMES DEWAR. In 1892 he was engaged in researches on liquid air, and required a vessel to maintain the air at the low temperature necessary to keep it liquefied. The type of vessel designed by Dewar for this purpose can also be used as a *thermos flask*, to keep a liquid hot. This is yet another instance of an invention in the course of pure research being applied for the benefit of the community.

The thermos flask consists of a double-walled glass container, with a vacuum between the walls, *A*, *B* (Fig. 168). The sides of *A*, *B* in the vacuum are silvered, and cork supports, *C*, are placed between the glass and the outside, *M*, of the flask. If a hot liquid is placed inside the flask it cannot lose heat by conduction or convection, as the space between the walls $A$, $B$ is a vacuum. The small amount of heat lost from the liquid by radiation is diminished by the silvering on the wall $A$, and any radiation striking the wall $B$ is reflected, as the latter is also silvered.

## Summary

1. In conduction, heat is transferred without any change in the average position of the particles of the substance concerned. In convection, heat is carried by the particles concerned to other places. In radiation, the particles appear to play no part in transfer of heat.

2. A miner's lamp is based on the principle that a metal gauze is a good *conductor* of heat. A hot water circulation system transfers heat by *convection*. A thermos flask has a double-walled container, with a vacuum between the walls, and any *radiated* heat is reflected back by the silvering on the outer wall.

## EXERCISES ON CHAPTER XVI

**1.** Why does an iron bar feel colder than a piece of wood on a cold day ?

**2.** Is water a good or a bad conductor of heat ? Describe an experiment to verify your answer.

**3.** A metal gauze is held over a gas jet which is turned on. Describe and explain what happens when a lighted match is applied to the gas (i) above the gauze, (ii) below the gauze. How is the principle of this experiment utilized in the construction of the miner's lamp ?

**4.** Is air a bad or a good conductor of heat ? Explain why a rug is warm to bare feet. Describe an experiment to show that copper is a good conductor of heat and wood is a bad conductor.

**5.** What is the difference between *conduction* and *convection* of heat ? Describe a simple convection system of hot water supply in a house.

**6.** (i) A man warms his hands in front of the fire; name the processes by which heat reaches him. (ii) Explain how a green-house becomes hot in the summer.

**7.** What is the difference between *convection* and *radiation* of heat ? Why are the roofs of metal huts often painted white and radiators in buildings painted black ?

**8.** Draw a sketch of a thermos flask, and explain why the liquid is kept hot.

**9.** Describe an experiment to show that a blackened surface is a good radiator of heat and a highly polished silvered surface is a bad radiator.

**10.** Give a diagram and explain the action of a form of hot water circulator. (*L.*)

**11.** Distinguish between the transference of heat by *conduction* and *convection*. How would you show that water is a bad conductor of heat ? (*L.*)

**12.** Name the three ways in which heat can be transferred from place to place, and give an example of *one* of them, stating to which method your example refers. (*N.*)

**13.** Explain what is meant by *convection of heat*, and describe a common application of convection. (*C.W.B.*)

**14.** It is stated that an oil stove heats a room mainly by convection and an electric fire mainly by radiation. What experiments or observations would you make to test this statement ? (*C.*)

**15.** How would you show that a black surface is a better absorber of heat radiation than a polished surface ? Give a diagram of the necessary apparatus. (*L.*)

**16.** Explain (*a*) how some of the heat is wasted when a room is heated by means of a coal fire burning in an open grate, (*b*) how an occupant of the room sitting well away from the fire is warmed by it, (*c*) why an eiderdown is very effective in keeping a person warm in bed, (*d*) why it feels warmer to the feet to step out of bed on to a rug rather than on to linoleum. (*N.*)

**17.** Explain the function of the silvering on the hotter wall of a thermos flask containing a hot liquid. Explain briefly why the flask would become ineffective if the sealing point were broken. (*N.*)

**18.** A room is heated by hot water radiators. Describe the process by which heat is conveyed from the water in the radiator to a person on the other side of the room. Apart from its cheapness, give a reason why water is a very good liquid to choose for use in a heating system. Why should water in a heating system be changed as infrequently as possible ? (*C.*)

**19.** (*a*) Explain why heat losses are reduced but not completely eliminated when a warm object is (i) wrapped in cotton wool, (ii) separated from the surrounding air by an evacuated space. (*b*) Explain and compare the various means by which heat may be transferred from the white-hot filament of a gas-filled electric lamp to a nearby black surface vertically above it. (*N.*)

**20.** What are the methods by which heat is transferred from one place to another ? Describe a laboratory experiment to illustrate each of these modes of transference. Describe how, in the hot-water method of heating a house, heat is transferred from the boiler to a person in a room on an upper floor. (*O. & C.*)

**21.** Describe the various ways in which heat is lost by a red-hot metal sphere suspended from the ceiling of a room by a metal wire. What would be the effect of using a suspension of quartz instead of a metal ? Explain the cause of convection in a gas or liquid and describe an experiment which shows the process in operation. (*O. & C.*)

**22.** Give a short account of the methods by which a vessel containing a hot liquid loses its heat.

Draw a diagram of a vacuum flask and explain how the rate of loss of heat by a hot liquid placed in a flask is reduced to a minimum. (*L.*)

**23.** How could it be shown that water is a bad conductor of heat ?

Show by means of a diagram the essentials of a domestic hot water system and explain how it works. (*L.*)

# ANSWERS TO NUMERICAL EXERCISES

## CHAPTER X (p. 169)

**2.** (i) 145·4, 248, − 0·4° F. (ii) 100,
    61·1, − 12·2, − 37·8° C.
**7.** 80° F.

**6.** 172·4° F.
**12.** 10·6 mm.
**15.** (i) − 40° C. (ii) 90

## CHAPTER XI (p. 189)

**1.** 0·000027 per °C.
**2.** 0·00001 per °F.
**3.** 100·045 cm.
**6.** 0·000024, 0·000036 per °C.
    (i) 150·02, 149·96 sq. ft.
**8.** 247° C.
**9.** 0·0000105 per °F.
**10.** (a) 4·8 sq. ft. (b) 216 cu. ft.
**11.** 1·0008 : 1.

**12.** 53·4° C.
**13.** (a) 2·31 sq. cm. (b) 40·4 c.c.
**14.** (a) 0·042 cm. (b) 0·0396 cu. mm.
**15.** 80·48 c.c.
**20.** 15·97 c.c.
**24.** 0·000023 per °C.
**25.** 16·5 cm. from 0° C.
**27.** (a) 0·0048 c.c. (b) 19·2 cm.
**28.** 9·75 c.c., 172° C.

## CHAPTER XII (p. 207)

**1.** s = 1.
**3.** 5280 cal., 19·8° C.
**4.** 3440 B.Th.U.
**5.** 15 gm., 4500 cal.
**6.** 32·9° C.
**8.** 152° C.
**9.** 80 cal. per gm.
**10.** 80 gm.

**11.** (a) 1 cal. per sec. (b) ⅙.
**12.** 8·6 min.
**13.** 0·21, 6·04 gm.
**15.** 30 gm.
**16.** 10 min.
**17.** (a) 7½ min. (b) 0·074 lb. per min.
**18.** 43·75 min.

## CHAPTER XIII (p. 214)

**2.** (i) 4·2 joules per cal.
    (ii) 780 ft. lb. wt. per B.Th.U.
**3.** 195,000 ft. lb. wt.
**4.** 10·26 B.Th.U.
**5.** 0·12° C.
**7.** 880 cal.

**9.** 3·06° C.
**10.** 140 min.
**11.** 4·42 × 10⁷ ergs per cal.
**14.** 1 hr. 33·8 min.
**15.** 10·6 gm.
**16.** 0·22° C.
**17.** 9⅙%.

CHAPTER XIV (p. 229)

2. $\frac{1}{273}$ per °C.
4. 353° C., 71·6 c.c.
5. 81·96 cm.
6. 25, 87·5, 162·5° C.
7. 327·5, 342·5 c.c.
8. 1·18 atmos.
9. 26⅔ c.c.
10. 1.5 lb. wt. per sq. in.
12. 0·0012 gm. per c.c.
13. 1·5 atmos.

14. 6 atmos.
16. 24 per min.
17. 1076 mm.; 0·00352 per °C.
    − 284° C.
18. (a) 20 ft. lb. wt.
    (b) 0·0256 B.Th.U.
19. 0·0037 per °C.
20. 273, 546° K.
21. 1092° C., 15 litres, − 91° C.

CHAPTER XV (p. 244)

9. 67% Higher.
17. Alcohol.

19. 99·63° C., 0·43° C.

# LIGHT ENERGY

LIGHT THICKENS

# CHAPTER XVII

## LIGHT BEAMS AND RAYS

WHEN we look at an object, some energy passes from it to our eyes and stimulates our sense of vision. The energy which stimulates vision is called **light energy**, and travels at the enormous speed of 186,000 miles per second in air. Some substances, such as glass, allow a high proportion of light energy to pass through them, and they are known as **transparent** substances. Substances which completely prevent the light energy from passing through are known as **opaque** substances.

**Light travels in straight lines.** Suppose that two screens, $P$, $Q$, with pin holes $A$, $B$ in them, are arranged as shown in Fig. 169, with a source of light $S$ directly behind $A$ so that light passes from $A$ to $B$. Experiment shows that an observer $E$ can see the light through a pin-hole $C$ in a third screen, $R$, only if $A$, $B$, $C$ are *placed in the same straight line.* We thus conclude that the light travels in a straight line.

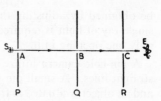

FIG. 169. Light experiment

More evidence that light travels in straight lines is obtained when sunlight is observed streaming through an opening into a darkened room. The dust particles are then illuminated by the light, and a straight edge is observed at the light boundary. Shadows, too, have sharp edges because light travels in straight lines (see p. 263).

**The ray box.** The path taken by light, represented by the line $ABC$ in Fig. 169 is known as a **ray**. A **beam** of light is the name given to a collection of rays of light, such as those given out by a lamp, a torch, or a searchlight. A *parallel* beam is shown in

Fig. 170 (*a*), a *divergent* (spreading) beam in Fig. 170 (*b*), and a *convergent* (concentrating) beam in Fig. 170 (*c*).

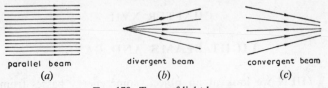

parallel beam          divergent beam          convergent beam
(*a*)                    (*b*)                    (*c*)

FIG. 170. Types of light beams

A *ray box* is an apparatus which provides separated rays (or narrow beams) of light (Fig. 171). It consists of a wooden box *B*

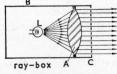

FIG. 171. Ray box

containing a small lamp *L* in front of a lens *A*. Beyond *A*, and covering the width of the box, is a strip *C* of metal with a number of narrow openings in it like a comb, so that separated rays of light pass through *C*. A divergent, convergent, or parallel beam of rays can be obtained by adjusting the position of *L* from the lens. If a single ray of light is required, *C* is replaced by a metal strip with only one opening in it.

**A pin-hole camera** utilizes the fact that rays of light travel in straight lines. A small pin-hole *H* is made in one side of a box, and an object is situated in front of *H*. A ray from the top point *A* of the object then passes straight through *H* and meets the back of

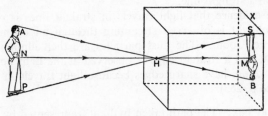

FIG. 172. Pin-hole camera

the camera at *B*. Fig. 172 illustrates other rays from the object passing straight through *H*, and building up an image of the object at $X$ on the back of the camera. The image is a clear one if each point on the object gives rise to *one* point on $X$, but this can only happen if the pin-hole is small. A wider hole allows rays from a point such as $A$ on the object to spread over a small *area* round $B$, thus blurring the image. The pin-hole camera is sometimes preferred to the photographic camera by surveyors, as the lens in the latter may produce distortion when a building, for example, is photographed.

*The size of the image.* In Fig. 172, suppose $HN$ is the perpendicular from $H$ to the object $AP$, and $HM$ is the perpendicular from $H$ to the image $BS$. Then, since triangles $BMH$, $ANH$ are similar, $BH/AH = MH/NH$; now triangles $BSH$, $APH$ are similar, and hence $BH/AH = BS/AP$.

$$\therefore \quad BS/AP = MH/NH$$

$$\therefore \quad \frac{\text{length of image}}{\text{length of object}} = \frac{\text{image distance from } H}{\text{object distance from } H}.$$

Suppose that the length of the camera from hole to plate is 3 in. and a 6 ft. man stands 8 ft. from the hole. The length, $x$ in., of the picture is then given, from above, by

$$\frac{x \text{ in.}}{6 \text{ ft.}} = \frac{3 \text{ in.}}{8 \text{ ft.}}$$

$$\therefore \quad x = \frac{6 \times 3}{8} = 2\tfrac{1}{4} \text{ in.}$$

**Eclipses.** Consider two solid opaque objects, $M$, $E$, arranged as shown in Fig. 173 and illuminated by a large source of light $S$. $E$ is illuminated by the rays of light from $S$ which reach it, but

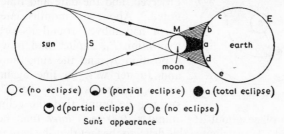

Fig. 173. Eclipses (*not to scale*)

owing to the presence of $M$ between $S$ and $E$, shadows are obtained on parts of $E$. By drawing the extreme rays from $S$ it can be seen that the region $a$ on $E$ receives no light, and this region,

known as the *umbra*, is therefore in total darkness. The region on
E between a, b, and a, d receives light from only part of S, and is
hence in partial shadow; such a region is known as a *penumbra*.
Beyond b to c, and beyond d to e, the region is illuminated by rays
from the whole of S, and is therefore bright.

Fig. 173 illustrates the *eclipse of the sun*. At some stage in
their movement round the sun, S, the moon, M, comes between it
and the earth, E. A *total eclipse* is then observed in the region a on
the earth, and a *partial eclipse* is observed at the regions b and d.
Observers at c and e on the earth receive rays from the whole of
the sun, so that no eclipse is seen. The sun's appearance at
different points on the earth is illustrated by the small circles
below the main figure.

The moon is not a luminous object; it merely reflects light from
the sun. An eclipse of the moon happens when the earth comes
between it and the sun, since the moon is then in the shadow of the
earth and is no longer illuminated by the sun.

**The velocity of light.** In 1673, a Danish astronomer called
RÖMER first calculated the velocity of light in air. Römer was
engaged in making regular observations on the eclipses of one of
Jupiter's satellites. At a certain time in the year, when the sun, S,

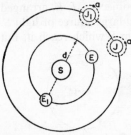

FIG. 174. Römer's method

the earth, E, and Jupiter, J, were in
line with each other (Fig. 174), the
eclipse of Jupiter's satellite, a, was
observed, and the date and time were
noted. About 6½ months later the
earth had moved round its orbit to
a position $E_1$, Jupiter had moved to
a position $J_1$, and the sun, the earth,
and Jupiter were in line again, but
on opposite sides of S. To Römer's
surprise he found that the eclipse of
a took place about 16½ minutes (990 sec.) *later* than he had
expected. He attributed this delay to the fact that the light from a
travels a greater distance to the earth when the earth, sun, and
Jupiter are in the positions $E_1$, S, $J_1$, than when in the positions S, E, J (Fig. 174). This *extra* distance is the diameter
of the earth's orbit round the sun, about 186,000,000 miles.

Consequently, Römer argued that the velocity of light is equal to $\dfrac{186,000,000 \text{ miles}}{990 \text{ seconds}}$, or about **186,000 miles per second.**

**Fizeau's experiment.** This speed is so enormous, that it was not until 1849 that a method was devised of measuring it by using light travelling from one place to another on the earth. Fig. 176 illustrates the essential features of the "toothed wheel" method then used by FIZEAU.

Light from a bright source $S$ is reflected from a plane glass surface $A_1$, and comes to a focus at the edge of a toothed wheel $W$. From here the light passes through a lens $M$ and travels in the form of a parallel beam for a distance of about 5 miles before it is brought to a focus by a lens $N$. A mirror $B$, placed at this focus, reflects the light back along its original path. On arriving at the glass plate $A_1$, some of the light passes directly through, and an image of the source $S$ is thus observed at $E$.

FIG. 175. TOOLMAKER'S MICROSCOPE
The shadow of a magnified screw thread is compared on the screen with the drawing of an accurate thread.

The wheel $W$ is rotated about the axle $O$, its speed being increased slowly from zero. The teeth of the wheel interrupt the light from $S$, but, when the succession of images reaches more than about 10 per second, the observer sees a steady image of S. As the speed of $W$ is increased further, a point is reached at which the image of $S$ disappears. This is because the light now travels through a gap in the rim $R$ to $B$ and back, and then meets $R$ on a tooth which has moved into the place where the gap had been; the light is thus unable to reach $E$.

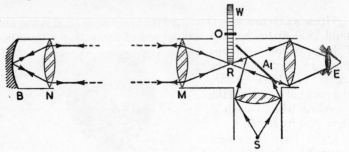

FIG. 176. Principle of Fizeau's experiment

In Fizeau's experiment, the number of teeth on the wheel was 720 and the image of $S$ first disappeared when the wheel rotated 12·6 times per second. The time taken for a tooth to move into the place of the next gap was thus $\dfrac{1}{2 \times 720 \times 12 \cdot 6}$ seconds, as there are $2 \times 720$ teeth-plus-gaps round the wheel. The distance between the wheel and the reflecting mirror was 8633 metres, and as the light travelled to-and-fro, the velocity

$$= 2 \times 8633 \div \frac{1}{2 \times 720 \times 12 \cdot 6} \text{ metres per sec.}$$
$$= 2 \times 8633 \times 2 \times 720 \times 12 \cdot 6 \text{ metres per sec.}$$
$$= 3 \cdot 13 \times 10^{8} \text{ metres per sec.}$$

---
### Summary

1. Light travels in straight lines.

2. Eclipses of the sun are due to the shadow formed when the moon passes between the sun and earth. An **umbra** is a region of total shadow; a **penumbra** is a region of partial shadow.

3. The velocity of light in air is about 186,000 miles per second.

## EXERCISES ON CHAPTER XVII

**1.** Name five appliances which utilize lenses or mirrors.

**2.** Draw sketches of a converging beam, a parallel beam, and a diverging beam of light. Which type of beam is obtained in a searchlight?

**3.** Explain the total eclipse of the sun by the moon, drawing a diagram. Explain also the appearance of the sun at different parts of the earth's surface.

**4.** Why are you unable to see round corners? Describe an experiment to show how light rays travel in air.

**5.** Define the terms *umbra* and *penumbra*, and draw sketches showing how they are formed.

**6.** Describe how a pin-hole camera works.

**7.** The distance between the pin-hole and screen of a pin-hole camera is 5 in., and the plate is 8 in. long. At what minimum distance from the pin-hole must a 6ft. man stand if a full-length photo is required?

**8.** Draw a diagram showing how an eclipse of the sun is caused. (*N.*)

**9.** Describe an experiment to demonstrate the rectilinear propagation of light, and explain the formation of a total eclipse of the sun. (*C.W.B.*)

**10.** What effect, if any, has an increase in the size of the hole of a pin-hole camera on: (*a*) the size of the image; (*b*) the brightness of the image; (*c*) the sharpness of the image? What is the chief disadvantage of the pin-hole camera in use? (*N.*)

# CHAPTER XVIII

## REFLECTION AT PLANE SURFACES

PLANE mirrors, consisting of plane glass usually silvered on the back, reflect a high percentage of the light energy incident on them. They are therefore commonly used as looking-glasses, for signalling at sea by using the sun's rays, as car mirrors, and for making simple periscopes. Plane mirrors are also used in the sextant, an instrument employed at sea for navigational purposes (p. 275). The working of plane mirrors is explained by the regular way in which they reflect light, as we shall now see.

**Laws of reflection.** If a ray of light $ON$, obtained from a ray box, is incident on a plane mirror $M$ at $N$ (Fig. 177), experiment shows

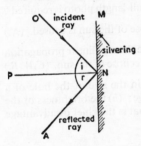

FIG. 177. Law of reflection

that the ray is reflected in a definite direction $NA$. By marking two points on $NA$ and on $ON$, and joining them with a ruler, the reflected and incident directions can be drawn to intersect the mirror at $N$; and, if the line $NP$ through $N$ perpendicular to the mirror surface is drawn, measurement of the angles $ONP$, $ANP$ shows they are equal. When the experiment is repeated with rays incident in different directions on the mirror at $N$, observation shows

that in each case the reflected ray and the incident ray make the same angle with the *normal* at $N$, as $PN$ is called.

The *angle of incidence* ($i$) of a ray of light on a surface is defined as the angle between the incident ray and the normal to the surface at the point of incidence; the *angle of reflection* ($r$) is the angle between the reflected ray and the normal at the point of incidence. In Fig. 177, $i$ = angle $ONP$ and $r$ = angle $ANP$. The **laws of reflection,** obtained from experiment, state that:

(1) **The incident ray, the normal, and the reflected ray all lie in the same plane.**

(2) **The angle of incidence = the angle of reflection.**

**Diffusion of light.** A plane mirror thus reflects a parallel beam of light in a definite direction, and in this case *regular reflection* is said to take place. When a parallel beam of light arrives at a surface *S* which is not highly polished, the individual rays are reflected in different directions (*R*) from the surface (Fig. 178). This is due to the irregular nature of the surface (which can be seen under a high-power microscope) and the incident rays are said to be scattered or *diffusely reflected* from the surface. The great majority of objects in everyday life are seen by light diffusely reflected from them. Paper diffuses light.

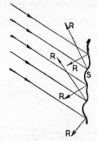

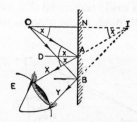

FIG. 178. Diffusion          FIG. 179. Position of image

**The formation of images in a plane mirror.** Consider a small object *O* placed in front of a plane mirror *NB* (Fig. 179). Rays from *O*, such as *OA* and *OB*, are reflected along *AX* and *BY* respectively to make equal angles on opposite sides of the normals at *A* and *B*, and an observer at *E* will receive the reflected rays. Now the eye sees rays of light *in the direction in which they enter it*. Hence an observer sees the rays as if they came from a point *I*, which is the point of intersection of *XA* and *YB* produced backwards. Any other ray from *O*, such as a ray *ON* incident normally on the mirror, is also reflected as if it came from *I*, and hence an observer sees an *image* of *O* at *I*.

If a screen were placed at *I*, no image would appear on it, for the light rays do not actually pass through *I*. An image of this sort, which cannot be received on a screen, is termed a **virtual** image. An image through which light rays actually pass, and which can be received on a screen, is known as a **real** image.

**The position of the image in a plane mirror.** *The image I in a plane mirror is as far behind the mirror as the object O is in front* (Fig. 179). The proof is as follows:

$\angle OAD = \angle XAD = x$, say (law of reflection)

But $\angle OAD =$ alternate $\angle AON$, $\angle XAD =$ corresponding $\angle AIN$

$$\therefore \quad \angle AON = \angle AIN$$

Since $\angle ONA = 90° = \angle INA$, $NA$ is common, and $\angle AON = \angle AIN$, the triangles $ONA$, $INA$ are congruent.

Thus $IN = ON$, and hence

the image distance from the mirror

= the object distance from the mirror.

It can easily be proved that any other reflected ray, such as $OB$, passes through $I$; the proof is left to the reader.

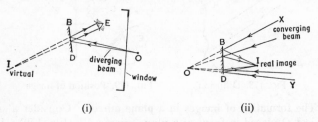

(i)                                    (ii)

FIG. 180. Virtual and real images by plane mirror

**Virtual and real images by plane mirror.** A car-driver, whose eye is represented by $E$ in Fig. 180 (i), sees traffic behind as *virtual images* when he looks into the plane driving mirror $BD$. An object $O$ outside sends a diverging beam of light to the mirror, and this is reflected as a diverging beam which appears to the eye to come from $I$.

A *real image* can also be formed by a plane mirror. This time, however, a *converging* beam is needed, obtained by using the ray-box (p. 262) for example. Fig. 180 (ii) shows what happens when the beam converges to a point $O$ behind the plane mirror $BD$. Incident rays, such as $XB$ and $DY$, are all reflected by the mirror

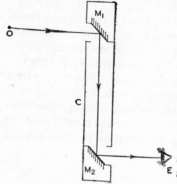

to a point $I$ in front of it, where $I$ and $O$ are equidistant from $BD$. Thus a real image, $I$, is formed.

**The simple periscope** consists of two mirrors $M_1$, $M_2$ inclined at an angle of 45° to the horizontal and placed in a tube $C$ (Fig. 181). Rays from an object at $O$ are reflected by $M_1$ and then by $M_2$ to reach an observer at $E$. Thus objects in front of a crowd can be viewed at $E$ by raising the periscope above the heads of the crowd.

FIG. 181. Simple periscope

**Lateral inversion.** Consider an $E$-shaped object placed in front of a plane mirror $M$ (Fig. 182). The image $a'$ of the point $a$ on the object is the same distance behind the mirror as $a$ is in front of the mirror (see p. 270), and similarly for other points on the object, such as $b$, $c$. The image formed can be obtained by finding the images of all the points which make up the object, and it will be seen that the image obtained is the same size as the object but *laterally inverted*. Thus a right-handed batsman appears to be left-handed if his stance is observed in a mirror, and the imprint of words on a blotting-paper

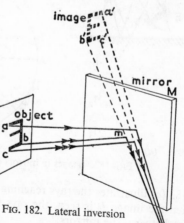

FIG. 182. Lateral inversion

can be seen as they were originally written by holding the blotting-paper up to a mirror (see Fig. 182).

### Images in pairs of mirrors.

*Parallel mirrors.* If an object $O$ is placed between two parallel mirrors $M$, $m$, an observer at $E$ sees a large number of images in the mirror into which he is looking. These images arise in the following way: $O$ forms an image $I_1$ in the mirror $M$, $I_1$ forms an image $I_2$ in $m$, $I_2$ again forms an image $I_3$ in $M$, and so on. Other images $i_1$, $i_2$, $i_3$, etc., are formed, starting with the image $i_1$ of $O$ in the mirror $m$. The number of images seen, although theoretically infinite, is in fact limited and depends on the condition of the mirrors, since some light energy is absorbed in each mirror.

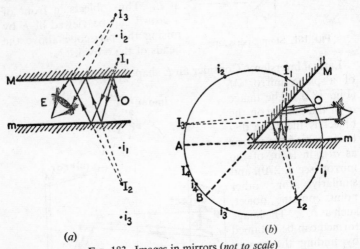

(a)                        (b)

Fig. 183. Images in mirrors (*not to scale*)

Fig. 183 (*a*) illustrates the rays reaching the observer's eye $E$ from $O$ when the image $I_3$ is seen, the rays from $O$ being reflected to and fro between the mirrors.

With *inclined mirrors,* a finite number of images is observed. Suppose $O$ is an object placed between the two mirrors $M$, $m$ (Fig. 183 (*b*)). The image of $O$ in $M$ is $I_1$, the image of $I_1$ in $m$ is $I_2$, and so on. As long as an image is formed *in front* of the silvering of one mirror, another image is formed. The last image, $I_4$, is therefore obtained in the sector $XAB$ opposite to that enclosed by the two mirrors.

Since the image is as far behind a plane mirror as the object is in front, the perpendicular distances of $O$ and $I_1$ from the mirror $M$ are equal. Consequently, by congruent triangles, $XI_1 = XO$, where $X$ is the point of intersection of the sections of the mirrors. Similarly, $XI_1 = XI_2$, considering the image $I_2$ of $I_1$ in the mirror $m$. In this way, it can be seen that the images of $O$ are equidistant from $X$, and hence they all lie on a circle of centre $X$ and radius $XO$. This is a useful method of drawing the images of $O$ quickly. Other images $i_1 i_2$, $i_3$, $i_4$ are formed starting with the image $i_1$ of $O$ in the mirror $m$. Fig. 183 $(b)$ illustrates the rays from $O$ which reach the eye $E$ when the image $I_3$ is observed.

**The kaleidoscope** utilizes the images formed in two inclined mirrors or glass surfaces, usually at an angle of 60° to each other. At one end of the tube containing the mirrors are differently coloured pieces of tinsel, and looking through the other end of the tube the observer sees five images of the tinsel pieces, in addition to the pieces themselves, arranged symmetrically in a circle. When the tube is shaken the tinsel pieces rearrange themselves, and a new symmetrical pattern is seen. The kaleidoscope is not merely a toy, but is used by designers to obtain fresh ideas for colour patterns.

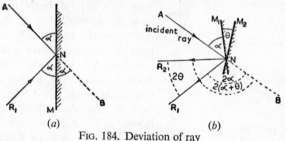

(a)                                             (b)

FIG. 184. Deviation of ray

**The deviation of a ray by a rotated mirror.** Consider a ray of light $AN$ reflected along a direction $NR_1$ by a plane mirror (Fig. 184 $(a)$). The **angle of deviation** of the ray caused by the mirror is angle $BNR_1$, the angle between the initial ($NB$) and the final direction $(NR_1)$ of the ray. Suppose that $a$ is the angle made by $AN$ with the mirror surface ($a$ is known as the **glancing angle** with the surface, and must not be confused with the incident angle, which is $(90° - a)$. Then, since vertically opposite angle $MNB$ is $a$ and angle $R_1NM$ is $a$, from the law of reflection, it follows that angle $BNR_1$ is $2a$. Thus the mirror deviates the ray through *twice* the glancing angle $a$.

Suppose the direction of the incident ray $AN$ remains unaltered and the mirror is rotated from the position $M_1$ to a position $M_2$ through an angle $\theta$ (Fig. 184 (*b*) ). The reflected ray is then deviated by an angle $R_1NR_2$ from its original direction $NR_1$ to a direction $NR_2$. With the mirror in the position $M_2$, the glancing angle made by the ray $AN$ with the mirror is $(a + \theta)$. Hence, from our result above,

the angle of deviation $BNR_2 = 2(a + \theta)$

But the angle $BNR_1 = 2a$, already proved

∴   angle $BNR_2$ − angle $BNR_1 = 2(a + \theta) - 2a$

∴     angle $R_1NR_2 = 2\theta$

Thus *the angle of deviation of the deflected ray* when the mirror is rotated is *twice the angle of rotation of the mirror.*

In many sensitive electrical measuring instruments a small mirror is used to measure the deflection produced by the passage of a current through the instrument (p. 588). A beam of light is incident normally in a direction $AN$ upon the mirror $M_1$, and the light is reflected back along its own path $NA$ (Fig. 185). When the current passes, the mirror rotates through a small angle $\theta$ to a position $M_2$, and the reflected ray $NA$ is rotated through an angle $2\theta$ to a direction $NR$. A scale $S$ placed as shown enables the deflection of the light from $A$ to $R$ to be measured.

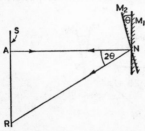

FIG. 185. Rotating mirror

Since $\tan 2\theta = \dfrac{AR}{AN}$, $AR = AN \tan 2\theta$.

Thus $AR$ is related to the angle of rotation ($\theta$) of the mirror for a given distance $AN$, and hence the deflection of the light along the scale $S$ is a measure of the current.

It should be noted that the reflected ray constitutes a pointer of negligible mass, and is therefore ideal in this respect. The arrangement of mirror and reflected ray is known as an *optical lever*, and a magnification of two is obtained by using the rotation of the reflected ray instead of the actual rotation of the mirror.

**The sextant.** The sextant is an instrument for finding the angle of elevation of the sun (or a star), and is used in ships to assist

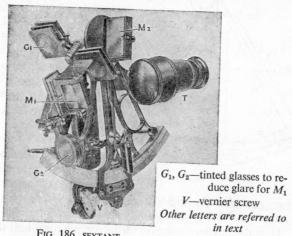

$G_1$, $G_2$—tinted glasses to reduce glare for $M_1$
$V$—vernier screw
*Other letters are referred to
in text*

FIG. 186. SEXTANT

navigation. It consists essentially of two mirrors $M_1$, $M_2$, the former being fixed and the latter movable about a horizontal axis through its middle (Fig. 187). The lower mirror $M$ is made so that one half is clear glass and the other half is silvered. A telescope $T$ sighted on $M_1$ is thus able to observe the horizon through the clear glass, as well as objects sending rays reflected by $M_2$ to the silvered part of $M_1$.

When $M_2$ is parallel to $M_1$, an observer through $T$ sees the horizon through the clear glass in $M_1$ coincident wth the horizon seen by reflection from $M_2$. The latter is illustrated in Fig. 187 by the reflection of $HO$ along $OA$ and $AT$. As $M_2$ is rotated the horizon reflected from it disappears from view in $T$, and the rotation is continued to a position $m_2$ until the sun is seen on

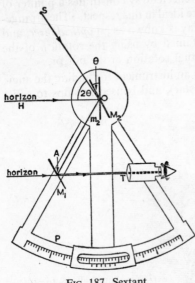

the horizon viewed directly through $M_1$. The sun's rays are then reflected from the top mirror downward along $OA$ to $M_1$ and then to $T$. *The angle of rotation, $\theta$, of the mirror is half the angular elevation of the sun.*

To prove this, imagine the ray *OA reversed.* Then, in the position $M_2$, the top mirror reflects $AO$ along the horizon direction, $OH$. When the mirror is rotated through an angle $\theta$ to $m_2$, the ray $AO$ is reflected along $OS$ in the direction of the sun $S$. Now it was proved on p. 274 that a reflected ray is deviated through twice the angle of rotation of a

FIG. 187. Sextant

mirror, provided the incident ray direction is constant. *Since the latter condition is fulfilled here,* it follows that angle $SOH$ is $2\theta$. Thus the angle of rotation of the top mirror is half the angle of elevation (angle $SOH$) of the sun.

─────Summary─────

1. The laws of reflection state: (*i*) The incident ray, the normal, and the reflected ray all lie in the same plane; (*ii*) The angle of incidence = the angle of reflection.

2. A *real* image is one which can be formed on a screen; a *virtual* image is one which cannot be formed on a screen.

3. The respective distances of the object and image from a plane mirror are equal: the image is virtual and laterally inverted.

## WORKED EXAMPLE

*A plane mirror* 4 in. *long is placed* 18 in. *in front of the driver of a car. Through what width of rear window,* 4 ft. 6 in. *behind him, can he see when looking into the mirror?* (*N.*)

In Fig. 180, p. 270 *BD*, the mirror, is 4 in. long, *PE* is 18 in., *EW* is 4 ft. 6 in. Then since $OP = EP$, $OP = 18$ in. Now triangles, $OAC$, $OBD$ are similar, since $AC$ is parallel to $BD$.

$$\therefore \quad \frac{AC}{BD} = \frac{OA}{OB} = \frac{OW}{OP}.$$

But $\quad OW = OP + PE + EW = 18 + 18 + 54 = 90$ in.,

$OP = 18$ in., and $BD = 4$ in.

$$\therefore \quad \frac{AC}{4} = \frac{90}{18}$$

$$\therefore \quad AC = \frac{90}{18} \times 4 = 20 \text{ in.}$$

Thus the width of the rear window is 20 in.

## EXERCISES ON CHAPTER XVIII

**1.** Name three applications of plane mirrors.

**2.** State the *laws of reflection* of light. A person stands 3 ft. from a tall looking-glass. Draw a sketch showing accurately the image.

**3.** Is the image of an object in a plane mirror virtual or real? Explain your answer with the aid of a diagram which shows the rays forming the image.

**4.** Explain with the aid of a diagram why a right-handed batsman appears left-handed when viewing his stance in a plane mirror.

**5.** Prove that the image in a plane mirror and the object are at equal distances from the mirror.

**6.** A ray of light is incident at an angle of 30° on a plane mirror. If the mirror is turned through 10° while the direction of the incident light is kept constant, calculate the rotation of the reflected ray. What is the rotation if the ray is originally at an angle of 40°, instead of 30°?

**7.** Describe a *sextant*. Explain how it is used to measure the angle of elevation of a star.

**8.** A small object is placed 2 ft. from a plane mirror, and a person 3 ft. from the mirror looks at the image. Draw to scale the beam of rays entering the eye of the observer.

**9.** State the laws of reflection of light. Two plane mirrors are placed

10*

at right angles with their reflecting surfaces inward. Show on a plan drawn to scale the positions of the images of an object situated between the mirrors, 4 in. from one mirror and 2 in. from the other. Also draw a pencil of rays to show how light from the object reaches an eye, suitably placed, after two reflections. (*L.*)

**10.** Explain (*a*) a sheet of ground glass becomes almost transparent when wet. (*b*) The image of a right hand formed by a plane mirror looks like a left hand. (*C.*)

**11.** State the laws of reflection of light at a plane surface, illustrating by a labelled diagram. Describe how you would verify the relation between the angles of incidence and reflection. Explain, with the aid of a ray diagram, how you would use two plane mirrors as a periscope. (*C.W.B.*)

**12.** Two plane mirrors are inclined at an angle to each other. A ray of light parallel to one of the mirrors travels, after two reflections, parallel to the other. Find the angle between the mirrors. (*O. & C.*)

**13.** Two parallel plane mirrors, facing each other, are 11 in. apart. A lighted candle is placed between them and 4 in. from one mirror. Draw a ray diagram showing the path of a diverging beam from the tip of the flame to the eye of a person looking at the second distant image seen in one mirror. On your diagram mark out relevant distances and state the distance of this image from the mirror on which it is seen. (*C.W.B.*)

**14.** Describe an experiment to show that the angle of reflection for a ray of light incident on a plane mirror is equal to the angle of incidence. A ray of light falls normally on a plane mirror. If the mirror is then turned through an angle of 30°, what is the angle through which the reflected ray turns ? (*C.*)

**15.** State the laws of reflection of light and mark the angles of incidence and reflection on a diagram. Prove the relation between the distance of the object from a plane mirror and the distance of the image. Explain how a mirror may be used to decipher marks of writing left on a blotting paper. (*O. & C.*)

**16.** State the laws of reflection of light. Two vertical plane mirrors intersect at right angles and a horizontal ray of light is reflected by the two mirrors in turn. Show that the final direction of the ray is parallel to the initial direction. Describe, with the aid of a diagram, the construction of **either** (*a*) the sextant, **or** (*b*) the periscope, and state the purpose for which it was designed. (*O. & C.*)

# CHAPTER XIX

## REFLECTION AT CURVED SURFACES

DRIVERS of motor vehicles often use a curved mirror (a mirror "bulging outward") to see traffic behind. The dentist employs a curved mirror, of a different type, for examining teeth closely, and sometimes shaving mirrors are of the curved type. Searchlights, such as those which illuminate the front of buildings for display purposes, utilize curved mirrors of a special design; while the largest telescope in the world, at Mount Palomar in California, has a curved mirror, 200 inches in diameter. (See Fig. 242.) Thus, curved mirrors are of great practical importance, and it is of interest to study how they work.

**Concave (converging) mirrors. Definitions.** Fig. 188 (*a*) illustrates a *concave* spherical mirror, silvered on the back at *S*. The mirror surface is a small part of a sphere of centre *C*, and *C* is therefore known as the **centre of curvature** of the concave mirror. The middle point *P* is known as the **pole** of the mirror. The mirror is symmetrical about the line *CP*, which is termed its **principal axis**.

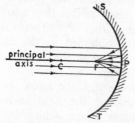

FIG. 188 (*a*). Narrow parallel beam reflected to single focus, *F*

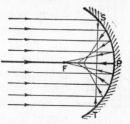

FIG. 188 (*b*). Wide parallel beam not reflected to single focus

The radius of the sphere of which the mirror is part is known as the **radius of curvature**, and is the distance from *C* to *P*. The width *ST* of the mirror is known as its **aperture**.

Experiment with a ray box shows that a parallel beam of light, incident on a mirror of small aperture parallel to the principal axis, is reflected to pass through a point, $F$, on the axis. $F$ is known as the **principal focus** of the mirror. Rays from the sun or stars, which are parallel when they arrive at the earth, can be concentrated at the focus by a concave mirror, and if a paper is placed at the focus, an image of the sun can be received at this point which can burn a hole in the paper on a bright sunny day. Archimedes is reputed to have set fire to enemy warships at anchor by concentrating the light of the sun on them with the aid of huge concave mirrors. All the rays reflected by the mirror in Fig. 188 (a) actually pass through $F$ in front of the mirror, and a concave mirror has therefore a **real** focus (see p. 270). The distance $PF$ from the pole to the focus is known as the **focal length** ($f$) of the mirror.

**Spherical aberration.**  If a wide parallel beam of light is incident on a concave mirror, experiment shows that the reflected rays are brought to a focus at different points (Fig. 188 (b) ). The phenomenon is known as *spherical aberration*, and the image on a screen is blurred. In practice, therefore, concave mirrors are made of small aperture, and we shall assume this is always the case in what follows, unless otherwise is specified.

**Luminous object at focus.  Searchlight principle.**  If a small lamp is placed at the focus $F$ of a spherical mirror, observation shows

parallel beam

parabolic mirror

FIG. 189.
Searchlight

that the rays reflected by the mirror are parallel if they strike it near the pole, $P$, and nearly parallel if they strike the mirror at some distance away from $P$. These results are similar to those obtained in the case of spherical aberration, except that the incident light now comes from the focus. As the light reflected at some distance from the pole is not parallel, most of the light leaves the mirror as a divergent beam, which becomes weaker as its distance from the mirror increases. The curved spherical mirror is therefore not used in searchlights.

A searchlight mirror is shaped in the form of a *parabola*, the shape of the path taken by a ball when it is thrown in the air. This mirror has the property of reflecting rays parallel to the principal axis when a lamp is placed at its focus $F$, no matter where they are incident on the mirror, and hence a bright parallel beam is obtained (Fig. 189). The same type of curved mirror is frequently used in an electric radiator, with the burning filament at its focus, so that the incident heat rays (infra-red rays, p. 353) are reflected as a parallel beam back into the room. The reflecting surface behind the headlamp of a car has also a parabolic shape.

**Convex (diverging) mirrors.** We have now to consider a *convex* spherical mirror, illustrated in Fig. 190. The reflecting surface

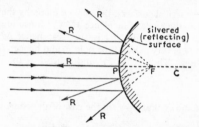

FIG. 190. Action of convex (diverging) mirror

of the mirror is part of a sphere of centre $C$. The pole $P$ of the mirror is at the middle, $C$ is the centre of curvature of the mirror, and $CP$ is the direction of the principal axis.

When a narrow beam of parallel rays is incident on the middle of the mirror, experiment shows that the reflected rays, $R$, form a *divergent* beam (Fig. 190). Nevertheless, all the rays appear to come from a point $F$ when they are produced back behind the mirror; and, by analogy with the action of a concave mirror on parallel rays, $F$ is known as the principal focus of the mirror. A screen placed at $F$ receives no rays since it is behind the mirror, and hence $F$ is a *virtual focus* (see p. 270).

The focal length ($f$) of the convex mirror is the distance from $F$ to the pole $P$; the radius of curvature ($r$) of the mirror is the distance $CP$.

**Relation between $f$ and $r$.** The focal length ($f$) of a concave or convex mirror is always *half* the radius of curvature ($r$),

i.e. $$f = \frac{r}{2}.$$

To prove this, consider a ray $MN$ parallel to the principal axis of a concave mirror incident at a point $N$ (Fig. 191 (a) ). The small region round $N$ may be considered a very tiny *plane* mirror, and as the normal at $N$ is $CN$ (since $CN$

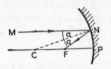

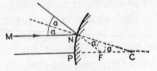

FIG. 191 (a). Concave mirror          FIG. 191 (b). Convex mirror

is a radius of the spherical surface), the angle $a$ made by the reflected ray $NF$ with $CN$ is equal to the angle of incidence $MNC$. Since angle $NCF$ = angle $MNC$ (alternate angles), triangle $NCF$ is isosceles; hence $NF = FC$. Now in practice the point $N$ is very close to the pole $P$, and hence $NF = FP$ to a very good approximation. Thus $FC = FP$, and hence $F$ is the mid-point of $CP$.

Consequently $f = FP = \dfrac{CP}{2} = \dfrac{r}{2}$.

A similar proof shows that $f = \dfrac{r}{2}$ in the case of the convex mirror (Fig. 191

(b) ); the proof is left as an exercise for the student.

**Drawing of images in concave mirrors.** When an optical diagram is drawn to show the formation of an image, the object is usually

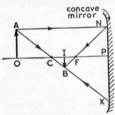

FIG. 192.
Drawing of image

represented by a straight line $OA$ perpendicular to the principal axis (Fig. 192). An arrow is drawn to represent the top, $A$, of the object, and the foot $O$ of the object is placed on the principal axis, as shown. As a small central portion of the mirror is used to obtain a clear image, the reflecting mirror surface should be represented as a *straight* line in drawing images accurately.

(1) *Object further from mirror than centre of curvature.* Consider a ray $AN$ from $A$ parallel to the principal axis and incident on the mirror. Since a beam of light parallel to the principal axis passes through the focus $F$ after reflection, it follows that the ray $AN$ passes through $F$ after reflection at $N$. Suppose a ray $AC$ passing from $A$ through the centre of curvature is incident on the mirror. Since $CX$ is a radius of the sphere of

which the mirror is part, the ray $ACX$ is a *normal* to the mirror at the point of incidence $X$. Hence the ray is reflected back along its incident path.

We now know the directions, $XC$ and $NF$, of two rays from $A$ reflected from the mirror. If we take any other ray from $A$ incident on the mirror, the reflected ray passes through the point of intersection, $B$, of $NF$ and $XC$ if the aperture of the mirror is small, and hence $B$ is the image of $A$. As far as a drawing is concerned, $B$ can be found from the point of intersection of *two* reflected rays, and in general two of the following are the easiest to draw: (1) A ray parallel to the principal axis, which is reflected to pass through the principal focus; (2) a ray passing through (or coming from) the centre of curvature, which is reflected back along its own path; (3) a ray passing through the focus which is reflected parallel to the principal axis.

The foot $O$ of the object must give rise to an image $I$ also on the principal axis, since the ray $OP$ is a normal to the mirror at $P$ and is hence reflected back along its incident direction. Now a mirror of small aperture does not distort an object. Hence, as the object $OA$ is drawn as a straight line perpendicular to the principal axis in the diagram, the image must be straight and perpendicular to the axis. The image is therefore completed, once $B$ is obtained, by drawing a perpendicular from $B$ to the axis.

It will be noted from Fig. 192 that the image is *inverted* compared with the object, and is smaller than the object, or *diminished* in size. Further, the reflected rays forming the points on the image, such as $B$, actually pass through these points in front of the mirror, and hence the inverted image can be obtained on a screen. The image is therefore *real*. (See also Fig. 196, p. 288.)

(2) *Object at centre of curvature.* When an object approaches a concave mirror from a long distance away, the image formed is real and inverted and increases in size. When the centre of curvature $C$ is reached, the inverted real image is formed at the same place, and is the same size as the object, Fig. 193 (a).

(3) *Object between centre of curvature and focus.* When the object lies between $C$ and $F$, the image is real, inverted, and larger than the object, as the reader should verify; and the image recedes away from the mirror until the object reaches the focus.

(4) **Object at focus.** When the object reaches the focus, the image is a very long way off, at infinity. The rays from the object are then reflected parallel from the mirror (see Fig. 188 (*a*)).

(5) **Object between focus and mirror.** When the object is between $F$ and the pole $P$, the rays from the top $A$ of the object are reflected in a *divergent* beam from the mirror. Fig. 193 (*b*)

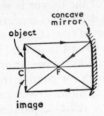

Fig. 193 (*a*). Object and image coincide at $C$

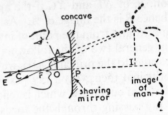

Fig. 193 (*b*). Object between $P$ and $F$—virtual magnified image

illustrates the reflection of a ray from $A$, which appears to come from the centre of curvature, and a ray which is parallel to the principal axis. An observer at $E$ sees the reflected rays as if they came from a point $B$ situated *behind* the mirror, and hence a *virtual* image is obtained in this case. The image $BI$, it will be carefully noted, is the same way up as the object (or *erect*) and is larger than the object. Fig. 193 (*b*) explains the use of a concave mirror as a shaving mirror. The face is placed close to the mirror,

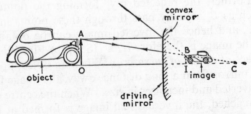

Fig. 194. Virtual diminished image in convex mirror

so that it is between the focus and the pole, when a large erect image of the face is seen. If the face is further away from the mirror than the focus, an inverted image is seen.

**Images in a convex mirror.** Consider an object *AO* in front of a mirror of the convex type (Fig. 194). A ray from *A* parallel to the principal axis appears to come from the virtual focus behind the mirror after reflection, and a ray incident towards the centre of curvature is reflected back along its own path. The two reflected rays are *divergent*, and appear to come from a point *B* *behind* the mirror. Thus the image *BI* is virtual and erect. As the reader should verify, this is true no matter where the object is positioned in front of the mirror. It will also be found that the image is always smaller than the object.

The convex mirror is thus used as a front driving mirror for vehicles; it will never produce an inverted image, unlike a concave mirror. Moreover, (*a*) the images are diminished and many objects can therefore be seen in a small mirror, (*b*) objects such as *O*, *P* at a wide angle can be observed, so that a convex mirror has a wide field of view (Fig. 195). A plane mirror is used to observe objects directly behind a car, and has a limited field of view.

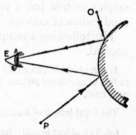

Fig. 195. Wide field of view

The reader is advised to memorize the different types of images formed by a concave and a convex mirror.

**Formulae with mirrors. The sign convention.** When considering the use of mirrors for various purposes scientists adopt a *formula method* for finding the position, size and nature of the images, instead of using the drawing method.

It can be proved that

$$\frac{1}{v} + \frac{1}{u} = \frac{1}{f},$$

where *u* and *v* represent the distances of object and image from a concave or convex mirror and *f* represents its focal length (see page 288).

In using this formula, however, it is necessary to adhere strictly to an agreement or "convention" about which distances are to be considered positive and which negative. The convention used here is this:

*Real objects and real images are regarded as being at a positive (+) distance from the mirror; virtual images are regarded as being at a negative (−) distance from the mirror.*

Thus, in brief, "real is positive, virtual is negative in sign." A concave mirror has a real focus (p. 279), and hence the focal length and radius of curvature of a concave mirror are positive. Mirrors bought from manufacturers are sometimes contained in packages labelled "+15 cm.", for example, and from the + sign, without opening the package, the mirror is known to be a concave one. A convex mirror has a virtual focus (p. 281), and its focal length (and radius of curvature) are therefore negative.

The following examples illustrate how the sign convention is applied.

1. An object is placed (*a*) 20 cm., (*b*) 5 cm., in front of a concave mirror of radius of curvature 30 cm. Calculate the position and nature of the image in each case.

The focal length of a concave mirror is positive. Hence $f = \dfrac{r}{2} = +15$ cm.

(*a*) The object is real; hence $u = +20$ cm.

Since
$$\frac{1}{v} + \frac{1}{u} = \frac{1}{f}$$

$$\therefore \quad \frac{1}{v} + \frac{1}{(+20)} = \frac{1}{(+15)}$$

$$\therefore \quad \frac{1}{v} + \frac{1}{20} = \frac{1}{15}$$

$$\therefore \quad \frac{1}{v} = \frac{1}{15} - \frac{1}{20} = \frac{1}{60}.$$

Inverting, $\therefore \quad v = 60.$

Thus the image is 60 cm. from the mirror, and since the sign of *v* is positive the image is real.

(*b*) In this case $u = +5$ cm., $f = +15$ cm.

Since
$$\frac{1}{v} + \frac{1}{u} = \frac{1}{f}$$

$$\therefore \quad \frac{1}{v} + \frac{1}{(+5)} = \frac{1}{(+15)}$$

$$\therefore \quad \frac{1}{v} + \frac{1}{5} = \frac{1}{15}$$

$$\therefore \quad \frac{1}{v} = \frac{1}{15} - \frac{1}{5} = -\frac{2}{15}$$

$$\therefore \quad v = -\frac{15}{2} = -7\tfrac{1}{2}$$

Thus the image is $7\tfrac{1}{2}$ cm. from the mirror, and since the sign of $v$ is negative the image is a virtual one and behind the mirror. See p. 284.

2. The image in a convex mirror of radius of curvature 12 cm. is 4 cm. from the mirror. Calculate the position of the object.

Since the mirror is convex, $f = \dfrac{r}{2} = \dfrac{(-12)}{2} = -6$ cm. Now the image in a convex mirror is always virtual (p. 285); hence the image distance $v = -4$ cm.

Since
$$\frac{1}{v} + \frac{1}{u} = \frac{1}{f}$$

$$\therefore \quad \frac{1}{(-4)} + \frac{1}{u} = \frac{1}{(-6)} \quad . \quad . \quad . \quad . \quad . \quad (a)$$

$$\therefore \quad -\frac{1}{4} + \frac{1}{u} = -\frac{1}{6} \quad . \quad . \quad . \quad . \quad . \quad (b)$$

$$\therefore \quad \frac{1}{u} = -\frac{1}{6} + \frac{1}{4} = \frac{1}{12}$$

$$\therefore \quad u = 12.$$

Thus the object is 12 cm. from the mirror, and as $u$ is positive in sign it confirms that the object is real.

In the examples it will be noted that the $+$ and $-$ signs are first substituted in the formula, as in equation $(a)$ above, and the corresponding fractions are simplified in sign in the next line, as in equation $(b)$.

**The magnification** produced by a mirror can be calculated from a formula, as well as obtained by drawing. It is shown on p. 289 that, numerically,

$$\frac{\text{length of the image } (IB)}{\text{length of the object } (OA)} = \frac{\text{image distance } (v)}{\text{object distance } (u)},$$

and since the *magnification*, $m$, produced by the mirror is given by $m = \dfrac{IB}{OA}$, it follows that, numerically,

$$m = \frac{v}{u}.$$

From example 1 (a) p. 286, the image distance $v = + 60$ cm., and the object distance $u = + 20$ cm. Hence $\dfrac{v}{u} = \dfrac{+60}{+20} = 3$, and the object is magnified 3 times by the mirror. From example 1 (b), the image distance $v = -7\frac{1}{2}$ cm.,

FIG. 196. CONCAVE MIRRORS OF VARIOUS SIZES

Note that the image of the back of the room is inverted

and the object distance $u = + 5$ cm.; hence the magnification produced, $m = -\dfrac{7\frac{1}{2}}{5} = -1\frac{1}{2}$. Thus the image is $1\frac{1}{2}$ times as long as the object and is virtual.

A positive sign in the result for $m$ denotes a real image, and a negative sign denotes a virtual image; but from results of the drawings on pp. 282, 284, the reader should be in a position to know whether a real or a virtual image is obtained with a concave or a convex mirror *before* embarking on the calculation.

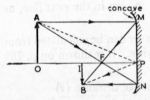

FIG. 197. Mirror formulae

**Proof that** $\dfrac{1}{v} + \dfrac{1}{u} = \dfrac{1}{f}$ **and** $m = \dfrac{v}{u}$. Fig. 197 illustrates an object $OA$ in front of a concave mirror, forming an inverted real image $IB$. If we assume that rays are incident on the mirror near the principal axis $OP$, we can take $PN$ and $MP$ to be straight lines without much error. Then, since triangles $AFO$, $FPN$ are similar triangles, $\dfrac{AO}{PN} = \dfrac{OF}{FP}$. But $OF = OP - FP = u - f$.

$$\therefore \quad \frac{AO}{PN} = \frac{u - f}{f} \qquad . \qquad . \qquad . \qquad (1)$$

Now triangles $MPF$, $IBF$ are similar. Hence $MP/IB = FP/IF$. But $MP = AO$, and $IB = PN$. Thus $AO/PN = FP/IF$. All the distances here are positive as they are real.

Since $FP = f$, $IF = IP - FP = v - f$.

$$\therefore \quad \frac{AO}{PN} = \frac{f}{v - f} \qquad . \qquad . \qquad . \qquad (2)$$

From (1) and (2),
$$\frac{u - f}{f} = \frac{f}{v - f}$$
$$\therefore \quad uv - uf - vf + f^2 = f^2$$
$$\therefore \quad uv - uf - vf = 0$$
$$\therefore \quad uf + vf = uv$$

Dividing each term by $uvf$,

$$\therefore \quad \frac{1}{v} + \frac{1}{u} = \frac{1}{f} \qquad . \qquad . \qquad . \qquad (3)$$

A similar procedure with convex mirrors, taking into account the sign convention, results in the same formula.

The *magnification*, $m = \dfrac{\text{image length}}{\text{object length}} = \dfrac{IB}{OA}$. A ray $AP$ from the top point $A$ of the object is reflected at $P$ to pass through the top point $B$ of the image; and, since $PO$ is the normal to the mirror at $P$, angle $APO$ = angle $BPI$. Thus triangles $APO$, $BPI$ are similar, and hence $\dfrac{IB}{OA} = \dfrac{IP}{OP} = \dfrac{v}{u}$.

$$\therefore \quad \text{magnification}, \; m = \frac{v}{u} \qquad . \qquad . \qquad . \qquad (4)$$

**To show that object and image coincide at the centre of curvature of a concave mirror.** In Fig. 193 (a), p. 284, it was shown by drawing that the image of an object in a concave mirror coincides in position with it when the latter is situated at the centre of curvature. To prove this result by the formula method, we substitute the object distance, $u$, which is $+ r$ in this case, in $\frac{1}{v} + \frac{1}{u} = \frac{1}{f} = \frac{2}{r}$.

The image distance, $v$, is thus given by

$$\frac{1}{v} + \frac{1}{(+r)} = \frac{2}{(+r)}$$

i.e.

$$\frac{1}{v} = \frac{2}{r} - \frac{1}{r} = \frac{1}{r}$$

$$\therefore \quad v = r.$$

Consequently the image is also at the centre of curvature, and the object and image thus coincide in position at this point.

### Experimental determination of $r$ and $f$ for a concave mirror.

**1. Approximate method.** A quick method of finding roughly the focal length of a concave mirror consists of holding it at one end of a long room to face the window. An inverted clear image of the latter is then obtained on a screen in front of the mirror, and the distance $d$ from the screen to the mirror is measured. Since the window is a long way from the mirror, the rays reaching the latter from any point on the window are fairly *parallel*, and hence $d$ is roughly equal to the focal length (see Fig. 188 $(a)$ ).

### 2. Centre of curvature method.

$(a)$ **Using illuminated object.** The radius of curvature, $r$, of a concave mirror $M$ can be found fairly accurately by using

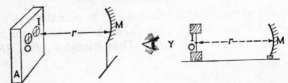

FIG. 198 $(a)$. Illuminated object method

FIG. 198 $(b)$. Pin (no parallax) method

illuminated cross wires $O$ in a box $A$ as an object (Fig. 198 $(a)$ ). When $A$ is moved to or fro, a clear inverted image $I$ is obtained beside $O$ at some stage. In this position, as illustrated on p. 284,

Fig. 193 (a), the distance from O to M is the radius of curvature, r, of the mirror. The focal length, f, of the mirror is half the radius of curvature (p. 281), and hence f can be calculated.

(b) *Using a pin.* Instead of using illuminated cross wires as an object, a pin O in a wooden block can be set up in front of M. An inverted image of the pin is observed when O is farther from the mirror than its focal length (p. 284); and, as O is shifted to or from the mirror between Y and M, and the observer's head is moved slightly from side to side, a position of O is reached when the inverted image I of the pin keeps perfectly in line with the object pin O as they both move. The image and object then coincide exactly in position, just as they do in Fig. 198 (b), and the distance from O to M is the radius of curvature of the mirror. The focal length is half the radius of curvature, and can thus be calculated.

The above method of locating the position of an image by a pin can be illustrated by holding up the forefinger of each hand and placing one a few inches behind the other. When the eye is moved from side to side, it will be observed that the finger farther away moves always in the same direction as the eye, and that the nearer finger always moves in the opposite direction. The two fingers are said to have an amount of **parallax** between them in this case; but as they are brought nearer to each other the amount of parallax diminishes. *When they are coincident there is no parallax between them*, as they both appear to move together when the eye moves from side to side. Thus the image I in Fig. 198 (b) coincides in position with the pin O when no parallax is obtained between them. The method of no parallax is a more accurate method of locating an image than the triangular gauze method, as the clearest position of the image of the triangular gauze cannot be obtained with such a high degree of accuracy as the position of no parallax.

**3. Formula method.** Another accurate method of measuring focal length is to use the formula $\frac{1}{v} + \frac{1}{u} = \frac{1}{f}$. Illuminated cross wires are set up as an object in front of the concave mirror M, and a screen B is then moved until the cross wires are shown as clearly as possible on it (Fig. 199 (a) ). The distance

*IM* or *v* of the image *I* from *M* is now measured; the object distance *OM*, or *u*, is also measured. Suppose $u = + 25 \cdot 1$ cm., $v = + 20 \cdot 9$ cm. Then since $\dfrac{1}{v} + \dfrac{1}{u} = \dfrac{1}{f}$,

$$\frac{1}{25 \cdot 1} + \frac{1}{20 \cdot 9} = \frac{1}{f}$$

$$\therefore \frac{20 \cdot 9 + 25 \cdot 1}{25 \cdot 1 \times 20 \cdot 9} = \frac{1}{f}$$

$$\therefore \frac{25 \cdot 1 \times 20 \cdot 9}{46} = f$$

$$\therefore \quad f = 11 \cdot 4 \text{ cm.}$$

The experiment is repeated for varying object distances, when varying image distances are obtained. The focal length is calculated from each pair of corresponding values of *u* and *v* as shown

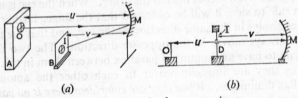

(a)　　　　　　　　　　　　(b)

FIG. 199. Focal length of concave mirror

above, and the average value of the focal length is calculated from the results obtained.

The 'no parallax' pin method can also be used to obtain values of *v* and *u*, as illustrated in Fig. 199 (*b*). The object pin *O* gives rise in an inverted image *I*, whose position is located by moving another pin, *D*, until no parallax is obtained between *O* and *I*.

**Focal length of convex mirror.** This can be found by placing an object pin *O* in front of the mirror *M*, and then moving a locator pin *L* behind the mirror until it coincides by no parallax with the virtual image *I* (Fig. 199c).

The distance from *L* to *M* is *v*, and *OM* is *u*. Substituting in $1/v + 1/u = 1/f$, and remembering *v* is negative, *f* can be found.

# Experimental verification of magnification and mirror formula.

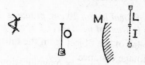

FIG. 199C. Focal length of convex mirror

The *magnification formula*, $m = v/u$, can be verified by measuring the height of the triangle $O$, the height of its image $I$, and the corresponding distances $u$, $v$ (Fig. 199 (*a*) ). It is then found that image height : object height $= v : u$. On varying the object position, the same result is obtained.

The *mirror formula* can be verified by measuring $u$ and $v$ in Fig. 199 (*a*), and calculating $1/v + 1/u$ in decimals. When $f$ is measured by the method illustrated on p. 290, it will be found that $1/f = 1/v + 1/u$. When $u$ and $v$ are varied, the same result is obtained.

=Summary=

1. The focal length of a curved spherical mirror is half the radius of curvature.

2. With a **concave (converging) mirror :** (*a*) The focus is real, (*b*) the image is real and inverted when the distance of the object from the mirror is greater than the focal length, (*c*) the image is virtual, erect and magnified when the distance of the object from the mirror is *less* than its focal length.

3. With a **convex (diverging) mirror,** the image is always erect, diminished, and virtual.

4. The formula for a curved mirror is $1/v + 1/u = 1/f$. Magnification, $m = v/u$.

5. "Real is positive, virtual is negative" is the sign convention. A concave mirror has a positive focal length; a convex mirror has a negative focal length.

6. The focal length of a concave mirror can be measured by using an illuminated object, finding the image on a screen, and then substituting in $1/v + 1/u = 1/f$.

## WORKED EXAMPLES

1. *A concave mirror of focal length 20 cm. produces a magnification of three when an object is placed in front of it. Where is the object if the image is erect?*

Let $x$ = the distance ($u$) of the object from the mirror in cm.

Then, since
$$\frac{\text{image distance}}{\text{object distance}} = \text{magnification,}$$

$3x$ = the distance ($v$) of the image from the mirror in cm.

Now a larger erect image than the object is obtained with a concave mirror when the image is *virtual*, see p. 284, and hence the distance $u = x$ is positive in sign and the distance $v = 3x$ is negative in sign.

Since $\dfrac{1}{v} + \dfrac{1}{u} = \dfrac{1}{f}$ and the mirror is concave,

then
$$\frac{1}{(-3x)} + \frac{1}{(+x)} = \frac{1}{(+20)}$$
$$\therefore \; -\frac{1}{3x} + \frac{1}{x} = \frac{1}{20}$$
$$\therefore \; \frac{2}{3x} = \frac{1}{20}$$
$$\therefore \; x = \tfrac{2}{3} \times 20 = 13\tfrac{1}{3}.$$

Thus the object is $13\tfrac{1}{3}$ cm. from the mirror.

2. *Distinguish between real and virtual images. In what circumstances is each formed by means of a concave spherical mirror with a radius of curvature of 12 in.? An object $1\tfrac{1}{2}$ in. high, with its base on the principal axis, is 9 in. in front of a concave mirror of 12 in. radius of curvature. Find the position, size, and nature of the image, preferably by means of a graphical construction depending directly on the laws of reflection.* (L.)

A real image is one which can be received on a screen; a virtual image is one which cannot be received on a screen. The focal length of the concave mirror $= \dfrac{r}{2} = \dfrac{12}{2} = 6$ in. Hence a real image is obtained when an object is farther than 6 in. from the mirror, and a virtual image when the object is nearer than 6 in. from the mirror.

A drawing is left to the reader (see Fig. 192). The position of the image is given by substituting $u = +9, f = +6$ in the mirror formula $\dfrac{1}{v} + \dfrac{1}{u} = \dfrac{1}{f}$, and hence

$$\frac{1}{v} + \frac{1}{(+9)} = \frac{1}{(+6)}$$
$$\therefore \; \frac{1}{v} = \frac{1}{6} - \frac{1}{9} = \frac{1}{18}$$
$$\therefore \; v = 18 \text{ cm.}$$

Since the sign of $v$ is positive, a real image is obtained 18 in. from the mirror, and it is inverted. The magnification, $m$, $= \dfrac{v}{u}$, and hence $m = \dfrac{18}{9} = 2$. Thus the size of the image $= 2 \times 1\frac{1}{2}$ in. $= 3$ in.

## EXERCISES ON CHAPTER XIX

**1.** Name three applications of curved mirrors.

**2.** Draw diagrams showing the effect of (i) a concave mirror, (ii) a convex mirror on a parallel beam of light, and mark the positions of the focus and centre of curvature in each case. State, with a reason, the *sign* of the focal lengths of the two mirrors.

**3.** An object 1 in. tall is placed 15 cm. from a concave mirror of focal length 10 cm. Draw the image accurately to scale, and find the magnification. (*Note.* The mirror must be represented by a straight line.)

**4.** Calculate the position of the image in question 3 and the magnification. Verify your drawing.

**5.** Describe how you would find quickly, but approximately, the focal length of a concave mirror.

**6.** What is the relation between the focal length of a concave or convex mirror and its radius of curvature? A real image of an object in a concave mirror of radius 24 cm. is formed 16 cm. from the mirror. Calculate the object position, and draw a sketch to illustrate the formation of the image.

**7.** An object 2 in. tall is placed 12 cm. in front of a concave mirror of radius of curvature 30 cm. Draw the image to scale, and find its position and the magnification. Verify your drawing by calculation.

**8.** Repeat question 7 if the mirror is a convex one of the same radius, 30 cm.

**9.** When is a *virtual* image formed by a concave mirror? What else do you know about this image? A magnified virtual image is formed 12 cm. from a concave mirror of focal length 18 cm. Calculate the object position and the magnification.

**10.** Why is a convex, and not a concave, mirror used for driving? An image in a convex mirror of radius of curvature 10 in. is formed 3 in. from the mirror. Find the position of the object and the magnification (i) by drawing, (ii) by calculation.

**11.** An object is magnified four times by a concave mirror of focal length 24 cm. Calculate the object position if the image is (i) real, (ii) virtual.

**12.** Distinguish between a *real* and a *virtual* image and, by means of ray diagrams, show how each type of image may be produced by a concave mirror. (*L*.)

**13.** Describe how you would find the focal length of a concave mirror. Describe the optical arrangement in a searchlight giving a parallel beam, and explain the adjustment necessary to give a diverging beam. By day, searchlights should never be left in such a way that they might point at the sun; explain this fact. (*N*.)

**14.** Define the principal focus of a convex spherical mirror, and state where it is situated. A convex mirror has a radius of curvature of 6 in.; an object 2 in. long stands 2 in. front of the mirror. Calculate the position and size of the image. Draw to scale a diagram showing the beam by which an eye, situated off the axis, sees one end of the object reflected in the mirror. (*C*.)

**15.** A luminous object is placed 6 in. from (*a*) a plane mirror, (*b*) a concave mirror of focal length 9 in. In each case, draw a ray diagram to illustrate the formation of the image, and give details of the nature, position, and magnification of the image. How would you verify experimentally the value given for the focal length of the concave mirror, (i) approximately and quickly, (ii) more precisely ? (*C.W.B.*)

**16.** Describe a method of measuring the focal length of a concave mirror. The image of an object placed in front of a concave mirror is one-half the size of the object, and is 45 cm. from the object. What is the focal length of the mirror? (*C*.)

**17.** Explain a quick method of determining the radius of curvature of a spherical concave mirror. How would you use a concave mirror, of radius of curvature 24 cm., to give a three-times magnified erect image of an object ? Give a ray diagram to explain the formation of the image. (*C.W.B.*)

**18.** Draw a ray diagram to illustrate the formation of an image by a concave *mirror* when an object is placed on and perpendicular to the principal axis at a distance from the mirror equal to one half its focal length. Find the magnification produced and state the nature of the image. The table below gives a series of experimental values for *u* and *v*, the distances of an object and its real image respectively from the pole of a concave mirror. Plot a graph to show the variation of *v* with *u*, deduce the value of $u$ for which $u = v$, and hence or otherwise deduce the focal length of the mirror.

| $u$ in cm. | 8·2 | 9·0 | 9·8 | 11·6 | 12·5 | 14·2 |
|---|---|---|---|---|---|---|
| $v$ in cm. | 15·8 | 13·6 | 12·0 | 10·1 | 9·6 | 8·7 |

(N.)

**19.** How would you determine the focal length of a concave mirror ? Deduce the relationship between the focal length and the radius of such a mirror. How do the nature and magnification of the image vary as an object is brought from a great distance up to such a mirror ? Illustrate your answer with diagrams. (O. & C.)

**20.** Distinguish between *real* and *virtual* images. Illustrate your answer by diagrams showing the formation of images by pencils of rays reflected from (a) a concave mirror, (b) a plane mirror.

A concave mirror, radius of curvature 20 cm., forms a *virtual* image of a small object placed at right angles to its axis, the image being four times as high as the object. What is the distance of the object from the mirror ? How far, and in what direction, must the object be moved in order that a *real* image, having a magnification of four, may be produced ? (L.)

**21.** Define *principal focus* of a concave mirror.

An object 2 cm. high is situated on and perpendicular to the axis of a concave mirror of radius of curvature 30 cm. and is 10 cm. from the mirror. Find, graphically or by calculation, the position and size of the image and indicate on a diagram, the paths of three rays from a point on the object to an eye viewing the image. (L.)

# CHAPTER XX

## REFRACTION AT PLANE SURFACES

IT is noticeable that the bottom of a swimming pool seems nearer to the surface than is actually the case. Similarly, a straight stick, partly immersed in water, appears bent at the surface into two straight parts. These observations show that light from objects in water does not travel in the same direction when passing into the air. This phenomenon, which was first studied about A.D. 100, occurs also when light passes through lenses, and a scientific study of it has paved the way for the manufacture of efficient eye-glasses, telescopes, and microscopes.

**Refraction (or "bending") of light.** If a ray of light $PA$ passes from air into a rectangular glass block, *the light travels in a new direction $AB$ in the glass*, and emerges along a direction $BL$ which is parallel to $PA$ (Fig. 200). We say that the ray of light is "bent" or *refracted* in the glass at $A$, and the phenomenon is called the *refraction* of light.

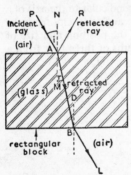

FIG. 200. Refraction

At this stage, it should be carefully noted that *reflection*, as well as refraction, of light takes place at the boundary between air and glass. Fig. 200, for example, illustrates the reflected ray $AR$ from a glass boundary, which can be observed faintly when the glass block is placed on a sheet of paper. The energy of the light in the incident ray $PA$ is equal to the light energy in the reflected ray $AR$ plus the light energy in the refracted ray $AB$. The ray $AB$ is much brighter than the reflected ray $AR$, as most of the incident light energy passes into the glass.

**Laws of refraction.** Experiments, such as those just described, show that a ray travelling from one medium to a *denser* medium (e.g. from air to water or glass) is refracted *towards* the normal at the point of incidence. Thus, in Fig. 200, $AB$, the refracted ray, is inclined at a smaller angle to the normal $NAM$ than the incident ray $PA$. Angle $PAN$ is the angle of incidence, $i$, and angle $BAM$ is the angle of refraction, $r$. As the angle of incidence is increased, the angle of refraction increases; but each time the latter is smaller than the angle of incidence. It was not until 1621 that the law governing refraction was discovered. In that year SNELL, a Dutch professor, found that *the ratio* $\dfrac{\sin i}{\sin r}$ *was always a constant* for a given pair of media, such as air and glass. Thus in one experiment with glass $i = 60°$, $r = 35°$ by measurement, and hence $\dfrac{\sin i}{\sin r} = \dfrac{\sin 60°}{\sin 35°} = \dfrac{0\cdot866}{0\cdot574} = 1\cdot52$. In another experiment with the same glass, $i = 70°$, $r = 38°$, and hence $\dfrac{\sin i}{\sin r} = \dfrac{\sin 70°}{\sin 38°} = \dfrac{0\cdot94}{0\cdot616} = 1\cdot52$. This constant value, $1\cdot52$, is known as the **refractive index** of glass, and we shall denote the latter by the symbol $\mu$. Thus, in general,

$$\frac{\sin i}{\sin r} = \mu \qquad . \qquad . \qquad . \qquad . \quad (1)$$

The value of the refractive index, $\mu$, from air to glass is about $1\cdot5$ (or $1\frac{1}{2}$); from air to water, $\mu$ is about $1\cdot33$ (or $1\frac{1}{3}$); from water to glass, $\mu$ is about $1\cdot13$. Although it is meaningless to talk about the refractive index of a medium (such as glass) without specifying the medium (such as air) in which the light was originally travelling before refraction occurred, scientists have compiled tables of the refractive indices of media on the assumption that the light travels originally in a *vacuum* before refraction takes place. The refractive index of air is only very slightly greater than 1; and, except where extreme accuracy is required, the refractive index of glass, for example, can be measured with light incident on it from air.

Lenses for eye-glasses, telescopes and microscopes are usually made from *crown glass* or *flint glass*, which have different refractive indices. The construction of the lenses of optical instruments

(a) Cracking up optical glass into lumps suitable for subsequent moulding

(b) Arranging slabs of optical glass on table prior to polishing opposite edges

FIG. 201. MANUFACTURE OF OPTICAL GLASS

depends on an accurate knowledge of the refractive indices of the glass used. (See Fig. 201).

All experiments on refraction, similar to those described on p. 298, show that—

**(1) The incident ray, the normal, and the refracted ray are all in the same plane.**

**(2) The ratio $\dfrac{\sin i}{\sin r}$ is always a constant for a given pair of media.**

These are known as the **two laws of refraction.** It should be noted that the normal to a plane boundary can easily be identified with a ray box, as it is the only direction along which an incident ray passes into the medium without change of direction. A ray along the normal $NA$ (Fig. 200), for example, passes into the glass along $AM$.

**Light travelling from a denser to a rarer medium.** When light passes through a telescope lens into air, it travels from one medium, glass, into a rarer (or less dense) medium, air. Observations with a ray box show that in this case the light is refracted into the air *away* from the normal at the point of incidence. In Fig. 200, for example, the ray $AB$ in passing from glass to air is refracted along the direction $BL$ which makes a greater angle with the normal $DB$. In Fig. 202, a ray $YX$ travelling in water emerges into air along a direction $XT$, and the angle of refraction, $r$, is greater than the angle of incidence, $i$.

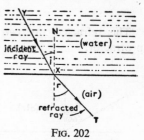

FIG. 202

Dense to rare medium

The refractive index from water to air can be written $_w\mu_a$, and is given by $\dfrac{\sin i}{\sin r}$ in Fig. 202. Since the refractive index from air to water, $_a\mu_w$, is $1\frac{1}{3}$, or $\frac{4}{3}$, the value of $_w\mu_a$ is $\frac{3}{4}$; this can be understood by imagining a ray of light travelling in air along $TX$, when it is refracted in the water along $XY$. Similarly, the refractive index from glass to air, $_g\mu_a$, is $\frac{2}{3}$, since the refractive index from air to glass, $_a\mu_g$, is $\frac{3}{2}$.

11

Light is refracted when it travels from one medium to another *because its speed is different in the two media.* The speed in air is about 186,000 miles per second, in glass about 124,000 miles per second, and in water about 140,000 miles per second.

**Obtaining the direction of the refracted ray.**

(1) *By calculation.* The calculation of the direction of the refracted ray may be illustrated by the following example. Consider a ray of light incident in air at an angle of 60° on a plane water surface (Fig. 203). If $x$ is the angle of refraction and $\mu$ for water is 1·33, then

$$\frac{\sin 60°}{\sin x} = \mu = 1\cdot33$$

$$\therefore \quad \frac{0\cdot866}{\sin x} = 1\cdot33$$

$$\sin x = \frac{0\cdot86}{1\cdot33} = 0\cdot645$$

$$\therefore \quad x = 40° \text{ (approx.)}$$

The refracted ray can now be drawn with the aid of a protractor.

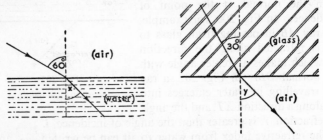

FIG. 203. Calculations on refraction   FIG. 204.

Again, suppose a ray of light in glass is incident at an angle of 30° on the boundary with air (Fig. 204). Then, if $\mu$ for glass is 1·5, the angle of refraction, $y$, which is greater than 30°, is given by

$$\frac{\sin y}{\sin 30°} = 1\cdot5.$$

(It should be noted that, in order to obtain a value for $\mu$ greater than 1, sin $y$ must be the *numerator* of the ratio with sin 30°.)

Thus $\dfrac{\sin y}{0\cdot 5} = 1\cdot 5$, i.e. sin $y = 0\cdot 75$.   Consequently, $y = 49°$.

(2) *By drawing.*  The refracted ray may also be obtained by a drawing method.  Suppose a ray $AO$ is incident in air at a *known* angle $i$ on a plane glass surface.  If the value of the refractive index $\mu = 1\cdot 5$, construct two circles whose radii are respectively 1 unit and 1·5 units (Fig. 205 (*a*) ), and produce the incident ray

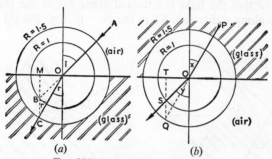

(*a*)                              (*b*)

Fig. 205. Refracted ray by drawing

to cut the smaller circle at $B$.  From $B$ draw the line $BM$ perpendicular to the surface, and produce $MB$ to meet the larger circle at $C$.  Join $OC$.  Then $OC$ is the refracted ray in the glass.

To prove this, we note that angle $MCO$ = the angle of refraction, $r$, and angle $MBO$ = the angle of incidence, $i$.  Thus $\dfrac{\sin i}{\sin r} = \dfrac{\sin \angle MBO}{\sin \angle MCO} = \dfrac{OM/OB}{OM/OC} = \dfrac{OC}{OB}$.  But $\dfrac{OC}{OB} = \dfrac{1\cdot 5}{1} = \mu$.  Consequently $OC$ must be the refracted ray.

When a ray $PO$ is incident *in glass* at a known angle $x$, the refracted ray in air can be drawn by a suitable change in the construction (Fig. 205 (*b*) ).  This time, produce $PO$ to cut the *larger* circle at $Q$, and draw a perpendicular $QT$ to the plane boundary, cutting the smaller circle at $S$.  Join $QS$.  Then $QS$ is the refracted ray in the air.  It is left to the reader as an exercise to show that $\dfrac{\sin y}{\sin x} = 1\cdot 5$.

If the perpendicular from $Q$ to the surface does not intersect the smaller circle, no refracted ray is possible. This case, which is important in practice, is considered on p. 308.

**The prism.** In the optical sense, a *prism* is the name given to a glass block whose sides are inclined to each other and whose edges are parallel. Fig. 206 (*a*) illustrates a prism whose sides meet along an edge *EH*, and *ABC* is a section of the prism at right angles to the edge so that *EH* passes through *A*. If a ray *PQ* is incident at *Q* on the prism (Fig. 206 (*b*) ), the laws of refraction show that the light is refracted along *QS* in the glass, and emerges along a direction *ST* in the air. In Fig. 206 (*b*) angle *a*

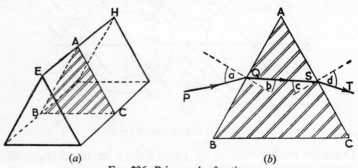

FIG. 206. Prism and refraction

is the angle of incidence, *b* is the angle of refraction, *c* is the angle of incidence at *S* in the glass, and *d*, which is greater than *c*, is the angle of emergence of the ray. In general, as the reader can verify by actual drawing, the prism bends rays of light incident as shown on *AB towards the base and away from the edge at A*, and this property is common to all prisms.

As explained on pp. 310, 340, prisms serve a useful purpose in optical instruments such as prism binoculars and submarine periscopes.

**Applications of refraction.** (1) *Apparent depth.* It is a common-place observation that a pool of water appears shallower than is actually the case and that objects at the bottom of a glass block appear raised.

To explain this, consider an object $O$ at the bottom of a parallel-sided block of glass, viewed by an observer at $E$ vertically above it (Fig. 207). The eye sees $O$ by rays *from the object* which enter the eye. The ray $ON$ which is normal to the upper surface passes straight through into the air along $NR$. The ray $OA$ slightly inclined to $ON$ is refracted into the air along a direction $AP$ which depends on the refractive index of the glass. Thus the rays $NR$, $AP$ enter the eye (the inclinations of $OA$, $AP$ to the normal are exaggerated for clarity in Fig. 207), and the observer sees the object in the direction from which the rays appear to come. The image of $O$ is hence the point of intersection, $I$, of $PA$ and $RN$, and it can be seen that $I$ is nearer to the surface than $O$.

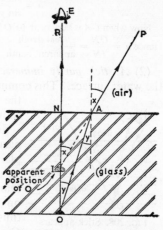

FIG. 207. Apparent depth

A similar argument shows that other points near $O$ at the bottom of the glass block appear raised when viewed by the observer. Thus, the bottom of a pool of water appears nearer to the surface than is actually the case.

In Fig. 207 $ON$ is the *real depth* of the object $O$, and $IN$ is termed the *apparent depth* of $O$. There is a simple relation between the refractive index ($\mu$) and the real and apparent depths which states that—

$$\mu = \frac{\text{real depth}}{\text{apparent depth}}.$$

Thus if the refractive index of water is $1\frac{1}{3}$ and the bottom of a pool is 4ft. below the surface, the apparent depth, $d$, is given by $1\frac{1}{3} = \frac{4}{d}$. Hence $d = 3$ ft.

The proof of the above relation is as follows—

Suppose the normal is drawn at $A$ and the angles at this point are $x$ and $y$

as shown. Then $\mu = \dfrac{\sin x}{\sin y}$. But angle $NOA = y$, angle $NIA = x$, and $\sin \angle NIA = NA/IA$, $\sin \angle NOA = NA/OA$.

$$\therefore \quad \mu = \frac{\sin x}{\sin y} = \frac{NA/IA}{NA/OA} = \frac{OA}{IA}.$$

Now when $OA$ is very near to $ON$, we may say $OA = ON$ and $IA = IN$. Hence $\mu = \dfrac{ON}{IN} = \dfrac{\text{real depth}}{\text{apparent depth}}$.

(2) *A stick partly immersed in water* appears to be bent at the water surface. This common observation is a consequence of the law of refraction, as we shall now show. (Fig. 208).

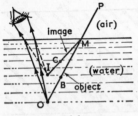

FIG. 208. Stick appears to be bent along *MI*

Suppose $PO$ is the stick, with the part $MO$ below the water. Then, as explained on p. 305, the end $O$ appears to be at $I$, a point nearer to the surface than $O$. Similarly, another point $B$ on the stick appears to be at $C$, and thus every part of the stick between $O$ and $M$ appears to be on a line $MI$, which is the image of $OM$ to an observer. Thus the stick appears to be bent towards the surface at $M$, where it enters the water.

(3) *Bringing objects into view.* An object $O$ hidden from an observer $E$ in a container $C$ can be brought into view by adding sufficient water to $C$. Fig. 209 (*a*) illustrates a ray from $O$ which

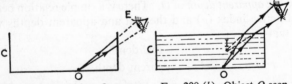

FIG. 209 (*a*). Object *O* unseen      FIG. 209 (*b*). Object *O* seen

just passes the edge of $C$, so that an observer at $E$ cannot see $O$. Fig. 209 (*b*) shows how a ray from $O$ is refracted at the water surface towards $E$, although the latter has not moved, so that the observer sees the object at $I$ in this case.

**Measurement of refractive index.** The accurate measurement of the refractive index of materials, expecially of glass, is of considerable commercial importance. The correct design of lenses for eyeglasses, microscopes, and telescopes, for example, involves the refractive index of the glass chosen. The refractive index may be measured in several ways, of which the most commonly used are described below.

(1) *Apparent depth method.* One of the best methods of measuring the refractive index of glass is illustrated in Fig. 210, the glass being made in the form of a rectangular block, $B$. A travelling microscope, $M$ (one which can move up and down along a vertical graduated scale), is first focused on an ink mark at $O$ on a sheet of paper, and the reading on the microscope scale, $x$ cm. say, is taken. The block is then placed on top of $O$, and the microscope is then raised to focus on $I$, which is the apparent position of $O$. Suppose the reading on the scale is $y$ cm. Some lycopodium dust particles are now sprinkled at $N$ on the top of the block, the microscope is raised again to focus on the particles, and the reading, $z$ cm., is taken. Then

$$\text{real depth of } O = ON = x - z \text{ cm.,}$$
$$\text{and apparent depth of } O = IN = y - z \text{ cm.}$$

Thus $\mu = \dfrac{x - z}{y - z}$, and can be calculated. The 'apparent depth' method can be used in the same way to measure the refractive index of water.

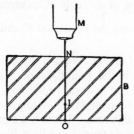

FIG. 210. Apparent depth

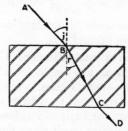

FIG. 211. Ray method

(2) *Ray path method.* The refractive index of a rectangular glass block can be found roughly by passing a ray of light through

it, using ray box or pin. The incident ray *AB* is refracted by the glass and emerges in the air along a parallel direction *CD* (Fig. 211), and the two rays *AB*, *CD* and the outline of the glass block are then drawn. The refracted ray is obtained by joining *B* and *C*, and after drawing the normal at *B* the angles of incidence ($i$) and refraction ($r$) at *B* are measured. The refractive index is then calculated by the formula $\mu = \dfrac{\sin i}{\sin r}$. The angles of incidence and refraction can also be measured at *C* if so desired.

*The refractive index of a glass prism* can be found by the method just discussed. A ray of light *AB* is incident on the prism at *B*, and the outline of the prism, the incident ray, and the ray *CD* emerging from the glass are drawn (Fig. 212). The refracted ray *BC* in the glass is obtained by joining *B* and *C* after the prism is removed, and the angle of incidence ($i$) and the angle of refraction

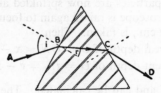

FIG. 212. Refractive index of prism

($r$) at *B* are measured. The refractive index of the glass is then given by $\mu = \dfrac{\sin i}{\sin r}$. Another value of an angle of incidence and refraction can be obtained by drawing the normal at *C* and using the rays *BC* and *CD*.

**Total internal reflection. Critical angle.** When a ray of light is incident at any angle from air on the boundary of a denser medium such as glass, a refracted ray is always obtained. This follows from our knowledge that the angle of refraction is always *less* than the angle of incidence in this case. There is a considerably different result when light travels from a medium like glass to a less dense medium, such as air.

Fig. 213 (*a*) illustrates the phenomena observed when light travelling *in glass* is incident on its plane boundary with air. When the angle of incidence in the glass is 15° say, the ray is refracted away from the normal in the air. As stated on p. 298, a reflected ray is also obtained, which is weak in intensity compared with

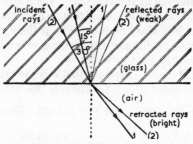

FIG. 213 (*a*). Refraction and reflection

the refracted ray. The energy of the light in the incident ray is equal to the sum of the energies of the light in the reflected and refracted rays. When the angle of incidence in the glass is increased to 30° say, a bright refracted ray is obtained which is further inclined to the normal than before, and a new weak reflected ray is observed. As the angle of incidence is increased further, a striking phenomenon is observed. At some stage, represented in Fig. 213 (*b*), the bright refracted ray just reaches the

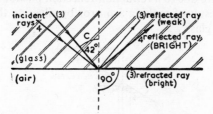

FIG. 213 (*b*). Critical angle phenomena

glass-air boundary and the reflected ray is weak; but when a slight increase in the angle of incidence takes place, *a bright reflected ray is observed in the glass*, and *no refracted ray is obtained.* As the angle of incidence is increased further, a bright reflected ray is still obtained and no refraction occurs.

The angle in the glass at which all the light first becomes

11*

reflected is known as the **critical angle** between the glass and air, and, as shown below, it is about 42° when $\mu = 1.5$ for the glass. For angles of incidence greater than 42° the light is said to undergo **total internal reflection** at the boundary, since no refraction then occurs. It should be carefully noted that no total internal reflection is possible at the boundary of two media if the light is travelling from the *less* dense medium of the two, e.g. when light travels in air and is incident on glass or water.

*Value of critical angle.* Suppose that $c$ is the critical angle for glass of refractive index $\mu = 1.5$. Then, at this angle, the angle of refraction in the air is 90°. See Fig. 213 (*b*). Since $\mu = \dfrac{\sin 90°}{\sin c}$, it follows that $1.5 = \dfrac{\sin 90°}{\sin c} = \dfrac{1}{\sin c}$, as $\sin 90° = 1$. Thus $\sin c = \dfrac{1}{1.5} = 0.0667$. Consequently, $c = 42°$ (approx.).

The value of the critical angle, $c$, for water of $\mu = 1.33$ is given by $1.33 = \dfrac{\sin 90°}{\sin c}$, from which $\sin c = 0.75$. Hence $c = 49°$.

**Applications of total internal reflection.** (1) *Submarine periscope.* If a plane mirror is used to observe an object, faint ("ghost") images can be seen besides a prominent image. The faint images are due to a refraction-reflection phenomenon, which occurs as a result of making a mirror by silvering the back of a thin block of glass.

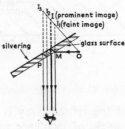

FIG. 214. Multiple images

Fig. 214 illustrates the formation of the images of an object $O$ in front of a plane mirror. A ray from $O$ strikes the glass at $M$, and part of the light is reflected along $MI_1$ to give rise to a faint image. Most of the light, however, is refracted at $M$, enters the glass, and is reflected at $N$ on the silvered surface. The light thus emerges at $P$ along $PI$, which gives rise to the prominent image observed in the mirror; but reflection takes place at $P$ once more, leading to the formation of a faint image $I_2$. Other faint images arise in the same way. We have seen that two parallel mirrors can be used to make a simple periscope (p. 271); owing to the many

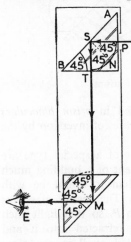

FIG. 215. (*Not to scale*)

faint images formed, mirrors are not used for submarine periscopes where precision of a high standard is necessary in sighting accurately on enemy ships. Instead of two mirrors, *two right-angled isosceles prisms are used* (Fig. 215).

Each prism has an acute angle of 45° since it is isosceles. Consider a ray $OP$ from an object $O$ incident normally on one of the sides containing the right angle $C$ of the prism. The ray passes straight through $AC$ and is incident at $S$ on the hypotenuse face $AB$ at an angle of 45°. This angle, it should be noted, is the angle of incidence in glass, and since 45° is greater than the critical angle (42°) if the glass has a refractive index 1·5, no light emerges beyond $S$ into the air. Total reflection thus takes place at $S$ and, in accordance with the law of reflection, a ray $ST$ is obtained making an angle of 45° with the normal at $S$. Thus the ray $ST$ is perpendicular to the ray $PS$ and is almost as bright as $PS$. From the geometry of the diagram it can be seen that $ST$ strikes the face $BN$ of the prism normally, and hence the ray emerges into the air in a direction perpendicular to its original direction $OS$. Thus the light emerging from the prism is practically as bright as the incident light from $O$, and the prism gives rise to only one image, unlike a plane mirror. To obtain a periscope action another right-angled isosceles prism is placed below the first one. The action of the light incident on this prism is exactly the same as that already described, and total internal reflection takes place in the glass at *M*. The ray thus finally emerges in air in a direction parallel to *OP*, and in this way an observer at *E* sees the object at *O*.

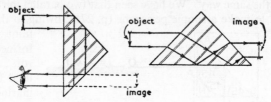

FIG. 216. Inversion effects of prisms

Right-angled isosceles prisms are used in *prism binoculars* (p. 341). Fig. 216 illustrates two other cases of inversion by the use of these prisms.

(2) *Basement lighting.* Glass windows of a special type are frequently employed to make the basement of a building much

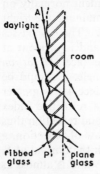

FIG. 217.
Basement lighting

lighter in daylight than it would be with ordinary plane glass. The special glass has "ribs" on the outside *P*, so that light such as *A* from above is refracted into it and is then totally reflected at *P* in the glass towards the interior of the room (Fig. 217). In this way much of the daylight is usefully employed. If a plane glass window had been used, most of the light from above would pass through to the floor of the room.

(3) *The mirage.* Total internal reflection can be obtained in gases, such as air, as well as in solid media like glass. The criterion again is that the light must be travelling from a denser to a less dense medium.

Fig. 218 illustrates how a mirage occurs in the desert. The layers of air near the ground are hotter than the layers higher up, because they are heated by the hot sand.    A ray from the top of a tree $T$ incident in a downward direction passes from a colder to a warmer layer of air (i.e. from a denser to a rarer medium), and

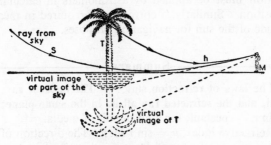

FIG. 218. Mirage

bends away more and more from the incident direction until it enters a layer of air, such as at $h$, where total internal reflection occurs.    The ray is then bent *upwards* at $h$, and as it now passes from one layer to another which is denser, the refraction continues to take place in a gradually increasing upward direction until it enters the eye of the observer $M$.    To him the ray appears to come from a place beneath the tree, and in general he sees an image of the tree below its actual position.    Similarly, he can also see an image of part of the sky, $S$, round the image of the tree, and as he can also see the tree directly the whole appearance is similar to that of a tree standing by a pool of water.

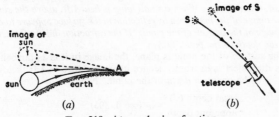

FIG. 219. Atmospheric refraction

**Other refractive phenomena.**    Atmospheric refraction is also responsible for the fact that the sun can often still be observed at a point such as *A* on the earth after it has disappeared below the horizon (Fig. 219 (*a*) ). For the same reason a star appears to be slightly more elevated than is actually the case (Fig. 219 (*b*) ), and a correction must be applied by astronomers in calculating its exact position. Similarly, a correction is required in measuring the altitude of the sun for navigation purposes.

---

## Summary

1. The **laws of refraction** state: (i) **The incident ray, the normal, and the refracted ray all lie in the same plane;** (ii) **sin *i*/sin *r* = constant,** for a given pair of media.

2. Refractive index, $\mu = \sin i / \sin r$. The direction of the refracted ray can be found by using $\mu = \sin i / \sin r$; or by drawing two concentric circles whose radii are in the ratio $\mu : 1$.

3. Real depth/apparent depth $= \mu$.

4. **Total internal reflection only occurs when light travels from a dense medium (like glass) to a less dense medium (like air).** The critical angle (*c*) can be found from $\sin c = \dfrac{1}{\mu}$.

---

### WORKED EXAMPLE

*Explain, with the aid of a suitable ray diagram, why a scratch on one face of a rectangular glass block, when viewed through the opposite face, appears to be nearer than it actually is. When a small lamp is held 4 ft. above the surface of a pond, the image of the lamp seen by reflection in the surface appears to coincide with the image of the bottom of the pond. If the refraction index of water is 4/3, calculate the depth of the pond. (L.)*

Since the surface of the pond is plane, the image in it is as far behind the surface as the object is in front. Hence image distance = 4 ft. below pond surface; this is the *apparent depth* of the pond. Now

$$\frac{\text{real depth } (d)}{\text{apparent depth}} = \mu \text{ (see p. 305)}$$

$$\therefore \quad \frac{d}{4 \text{ ft.}} = \frac{4}{3}$$

$$\therefore \quad 3d = 16 \text{ ft., i.e. } d = 5\tfrac{1}{3} \text{ ft.}$$

## EXERCISES ON CHAPTER XX

**1.** Draw sketches showing the incident and refracted rays when light is refracted (i) from air to glass, (ii) from water to air, (iii) from glass to air, (iv) from air to water, (v) from glass to water.

**2.** A ray of light is incident at an angle of 60° in air on a block of glass; the angle of refraction is 35°. Calculate the refractive index. Repeat the calculation if the angle of refraction is 31° and the angle of incidence is then 50°.

**3.** Define *refractive index*. A ray of light is incident at an angle of 65° on a block of glass of refractive index 1·5. Draw the refracted ray accurately, and verify your drawing by calculating the angle of refraction.

**4.** A ray of light *in water* is incident at an angle of 30° on the plane boundary with air. Draw accurately the refracted ray, assuming the refractive index of water is 1·33. Verify your answer by calculation.

**5.** Explain why a pool of water appears shallower than is actually the case. A pool of water is 4 ft. deep; what is the apparent depth if the refractive index of water is 4/3 ?

**6.** Describe a method of measuring the refractive index of glass in the form of (i) a rectangular block, (ii) a triangular prism, showing how your answer is calculated in each case.

**7.** A travelling microscope is focused on a piece of paper on a table, and the reading is then 8·54 cm. A glass block is now placed on the paper and the reading is 6·97 cm. when the microscope is focused again on the paper. The reading is 4·23 cm. when it is focused on top of the block. Calculate the refractive index of the glass from the three readings.

**8.** Explain the meaning of *critical angle, total internal reflection*; draw diagrams to illustrate your answer. Can a critical angle be obtained when light travels from (i) air to glass, (ii) water to air, (iii) glass to air, (iv) air to water ?

**9.** The refractive index of a specimen of glass is 1·51; calculate its critical angle. What is the critical angle from water to air if the refractive index of water is 4/3 ?

**10.** Draw a diagram of a glass prism which can act as a good reflector of light, and explain the action.

**11.** A ray of light is incident in glass at an angle of (i) 25°, (ii) 41°, (iii) 42°, (iv) 70°, on a plane boundary with air. The critical angle of the glass is 41°. Draw sketches showing the behaviour of the light in each case.

**12.** Explain how a mirage is formed.

**13.** Answer *two* of the following: (*a*) Explain what is meant by *total internal reflection*, stating clearly the conditions for it to occur. Give a practical example of its application. (*b*) Explain, with the help of a diagram, the bent appearance of a straight stick dipping into water at an angle to the surface. (*c*) Distinguish between a *real* and a *virtual* image, and by means of ray diagrams, show how each type of image may be produced by a concave mirror. (*L.*)

**14.** Describe an experiment to find the refractive index of a liquid. (*L.*)

**15.** With the help of a diagram explain the term *refractive index*. (*N.*)

**16.** To an observer in an aeroplane a submerged submarine seems to be nearer the surface than it actually is. Explain this with the help of a diagram. Describe how you would determine the refractive index of a sample of sea-water, using the method of real and apparent depth and proving any formula you use. (*N.*)

**17.** Under what conditions is light totally reflected at the surface separating two media ? Show, by means of a diagram, how a beam of light may be turned through a right angle by means of an isosceles right-angled glass prism. Explain why the same effect could not be produced with a similar prism made of ice. (Critical angle for glass-air surface = 41°; critical angle for ice-air surface = 50°.) (*C.*)

**18.** State the laws of refraction of light, illustrating your statements by a diagram. How would you verify one of the laws experimentally ? A point source of light is about 1 in. from one face of a rectangular slab of glass, of thickness about 3 in., and is observed from beyond the opposite face. Draw a ray diagram showing the path of a divergent beam of light from the point source to the eye of the observer. (*C.W.B.*)

**19.** Explain, by the aid of suitable diagrams, the meaning of *total internal reflection*, and *critical angle*. A man stands on a ladder in a swimming bath, so that his head is just above the water, and observes the light from a lamp held at the bottom of the bath. As the lamp is moved horizontally away from him he observes that the lamp cannot be seen when it is more than 7·5 ft away from him, measured horizontally. Explain this observation and calculate the depth of the water. Take the refractive index of the water as 4/3. (*L.*)

**20.** How would you determine experimentally the angle of deviation produced by a prism when light falls on one of its faces at an angle of incidence of 70 degrees ? A ray of monochromatic red light is incident perpendicularly on one of the faces of a glass prism of refracting angle 30 degrees. Find (a) the angles of incidence and refraction at the other face, and (b) the deviation produced by the prism. State in each case whether these angles would be greater or less or unchanged if the incident light had been blue. (Refractive index of the glass for red light 1·50.) (N.)

**21.** State the *laws of refraction* and explain the terms *critical angle, total internal reflection*. Show that the critical angle for glass of refractive index 1·5 is 42° approximately. Hence show, with the aid of a diagram, how a right-angled prism of refractive index 1·5 can be used to reverse the direction of a beam of light. (O. & C.)

**22.** Explain the terms *refractive index, critical angle, total internal reflection*. Deduce the relation between the refractive index and critical angle. Explain the appearance of a stick partially immersed obliquely in water. (O. & C.)

**23.** Sodium light falls normally upon one of the faces forming the refracting angle of a glass prism. Find the angle through which a ray is deviated in passing the prism, given that the refractive index of the glass for sodium light is 1·5, and that the refracting angle of the prism is 30°.

With the help of diagrams describe and explain what would happen at the second face if (a) white light were used instead of sodium light, (b) the refracting angle of the prism were 45° instead of 30°. (N.)

**24.** How would you find the apparent depth of water in a beaker ? The following results were taken from a school note-book:

| Actual depth of water, cm. | 6 | 8 | 10 | 12 | 14 | 16 |
|---|---|---|---|---|---|---|
| Apparent depth, cm. | | 4·8 | 5·7 | 7·8 | 9·0 | 10·2 | 12·1 |

Plot the results on the graph paper and draw the best straight line passing through the origin and these points. From the graph deduce the refractive index of water. (N.)

# CHAPTER XXI

# REFRACTION AT CURVED SURFACES

FROM ancient times, lenses have been known to possess the property of concentrating light on one point, and hundreds of years ago a glass sphere containing water was used as a "burning glass." The first sunshine recorder, made in 1857, utilized a glass globe to char a paper graduated in fractions of an hour, and a good modern example is shown in Fig. 223 on p. 320.

**Types of lens.** One or both of the surfaces of a glass lens are usually spherical in shape, and Fig. 220 illustrates six common types of lens. A lens thicker in the middle than at the edges is known as a **converging** or **convex** lens (Fig. 220 (*a*) ), while a lens thinner in the middle than at the edges is known as a **diverging** or **concave** lens (Fig. 220 (*b*) ).

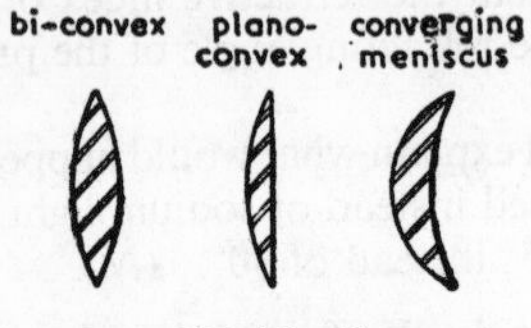

bi-convex    plano-    converging    bi-    plano-    diverging
             convex    meniscus     concave  concave   meniscus

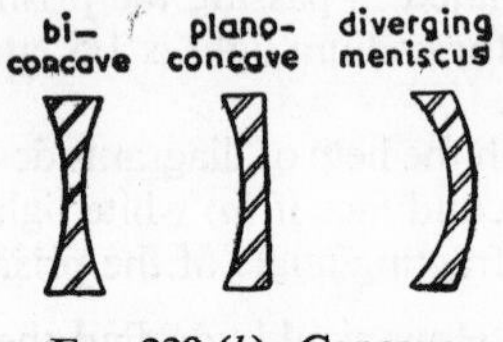

FIG. 220 (*a*). Convex         FIG. 220 (*b*). Concave
(converging) lenses             (diverging) lenses

Lenses play a vital part in everyday life. The eye has a natural lens inside it, and manufactured lenses are used in spectacles, cameras, telescopes, and microscopes. Some types of spectacles contain converging lenses, others contain diverging lenses; one type of telescope has a lens made by combining a converging and a diverging lens.

**The converging (convex) lens. Action on rays.** If a parallel beam of light from a ray box is incident on a small central portion of a convex lens, experiment shows that, after refraction through the

318

glass, the rays converge to a point $F$ on the principal axis of the lens (Fig. 221). $F$ is known as the **principal focus** of the lens, and in the case of a convex lens it is a *real* focus. The distance from $F$ to the lens, $C$, is known as the **focal length** of the lens, and its magnitude is governed by the curvature of the two surfaces of the lens and the refractive index of the glass.

As light rays are reversible, a very small lamp at $F$ gives out a divergent beam which emerges parallel after refraction through the lens. Further, a beam of parallel light incident on the central portion of the lens in Fig. 221 from right to left converges to a

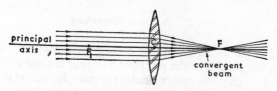

FIG. 221. Convex lens has REAL focus

focus at a point $F_1$ on the other side of the lens, and this point is also a principal focus of the lens. The distances of $F_1$ and $F$ from the lens are equal. A divergent beam from a small luminous source at $F_1$ emerges parallel after refraction through the lens.

Thus a ray parallel to the principal axis incident on the lens passes through a focus after refraction, and a ray through a focus incident on the lens emerges parallel to the principal axis.

**Spherical aberration.** If a *wide* beam of light is incident on a convex lens, experiment shows that the outer rays are brought to a focus nearer to the lens than the rays incident on the central portion of the lens, which are brought to a focus at the principal focus. This phenomenon is known as *spherical aberration*, and results in an imperfect focus (compare p. 280). In practice, therefore, a lens is usually small, and beams of light are incident only near its centre.

**How a convex lens brings parallel rays to a focus.** A lens can be regarded as being made up of a very large number of truncated prisms, whose sides are inclined at increasing angles as we proceed from the middle to the top or bottom of the lens. Fig. 222 illustrates a lens "cut" into a few of such prisms, the prisms at the top and bottom, $X$, $Y$, having the greatest refracting angle. Now, in general, a prism deviates a ray towards its base (see p. 304), and, if the direction of the incident ray is constant, the larger the

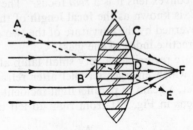

FIG. 222. Action of convex lens

refracting angle of the prism, the greater the deviation. Consequently a ray parallel to the principal axis near the top of the lens is deviated much more than a similar ray near the central portion; applying this result to the deviation caused by every part of the lens, it can be shown that the parallel beam of rays is brought to a focus at one point, *F*.

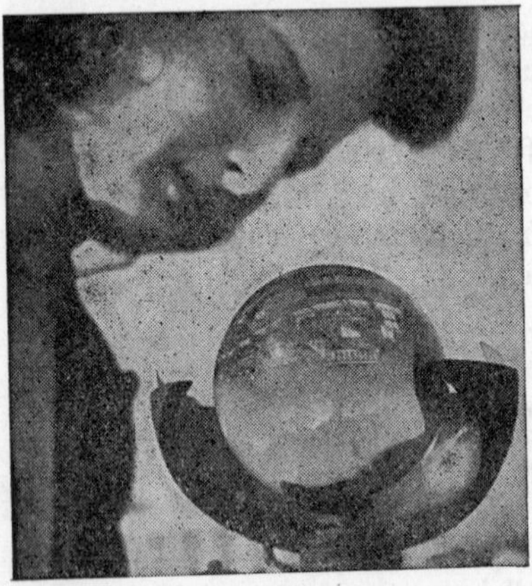

In drawing images of objects in lenses we often utilize a ray incident on the central portion, *C*, of the lens, such as the ray *AB* in Fig. 222. If we imagine it apart from the remainder of the lens, *C* is a *plane-parallel* piece of glass, and hence *AB* emerges in a parallel direction *DE* (p. 298). Now in practice lenses are made thin, so that *BD* is very small, and hence the incident ray *AB* is only very slightly displaced along *DE*. We therefore consider that the

FIG. 223. SUNSHINE RECORDER

—on the roof of the Air Ministry, Kingsway, London. It consists of a 4 in. glass sphere which brings the sun's rays to a focus on a card graduated to cover 12 hours, mounted in a metal bowl.

ray *AB* passes *straight through* the lens, and this is true for any ray incident on the middle of the lens. We shall denote the middle of the lens by *C* in sketches.

**Drawing of images in convex lenses.** As with mirrors, an object in front of a lens will be drawn as a straight line perpendicular to the principal axis, with an arrow at its head. We shall assume that a *thin* lens is used, and that rays are incident close to the principal axis, in which case we must draw the lens as a straight line to obtain accurate images (compare p. 282).

(1) *Object further from lens than the focal length.* A ray from the top point *A* of the object parallel to the principal axis is refracted so that it passes through the focus *F* after refraction (Fig. 224 (*a*)). Another ray from *A* incident on the middle, *C*, of the lens passes straight through, and the two rays emerging from the lens intersect on the side away from the object at *B*, which is therefore the image of *A*. Although we utilize two rays from *A*, any other ray from the latter would pass through *B* after refraction through the lens. For example, a ray *AF₁* through the focus *F₁* on the left side of the lens emerges parallel to the principal axis and passes through *B*.

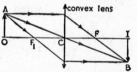

FIG. 224 (*a*). Object OUTSIDE focus—real, inverted image

The point *B* is below the axis of the lens. By drawing a perpendicular to the axis from *B* to obtain the rest of the image, it can be seen that the latter is *real* and *inverted*. As the object is moved nearer to the lens the size and distance of the image increases, but the image remains real and inverted until the object reaches the focus.

(2) *Object at focus.* Rays from an object at the focus give rise to parallel rays after they are refracted through the lens

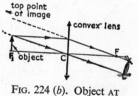

FIG. 224 (*b*). Object AT focus

Fig. 224 (*b*) illustrates two of the parallel emergent rays obtained from the top point of the object, whose image is therefore observed a very long way from the lens (theoretically at infinity).

(3) *Object nearer to the lens than the focal length.* In this case an erect virtual image is obtained (Fig. 225). A ray *AM* from the top point *A* parallel to the principal axis passes through the focus *F* after refraction, and the ray *AC* through the centre of the lens passes straight through along *CY*. The two emergent rays *MF*, *CY* are *divergent*, however, and hence cannot intersect on the side of the lens opposite to the object. Consequently the image of the point *A* is *virtual*, and is located at *B*, which is on the same side of the lens as the object; the remainder of the image is obtained by drawing the perpendicular from *B* to the principal axis.

**The simple magnifying glass.** The virtual image obtained above is much larger than the object and is erect; and, although it cannot be received on a screen, an observer at *E* sees the image by the

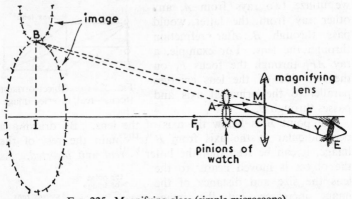

Fig. 225. Magnifying glass (simple microscope)

divergent cone of rays entering the eye (Fig. 225). Thus, if a page of this book is examined by a convex lens held nearer to it than its focal length, a magnified image is observed which is the same way up as the words, the object in this case. If the lens is held too far away from the object examined, so that the latter is further away from the lens than its focal length, an *inverted* image of the object is formed, as illustrated in Fig. 224 (*a*).

**Summary : converging (convex) lens.** When an object is a very long way from a convex lens a real inverted image is formed at the focus. As the object approaches the lens the image becomes larger, but is real and inverted until the object reaches a focus, when the image is formed a very long way from the lens. Beyond the focus, nearer to the lens, a magnified, erect, virtual image is obtained.

**The action of a diverging (concave) lens.** If a beam of parallel rays is incident on the central portion of a thin concave (diverging) lens, experiment shows that a *divergent* beam emerges from the lens after refraction (Fig. 226). Moreover, all the rays pass through

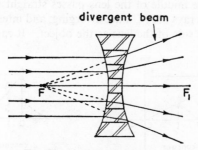

Fig. 226. Action of concave lens

a point $F$ on the principal axis after they are produced back and, by analogy with the convex lens, $F$ is termed the *principal focus*. The distance from $F$ to the lens is the focal length. It can now be seen that the concave lens has a *virtual* focus, whereas the convex lens has a real focus.

If parallel rays are incident on the lens from right to left in Fig. 226, the rays diverge after refraction and appear to come from the point $F_1$, which is at the same distance from the lens as $F$. The point $F_1$ is also a focus of the lens. As shown in the case of the convex lens, the action of a concave lens on a parallel beam of rays can be explained by considering the whole lens as made up of a large number of truncated prisms whose refracting angles increase away from the centre of the lens. The parallel rays are deviated towards the bases of the prisms by the refraction, and are hence bent away from the principal axis (compare the action of the convex lens, p. 320). The greatest deviation occurs for rays incident on the prism furthest away from the centre of the lens, and a divergent beam is thus obtained.

A ray incident on the middle of the thin lens is refracted so that it passes straight through, as the central portion of the lens is a thin parallel-sided block of glass (see p. 320).

**Images formed by a concave lens.** Consider an object $OA$ in front of a concave lens (Fig. 227 (*a*) ). A ray $AM$ from the top point $A$ parallel to the axis of the lens is refracted along $MY$ so that the latter appears to come from the focus, $F$, and a ray $AC$ incident on the middle of the lens passes straight through along $CX$. Thus the rays $MY$, $CX$ are diverging, and intersect at a point $B$ on the same side of the lens as the object. It can now be seen

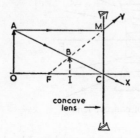

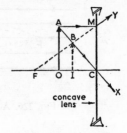

FIG. 227 (*a*). Virtual erect image

FIG. 227 (*b*). Virtual erect image

that a *virtual* erect image, $BI$, is obtained, which is smaller than the object. Unlike that obtained with a convex lens, the image obtained with a concave lens is always virtual, erect, and diminished, *no matter where the object is placed*. Fig. 227 (*b*) illustrates the formation of the image when the object is nearer to the lens than the focus.

**The lens formula.** Instead of drawing the paths of rays, the position of the image formed by a lens may be found by applying the formula

$$\frac{1}{v} + \frac{1}{u} = \frac{1}{f} \qquad . \qquad . \qquad . \qquad . \qquad (1)$$

where $u$, $v$ are the respective distances of the object and image from the lens and $f$ is the focal length. This is the same formula as was used with curved mirrors (p. 285).

The sign convention must again be observed when numerical values are substituted in the formula, i.e. real object and real image distances are positive in sign, but a virtual image distance is negative in sign. Thus, since a convex lens has a real focus and a concave lens has an imaginary focus,

*the focal length of a convex lens is positive,*

*the focal length of a concave lens is negative.*

From Fig. 224 (*a*), with a convex lens, or Fig. 227 (*a*) with a concave lens, it can be seen that

$$\frac{\text{length of image } (BI)}{\text{length of object } (AO)} = \frac{CI}{CO} = \frac{\text{image distance}}{\text{object distance}}.$$

Hence the magnification ($m$) produced by a lens is given numerically by

$$m = \frac{v}{u} \quad . \quad . \quad . \quad . \quad . \quad (2)$$

The following examples on lenses should be carefully studied.

1. A convex lens has a focal length of 20 cm., and an object is placed (*a*) 25 cm., (*b*) 15 cm. from it. Calculate the position of the image, and magnification, in each case.

The lens is convex; hence $f = +20$.

(*a*) Substituting $u = +25$ in $\dfrac{1}{v} + \dfrac{1}{u} = \dfrac{1}{f}$, we have

$$\frac{1}{v} + \frac{1}{(+25)} = \frac{1}{(+20)}$$

$$\therefore \frac{1}{v} + \frac{1}{25} = \frac{1}{20}, \text{ i.e. } \frac{1}{v} = \frac{1}{20} - \frac{1}{25}$$

$$\therefore \frac{1}{v} = \frac{5-4}{100} = \frac{1}{100}, \text{ i.e. } v = 100.$$

Since $v$ is positive, the image is real and 100 cm. from the lens (see Fig. 224 (*a*)).
The magnification, $m$, $= \dfrac{v}{u} = \dfrac{100}{25} = 4$.

(b) Substituting $u = +15$ in $\dfrac{1}{v} + \dfrac{1}{u} = \dfrac{1}{f}$, we have

$$\frac{1}{v} + \frac{1}{(+15)} = \frac{1}{(+20)}$$

$$\therefore \frac{1}{v} = \frac{1}{20} - \frac{1}{15} = \frac{3-4}{60} = -\frac{1}{60}$$

$$\therefore v = -60$$

Since $v$ is negative, the image is virtual and 60 cm. from the lens (see Fig. 225). The magnification, $m$, $= \dfrac{v}{u} = \dfrac{60}{15} = 4$.

2. A concave lens has a focal length of 12 cm., and an object is placed 60 cm. in front of it. Find the position of the image and the magnification.

The lens is concave; hence $f = -12$. Substituting $u = +60$ in the formula $\dfrac{1}{v} + \dfrac{1}{u} = \dfrac{1}{f}$, we have

$$\frac{1}{v} + \frac{1}{(+60)} = \frac{1}{(-12)}$$

$$\therefore \frac{1}{v} + \frac{1}{60} = -\frac{1}{12}$$

$$\therefore \frac{1}{v} = -\frac{1}{12} - \frac{1}{60} = -\frac{5+1}{60} = -\frac{6}{60}$$

$$\therefore v = -\frac{60}{6} = -10$$

The image is thus virtual and 10 cm. from the lens (see Fig. 227). The magnification, $m = \dfrac{v}{u} = \dfrac{10}{60} = \dfrac{1}{6}$.

## Experimental determination of $f$ for a convex lens.

(1) *Approximate method.* A quick method of finding the focal length of a convex lens consists of holding the lens far away from a window in a room, and then moving the lens until the inverted image, $A$, of the window and its frame is shown as clearly as possible on a screen in front of the lens (Fig. 228 (a) ). Since the window acts as a "distant" object, the distance from $A$ to the lens is approximately the focal length, $f$.

(2) *Formula method.* This consists of setting up an illuminated gauze triangle in front of the lens, and then obtaining the

inverted image of the triangle clearly on a screen placed on the other side of the lens (Fig. 228 (b) ). Since the object distance, $u$, and the

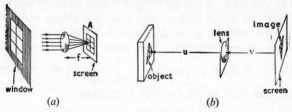

FIG. 228. Focal length by (a) approximate (b) accurate method

image distance, $v$, from the lens are known, it is possible to calculate the focal length, $f$, by substituting their values in the lens formula $\dfrac{1}{v} + \dfrac{1}{u} = \dfrac{1}{f}$. The experiment is repeated with varying values of $u$ and $v$, and the average of the results for $f$ is calculated. This method gives accurate results.

(3) *Plane mirror method.* This is illustrated in Fig. 229 (a). The lens $L$ is placed in front of a plane mirror $M$, and an object $O$, such as an illuminated gauze triangle, is moved in front of the lens until the image $I$ is formed at the same place. The rays

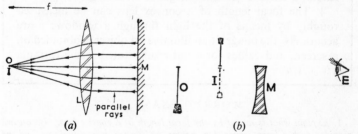

FIG. 229. Focal length of (a) convex (b) concave lens

striking $M$ must now be returning along their original path, and hence all these rays must be perpendicular to $M$. Consequently the rays incident on the lens $L$ from $M$ are parallel, and $O$ and $I$ are situated at the focus of $L$. Thus the focal length is the distance from $O$ to the lens. This method also is accurate.

**Method for focal length of concave lens.** The focal length of a concave lens $M$ can be found by placing an object pin $O$ in front of it (Fig. 229 $(b)$). A virtual image $I$ is then seen through the lens by an observer $E$. A locator pin, held in a clamp and stand, is then moved until it coincides with $I$ by no parallax. Then $IM = v =$ image distance, and $OM = u =$ object distance. Substituting in the lens formula $\frac{1}{v} + \frac{1}{u} = \frac{1}{f}$ and remembering $v$ is negative, the focal length $f$ can be calculated.

---Summary---

1. A convex (converging) lens has a real focus; a concave (diverging) lens has a virtual focus.

2. With a convex lens: $(a)$ The image is inverted and real when the object is further from the lens than its focal length, $(b)$ the image is magnified, erect and virtual when the object is closer to the lens than its focal length (magnifying glass principle).

3. **The formula for a lens is $1/v + 1/u = 1/f$; the magnification, $m = v/u$.** "Real is positive, virtual is negative" is the sign convention. **A convex lens has a positive focal length, a concave lens has a negative focal length.**

4. The focal length of a convex lens can be measured roughly by means of the light through a window; more accurately, the image of an illuminated object is obtained on a screen, and values for $u$ and $v$ substituted in the formula $1/v + 1/u = 1/f$.

### WORKED EXAMPLES

1. *Explain what is meant by the focal length of a convex lens. If you are provided with a lamp, gauze, lens-holder, and screen, how would you find the focal length of a convex lens? A convex lens of focal length 5 cm. is used as a magnifying glass. How far from the object must the lens be placed so that the size of the virtual image is four times that of the object?* (L.)

Let $x$ cm. be the distance of the lens from the object. Then, since $m = v/u = 4$, $v = -4x$ cm.; the image is virtual. Substituting in $\frac{1}{v} + \frac{1}{u} = \frac{1}{f}$, we have, since $f$ is $+5$ cm., $\frac{1}{(-4x)} + \frac{1}{(+x)} = \frac{1}{+5}$

$$\therefore \quad -\frac{1}{4x} + \frac{1}{x} = \frac{1}{5}$$

$$\therefore \quad \frac{-1+4}{4x} = \frac{1}{5}, \text{ i.e. } \frac{3}{4x} = \frac{1}{5}$$

$$\therefore \quad 4x = 15, \text{ i.e. } x = \frac{15}{4} = 3\tfrac{3}{4} \text{ cm.}$$

2. *Describe how you would proceed in order to find the diameter of a ball-bearing with the aid of a convex lens, a screen, a metre stick, and an electric lamp. A small object, $\frac{1}{2}$ in. high, stands on and perpendicular to the axis of a convex lens and a real image $1\frac{1}{2}$ in. high is formed on a screen 12 in. from the object. Make a scale diagram of the arrangement and from it (or otherwise) find the distances of the object and image from the lens, and the focal length of the lens.* (L.)

Arrange the lamp $A$ and the ball-bearing $O$ in front of the lens $L$ so that a *magnified real* image is formed on the screen $S$ on the other side of $L$ (Fig. 230). The diameter, $d$, of the image is then measured, together with

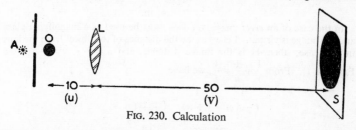

Fig. 230. Calculation

the distances of the image and object, $O$, from $L$. The magnification, $m$, is given by $\frac{v}{u}$ with the usual notation; so that if $v = 50$ cm., and $u = 10$ cm., $m = \frac{50}{10} = 5$. It follows that the diameter, $d$, of the ball-bearing is $\frac{1}{5}$ of the measured diameter of the image, and thus $d$ can be calculated.

A scale drawing is left to the reader. The magnification $= \frac{1\frac{1}{2}}{\frac{1}{2}} = 3$, and hence the object distance: image distance (measured from lens) $= 1 : 3$. Thus the object distance $= \frac{1}{4} \times 12$ in. $= 3$ in., and the image distance $= \frac{3}{4} \times 12$ in. $= 9$ in. The focal length, $f$, is given by

$$\frac{1}{v} + \frac{1}{u} = \frac{1}{f}$$

$$\therefore \quad \frac{1}{9} + \frac{1}{3} = \frac{1}{f}, \text{ i.e. } \frac{1}{f} = \frac{4}{9}$$

$$\therefore \quad f = \frac{9}{4} = 2\tfrac{1}{4} \text{ in.}$$

**3.** *What is meant by the focal length of a convex lens? Describe an experiment to determine its value. A luminous object is held 6 in. in front of a convex lens, and an inverted image magnified three times is formed. How far from the object must the same lens be placed in order to form an erect image magnified four times?* (*C.W.B.*)

Since magnification $= \dfrac{v}{u} = 3$, the image distance from the lens $= 3 \times 6$ in.

$= 18$ in. The image is real, as it is inverted (see p. 322); hence $u = +6$ in. $v = +18$ in. Substituting in $\dfrac{1}{v} + \dfrac{1}{u} = \dfrac{1}{f}$, we have

$$\frac{1}{(+18)} + \frac{1}{(+6)} = \frac{1}{f}$$

$$\therefore \quad \frac{1}{18} + \frac{1}{6} = \frac{1}{f}$$

$$\therefore \quad \frac{4}{18} = \frac{1}{f}, \text{ i.e. } 4f = 18$$

$$\therefore \quad f = 4\tfrac{1}{2} \text{ in.}$$

In the case of an *erect* image, the lens must be used as a magnifying glass and the image is virtual. Let $x$ cm. be the distance of the object from the lens in this case; then $4x$ is the image distance, and $u = +x$, $v = -4x$ (see Example 1). From $\dfrac{1}{v} + \dfrac{1}{u} = \dfrac{1}{f}$, we have

$$\frac{1}{(-4x)} + \frac{1}{(+x)} = \frac{1}{(+4\tfrac{1}{2})} = \frac{2}{9}$$

$$\therefore \quad -\frac{1}{4x} + \frac{1}{x} = \frac{2}{9}, \text{ i.e. } \frac{3}{4x} = \frac{2}{9}$$

$$\therefore \quad 8x = 27, \text{ i.e. } x = 3\tfrac{3}{8} \text{ in.}$$

## EXERCISES ON CHAPTER XXI

**1.** Name four applications of lenses.

**2.** Draw diagrams illustrating the action of a parallel beam of light on (i) a convex lens, (ii) a concave lens. Mark clearly the focus in each case. What is the *sign* of the focal lengths of a concave and a convex lens?

**3.** Why is a convex lens usually called a *converging lens* and a concave lens called a *diverging lens*? An object 2 in. high is placed 12 cm. in front of a converging lens of focal length 8 cm. Draw accurately to scale the image formed, and obtain the position and magnification from your drawing. (*Note.* A lens must be represented by a straight line in accurate drawings of images.)

**4.** By using the lens and magnification formulae, calculate the image position in question 3 and the magnification.

**5.** Where must an object be positioned to form an erect magnified image with a convex lens ? An object 2 cm. high is placed 5 in. in front of a convex lens of focal length 6 in. Draw the image formed to scale, and obtain its position and magnification from the drawing.

**6.** Verify your answers in question 5 by calculation.

**7.** What kind of image is formed by a concave or diverging lens ? An object 1·5 in. high is placed 24 cm. from a concave lens of focal length 16 cm. Find by drawing the image position and its size.

**8.** Verify your answer to question 7 by using formulae.

**9.** Draw a diagram showing how a convex lens is used as a magnifying glass.

**10.** An inverted image, three times the size of the object, is obtained with a convex lens of focal length 20 cm. Calculate the position of the object. What would be the new position if the magnified image was erect instead of inverted ?

**11.** Define (a) principal focus, (b) focal length, of a convex lens. Describe how you would make an accurate determination of the focal length, showing how the result is calculated. Draw a labelled ray diagram to show how a convex lens is used as a magnifying glass. (L.)

**12.** A converging (convex) lens forms a real image of a small object magnified four times. The distance between object and image is 100 cm. By means of a scale drawing (only) deduce the position of the lens and its focal length. (L.)

**13.** Describe *two* methods of producing a parallel beam of light. A parallel beam of light falls on a concave lens in a direction parallel to its principal axis, and after passing through the concave lens is again made parallel with the aid of a convex lens. Show with a diagram how this may be done, and explain the fact that any convex lens will serve the purpose provided that its focal length is numerically greater than that of the concave lens. Find the separation of the lenses if the focal lengths of the lenses are 12·5 cm. and 50·0 cm. (N.)

**14.** Draw diagrams to illustrate how a convex lens may produce a magnified image (a) which is real, (b) which is virtual. A convex lens situated 6 in. from an object produces a real image four times the size of the object. Where must the lens be placed to give a virtual image three times the size of the object ? (*C*.)

**15.** What is meant by the focal length of a lens ? Calculate at what distance from a converging lens of focal length 20 cm. an object must be placed in order that the image may be (*a*) real, and three times the size of the object; (*b*) virtual, and three times the size of the object. Draw diagrams to illustrate the two cases. (*O. & C.*)

**16.** A luminous object is placed at a given distance from a convex lens of 15 cm. focal length and a real image is produced. If the object distance is equal to the image distance, calculate its value. Give a ray diagram to explain the formation of the image. (*C.W.B.*)

**17.** What is the *focal length* of a lens ? Describe an experiment to determine its value for a convex lens. When a luminous object is held 24 cm. from a convex lens an inverted image is formed 24 cm. from the lens. What is the focal length of the lens ? How must the object be moved in order that a virtual image may be produced at the position from which the object has moved ? (*L.*)

**18.** What do you understand by the terms *principal focus* and *focal length* of a lens ? Illustrate your answer with labelled ray diagrams for converging (convex) and diverging (concave) lenses. An object is placed on the principal axis of a convex lens of focal length 5 cm. Find the magnification produced by the lens when the distance of the object from it is (*a*) 6 cm., (*b*) 4 cm. Distinguish between the natures of the images found in the two cases. (*N.*)

**19.** State the formula relating the position of object and image for a convex lens. State clearly the sign convention you use. How would you demonstrate experimentally the relation you give ? An object is placed between a convex lens of focal length 10 cm. and a plane mirror, being 15 cm. distant from each. Two real images will be formed by the lens on the opposite side from the object. What will be the positions of these images ? (*O. & C.*)

**20.** State the formula connecting the distance of an object from a lens with the distance of its image from the lens, stating the sign convention which you use. Two convex lenses have different focal lengths. When a pin is held at a distance of 60 cm. from one lens, a real image is produced 120 cm. from the lens, while if the pin is held at 60 cm. from the other lens, a virtual image is produced 120 cm. from the lens. Find the focal lengths of the two lenses. Draw diagrams illustrating the formation of the images in the two cases. (*O. & C.*)

# CHAPTER XXII

## PRINCIPLES OF OPTICAL INSTRUMENTS

**T**HE eye. No mechanic could fashion a lens to compete in range and action with the human eye, which can adjust itself easily to objects at varying distances and has a natural "stop" for protecting itself from light of too great intensity. The eye has a lens ($L$) made of tough gelatinous material (Fig. 231). The lens forms an image of the object on the "screen", known as the *retina* ($R$), at the back of the eye. The most light-sensitive spot on the retina is called the *yellow spot* ($S$) or *fovea centralis*. The light impulses here are conveyed to the brain by the optic nerve $N$ and give rise to the sensation of vision. Although the image is inverted on the retina because $L$ is a convex lens, the mind interprets the image as that of an upright object. The ability of the ciliary

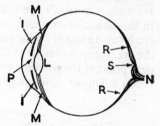

FIG. 231. The eye

muscles ($M$) to alter the focal length of the lens $L$ enables the eye to adjust itself to objects at different distances; we call this action the *power of accommodation* of the eye.

The *pupil* is a hole in the middle of the *iris* ($I$) or coloured diaphragm, and appears black because no light is reflected from it.

If the amount of light received by the eye is great, the pupil $P$ in front of the lens $L$ contracts, and reduces the amount of light entering the eye. If the amount of light received by the eye is small, the pupil expands.

If objects in a room are viewed with one eye closed, they appear to be "flatter" than if both eyes are open. Similarly, the action of touching an object is performed more easily with both eyes open than with one eye closed. The use of two eyes enables two slightly different impressions of a scene to be obtained. This gives a sense

12                                    333

of distance and perspective which is lacking if one eye is used, and objects stand out in relief, giving a *stereoscopic effect*.

**Accommodation.** When a very distant object (an object at "infinity") is observed, the ciliary muscles are fully relaxed, and the lens's surfaces are flattened. The eye is then said to be *unaccommodated*. When the eye is focussed on an object at a finite distance from it, the ciliary muscles are tense, and the lens is more convex. The eye is then said to be *accommodated*. An object cannot be observed without strain by most people at a distance of less than about 25 cm., or 10 in., which is known as the *least distance of distinct vision*. The point about 25 cm. from the eye is then known as its *near point*. The *far point* of a normal eye is "infinity."

**Some defects of vision, and how they are corrected.**

**Long sight.** Due to defects of vision such as a short eyeball for example, a person's eye may not be able to focus near objects. In this case, for example, an object at $B$ distant 50 cm. from the eye forms an image on the retina $R$, but an object at $A$, 25 cm. from the eye, forms an image at $C$ behind the retina (Fig. 232 (*a*)). The near point of the person's eye is thus 50 cm. away. This defect of vision is known as **long sight.**

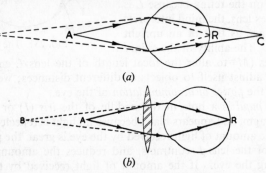

Fig. 232. Long sight and its correction

A *convex* lens, $L$, in front of the eye corrects this defect. As illustrated in Fig. 232 (*b*), light from an object at $A$ now appears to be coming from the point $B$, so that an image is formed on the retina.

*Presbyopia.* Due to old age, when the eye lens becomes inelastic for example, some people are unable to see distant objects and parallel rays then converge to a point C behind the retina R Fig. 233 (*a*)). Using a convex lens M (Fig. 233 (*b*)), they are able to bring the light to a focus on the retina and and thus see distant

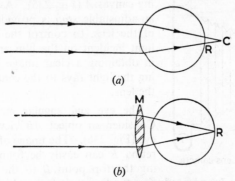

(*a*)

(*b*)

Fig. 233  Presbyopia and correction

objects clearly. Also, they may not be able to see near objects clearly. This defect of vision is known as *presbyopia*. It may be corrected both for near and far vision by wearing *bifocal* spectacles, whose lenses have a different upper and lower part.

**Short sight.** In some cases, due to a long eyeball, parallel light is brought to a focus at a point C in front of the retina R, and distant objects cannot be clearly seen. This defect is known as "short sight". The furthest point A from the eye which can now

(*a*)                                   (*b*)

Fig. 234.  Short sight and its correction

be focused on the retina (Fig. 234 (*a*)), is called the person's "far point". The defect can be compensated by a suitable *concave* lens L, as illustrated in Fig. 234 (*b*), in which case parallel light incident on $L$ appears to diverge from $A$ and hence is brought to a focus on the retina $R$.

**The photographic camera** is very similar in action to the eye. It has a lens $L$ in front, whose distance from the light-sensitive film

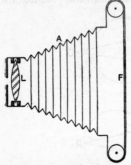

F can be adjusted by means of a folding canvas $A$ (Fig. 235). As in the eye, an adjustable *stop* is provided in front of the lens, to control the amount of light incident on the film and to assist in obtaining a clear image by restricting the light rays to the central part of the lens.

**The eye and angular magnification.** Consider an object $AB$ viewed by the eye (Fig. 236). The image of $AB$ on the retina $R$ can easily be found by joining the top point $B$ to the middle of

FIG. 235. Lens camera

the eye-lens $L$ and producing the line to $b$, from which it can be seen that $ab$ is the length of the image. Now the distance from $L$ to $R$ is practically constant for a given eye; hence the length of the image $ab$ is proportional to the angle $aLb$ ($a$). But vertically

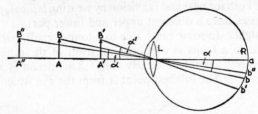

FIG. 236. Angular magnification

opposite angle $BLA = a$. Hence we arrive at the important result that *the length of the image on the retina is proportional to the angle subtended at the eye by the object.*

If the same object is moved to $A'B'$ nearer to the eye, the angle subtended by the object at $L$ increases, and the length of the image on the retina is now $ab'$. Thus, although the object has the same physical size in the two positions $AB$, $A'B'$, it appears to be bigger in the position $A'B'$. If the object is moved back to $A''B''$ the angle subtended at the eye is reduced, and hence the object appears smaller. The purpose of optical instruments such as the microscope and telescope is to increase the angle subtended at the eye by the object concerned.

**The simple microscope, or magnifying glass.** The most comfortable position for viewing an object is about 10 in. from the eye (p. 334), and we shall denote this distance by $D$. Suppose that an object $AO$ of height $h$ is at the distance $D$ from the eye and that the angle subtended at the latter is $\alpha$ (Fig. 237 (a)). Now suppose that a convex lens $L$ is placed in front of the eye so that a virtual erect image of the same object is observed again at a distance $D$

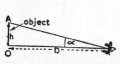

FIG. 237 (a). Unaided eye

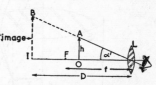

FIG. 237 (b). Eye aided by lens

from the eye, in which case the object must be nearer to the lens than its focus $F$ (see p. 322), as in Fig. 237 (b). The angle subtended by the virtual image $BI$ at the eye is now $\alpha'$, say, which is *greater* than the angle $\alpha$ subtended at the unaided eye (Fig. 237 (a)), so that the image on the retina is bigger when the lens is used.

The *angular magnification* ($m$), or, briefly, the *magnification*, produced by the lens is defined as the ratio $\dfrac{\alpha'}{\alpha}$, where $\alpha'$, $\alpha$ are the respective angles subtended at the eye in Fig. 237 (a), (b). It should be carefully noted that the image $BI$ in Fig. 237 (b) and the object $AO$ in Fig. 237 (a) are each at a distance $D$ from the eye, where they are most comfortably seen. The ratio $\dfrac{\alpha'}{\alpha}$ is thus $\dfrac{BI}{D} \Big/ \dfrac{AO}{D}$, or $\dfrac{BI}{AO}$; when this ratio is calculated in terms of $D$ and the focal length $f$ of the lens $L$, which is beyond the scope of this book, the result is

$$\text{angular magnification, } m = 1 - \frac{D}{f}.$$

If the focal length of the lens is $+ 2$ cm., and $D$ is $- 25$ cm. ($D$ is left of the lens), $m = 1 + \frac{25}{2} = 13.5$. It can be seen from the formula for $m$ that the magnification is greater the shorter the focal length $f$ of the convex lens used; but there is a limit to the magnification obtainable with a single lens, since the lens surfaces cannot be ground to make $f$ extremely small.

**The principle of the compound microscope.** The compound microscope is used extensively in biological researches, and in

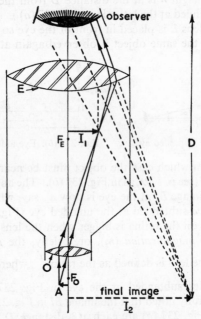

FIG. 238. Compound microscope

some researches in physics. Under the microscope, objects as small as bacteria and other carriers of disease have been magnified sufficiently for detailed examination.

Fig. 238 illustrates the principle of a microscope made from two convex lenses. The lens $O$ nearer to the object is known as the *objective lens*, and the lens $E$ near to the eye is known as the *eye-*

*piece lens.* The objective and eyepiece are contained inside a tube so that external light is excluded.

The microscope tube is moved so that the object $A$ is just *outside* the focal length of the objective $O$. An inverted real image, $I_1$, of $A$ is then obtained, and this is arranged to be at a distance from the eyepiece lens $E$ *smaller* than its focal length, so that the latter now acts as a simple magnifying glass for which $I_1$ is the "object." A large virtual image is therefore formed at $I_2$, and is normally at the near point, distant $D$ from the eye. It can be seen from the diagram that the angle subtended at the eye by the final image $I_2$ is much larger than the angle subtended at it by the object $A$ if the latter were placed at a distance $D$ from the *unaided* eye. Hence a large angular magnification is obtained by the use of the two lenses, which usually have short focal lengths.

**The principle of the telescope.** The microscope is used for viewing near objects, whereas the telescope is normally used for viewing distant objects.

A simple type of astronomical telescope consists of a convex objective lens $O$ and a convex eyepiece lens $E$ (Fig. 239). Suppose that a distant object is viewed, so that rays from a particular point on the object arrive *parallel* at the objective lens. Fig. 239 (*a*) illustrates three rays $A$ from the top point of the object, and three rays $B$ from the foot of the object which may be imagined on the axis of the lenses. It will be noted that the angle $\alpha$ between the two sets of parallel rays is the angle subtended at the lens $O$ by the distant object.

The image $I_1$, formed is at the focus $F$ of the objective lens; and if $F$ is also the focus of the eyepiece lens a final virtual image is observed through $E$ a very long way off, i.e. at infinity. Fig. 239 (*a*) traces the path of the rays forming the top point of the image of $I_1$ in $E$.

Since the foot of the final image is on the axis, the angle subtended at the eye by the final image is $\alpha'$ in Fig. 239 (*a*). The angle subtended at the unaided eye by the distant object is $\alpha$ to an excellent approximation, as the distance from the far object to $E$ is practically the same as the distance from the object to $O$. The angular magnification $m$ is hence given by $\dfrac{\alpha'}{\alpha}$, and in practice the focal length of the objective lens $O$ must be large and that of the eyepiece lens $E$ must be small to achieve a high angular magnification. From Fig. 239 (a), it can be seen that the distance between the two lenses is equal to the sum of their focal lengths.

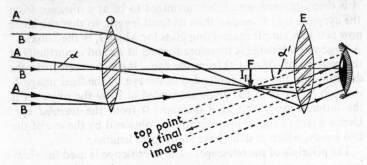

FIG. 239 (a). Telescope in normal use

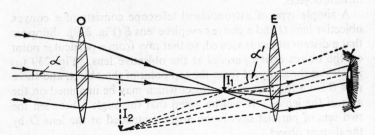

FIG. 239 (b). Final image at distance $D$ from eye

Fig. 239 (b) illustrates the formation of the final image $I_2$ in a telescope when it is viewed at the distance $D$ from the eye. This is not considered the "normal" use of the telescope, although the magnification is greatest and the distant object is observed most distinctly in this case.

**Prism binoculars** are short telescopes which have a convex objective lens and a convex eyepiece lens, as in the case of the telescope just described. Their special feature is a pair of right-angle isosceles prisms with their edges at right angles to each other, which are contained in the tube between the objective and eyepiece (Fig. 240). Light from the object enters the first lens, and the two prisms *invert* the image formed so that an upright image is finally viewed. In the absence of the prisms an inverted image of the object would be seen and the tube would have to be about three times the length of the binoculars tube to give the same magnification.

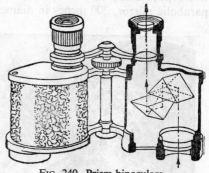

**Galileo's telescope.** About 300 years ago Galileo invented a telescope which is known by his name, and the principle of which is used in

FIG. 240. Prism binoculars

*opera glasses.* Unlike the telescope just described, which gives an inverted image (see Fig. 239 (*b*) ), Galileo's telescope gives an upright image.

The telescope consists of a convex objective lens *O* and a *concave* eyepiece lens *E* (Fig. 241). For simplicity, consider rays *A* coming

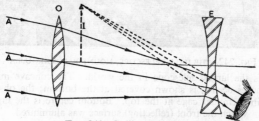

FIG. 241. Galilean telescope

from the top part of the distant object and incident on *O*. The rays are refracted, and are incident on *E* before they come to a focus, with the result that the rays are rendered divergent (this is the action of a concave lens on any beam of light). The observer thus receives rays which appear to come from the top point *I* of the virtual image in *E*, and since *I* and the top point of the distant object are both above the axis of the lenses the image is upright.

**The reflecting telescope. The Hale telescope.** The largest telescope in the world has been mounted for astronomical observations at Mount Palomar in California (Fig. 242). Unlike the telescopes already described, this one consists of a huge concave parabolic mirror, 200 inches in diameter, with a polished coating

FIG. 242. THE HALE TELESCOPE, MOUNT PALOMAR, U.S.A.

This is the largest telescope in the world. The concave mirror, 200 inches across, is shown covered at the bottom; the observer rides in a circular cage at the top. Bottom centre is the tank in which the front (reflecting) surface was aluminized

of aluminium on the *front* surface which acts as a reflector. The mirror is made large in order to collect as much light as possible from the stars and planets to be observed, and the absence of lenses eliminates chromatic aberration (p. 351) and other defects of images inherent in the use of lenses, such as the absorption of light by the glass. The idea of using a mirror for an astronomical telescope was first put forward by Newton.

Fig. 243 (*a*) illustrates the path of rays from a star, for example, incident on the concave paraboloid mirror, *P*. The light is reflected towards the focus, and is intercepted by a plane mirror, *M*, so that the light enters an eyepiece *E* through which it can be observed. Fig. 243 (*b*) illustrates the use of a convex mirror, C, instead of a plane mirror, a method suggested by Cassegrain.

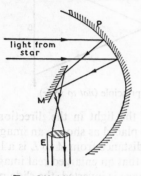

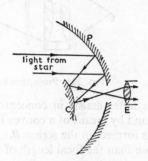

FIG. 243 (*a*). Newton reflector

FIG. 243 (*b*). Cassegrain reflector

The whole telescope system at Palomar has a special clockwork control for keeping the eyepiece trained on a particular object while the earth rotates on its axis. Photographic plates are also to be used in place of the eyepiece *E* to photograph the stars and planets. Many astronomical theories will be tested by the use of the 200-inch telescope. It is hoped to photograph Mars and to find out whether or not canals, for example, exist on the planet, as some scientists suggest. A knowledge of the relative abundance of the chemical elements in the stars, which may be obtained from photographs, may be of assistance in finding out how the sun derives its energy.

The glass of the Palomar mirror is made of a special pyrex material, like that used for cooking dishes, which has an extremely small coefficient of expansion. Cracking of the glass as it cooled after its heat treatment was thus avoided (see p. 171). The grinding and polishing of the glass was begun in 1936, and the telescope was ready for use in 1949.

**The principle of the projection lantern.** The essential features of a projection lantern are illustrated in Fig. 244. $S$ is a small incandescent lamp of very high candle-power, which illuminates the slide $A$ placed in front of a convex lens arrangement, $C$.

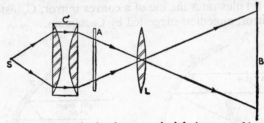

FIG. 244. Projection lantern principle (*not to scale*)

The latter assists in concentrating the light in the direction of $A$, and by means of a convex lens $L$ placed as shown, an image of $A$ is formed on the screen $B$. The distance from $A$ to $L$ is a little more than the focal length of $L$, so that an enlarged real image is formed on the screen. Since the image is inverted, the slide must be placed in an inverted position in order to obtain the picture the right way up.

---
**Summary**
---

1. The human eye has a "lens" and a "screen" known as the retina; the pupil acts as a "stop" controlling the amount of light entering the eye. The ciliary muscles alter the focal length of the lens ("power of accommodation").

2. The near point of a normal eye is about 10 in. or 25 cm. from the eye; the far point of a normal eye is at infinity.

Long sight (hypermetropia) is corrected by a convex lens. Short sight (myopia) is corrected by a concave lens.

3. **The length of the image on the retina is proportional to the angle subtended at the eye by the object.**

4. The **compound microscope** has two convex lenses of short focal length. The **simple astronomical telescope** has an objective convex lens of long focal length and an eyepiece convex lens of short focal length, the distance apart being approximately equal to the sum of the focal lengths.

## EXERCISES ON CHAPTER XXII

**1.** Where is the *near point* and *far point* of a normal eye ? What is the name of the "screen" on which an image is formed by the eye ? How does the eye focus objects at different distances from it ?

**2.** Describe why the eye and the photographic camera are similar. Point out two differences between them.

**3.** What is *short sight* ? How is it corrected ? Draw diagrams to illustrate your answers.

**4.** A picture 3 ft. from the eye appears smaller when viewed 30 ft. away. Explain this with diagrams.

**5.** On what does the apparent size of an object depend when it is viewed ? Draw a diagram of a lens acting as a magnifying glass, and explain the magnification by using the angles subtended at the eye.

**6.** How would you arrange two convex lenses to act as a telescope if their focal lengths are 2 cm. and 30 cm. respectively ? Draw a ray diagram showing how the final image is formed.

**7.** How would you arrange two convex lenses to act as a microscope ? Draw a diagram showing how the final image is formed.

**8.** How can a convex and a concave lens be set up to act as a telescope ? Draw a ray diagram showing the formation of the image, and state the advantage of this type of telescope.

**9.** Write an account of the largest telescope in the world.

**10.** Explain how a compound microscope may be made from two suitable thin lenses. Draw a diagram tracing the paths through the system of two rays from a point on the object not on the principal axis and show the positions of the intermediate and final images. The 2 in. objective of a microscope is replaced by one of $\frac{1}{4}$ in. focal length. State (*without attempting numerical calculation*) what must be done to bring the image into focus again, and what difference the exchange makes to the appearance of the final image. (*N.*)

**11.** What is meant by long sight and short sight, and how may they be corrected ? A long-sighted person cannot see distinctly objects nearer than 50 cm. What kind of lens must he use in order that he may be able to read a book at distances down to 25 cm. ? (*C.*)

**12.** Illustrate by a labelled diagram the use of two lenses as a simple telescope. (*C.W.B.*)

**13.** Describe a simple photographic camera with a single movable lens and explain how it works. The camera of a reconnaissance aeroplane has a lens of 9 in. focal length. A picture, enlarged 10 times from the original negative which was taken from a height of 10,000 ft., shows marshalling yards 27 in. long. What is the actual length of this target ? (N.)

**14.** Describe the optical system of the projection lantern, and explain (a) where the slide must be placed, relative to the principal focus of the projection lens, in order to give a sharply focused image on the screen, (b) how the slide must be arranged so that things depicted on the screen appear the right way up and the right way round. An object placed 8 cm. from a certain lens gives a virtual image at a distance of 12 cm. from the lens. Find the value of the focal length, and state whether the lens is concave or convex. (N.)

**15.** Describe how to arrange two lenses to form a compound microscope. What is the nature of the final image ? The objective lens of such a microscope has a focal length of 1 in., and is used to view an object $1\frac{1}{4}$ in. from it. Find the position of the image. This image is then viewed with the eye-lens which is placed $3\frac{1}{8}$ in. from it. If the eye-lens has a focal length of 5 in., where will the final image be formed ? (C.)

**16.** In a simple astronomical telescope, the objective has a focal length of 75 cm. and that of the eyepiece is 5 cm. If the lenses are $79\frac{1}{6}$ cm. apart, where will the final image of a distant star be formed ? (C.)

**17.** Draw a diagram to show the arrangement of the lenses in a compound microscope, stating clearly the kind of lens used. Trace the paths to the eye of **two** rays from a point of the object not on the axis of the microscope and mark on the ray diagram the relative positions of the *object* and *the focal point of the object lens*. If you were supplied with four lenses of focal lengths 100 cm., 25 cm., 3 cm., 2 cm., respectively, state, giving a reason, which lenses you would select as most suitable for constructing a compound microscope. (O. & C.)

**18.** State the theoretical formula connecting the distances of the image and object from a convex lens with its focal length. How may it be proved experimentally for real images ?
If you were desiring to make an astronomical telescope of magnifying power 8 what kind of lens would you purchase for the object glass and the eyepiece respectively ? What focal lengths would you select ? Draw a typical ray diagram showing the path through this instrument of two parallel rays incident at an angle with the principal axis. (O. & C.)

**19.** Illustrate, by ray diagrams, the effects of both *converging* and *diverging* lenses on (*a*) a parallel beam, (*b*) a converging beam of light. A person whose greatest distance of distinct vision is 4 ft. wishes to see a person distinctly at a distance of 20 ft. What *type* of spectacle lens will he require and of what numerical focal length? Draw a ray diagram to illustrate your answer. (*O. & C.*)

**20.** Describe, giving ray diagrams, how (*a*) a simple microscope, (*b*) a camera, produce their images.

A small object is viewed through a converging lens held close to the eye. An image 1·5 cm. long is formed 24 cm. from the lens whose focal length is 6 cm. Find, by calculation or a scale diagram, the size of the object and its distance from the lens. (*L.*)

**21.** Draw a large diagram (not less than 4 in. across) of a human eye. Label *eight* important parts.

An eye is often compared with a camera. Discuss the resemblances. (*L.*)

**22.** Give a labelled diagram of the human eye (no description is necessary). How does the eye adjust itself to deal with (*a*) light of varying intensity, (*b*) objects at different distances?

Young children often read comfortably with the book very close to the face. Why is this? Why does the sight of such children often become normal as they grow older? (*L.*)

**23.** What lenses are used and how are they arranged in a compound microscope? Give the purpose of each lens.

Draw a diagram showing the paths of two rays from a non-axial point on an object to an eye looking at it through the microscope. (*L.*)

# CHAPTER XXIII

## COLOURS OF LIGHT. DISPERSION

IN 1666 Newton made one of his greatest discoveries; he found that sunlight (white light) was made up of many different colours. Newton records that one day he noticed sunlight streaming through a hole $H$ in a shutter, and placed a glass prism $P$ in the path of the light to observe what would happen (Fig. 245). To his great surprise a number of *differently coloured* images were obtained on a screen $S$ placed behind the prism. He immediately

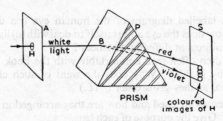

FIG. 245. Dispersion of white light

experimented further to find out whether the colouring was imposed on the white light by the prism, or whether the colouring was inherent in the white light itself.

To this end he placed another prism $Q$ of equal angle between $P$ and the screen, with the apex pointing the opposite way; he found that most of the colour effect disappeared from the image now received on the screen $S$, i.e. the image was nearly white again (Fig. 246 (*a*)). When the prism $Q$ was turned round, however, the coloured images were again on the screen, but in a different position (Fig. 246 (*b*)).

Newton proceeded to isolate a particular colour by means of a slit in a screen, and found that the *same* colour was obtained on a screen when another prism was placed in the path of the light, showing that no further decomposition of the colour takes place. These, and other experiments, led him to the conclusion that *white light contains a mixture of colours which are red, orange, yellow, green, blue, indigo, and violet.*

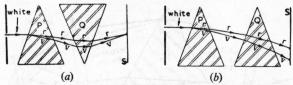

FIG. 246. (*a*) Recombination of colours (*b*) More dispersion

**How the prism separates the colours in white light.** These seven colours are known as the **spectrum of white light**; and when the apex of the prism is pointing upwards, as in Fig. 245, the red is at the top of the spectrum and the violet is at the bottom. To explain the production of the spectrum, consider a ray *AB* of white light incident on the prism. The ray contains a mixture of all colours, ranging from red to violet, and the prism experiment shows that *the refractive index of the prism material is different for different colours* (e.g. the refractive index ($\mu$) may be 1·51 for red light and 1·53 for violet light). Now the angle of incidence ($i$) on the prism is the same for all colours, since the latter are contained in the ray *AB*, and $\mu = \sin i / \sin r$. Thus the red and violet rays are refracted in different directions in the prism; the same is true for the orange, yellow, green, blue and indigo colours. Consequently the different colours appear on different places on the screen, and the white light is said to be **dispersed** by the prism.

**Obtaining a pure spectrum.** The spectrum of white light obtained by Newton was an "impure" one, i.e. one of the coloured images of the hole illuminated by sunlight was actually made up of a mixture of colours, although a particular colour was predominant. Light of one colour, for example red, is termed **monochromatic light**, and a "pure spectrum" consists of a series of differently coloured monochromatic images, ranging from the red to the violet.

The impure spectrum occurs mainly because (*i*) the hole illuminated by white light is large, thus tending to make the images of the hole overlap; (*ii*) the differently coloured rays are not brought to a focus, thus allowing further overlapping of the colours. In order to obtain a pure spectrum, the arrangement of Fig. 247 can be used. *S* is a narrow slit illuminated by a source of white light,

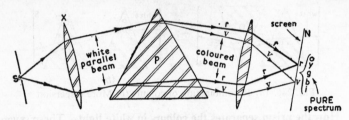

FIG. 247. Pure spectrum (*not to scale*)

such as an electric lamp, and is situated at the focus of a lens *X* to provide a parallel beam incident on the prism *P*. Refraction takes place at the prism, and the different colours travel along different paths; it should be carefully noted, however, that the rays of a *particular* colour are parallel inside the prism, and remain parallel on emerging. The rays are then incident on a lens, which brings them to a focus on a screen *N*; and since the red rays, for example, are all parallel, they come to a focus at *one* place on *N*. The parallel violet rays are all brought to a focus at a *different* place on the screen (Fig. 247), and a similar result is obtained for the orange, yellow, green blue, and indigo colours. Thus the band of light on the screen is made up of a series of images of the slit each of which contains light of one colour only.

**The achromatic lens.** If a parallel beam of white light is incident on a convex lens *L*, observation shows that a blurred coloured image is formed on a screen near the focus of the lens. The colouring of the image can be explained by considering the lens to be made up of many prisms (see p. 320), so that the colours in a ray *AD* of white light are refracted in different directions through the lens. As a result, the red rays in the parallel beam of white light are all brought to a focus at *R*, while the blue rays are brought to a focus at a different place *B*, nearer to the lens *L* (Fig. 248 (*a*) ). A coloured image is hence obtained on a screen placed between *B* and *R*, and in this case we cannot refer to a definite focus of the lens $L$. The lens is said to have caused **dispersion** between the red and blue rays, and the defect is known as **chromatic aberration.**

Newton had observed a coloured image when using lenses for a telescope, but he considered that there was no way of overcoming this defect. The problem was solved years later, however, by joining together two lenses made of different glasses, which eliminate the colour effect between the red and blue rays. The combination of lenses is known as an *achromatic* lens for these two colours.

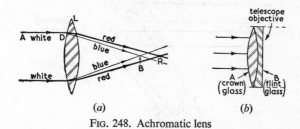

(a)                    (b)

FIG. 248. Achromatic lens

Fig. 248 (*b*) illustrates a common type of achromatic lens. A concave lens $B$ of flint glass is attached to a convex lens $A$ of crown glass, and when the focal lengths of the lenses are suitably chosen, the dispersion between the red and blue rays caused by $A$ is cancelled out by an equal dispersion by $B$ in the opposite direction. Achromatic lenses are used in cameras, microscopes and telescopes.

**The rainbow.** The colours of the rainbow are obtained when sunlight is refracted by water-drops and split into its colours.

Fig. 249 illustrates the essential features of the formation of the rainbow, which can only be seen by an observer standing with his *back* to the sun. The sun's rays, falling on a water-drop such as $A$ in the air, are refracted at the surface, then undergo total reflection inside the drop, and emerge into the air again. The emergent rays which make an angle of about 42° with the line joining the sun to the observer's head are red, while those making an angle of about 40° are violet. Thus the observer sees red ($R$) and violet ($V$) bows spread over the sky, which correspond in position respectively to all the water-drops at angles of 42° and 40° (Fig. 249);

the other spectral colours form bows between *R* and *V*, and the whole is known as the *primary rainbow*.

Other rainbows are sometimes seen. In particular, the *secondary rainbow* is formed by two successive internal reflections inside the water-drop, and its inner and outer bows are red and violet respectively, which is the reverse of the primary rainbow. Fig. 249 also shows the angles made in this case.

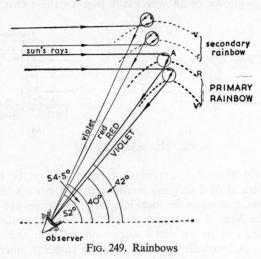

FIG. 249. Rainbows

**The colours of objects.** After Newton had obtained the spectrum of white light, he painted the different colours on sectors of a circular disc and revolved it rapidly. The disc then appeared greyish-white in colour, showing that successive impacts on the eye of the different colours blend into a whitish colour. When the blue sector was covered with black paper and the disc was rotated, a yellowish colour was observed; yellow is termed the **complementary colour** to blue. When the red sector was covered and the disc was rotated a greenish-blue colour was obtained, which is the complementary colour to red. Red, green, and blue are termed **primary colours,** as any colour can be obtained by mixing these colours in suitable proportions. This mixing effect applies to coloured lights, but does not apply to red, green, and blue pigments (paints), because, as explained below, the colours are not emitted by the pigments themselves.

An object which is white in daylight reflects all the different colours of the spectrum.  In blue light, therefore, the object appears blue; in red light it appears red.  An object which is black in daylight absorbs practically all the colours, and thus no reflection takes place.  A transparent red filter paper transmits only red light and absorbs the rest of the colours.  Hence a white sheet of paper appears red when viewed through it, and a bluish object appears black.  A red rose has this colour because it absorbs all the other colours of white light, and hence appears black if illuminated by blue or green light.  In the same way, a blue object appears black in all parts of the pure spectrum except the blue.

A blue pigment (paint) absorbs the red and yellow colours in white light; a yellow pigment absorbs the blue and violet colours in white light.  Green is the only colour which neither pigment absorbs; hence, when the pigments are mixed, a green colour is obtained.

**Infra-red and ultra-violet rays.  Electromagnetic waves.**  On p. 348 we discussed the production of the spectrum of colours, ranging from the red to the violet by means of a prism.  In 1800 SIR WILLIAM HERSCHEL discovered that the temperature on a thermometer rose as its blackened bulb was placed in the darkened part near the red rays of the solar spectrum, and came to the conclusion that the sun emitted rays which were invisible to the eye but which caused the sensation of heat.  These rays are known as *infra-red* rays.  It was also soon discovered that the sun emitted invisible rays beyond the violet part of the spectrum.  The *ultra-violet* rays, as they were called, had practically no heating effect, but certain glass materials became coloured with a bluish tint (fluoresced) when they were placed in the path of these rays, and the rays also affected a photographic plate.

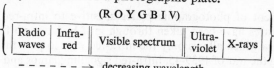

| (R O Y G B I V) | | | | |
| Radio waves | Infra-red | Visible spectrum | Ultra-violet | X-rays |

------ → decreasing wavelength

E L E C T R O M A G N E T I C   W A V E S

Radio waves, infra-red rays, the visible rays from red to violet, ultra-violet rays, and X-rays are all radiations of the same nature although they have different effects. They are called *electromagnetic radiations* or *waves*, because experiments have shown that each of the radiations must have electric and magnetic components, and different electromagnetic waves are characterized by different *wavelengths*. The B.B.C., for example, has transmitters sending out radio (electromagnetic) waves of different wavelength, such as those of 1500 metres (a long wavelength) and 30 metres (a short wavelength). Radio waves of a few centimetres have also been generated. The infra-red rays have wavelengths of about $\frac{1}{1000}$ centimetre, the yellow in the visible spectrum has a wavelength of about $\frac{1}{20000}$ centimetre, and the ultra-violet rays have a wavelength of about $\frac{1}{200000}$ centimetre. X-rays have wavelengths hundreds of times shorter than ultra-violet rays.

**Using infra-red rays.** Infra-red rays are scattered by haze and mist much less than rays in the visible spectrum, and with the

Ordinary photograph          Infra-red photograph

FIG. 250. INFRA-RED PHOTOGRAPHY
used to decipher ancient writing on leather

development of photographic plates sensitive to infra-red rays, clear pictures have been taken through mist by having a filter in the camera which allows only the infra-red rays to pass through.

During the war of 1939-45, infra-red signalling lamps and infra-red telescopes were developed, which enabled transport to deploy in the darkness. The invisible infra-red rays from the lamp were picked up by the telescope, which contained a concave mirror $M$ reflecting the rays to a similar mirror $N$, which in turn formed an image on a surface $A$ (Fig. 251). This surface was made of a caesium compound which emitted electrons when infra-red rays

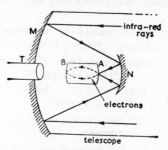

FIG. 251. Infra-red telescope

were incident on it (see "Photo-electricity," p. 630), and the electrons were focused on to a screen at $B$ coated with zinc sulphide, which fluoresced when the electrons impinged on it (see "Cathode-ray tube," p. 626). A "picture" was thus obtained on $B$, and was observed through an eyepiece $T$. Towards the end of the war bombers and fighter aeroplanes carried an infra-red lamp which could be seen and identified in the dark by the infra-red telescope without detection by the Germans.

―――Summary―――

1. The colours in the spectrum of white light are: **red, orange, yellow, green, blue, indigo, violet (ROYGBIV)**. A rainbow is due to refraction of sunlight by water droplets in the air.

2. **A pure spectrum can be obtained by using a narrow slit at the focus of a lens, a prism, another lens, and a screen at the latter's focus.**

3. Sunlight contains infra-red and ultra-violet rays, which are invisible. Infra-red rays produce a sensation of heat. Ultra-violet rays produce fluorescence in certain substances.

## EXERCISES ON CHAPTER XXIII

**1.** Draw a clear diagram showing how a ray of white light is dispersed into different colours by a prism. Name the colours in the spectrum of white light, and explain why they are dispersed by the prism.

**2.** Where must an observer stand to see a rainbow? Describe how a rainbow is formed, and draw a diagram to illustrate your answer.

**3.** The refractive index of a rectangular glass block for blue light is 1·523, and for red light is 1·511. Calculate the angle of refraction in the glass for the two colours when a ray of white light is incident on the glass at an angle of 60°. Draw a diagram.

**4.** Describe how a pure spectrum of white light can be obtained. Draw a diagram of the arrangement.

**5.** Describe and explain the colour effects observed when (i) a blue cap is illuminated by red light, (ii) a piece of coal is illuminated by white light, (iii) a violet flower is illuminated by white light, (iv) a red poppy is moved slowly through a pure spectrum of white light.

**6.** Describe an experiment to show how the colours of the spectrum of white light can be combined together to form white light again.

**7.** Does a lens disperse white light? State the meaning of *chromatic aberration*, and draw a sketch of a lens arrangement which provides an image fairly free of colour.

**8.** State the apparatus needed to form, on a screen, a pure spectrum of the light from an electric filament lamp. Give a diagram to show how it would be arranged and mark the paths of *two* rays of light from the lamp. How would the appearance of the spectrum alter if (a) a green gelatine or filter is placed in the path of the light, (b) the screen is painted green, (c) the lamp is replaced by a mercury vapour lamp such as is used for street lighting? Give reasons in each case. (*L.*)

**9.** Draw a labelled diagram to illustrate the *deviation* and *dispersion* of a narrow beam of white light by a triangular glass prism. (*N.*)

**10.** Explain fully how you would demonstrate that the refractive index of glass is less for red than for blue light. Give a reasoned opinion of the effect of this fact when first red and then blue light is used to find (a) the apparent thickness of a glass block, (b) the focal length of a converging lens. (*N.*)

**11.** State the laws of refraction of light at a plane surface. By a graphical construction, trace the path of a ray of light through an equilateral glass prism of refractive index 1·5, when the original angle of incidence is 45°. Measure the angle of deviation of the ray passing through the prism. Illustrate by means of a diagram the difference in the deviation, as compared with the mean value for white light, of (a) a red ray, (b) a blue ray, when passing through the same prism. Give reasons for your answer. (C.W.B.)

12. How would you project a pure spectrum on to a screen, using a source of white light? Give a diagram showing the arrangement. State and explain what would be observed if (a) a piece of red glass was put immediately in front of the source, (b) a burner with a lump of salt in the flame was substituted for the white source. (N.)

13. Define refractive index, and explain what is meant by refractive dispersion. Describe the appearance of the spectrum produced by a prism spectroscope when the source of light is (i) the sun, (ii) a bunsen flame containing some common salt, (iii) a filament lamp in which the filament is only red (and not white) hot. (C.)

14. When white light strikes a glass prism it gives rise to an impure spectrum. Explain what is meant by this, and show by a drawing how lenses may be used with the prism to obtain a pure spectrum. A flag has stripes of red, white, and blue. How will it appear if viewed in (a) red light, (b) green light? Give reasons. (O. & C.)

15. Describe an experiment to establish the relation between angles of incidence and the corresponding angles of refraction for glass. State the conclusion to be drawn from the experiment.

A parallel beam of white light is passed through a 60° glass prism and is received on a white screen. Draw a diagram to illustrate the experiment. What changes would you make to produce a pure spectrum on the screen and how would the screen differ in appearance in the two instances? (L.)

16. A triangular glass prism is placed between a screen and a narrow slit illuminated by an electric filament lamp. Describe the appearance of the coloured band on the screen when the light meets the prism at a suitable angle and explain its formation.

What additional apparatus would be required to obtain the least possible overlapping of the colours? Give a diagram of the new arrangement showing the paths of two rays from a point on the slit to the screen.

How would the appearance on the screen be altered if a piece of red glass were placed between the lamp and the slit? (L.)

# CHAPTER XXIV

## PRINCIPLES OF PHOTOMETRY

WE have already mentioned (p. 261) that light is a form of energy which stimulates the sense of vision, and it is obvious that the amount of light energy which reaches the eye is an important factor in everyday life. We can read at night, and often in the day-time in winter, only by light from lamps. Motorists and the general public are dependent for safety on proper lighting at night. The light coming from the chalk on the blackboard in a classroom depends mainly on the light falling on the board, either from daylight or the lamps in the room; workers engaged in the factory often require special lighting. The *illumination engineer* now plays a very important part in lighting design, both outdoor and indoor. This type of engineer deals with measurements of *light* or *luminous energy*, a subject known as **photometry.**

**Luminous intensity (candle power) of a source.** The filament of a lighted electric lamp emits a continuous stream of luminous (light) energy; the paper of this page also emits a continuous stream of luminous energy reflected from daylight or lamp. The *luminous intensity of a source of light* is a measure of the stream of light energy it emits per second, and the brighter the light from a lamp, the greater is its luminous intensity.

Years ago, before photometry was placed on a proper scientific basis, the luminous intensity of a source of light was expressed in terms of a unit defined by reference to the light energy from the flame of a candle of a certain size. There was an appreciable uncertainty in this standard, as the candle could not be reproduced to a high degree of accuracy; nevertheless, the unit was called one **candle-power,** and the name persists to this day. The meaning of the unit has altered, however; it is now defined by reference to the luminous energy emitted per second by an electric lamp at the National Physical Laboratory (N.P.L.), operated under certain conditions. These conditions can be reproduced to a high degree of accuracy owing to the reliability of present-day electrical instruments. Manufacturers of electric lamps send them to the N.P.L.,

FIG. 252. PHOTOMETRY

Light from the standard lamp shown is being compared by a photometer with light from a test lamp inside a large sphere, which collects all the light

where their candle-power (c.p.) is measured in terms of the unit candle-power; a given lamp may be classified as 50 c.p., and one which burns more brightly as 80 c.p. The total candle-power of a searchlight lamp is of the order of many thousands of candle-power.

**The intensity of illumination of a surface; the foot-candle.** We now turn from the source of light to consider the *area* or place on which the light falls; for example, the desk, the table, the walls of a room, the stage, the road, which are all illuminated at night by lamps. Suppose, for convenience, that we consider a small area on the top right-hand corner of this page. Light is falling on the area from all directions, and its **intensity of illumination** is defined as the *total amount of energy incident on it per unit area per second*. The "intensity of illumination" round a point on a surface is thus a different quantity from "luminous intensity," which

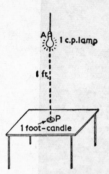

refers to a *source* of light, and the distinction must be borne in mind by the reader. We must also expect different units for the two quantities.

The **foot-candle** is defined as *the illumination round a point P on a surface when a lamp of 1 c.p. is placed 1 foot away from P in a direction perpendicular to the surface*, Fig. 253. The *metre-candle* (known also as a *lux*) and the *centimetre-candle* are other units of illumination; their definitions are the same as the foot-candle, except that

FIG. 253. Foot-candle the distances are now 1 metre and 1 centimetre respectively. An illumination of about 15 ft.-candles is required for sewing, and one of about 10 ft.-candles is recommended for an office.

**Variation of intensity of illumination ($E$) with distance ($d$).** Consider a small lamp $A$ of 80 c.p., and suppose $B$ is 1 sq. ft. of a surface of a sphere of centre $A$ (Fig. 254). The luminous energy from $A$ illuminating the surface is that contained in a cone $AMN$, and the same amount of luminous energy per second illuminates the area $C$ on a sphere of radius 2 ft. having $A$ as its centre. But the area of $C$ is *four* times the area of $B$, since the areas are proportional to the *squares* of their distances from $A$. Consequently the amount of

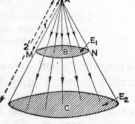

FIG. 254.

Intensity of illumination

luminous energy falling per second on a square inch of $C$ is four times as small as that falling on a square inch of $B$, i.e. the intensity of illumination over $C$ is four times as small as that over $B$. It can be seen that, in general, if the direction from the source is the same, **the intensity of illumination**

($E$) over a surface due to a small lamp decreases as the square of the distance ($d$) of the surface from the lamp. Thus

$$E \propto \frac{1}{d^2}, \qquad . \qquad . \qquad . \qquad . \qquad (1)$$

which is the "inverse-square law" of illumination.

**Verifying the inverse-square law.** The inverse-square law of illumination can be verified by means of the *grease-spot photometer* (Fig. 255). This consists of a white paper screen $S$ with a grease-spot at $P$ in the middle, so that light can pass only through the latter part of the screen. When a lamp $A$ of c.p. $I_1$ and a lamp $X$ of different c.p., $I_2$, are placed on the opposite sides of $S$, an observer at $Q$ sees the grease-spot lighter, or darker, than the surrounding screen; but by moving one lamp it is possible to

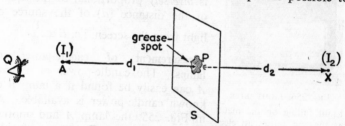

FIG. 255. Grease-spot photometer

make the grease-spot and screen of uniform brightness, *so that the spot is difficult to distinguish from the surrounding screen.* The luminous energy per second from $A$ reflected by the grease-spot, together with that from $X$ transmitted through the grease-spot, is then equal to the luminous energy per second reflected by the surrounding opaque screen facing $A$, and a "photometric balance" is now said to be obtained.

First, then, a lamp $A$ is arranged so that a photometric balance is obtained with a lamp $X$ on the other side. The distance, $d_1$, of $A$ from the screen is then measured. Suppose it is 20 cm. *Keeping the lamp $X$ fixed in position*, another three lamps of exactly the same type as $A$ are placed beside the latter, so that the total c.p. on this side of the screen is increased four times. Then all the lamps are moved back *together* to restore the photometric balance, and their distance from the screen is measured. This will be found to be about 40 cm., so that the distance is doubled to keep the intensity of illumination of the screen the same. If five more lamps of exactly the same type as $A$ are placed beside the four lamps, it will be found that the nine lamps together require to be moved back to about 60 cm. to restore the photometric balance, $X$ still being fixed in position. The c.p. of the lamps had thus increased in the ratio $1 : 4 : 9$, and, to keep the intensity of illumination of the screen constant, the square of the distance increased in the ratio $20^2 : 40^2 : 60^2$, or $1 : 4 : 9$. It follows that the intersity of illumination ($E$) of the screen is *inversely* proportional to the square of the distance ($d$) of the source of light from the screen, i.e. $E \propto \dfrac{1}{d^2}$.

FIG. 256. LIGHTMETER

Light falling on the metal at the top causes an electric current to flow whose magnitude is proportional to the intensity of illumination

**Measurement of candle-power of lamps.** The candle-power of a lamp $A$ can easily be found if a lamp $B$ of known candle-power is available. As in Fig. 255, the lamp $A$ and another lamp $X$ are arranged on opposite sides of a grease-spot photometer until a "balance" is obtained. The distance $d_1$ of $A$ from the screen is then measured. The lamp $A$ is then taken away, and a lamp $B$ of known c.p. is placed on the same side of the screen, the position of $X$ remaining constant. $B$ is moved until the photometric balance is restored, and its distance $d$ from the screen is then measured. Now the intensity of illumination of the screen is $\dfrac{I_1}{d_1{}^2}$ in the case of $A$, and $\dfrac{I}{d^2}$ in the case of $B$ (see p. 361), where $I_1$, $I$ are the respective candle-powers of $A$ and $B$. Since the intensities of illumination due to $A$ and $B$ are the same when the "balance" is obtained, it follows that $\dfrac{I_1}{d_1{}^2} = \dfrac{I}{d^2}$. Hence $\dfrac{I_1}{I} = \dfrac{d_1{}^2}{d^2}$.

Thus if $I = 50$ c.p., $d_1 = 25\cdot4$ cm., $d = 15\cdot2$ cm., the unknown c.p. $I_1$ is given by

$$\frac{I_1}{50} = \frac{25\cdot4^2}{15\cdot2^2}$$

$$\therefore \quad I_1 = \frac{50 \times 25\cdot4^2}{15\cdot2^2} = 139\cdot5 \text{ c.p.}$$

The same principle as the above is adopted at the N.P.L. for measuring candle-power, although the apparatus and the photometer used are much more elaborate (see Fig. 252).

If two lamps of candle-power $I$, $I'$ respectively are placed on *opposite* sides of the grease-spot, and moved until a "balance" is obtained, then, approximately, the intensity of illumination of the two sides of the spot are respectively equal. Thus $\dfrac{I}{d^2} = \dfrac{I'}{d'^2}$ approximately, where $d$, $d'$ are the respective distances of the two lamps from the grease-spot.

---

**Summary**

1. The "candle-power" or "luminous intensity" of a lamp is a measure of the light (luminous) energy it emits per second. "Candle-power" can be measured by a grease-spot photometer, using a lamp of known candle-power.

2. The "intensity of illumination" of a surface is the total amount of luminous energy falling on unit area per second. The foot-candle is defined as the illumination round a point on a surface when a lamp of 1 c.p. is placed 1 ft. away from the surface in a direction perpendicular to it.

3. **Intensity of illumination, $E = \dfrac{\text{Candle-power, } I}{d^2}$**, where $d$ is the distance, if the light is incident normally.

---

## WORKED EXAMPLES

1. *How would you show by experiment or otherwise that the intensity of illumination due to a small source of light varies inversely as the square of the distance from the source? Explain how this fact is used in order to compare the illuminating powers of two sources. Two sources, whose illuminating powers are 100 c.p. and 16 c.p. respectively, are separated by a distance of 70 cm. Where*

*must a screen be placed between the two sources in order that the intensities of illuminations on both sides may be equal? (L.)*

Suppose that the screen is $x$ cm. from the source of 100 c.p., and hence $(70 - x)$ cm. from the source of 16 c.p. Since the intensity of illumination, $E = \dfrac{I}{d^2}$, it follows that

$$\frac{100}{x^2} = \frac{16}{(70 - x)^2}.$$

Taking the square root of both sides,

$$\therefore \quad \frac{10}{x} = \frac{4}{70 - x}$$

$$\therefore \quad 10(70 - x) = 4x$$

$$\therefore \quad x = \frac{700}{14} = 50 \text{ cm.}$$

Thus the screen is 50 cm. from the source of 100 c.p.

2. *Define a foot-candle. A 40 c.p. lamp is placed 9 ft. away from another lamp L, and a screen placed between them at 5 ft. from the 40 c.p. lamp is found to be equally illuminated on both sides. If the 40 c.p. lamp is replaced by one of twice its power, the screen being kept in the same place, where must the lamp L be placed if the screen is still to be equally illuminated on both sides? (C.W.B.)*

The intensity of illumination at the screen due to the lamp of 40 c.p. ($I$) is $\dfrac{I}{d^2}$, or $\dfrac{40}{5^2}$. If $I_1$ is the c.p. of the lamp $L$, we have

$$\frac{I_1}{4^2} = \frac{40}{5^2}$$

$$\therefore \quad I_1 = \frac{16}{25} \times 40 = 25 \cdot 6 \text{ c.p.}$$

When the 40 c.p. lamp is replaced by one of 80 c.p., the intensity of illumination at the screen $= \dfrac{80}{5^2}$. Suppose $L$ is moved to a distance $x$ ft. from the screen to make the illumination the same on both sides. Then

$$\frac{25 \cdot 6}{x^2} = \frac{80}{5^2}$$

$$\therefore \quad 80x^2 = 5^2 \times 25 \cdot 6 = 640$$

$$\therefore \quad x^2 = 8$$

$$\therefore \quad x = \sqrt{8} = 2 \cdot 83 \text{ ft.}$$

*Alternatively.* The lamp $L$ is originally 4 ft. from the screen. When the 40 c.p. lamp is replaced by one of twice its power, the lamp $L$ must be moved back to a new distance $x$ ft. from the screen to make the illumination the same on both sides. Since the intensity of illumination of the screen is now twice as much, and the intensity varies inversely as the square of the distance, $\dfrac{4^2}{x^2} = 2$.

$$\therefore \quad 2x^2 = 16, \text{ i.e. } x^2 = 8$$

$$\therefore \quad x = \sqrt{8} = 2 \cdot 83 \text{ ft.}$$

**3.** *The intensity of illumination due to a point source of light varies inversely as the square of the distance from source to object, while that due to a search-light is to a large extent independent of the distance. With the help of diagrams explain the reason for this difference. A small 50 c.p. lamp is surrounded by a globe which cuts off 40 per cent. of the light falling on it. What is the intensity of illumination received by a book on which the light falls normally at a distance of 5 ft. from the lamp?* (N.)

Since the light falls *normally* on the book,

$$E = \frac{I}{d^2} \text{ (see p. 363)}$$

$$\therefore \quad E = \frac{\frac{60}{100} \times 50}{5^2} = 1 \cdot 2 \text{ ft.-candle.}$$

## EXERCISES ON CHAPTER XXIV

**1.** Name two applications of photometry. A 100 c.p. lamp is suspended directly over a point $X$ on a table. If $X$ is 4 ft. from the lamp, calculate the intensity of illumination at $X$, giving the units.

**2.** What is meant by the *candle-power* of a lamp? The intensity of illumination at a point on a wall 3 ft. directly in front of a lamp is 8 ft.-candles. Calculate the candle-power of the lamp.

**3.** Describe a photometer, and explain how you would use it to compare the candle-power of two lamps.

**4.** A lamp $X$ of 40 c.p. is placed 24 cm. from a photometer; another lamp $Y$ is situated 36 cm. from the photometer on the other side when a "balance" is reached. Calculate the c.p. of $Y$. If $X$ is moved back 6 cm. from the photometer, where must $Y$ be positioned to restore the "balance"?

**5.** What is the *inverse-square law* in photometry? Describe an experiment to verify the law, giving a table of imaginary results and showing how they are used to prove the law.

**6.** Find where a lamp of 36 c.p. gives the same intensity of illumination as a lamp of 25 c.p. gives at a point 4 ft. from the latter.

**7.** How far away from a 200 c.p. lamp must a screen be placed, normal to the incident light, in order to receive an illumination (or intensity of illumination) of 8 metre-candles? (N.)

13

**8.** Explain carefully why you would expect the intensity of illumination of a screen, illuminated by a small light source, to vary inversely with the square of its distance from the source. If a photographic print can be made with 4 sec. exposure at a distance of 4 ft. from a 32 c.p. lamp, what exposure will be required if the negative is held at 2 ft. from a 16 c.p. lamp ? (*O. & C.*)

**9.** A 40 c.p. lamp illuminates a screen which is 4 ft. away and perpendicular to the incident light. The intensity of illumination of the screen can be doubled (i) by moving the lamp, or (ii) by replacing the 40 c.p. lamp by another of suitable power. Calculate in each case how this can be done. (*C.W.B.*)

**10.** Define *illuminating power* and *intensity of illumination*. If a source of light of known power is supplied, describe how you would find the illuminating power of an electric torch. In printing from a photographic negative, an illumination of 20 ft.-candles is required for 5 sec. At what distance from the printing paper must a lamp of 100 c.p. be placed to give the correct exposure in 10 sec ? (*L.*)

**11.** Describe any form of photometer. Explain how it could be used to compare the illuminating power of a lamp with that of an identical lamp covered with a sheet of paper. A bright light gives an illumination of 2 foot-candles 40 yd. away. Calculate the illumination at 20 yd. distance if a sheet of paper which absorbs half the light incident on it is placed in front of the lamp. (*N.*)

**12.** Give an account of a method you would use for comparing the candle-powers of two electric lamps. Equal illumination is produced at a photometer by two lamps, *A* and *B*, when their respective distances from the photometer are 50 cm. and 75 cm. ; if the candle-pow er of *A* is 20, what is the candle-power of *B* ? A sudden decrease in the candle-power of *B* still gives equality of illumination at the photometer if *B* is moved 5 cm. from its original position. Find the ratio of the candle-powers of *B* before and after the sudden change. (*N.*)

**13.** What do you understand by the *inverse square law* as applied to illumination ? Describe an accurate form of photometer and explain how to use it to compare the candle-powers of two similar lamps. Three flash lamps of identical candle-power are placed close together and face a white vertical screen 10 ft. away. Calculate the distance the screen must have moved when *one* lamp is switched off in order to have the same intensity of illumination. (*O. & C.*)

# ANSWERS TO NUMERICAL EXERCISES

## CHAPTER XVII (p. 267)

**7.** 3 ft. 9 in.

## CHAPTER XVIII (p. 277)

**6.** 20°, 20°.  **14.** 60°.
**12.** 60°.

## CHAPTER XIX (p. 295)

**3.** $m = 2$.

**4.** 30 cm. from mirror, $m = 2$.

**6.** 48 cm. from mirror.

**7.** 60 cm. from mirror (virtual), 5.

**8.** $6\frac{2}{3}$ cm. from mirror (virtual), 0·55.

**9.** 7·2 cm. from mirror, 0·6.

**10.** 7·5 in. from mirror, 0·4

**11.** (i) 30, (ii) 18 cm.

**14.** 1·2 in. from mirror (virtual); 1·2 in.

**15.** (a) Virtual, 6 in. from mirror, $m = 1$, (b) virtual, 18 cm. from mirror, $m = 3$.

**16.** 30 cm.

**17.** Place object 8 cm. from mirror.

**18.** $u = 10·8$ cm., $f = 5·4$ cm.

**20.** 7·5 cm.; 5 cm. away from mirror.

**21.** 30 cm. behind mirror, 6 cm. high.

## CHAPTER XX (p. 315)

**2.** 1·51, 1·49.

**3.** 37·1°.

**4.** 41·7°.

**5.** 3 ft.

**7.** 1·57.

**9.** 41·5°, 48·6°.

**19.** 6·6 ft.

**20.** (a) 30°, 48·6°, (b) 18·6°.

**23.** 18·5°.

**24.** 1·3.

## CHAPTER XXI (p. 330)

**3.** 24 cm. from lens, 2.

**5.** 30 in. from lens (virtual), 6.

**7.** 9·6 cm. from lens, 0·6 in.

**10.** $26\frac{2}{3}$, $13\frac{1}{3}$ cm. from lens.

**12.** $f = 16$ cm.

**13.** 37·5 cm.

**14.** 3·2 in. from object.

**15.** (a) $26\frac{2}{3}$, (b) $13\frac{1}{3}$ cm.

**16.** 30 cm.

**17.** 12 cm.; 16 cm. towards lens.

**18.** (a) 5, (b) 5.

**19.** 30, $12\frac{6}{7}$ cm. from lens.

**20.** 40, 120 cm.

## CHAPTER XXII (p. 345)

**11.** Convex, $f = 50$ cm.

**13.** 3000 ft.

**14.** Convex, 24 cm.

**15.** 5 in., 10 in. from lens.

**16.** 25 cm. from eyepiece.

**19.** Concave, $f = 5$ ft.

**20.** 0·3, 4·8 cm.

CHAPTER XXIII (p. 356)

3. 34° 40′, 34° 58′.

CHAPTER XXIV (p. 365)

1. 6¼ ft. candles.
2. 72 c.p.
4. 90 c.p.;   45 cm. from photo-
    meter.
6. 4·8 ft.
7. 5 m.
8. 2 sec.

9. (i) 2·83 ft. from screen, (ii) 80 c.p.
    lamp.
10. 3·1 ft.
11. 4 ft. candles.
12. 45 c.p.; 1·15 : 1.
13. 1·83 ft.

# SOUND ENERGY

# CHAPTER XXV

## SOUND WAVES

WHEN a bicycle bell is rung and then lightly touched, it can be felt to be *vibrating*, i.e. it is moving to and fro continuously through a small distance. This vibratory movement is common to all objects emitting a sound, such as the violin, the piano, the organ-pipe, and the radio loud-speaker.

A sounding object vibrating in this way has a certain amount of energy. This can be demonstrated by touching a suspended piece of cork, X, with a prong of a sounding tuning fork F, when the cork moves through a surprising distance to Y as a result of the impact of the prong (Fig. 257 (a) ).

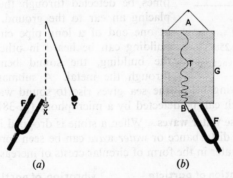

(a)                              (b)

FIG. 257. Effects of vibrating prong

**Sound vibrations.** Fig. 257 (b) illustrates a method of showing the vibrations due to a sounding tuning fork. The fork F has a bristle, B, attached by wax to one of its prongs, and B presses against the lower part of the face of a vertical glass plate, G, which has been covered with lamp-black. The fork F is fixed in position, and when it is sounded by means of a bow, the prongs vibrate horizontally and the bristle *B* marks a visible horizontal line on *G*. The plate *G* is suspended from a nail by a thread, *A*, and when the fork is next sounded, the thread is burnt at the top. The downward movement of the plate results in a visible trace, *T*, of the horizontal vibrations of the bristle, and the experiment has resulted in the "writing of the sound." (See also p. 401.)

**A medium is necessary for sound waves.** An important fact emerges when an electric bell *B* is placed in an air-tight vessel *V*

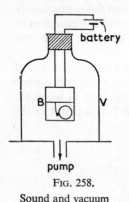

battery

connected to a pump (Fig. 258). As air is pumped out gradually from the vessel, the sound of the bell also dies gradually away. The clapper of the bell, however, can still be seen striking the gong, although no sound is heard. *Thus sound cannot pass through a vacuum*; a material medium is necessary to carry the sound.

The sound of horses' hooves can sometimes be detected through the earth by placing an ear to the ground. Tapping at one end of a long pipe circulating a building can be heard in other parts of the building, the sound being carried through the metal. A submarine's propellers moving below the sea gives rise to sound waves in the water, which can be detected by a microphone (p. 381).

pump

FIG. 258.

Sound and vacuum

**Transverse water waves.** When a stone is dropped into a pool of water, a disturbance or *water wave* can be seen spreading out along the water in the form of circular crests of increasing radius.

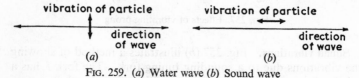

FIG. 259. (*a*) Water wave (*b*) Sound wave

A straw in the water moves up and down when the wave reaches it, but does not move across the surface. This is because the particles move up and down but do not move onwards in the direction of the waves. It may be asked: "What travels along the water?" The answer is "Energy," transferred from one moving part of the water to the next part, which in turn transfers it to the next part, and so on all along the water. As the particles of water move up and down, in a direction *perpendicular* to the horizontal direction along which the wave travels, the water wave is known as a *transverse wave* (Fig. 259 (a) ).

**Sound waves.** Sound travels through the air in waves, but, unlike water waves, these sound waves are not transverse but

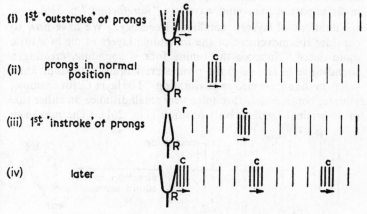

(i) 1st 'outstroke' of prongs

(ii) prongs in normal position

(iii) 1st 'instroke' of prongs

(iv) later

FIG. 260. Sound wave

longitudinal: the to-and-fro movement of the air particles is in the same direction as the movement of the waves (Fig. 259 (b) ). Suppose a tuning fork is sounding in air, so that both prongs are vibrating. When the prong R first moves to the right it disturbs the layer of air next to it (Fig. 260 (i) ). This layer pushes the layer next to it, and so on. Thus at the first "outstroke" of the prongs, the layers of air near R are compressed, as shown by c in Fig. 260 (i). When the prongs return to their normal position a short time later, the *compression c*, as the crowding of the layers may be termed, has travelled a short distance from the fork, Fig. 260 (ii). The "disturbance" has thus been passed on from one layer to the next; *the layers themselves have not moved* bodily

13*

with the disturbance. When the prongs swing inward a short time later, Fig. 260 (iii), the compression $c$ has travelled further away from $R$. As the vibrations of the prongs continue new compressions are started, and a number of compressions are present in the air at any instant, as represented in Fig. 260 (iv). When the diaphragm of the ear of an observer is reached, it undergoes a slight displacement due to the movement of the layer of air near to it, and a sound is heard. We shall see later that layers of air at a place are further apart than normal at some instants, e.g. at $r$ in Fig. 260 (iii), and this "condition" also travels in air as part of the sound wave (see "*Rarefaction*," p. 376).

**Vibrations of layers; amplitude frequency.** We have now to consider the movement of the individual layers of air in a little more detail. Suppose the tuning fork is sounding, creating a sound wave in air; since the prongs are vibrating, the individual layers in the air are also each vibrating. The layer $C$, for example, vibrates continuously through a very small distance on either side of its original (undisturbed) position (Fig. 261). The *maximum*

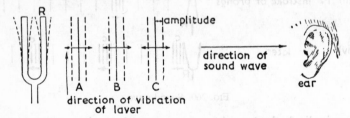

FIG. 261. Transmission of sound wave

distance moved on either side of the original position is known as the *amplitude* of the vibration of $C$, and in practice is a very small fraction of an inch. The number of complete to-and-fro oscillations of $C$ in 1 second is known as its *frequency* ($f$) of vibration. This is obviously the frequency of the tuning fork's vibration, so that if the latter is 256 per second, $C$ makes 256 complete oscillations or *cycles* in 1 second while the sound wave passes along. Other layers between the fork and the ear are also vibrating at the same frequency; but the vibrating layer $B$ near $C$ is a little "out of step" with the latter's movement, and a vibrating layer $A$ near

*B* is a little more "out of step" with *C*'s motion, since it takes time for the wave to travel from one place to another.

**Displacement and pressure curves; wavelength.** To illustrate the movement of the layers, consider Fig. 262 (i), in which the original (undisturbed) positions of the layers are shown exaggerated. Fig. 262 (ii) illustrates the new positions of these layers at some instant when a sound wave is travelling through the air. At this instant some of the layers are displaced to the right of their

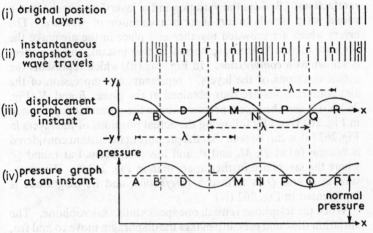

FIG. 262. Displacement and pressure graphs

original positions, others are just passing through their original positions, and some are displaced to the left of their original positions. If we represent the displacement, *y*, to the right as positive, and to the left as negative, the displacement graph at this instant is illustrated in Fig. 262 (iii). At *A*, the layer is at the end of its vibration to the right (Fig. 262 (ii) ); at *B* at this instant, the layer is just passing through its original position; at *D* the layer is at the end of its vibration to the left; and so on at *L*, *M*, *N*, *P*, *Q*. Corresponding to *A*, *M*, or *R* at this instant, we have *crests* of the wave; corresponding to *D* or *P* at the same instant, we have *troughs* of the wave. The *wavelength* (λ) of the sound wave in the air is the distance between successive crests or troughs, just like the wavelength of a water wave. Thus, from Fig. 262 (iii), the wavelength, $\lambda$, $= AM$, or $MR$, or $LQ$. As time goes on, the positions of the layers change, and the crests and troughs appear at another instant at different places from those shown in Fig. 262 (iii); but the displacement graph at any instant has always the same wave-form, and the wavelength, as defined above, is the same.

**Pressure variations due to a sound wave. Compressions and rarefactions.** Besides the variation in the layers' positions when a sound wave travels in air, there is a variation of *pressure*. The layers which are crowded together at a place in the air make the pressure there a little above normal at that instant, and this effect is known as a *compression*. In Fig. 262 (ii), which represents the actual positions of the layers, $c$ represents a compression of the air; compressions are thus obtained at the places $B$ and $N$ (Fig. 262 (iii) ), and the excess pressure above the normal is represented in Fig. 262 (iv). By following the actual positions of the layers in Fig. 262 (ii), it can be seen that the pressure at the instant considered is normal ($n$) at $D$, $M$, and $P$, and less than normal at $L$ and $Q$, where the layers are further apart than in Fig. 262 (i). The pressure at $L$ and $Q$ is termed a *rarefaction*, and its magnitude is represented in Fig. 262 (iv).

When the telephone is used, one speaks into a microphone. The variation of sound pressure makes the diaphragm move to and fro, and, as explained on p. 574, a varying electric current is produced in the telephone wires. At the other end of the "line" a person is listening with a *telephone earpiece*, an apparatus which changes the varying electrical currents back again into variation of air-pressure (see p. 573), so that a sound wave is again obtained. Essentially, then, a microphone converts sound energy to electrical energy, while a telephone earpiece does the reverse.

**Characteristics of sound.** A sound note can be completely identified from all other notes by three characteristics: (1) its *pitch*; (2) its *intensity*; (3) its *quality* (or *timbre*).

The **pitch** of a note is analogous to the colour of light. A note of high pitch has a high frequency, i.e. it is due to vibrations which are very many per second; 1000 cycles per second (1000 c.p.s.) is a high-pitched note, for example. A low note has a low frequency for example, 100 c.p.s. is a hum heard in a radio receiver prior to the valves becoming warm. Dogs and cats are capable of hearing notes beyond the upper limit of a human being, which is about 20,000 c.p.s.

The **intensity** of a note is a measure of the sound energy it produces, which controls the *loudness* of the note. As the volume control of a radio receiver is turned up, the note from the loud-speaker becomes louder and louder; the sound energy thus increases. At the same time the *amplitude* of vibration of the loud-speaker cone can be felt to increase. In general, *the intensity, or loudness, of a note is proportional to the square of the amplitude of vibration*. The amplitude of the air vibrations carrying a note decreases the further the distance from the source of sound, so that the loudness also diminishes.

The kinetic energy of a mass $m$ of air moving with an average velocity $v$ is given by $\frac{1}{2}mv^2$ (see p. 40). It follows that the sound energy from a vibrating object depends on the mass of air it sets into vibration, and the larger the mass of air, the louder is the sound obtained. A loudspeaker cone has a large surface area, so that the mass of air set into vibration by it is large, and the sound intensity is large. On the other hand, the vibrating diaphragm in a telephone earpiece has a small surface area, so that the mass of air set into vibration is small, and sound can only be heard when one's ear is close to the earpiece. By itself, the violin string sets into vibration a very small mass of air; but the hollow box to which it is attached has a comparatively large surface area, and a large mass of air vibrates when the box is set into vibration by the sounding strings. A loud note is thus obtained from the violin (see also "Sonometer," p. 390). Similarly, the note from a sound-ing tuning fork is a soft note, but when the base of the fork is placed on a table the note becomes loud, as a much larger mass of air, in contact with the table, is then set into vibration.

**The quality (or timbre) of a note.** If the same note is sounded on a piano, an organ, or a violin, the source of the sound can immediately be recognized by ear. In technical language, we say that the "quality" or "timbre" of the note is different when it is sounded on different instruments.

In practice, it is very difficult to get a pure note: one which contains only one frequency. The note from a tuning fork is a near approach to a pure note. The note from a piano, however, contains a "background" of other notes, of higher frequency than the one heard. For example, if a note of 256 cycles per second is sounded, notes of 512, 768, and 1024 cycles per second are also present; but the latter have a much smaller amplitude than the note of 256 cycles per second, so that their intensity, or loudness,

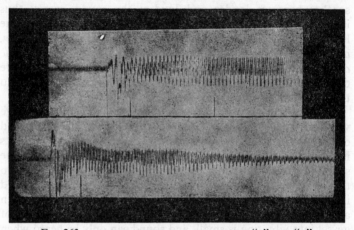

FIG. 263. SOUND WAVE FORMS OF THE LETTERS "T" AND "D"
as spoken by the human voice

is much lower. Similarly, an organ pipe emitting a note of 256 cycles per second also produces notes of 768 and 1280 cycles per second, at a much lower intensity. Notes which provide a "background" to the note heard are called *overtones*, and are registered subconsciously by the mind. The overtones of the same note vary with different instruments, and *the quality of a note is due to the overtones which accompany it*.

If a note $X$ has a frequency of 300 cycles per second, the note of frequency 600 cycles per second is known as the *second harmonic* of $X$. The third harmonic of $X$ has three times its frequency, i.e. 900 cycles per second, and so on. The note of 300 cycles per second is termed the "first harmonic." The overtones of a note are always harmonics of it, as illustrated by the figures given.

**The velocity of sound waves.** In 1864 REGNAULT, a famous French scientist, carried out an experiment to find the speed with which sound waves travel in air. Gunpowder was fired on the top of a mountain, and the flash of light which followed was observed on another mountain some miles away. A short time, $t$, afterwards, the sound was heard, and since light travels with an enormous speed compared with the speed of sound, $t$ may be taken as the time taken for the sound to travel the distance $d$ between the mountains. The velocity of sound, $v$, was then calculated from $v = \dfrac{d}{t}$. The experiment was not capable of yielding accurate results; not only were the instruments unable to measure the time $t$ accurately, but the time taken for the observer's reactions to the light and to the sound could not be determined and taken into account. Electrical instruments, such as a microphone and a cathode-ray tube, are now used to determine the velocity of sound accurately, but their use is outside the scope of this book (see also p. 381).

It has been shown that the velocity of sound in free air is about 1100 ft. per second, or about $\frac{1}{5}$ of a mile per second; the velocity of light in air is about 186,000 miles per second. The difference in velocity explains why the action of kicking a football, observed some distance away, is heard a little time after the football has left the kicker's foot. Another striking example is the observation that the whistle of a distant engine is heard after the steam is seen. The distance of a storm-centre can be determined by timing how long the thunder takes to reach one's ears after the lightning is seen, and using the fact that the velocity of sound in air is about $\frac{1}{5}$ of a mile per second.

We have seen that the particles of a medium are disturbed when a sound wave passes along it, and that compressions and rarefactions occur. Consequently, stresses are set up in the medium as the wave passes. Now different media, such as wood, water, iron, air, react differently to the same stresses imposed on them; it should therefore occasion no surprise that the velocity of sound varies with the medium concerned. The velocity of sound in water is about 4800 ft. per sec., in iron about 16,500 ft. per sec., in hydrogen about 4200 ft. per sec., and in carbon dioxide about 890 ft. per sec.

**The velocity of sound in air** varies with the temperature of the air. Experiment and theory show that *the velocity* (*v*) *of sound in a given gas is proportional to the square root of its absolute temperature* (*T*). In symbols,

$$v \propto \sqrt{T} \qquad . \qquad . \qquad . \qquad (1)$$

It follows that the speed in air is greater on a warm day than on a cold day, as the absolute temperature $T$ is then greater. The velocity of sound in air increases by approximately 2 ft. per sec. for each degree Centigrade rise in temperature.

On the other hand, experiment and theory show that *the velocity of sound in air is independent of the air pressure*. Thus the speed of sound is the same at the top of a mountain as at the foot of it.

**Relation between $v, f, \lambda$.** If a tuning fork of frequency 110 cycles per sec. is sounding in air, 110 complete waves are set up in the air in 1 second because the fork vibrates 110 times per second. Suppose that the wavelength is 10 ft. Then the distance travelled by the sound in 1 second = 110 × 10 = 1100 ft. But the velocity of sound is the distance travelled in 1 second. Hence the velocity = 1100 ft. per sec.

In general, the velocity of sound, $v$, is related to the frequency of vibration, $f$, and the corresponding wavelength, $\lambda$, by the relation

$$v = f\lambda.$$

**Reflection of sound.** Sound waves, like light waves, are reflected by surfaces such as walls. An *echo* is due to the reflection of sound waves back to the person shouting. The "whispering gallery" of St. Paul's Cathedral is circular in shape, and a person talking quietly on one side of the gallery can be heard on the other side by continuous reflection of the sound waves round the wall. The captain of a ship can roughly find the direction and distance of an iceberg by sounding the ship's foghorn and listening for the echo. If the interval between sounding the foghorn and hearing the echo is 2 seconds, the total distance travelled by the sound = 2 × 1100 ft. = 2200 ft., assuming the velocity of sound in air to be 1100 ft. per sec. The distance of the iceberg from the ship is thus 1100 ft.

By means of a method similar in principle, submarines have been detected when submerged below the water by an apparatus known as *Asdic*. These instruments send out very high-frequency sound waves, called **supersonics** because they have frequencies well above the audio-frequency range, and the waves are reflected back to the apparatus on meeting any object in water. The position of the object can be located from a knowledge of the time-interval elapsing before the return of the echo, and the velocity of the sound waves in water.

Many merchant vessels have an apparatus for finding the depth of the water. The *echo-sounder*, as the device is called, sends out a supersonic note from its position at $T$, which travels through the water to the sea bed at $A$ (Fig. 264). Here it is reflected back to the echo-sounder, and if the time of travel is 1 sec. and the speed of sound in water is 4800 ft. per sec., the distance travelled is 4800 ft. The depth of the sea bed, $d$, is thus 2400 ft.

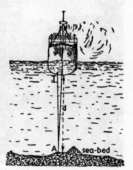

FIG. 264. Echo-sounding

An echo-sounder contains a stylus, or pen, which describes an arc of a circle as it moves to and fro; the arc is graduated in fathoms. A short burst of supersonic waves is sent towards the ocean bed and simultaneously the pen is released from the zero position and begins to swing across the arc. When the sound-echo is received back at the apparatus, the pen is made to record a brown mark on a sheet of recording paper impregnated with a chemical solution of starch and potassium iodide, and the depth is the reading on the scale corresponding to the mark on the paper. By making the paper move slowly downwards past the pen, successive echo marks appear one above the other. A continuous thick line is then obtained, which corresponds to the profile of the sea bed over which the ship is passing. Shoals of fish also are located nowadays by echo-sounding.

**Acoustics of rooms.** When a musician is playing in a hall, the sound reaches the ears of the audience (*a*) directly and (*b*) by reflection from the walls and ceilings a short time later. This repetition of sound makes the music appear indistinct to the audience, and the hall will be unsuitable for concerts (i.e. acoustically bad) if no steps are taken to minimize the reflection of the sound. Thick curtains round the walls assist reception because they are good absorbers of sound; an absorbent screen is used at

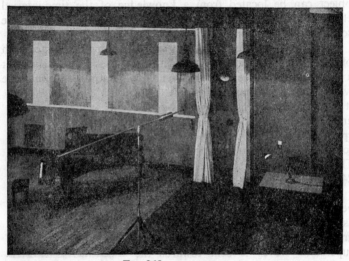

FIG. 265. B.B.C. STUDIO

This studio used for music, is acoustically "live" at one end (*plywood panelling and acoustic tiles*) and "dead" at the other end (*walls and ceiling covered with rock wood*).

the Royal Albert Hall at orchestral concerts. The same acoustical problem was present in the early days of broadcasting. When people in a play, for example, spoke a few feet away from the microphone, the sound reached it by reflection from the walls and ceiling as well as directly. B.B.C. engineers have spent many years in research into the *acoustics of rooms*, as the subject is called. The walls and ceilings are now faced with a special type of sound-absorbent wood. Other rooms are converted into suitable studios by means of heavy curtains or other materials on the walls and ceiling (Fig. 265).

Fig. 266 (*a*) illustrates the reflection of circular sound waves incident in a direction *AO* on a smooth plane *W*, such as a wall. The wall reflects the waves in a direction *OB* such that the angle of reflection is equal to the angle of incidence, which is the same as the case of light reflected from a plane mirror.

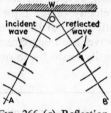

FIG. 266 (*a*) Reflection of sound

**Refraction of sound.** Just like light waves, sound waves can be refracted. TYNDALL performed an experiment in which a large soap bubble was filled with carbon dioxide, and a high-pitched whistle was sounded in front of it. A sensitive flame, which reacted sharply to a high-pitched sound, was moved about on the other side of the soap bubble, and was found to be affected at one position. The carbon dioxide gas in the bubble had thus acted towards the sound waves in the same way as a convex lens to light, and had refracted the waves spreading out from the whistle towards a single point on the other side.

It is a commonplace experience that a person shouting can be heard more easily some distance away if the wind is blowing

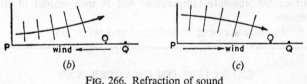

(*b*)                                   (*c*)

FIG. 266. Refraction of sound

towards the observer; if the wind is blowing in the opposite direction to the observer, the latter experiences more difficulty in hearing the sound. These observations are explained by the refraction of the sound by the wind. Fig. 266 (*b*) illustrates the case of the wind blowing in the opposite direction to a source of sound at *P*. The velocity of the air in contact with the ground is little influenced by the wind, whilst the velocity of the higher layers of air is diminished. The waves, which are represented by the straight lines in Fig. 266 (*b*), are thus refracted (bent) upwards, and an observer at *O* on the ground has difficulty in hearing the sound. A reverse effect is obtained when the wind blows in the same direction, *PQ*, as the sound from the source *P* (Fig. 266 (*c*) ). The velocity of the layer of air in contact with the ground is little affected, but the velocity of the upper layers is increased. The waves are thus refracted downwards, and the sound is then more easily heard by an observer at *O*.

─────Summary─────

1. Sound cannot pass through a vacuum; it requires a medium (such as air or water) to pass from one place to another.

2. A **longitudinal** wave travels in the same direction as the vibrations of the particles. A **transverse** wave travels in a perpendicular (transverse) direction to the vibrations of the particles.

3. Sound is produced by a vibrating object. The loudness of a note depends on the amplitude of vibration; the pitch depends on the frequency (number of vibrations per second; the timbre or quality depends on the overtones present.

4. **The velocity of sound in air is proportional to the square root of the absolute temperature but is independent of the pressure.**

5. Sound waves can be reflected and refracted.

## EXERCISES ON CHAPTER XXV

**1.** What does the *pitch* of a note depend on ? The note from a bicycle bell is lower in pitch when it is rung with one hand placed lightly on it. Explain this observation.

**2.** Can sound waves pass through iron and through water ? Describe an experiment to show that sound cannot pass through a vacuum.

**3.** Draw a sketch of a simple sound wave. Mark on it (i) the title of the axes, (ii) the amplitude of the wave, (iii) its wavelength. Draw a sketch of a note from a piano.

**4.** The volume control of a radio set is turned so that music played quietly suddenly becomes loud. What do you know about the sound waves before and after the volume control was turned? Draw sketches of the two waves to illustrate your answer.

**5.** Describe in detail how sound waves travel through air. State particularly what happens (i) at a given place in the air, (ii) at all the places in the air at given instants. Draw diagrams to illustrate your answer.

**6.** How does the pressure of the air at a given place vary when a sound wave is passing through it? Draw a sketch to illustrate the *compression and rarefaction* due to the wave, and explain the meaning of the two terms.

**7.** Describe an echo-sounder, and explain how it works.

**8.** Describe the Whispering Gallery of St. Paul's, and explain the phenomenon observed.

**9.** A sounding tuning fork gives a much louder sound when its end is placed on a table. Explain this observation. Why is a much louder sound obtained with a moving-coil loudspeaker than a telephone earpiece in the same set?

**10.** Draw a diagram showing why a distant sound is heard more easily when the wind is blowing towards the observer.

**11.** A note from a piano can be distinguished by ear from the same note played on an organ. What is the name given to this phenomenon? Explain why a person can detect a difference between the notes.

**12.** What is the relation between the velocity, wavelength, and frequency of a wave? What is the wavelength when the frequency of a note in air is 100 cycles per sec. and then 500 cycles of air? Calculate the frequency of a note in air when the wavelength is 170 cm. and then 6·8 metres. (Velocity of sound in air = 34,000 cm. per sec.)

**13.** (*a*) What is a *transverse* and a *longitudinal* wave? Give an example of each type of wave. (*b*) How does atmospheric pressure and temperature affect the velocity of sound in air?

**14.** State the factors which influence each of the following characteristics of a musical note: (*a*) its pitch; (*b*) its loudness; (*c*) its quality. (*N*.)

**15.** A piece of springy steel, clamped at one end, gives a musical note when set in vibration. The changes mentioned below are made in the system. State in *one* word, how the pitch of the note in each case compares with the original note. (a) The steel is clamped nearer the middle, (b) the amplitude of vibration is increased, (c) a pellet of wax is fixed on the free end. (N.)

**16.** A man standing between two parallel cliffs fires a rifle. He hears one echo after 1½ sec., one after 2½ sec., and one after 4 sec. Explain how these echoes reach him and calculate the distance apart of the two cliffs. The velocity of sound under the given conditions is 1120 ft. per sec. (L.)

**17.** Write a short account of the production of sound by a tuning fork and of the transmission of this sound through air. (N.)

**18.** Describe some form of siren. Why does its operation lead to the production of a musical sound? Explain how (a) the pitch, (b) the loudness of the note can be changed. If the frequency of the note is 300 vibrations per sec. and the velocity of sound in air is 1100 ft. per sec., calculate the wavelength of the sound waves produced. (N.)

**19.** Explain the terms *frequency*, *pitch*. A tuning fork vibrating in air at standard temperature and pressure produces waves of wavelength 4 ft., which travel with a velocity of 1100 ft. per sec. Calculate the frequency of the note emitted by the fork. (O. & C.)

**20.** Describe the motion of the particles of air between a loudspeaker, which is giving out a continuous steady note of constant pitch, and an observer.

How would this motion alter if (a) the pitch of the note were raised one octave; (b) the note were louder; (c) the temperature of the air increased? (N.)

# CHAPTER XXVI

## VIBRATIONS IN STRINGS AND PIPES

WHEN a note is obtained by bowing a violin string, the particles of the string vibrate in a special way, and the wave set up along the string is known as a *stationary wave*. Similarly, when a whistle is blown, the particles of air in it vibrate and a stationary wave is set up in the air.

**Stationary waves.** The essential features of a stationary wave can be demonstrated by attaching one end of a piece of thread to a vibrating clapper $C$ of an electric bell from which the gong is removed (Fig. 267). The thread passes over a pulley $P$, and a scale-pan is attached to the other end of the thread. By adjusting the value of the weight on the scale-pan, the thread can be seen to form loops between $P$ and $C$ as it vibrates up and down. A stationary wave is now set up along the thread. Owing to the vibration of the clapper a wave travels from $C$ to $P$, where it is

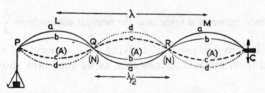

FIG. 267. $(A)$ = antinode, $(N)$ = node

reflected, and it then travels in the direction $PC$. Two waves travelling in opposite directions now travel along the thread, and have a combined effect on the latter which makes it vibrate in apparently stationary loops, as illustrated in Fig. 267. Thus the stationary wave along the thread is due to two waves of the same frequency and amplitude travelling in opposite directions.

**Nodes and antinodes.** Observation of the stationary wave in the thread shows that some points, $N$, are always at rest. These points are called the "nodes" of the wave. The particles of the string vibrate up and down with increasing amplitude as we proceed from $N$ to either side, and at a point $A$ exactly midway between consecutive nodes the amplitude of vibration is a maximum. These points ($A$) are known as the "antinodes" of the stationary wave. The appearance of the thread at different instants corresponds to $a$, $b$, $c$, $d$, in that order, so that the whole of the vibrating thread between $PC$ is horizontal at some instant.

**The wavelength, and its relation to the nodes and antinodes.** When the clapper $C$ vibrates up and down, a wave travels along the string and is reflected at $P$, as stated above. Now the wavelength ($\lambda$) of a wave is the distance between two successive crests, or two successive troughs (see p. 375); and it can be seen from the curve $a$ in Fig. 267 that the distance $LM$ is one wavelength, $\lambda$. The distance from $P$ to $R$ is also one wavelength. We therefore arrive at the following important results, which can now be verified by the reader: *In a stationary wave*,

*the distance between successive nodes (or antinodes)* $= \dfrac{\lambda}{2}$,

*the distance between a node and the nearest antinode* $= \dfrac{\lambda}{4}$.

These relations are used in studying the frequency of the notes obtained from musical instruments, as we shall now show.

## VIBRATIONS IN STRINGS

When a violin string is plucked in the middle a transverse wave travels along it and is reflected at the fixed ends, as in the case of the thread connected to the clapper in Fig. 267. A stationary wave is therefore set up along the violin string. Since the two ends of the string are fixed, the simplest or *fundamental* frequency, $f_o$, obtained is due to a stationary wave whose nodes $N$ are at the ends, and an antinode $A$ at the middle, where the string was plucked. (Fig. 268.) Thus the length $l$ of the string = the distance

$NN$ between two successive nodes $= \dfrac{\lambda}{2}$, where $\lambda$ is the wavelength of the transverse wave along the string (see p. 388).

If, for example, $l = 20$ cm. $= \dfrac{\lambda}{2}$, then $\lambda = 40$ cm.; if the velocity, $v$, of the wave along the string is 40,000 cm. per sec., the frequency $f$ of the note obtained is given by

$$f = \frac{v}{\lambda} = \frac{40,000}{40} = 1000 \text{ cycles per sec. (p. 380).}$$

Since $\dfrac{\lambda}{2} = l$, it follows that $\lambda = 2l$. Now the velocity $v$ of the wave set up along the string depends on (i) the tension $T$ in the

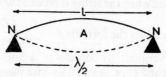

FIG. 268. Stationary wave in string

string, i.e. how tightly the string is stretched; (ii) the mass per unit length, $m$, of the string, i.e. the nature of the string. It can be proved, by methods outside the scope of this book, that $v = \sqrt{\dfrac{T}{m}}$. Hence, since the frequency of the wave, and note obtained, is given by $f = \dfrac{v}{\lambda}$, it follows that $f = \sqrt{\dfrac{T}{m}} \Big/ 2l$, or

$$f = \frac{1}{2l}\sqrt{\frac{T}{m}} \qquad . \qquad . \qquad . \qquad . \quad (1)$$

From this relation it can be seen that when the tension $T$ in a given string is kept constant, the frequency $f$ increases as the length of the string is reduced.

i.e. $$f \propto \frac{1}{l},$$

or *the frequency is inversely proportional to the length $l$ of the wire.*

Thus, if a length of string of 50 cm. gives a note of 800 cycles per sec. when plucked, a length of 100 cm. gives a note of 400 cycles per sec. (a lower frequency) when the same string is plucked, the tension being kept constant. Again, if a length of 35 cm. is plucked, the tension remaining constant, the frequency $f$ of the note obtained is given by $\dfrac{f}{800} = \dfrac{50 \text{ cm.}}{35 \text{ cm.}}$, since a note of *higher* frequency than 800 cycles per sec. is obtained. Thus, $f = \dfrac{50}{35} \times 800$ = 1143 cycles per sec.

From relation (1) it can also be seen that for a given length $l$ of string, the frequency of the note increases when the tension $T$ increases. More specifically, *the frequency is proportional to the square root of the tension in the wire.* Thus, if the same length of string is tightened so that the tension is four times as great, the frequency is doubled when the string is plucked with the same force.

**The sonometer** enables the formula $f = \dfrac{1}{2l}\sqrt{\dfrac{T}{m}}$ for the frequency of a string to be verified. It consists of a hollow box $S$, with a wire attached to it at $F$ (Fig. 269). The wire passes over bridges at $C$, $B$, and then over a fixed pulley $P$; and a weight $W$ is attached at the end to keep the wire under constant tension. A tuning fork of frequency 256 cycles per sec., for example, is struck, and with

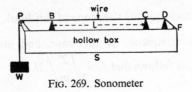

FIG. 269. Sonometer

$C$ kept fixed, the bridge $B$ is moved until the note obtained by plucking the midpoint of $BC$ is exactly the same as the tuning fork. The length of $BC$ is then noted. By using other forks of known frequency $f$ and noting the corresponding length $l$ of $BC$, the wire plucked, it can be shown that $f \times l$ is a constant, i.e. $f \propto \dfrac{1}{l}$. The best method is to plot the values of $\dfrac{1}{l}$, calculated in decimals, against the corresponding value of $f$, when a straight line graph is obtained.

The following results were obtained in an experiment—

| $f$ (c.p.s.) | $l$ (cm.) | $f \times l$ | $1/l$ |
|-----|------|------|-------|
| 256 | 35·0 | 9060 | 0·028 |
| 288 | 31·5 | 9070 | 0·032 |
| 320 | 28·0 | 8960 | 0·036 |
| 384 | 23·6 | 8960 | 0·042 |
| 426 | 21·2 | 9030 | 0·047 |
| 512 | 17·5 | 8960 | 0·057 |

The corresponding graph is shown in Fig. 270.

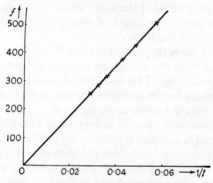

FIG. 270. Graph of $f$ v. $1/l$

If the scale of the frequency $f$, as well as $1/l$, is graduated from zero, the straight line graph should pass through the origin when produced back.

*To show that* $f \propto \sqrt{T}$ *when* $l$ *is constant*, the weight $W$ can be altered, keeping $B$ and $C$ fixed in position. Different notes are then produced as $W$ is altered. By calibrating a second wire (not shown) on the sonometer, under constant tension, the frequency $f$ of the notes can be found; when $f$ is plotted against $\sqrt{W}$, a straight line graph is obtained.

**How the sonometer can be "tuned." Beats.** If a person has not a musical ear, the length of wire on the sonometer can be tuned to the note from a sounding tuning fork by one of two methods.

(1) *Paper rider.* A small piece of paper, a paper rider, is bent in the form of an inverted V, and placed in the middle of the wire whenever the length of wire is altered. The tuning fork is then sounded, and placed upright with its end firmly on the sonometer box so that its prongs are free; whereupon the vibrations of the fork are communicated to the wood of the sonometer box and then to the wire attached. The wire itself vibrates slightly when its natural frequency is not the same as that of the sounding tuning fork, but when the natural frequency of the wire is exactly the same as that of the fork, the wire vibrates through the largest amplitude. At this stage the paper rider is observed to tremble and to fall off the middle of the wire, which is then an antinode of the wave in the wire. The length of wire is now tuned to the same frequency as that of the fork.

(2) *Beats.* In the early years of the last World War, a throbbing note was heard from the aeroplane engines of German bombers on raiding expeditions. This note was due to a combination of the sounds emitted by the multiple engines, and, from its sound, the throbbing note is known as a *beat* note.

The phenomenon of "beats" occurs when two notes of *nearly equal* frequency are sounded together. As an illustration, suppose a note $A$ of frequency 256 cycles per second is sounding at the same time as a note $B$ of frequency 260 cycles per second. The sound heard by a near observer will be due to a wave which is the *sum* of the waves due to the individual notes, whose displacement-time graphs are illustrated in Fig. 271 (i), (ii) (not to scale). Suppose at an instant that the displacements are a maximum and in the same direction. The sum, or resultant, of the two displacements is then a maximum, and a loud sound is heard, since the intensity of a note depends on the square of its amplitude (p. 377). Suppose this instant is represented by $O$ in Fig. 271 (iii). In $\frac{1}{8}$ sec., the note $A$ makes $\frac{1}{8} \times 256$, or 32, complete cycles; and in the same time the note $B$ makes $\frac{1}{8}$ of 260, or $32\frac{1}{2}$, complete cycles. Thus, at this instant, the displacement due to $A$ is a positive maximum, say, and the displacement due to $B$ is a negative maximum. The sum of the two displacements is consequently very small, and a much quieter sound is heard; this instant is represented by $M$ in Fig. 271 (iii). In $\frac{1}{4}$ of a second from the time $O$, corresponding to $N$, the note $A$ makes 64 complete cycles and the note $B$ makes 65 complete cycles. Thus the displacements due to each note are exactly the same at this instant as at the instant $O$, i.e. they are both positive maximum. A loud sound is again heard. Thus the frequency of the loud sounds or beats is 4 per second. In general, the frequency $f$ of the beats of two notes of nearly equal frequencies $f_1, f_2$ is given by $f = f_1 - f_2$.

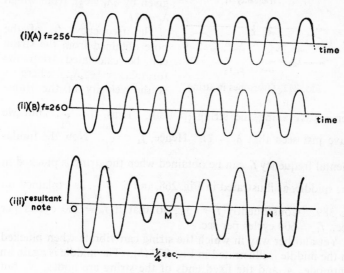

(i)(A) f=256

(ii)(B) f=260

(iii) resultant note

¼ sec.

FIG. 271. Beats

To tune the sonometer wire, the latter is plucked and the tuning fork sounded. The length of wire is altered until a throb or beat note is recognizable, in which case we know that the frequency of the wire is near to that of the fork. The length of wire is then adjusted until the beats are as slow as possible, and the wire can now be considered to be practically "tuned" to the fork.

**Overtones in a plucked string.** Notes of higher frequency can be obtained by plucking the middle of a violin string with increasing force. In each case a stationary wave is set up along the string, and Fig. 272 (a) illustrates the stationary wave set up along the string of length $l$ when the frequency is $f_1$. The middle of the stationary wave is an antinode, $A$, and the fixed ends of the string are nodes, $N$; and it can be seen that the distance from a node to the next node, which is half a wavelength, is $\frac{1}{3}l$. The wavelength, $\lambda_1$ is thus given by $\frac{1}{2}\lambda_1 = \frac{1}{3}l$, from which $\lambda_1 = \frac{2}{3}l$.

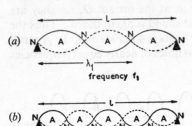

FIG. 272. Overtones in string

The frequency, $f_1$, of the note obtained from the string can be calculated from the formula $v = f_1\lambda_1$, where $v$ is the velocity of the transverse wave along the string (p. 380). Thus, $f_1 = \dfrac{v}{\lambda_1}$. But we have just seen that $\lambda_1 = \frac{2}{3}l$. Hence, $f_1 = \dfrac{3v}{2l}$. Now the fundamental frequency $f_0$ can be obtained when the string is plucked in the middle, as illustrated in Fig. 268, and $f_0 = \dfrac{v}{2l}$, as explained on p. 389. Consequently $f_1 = 3f_0$, so that if $f_0 = 200$ cycles per sec., $f_1 = 600$ cycles per sec.

Yet another way in which the string can vibrate when plucked in the middle is illustrated in Fig. 272 (b). The middle is again an antinode, $A$, and the fixed ends of the string are nodes, $N$; but the distance between two consecutive nodes is $\frac{1}{5}l$ in this case. Thus the new wavelength, $\lambda_2$, of the wave set up along the string is given by $\frac{1}{2}\lambda_2 = \frac{1}{5}l$, and hence $\lambda_2 = \frac{2}{5}l$. The frequency, $f_2$, of the note obtained is given by $f_2 = \dfrac{v}{\lambda_2} = \dfrac{5v}{2l}$, and since $f_0 = \dfrac{v}{2l}$, $f_2 = 5f_0$.

It can now be seen that higher notes of frequency $3f_0$, $5f_0$, $7f_0$, etc., may accompany a note of fundamental frequency $f_0$ obtained when the string is plucked in the middle. These higher notes are the *overtones* of the fundamental note, and, as explained on p. 378, they give the quality of the note.

*Even harmonics* are obtained by plucking the string one-quarter and one-eighth of the way along. Fig. 273 illustrates the stationary wave in the former case, as the antinode is then $\frac{1}{4}$ of the way along

FIG. 273. Frequency, $f, = 2f_0$

the string. Thus the length $l$ of the string is one wavelength, $\lambda$. The frequency, $f$, of the note is given by $f = \dfrac{v}{\lambda} = \dfrac{v}{l}$; and since the fundamental frequency $f_0 = \dfrac{v}{2l}$, from above, $f = 2f_0$. The reader should verify that the frequency is $4f_0$ when the string is plucked $\frac{1}{8}$ of the way along from one end.

## VIBRATIONS OF AIR IN PIPES

**The closed pipe.** A closed organ pipe consists basically of a pipe closed at one end $N$ and having an opening at the other end near $A$ (Fig. 274 (a) ). When air is blown into the pipe at $X$, a wave is excited at $A$, travels up the pipe, and is reflected at $N$, the closed end. Another wave is thus obtained in the direction $NA$, and so a *stationary wave* is formed in the air inside the pipe; just as a stationary wave was obtained in the string in Fig. 267 by two waves travelling in opposite directions.

$l = \lambda_0/4$    $l = 3\lambda_1/4$    $l = 5\lambda_2/4$

(a)    (b)    (c)

FIG. 274. Closed pipe

The simplest stationary wave in the pipe is one which has a node at the closed end $N$ of the pipe, where the air cannot move, and an antinode $A$ near the open end, where the air is free to move. Thus, roughly, the length $l$ of the pipe = distance from node $N$ to nearest antinode $A$.

$$\therefore \quad l = NA = \frac{\lambda}{4}$$

where $\lambda$ is the wavelength of the note obtained. See p. 388.

$$\lambda = 4l$$

But
$$v = f\lambda \text{ (p. 380)},$$

where $v$ is the velocity of sound in air and $f$ is the frequency of the note.

$$\therefore f = \frac{v}{\lambda} = \frac{v}{4l} \qquad . \qquad . \qquad . \qquad . \qquad (2)$$

**Overtones.** The frequency $v/4l$ is that of the lowest note obtained by blowing down the pipe, and is known as the *fundamental* frequency; its symbol is usually $f_0$. If air is blown slightly harder into the pipe, a note of higher frequency is heard, with a corresponding stationary wave-form shown in Fig. 274 (b). In this case, the length of the pipe, $l, = \frac{3\lambda_1}{4}$, where $\lambda_1$ is the wavelength, and hence $\lambda_1 = 4l/3$. Thus the higher frequency $f_1$ is given by

$$f_1 = \frac{v}{\lambda_1} = \frac{3v}{4l}$$

$$\therefore f_1 = 3f_0, \text{ from (2)}.$$

Thus the higher frequency ($f_1$) is three times the fundamental frequency ($f_0$).

By blowing harder, a note of higher frequency is obtained corresponding to the stationary wave shown in Fig. 274 (c). The frequency $f_2 = \frac{v}{\lambda_2} = \frac{5v}{4l}$, since $l = \frac{5\lambda_2}{4}$; thus $f_2 = 5f_0$. We now deduce that a closed pipe emits a fundamental note $f_0$ accompanied by notes of higher frequency $3f_0$, $5f_0$, $7f_0$, etc. The latter notes are the *overtones*, and accompany the fundamental note of frequency $f_0$; the overtones provide the "quality" or "timbre" of the fundamental note obtained from the pipe (see p. 378).

As an illustration, suppose the length of a closed pipe is 25 cm. Then, when the fundamental note of frequency $f_0$ is obtained, the wavelength $\lambda_0$ of the sound in air is given by $\lambda_0/4 = 25$ cm., from Fig 274 (a). Thus $\lambda_0 = 100$ cm. Now the velocity of sound $v$ in air is about 33,000 cm. per sec. Hence, as

$$\text{frequency} = \frac{\text{velocity}}{\text{wavelength}}$$

$$f_0 = \frac{v}{\lambda_0} = \frac{33,000}{100} = 330 \text{ cycles per sec.}$$

The first overtone corresponds to a wavelength $\lambda_1$ which is $\frac{1}{3}$ of that of the fundamental note, or 100/3 cm. (Fig. 274 (a), (b) ). Hence the frequency, $f_1$, of the overtone, which is equal to $\frac{v}{\lambda_1}$, is given by

$$f_1 = \frac{33,000}{100/3}.$$

Thus $f_1 = 990$ cycles per sec. Similarly, the second overtone is given by $f_2 = 5 \times 330$ cycles per sec., since the wavelength is $\frac{1}{5}$ of that of the fundamental frequency, i.e. $f_2 = 1650$ cycles per sec.

**The open pipe.** Fig. 275 (a) illustrates an organ pipe open at both ends. When the lowest (fundamental) note of frequency $f_0$ is obtained by blowing air into the pipe, a stationary wave is set up by reflection of the downward wave at the lower (open) end of the pipe. An antinode is obtained at both ends of the pipe, where the air is free to move; hence, since the distance between successive antinodes $= \frac{\lambda}{2}$ (p. 388), the length, $l$, of the pipe $= \frac{\lambda_0}{2}$, where $\lambda_0$ is the wavelength of the sound wave.

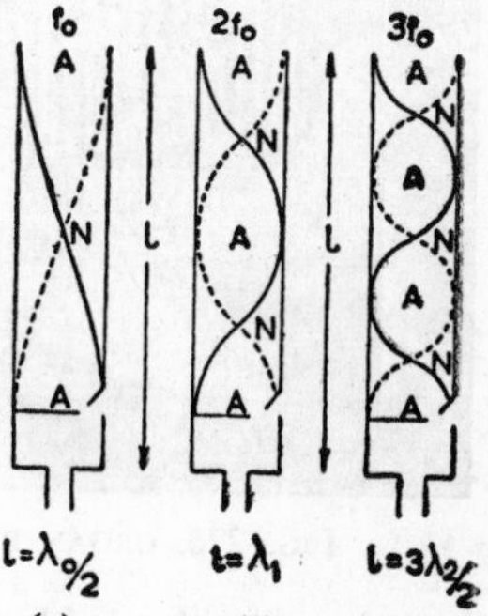

Thus                    $\lambda_0 = 2l$

$$\therefore f_0 = \frac{v}{\lambda_0} = \frac{v}{2l} \qquad . \qquad . \qquad . \qquad . \quad (3)$$

FIG. 275. Open pipe

By blowing harder into the pipe, a note of higher frequency, $f_1$, is obtained, corresponding to a stationary wave shown in Fig. 275 (b). Again, it should be noted, an antinode is obtained at both ends of the pipe. The length $l = \lambda_1$, the new wavelength. Hence the new frequency, $f_1 = \frac{v}{\lambda_1} = \frac{v}{l}$. Consequently, $f_1 = 2f_0$, from (3). By blowing harder into the pipe a note of higher frequency, $f_2$, is obtained, corresponding to Fig. 275 (c). It can be seen that $l = 3\lambda_2/2$, where $\lambda_2$ is the wavelength, and hence $f_2 = \frac{v}{\lambda_2} = \frac{3v}{2l}$. Consequently, $f_2 = 3f_0$, from (3). Thus,

14

accompanying the fundamental note $f_0$, are *overtones* of frequency $2f_0$, $3f_0$, $4f_0$, etc. This is different from the case of the closed pipe, in which the overtones are odd multiples of $f_0$ (see p. 396).

**The cinema organ** has sets of pipes of different shapes and sizes (see Fig. 276). Each set is designed to produce special overtones to the fundamental, so that one set emits notes sounding like those from a flute, another set emits notes sounding like those from a clarinet, and so on. In Fig. 276 the set of pipes doubled over near the base produce notes like those from a trombone.

FIG. 276. ORGAN PIPES

**Resonance.** A wooden bridge has a *natural frequency* of vibration which depends on its mechanical construction. If the movement of traffic and people should happen to give rise to regular impulses on the bridge, the latter will be set into vibration of small amplitude. But, if by chance, the frequency of the impulses were exactly equal to the natural frequency of the bridge, the latter would vibrate through a large amplitude, and the bridge would then be said to be in *resonance* with the applied impulses. The comparatively violent motion of the bridge would be dangerous to its stability. An army order actually forbids soldiers to keep in step when crossing a bridge for fear of accidents, although the danger is in fact slight.

An excellent example of resonance is obtained with a piano wire when the key of a note, G say, is depressed gently and held down. If one sings the note G and then stops, the string can be heard sounding the same note; it has been set into resonance by the note sung, which has the same natural frequency as the string. The newspapers have reported cases where a soprano on the wireless, singing a high note, has caused a glass tumbler in a listener's home to splinter into thousands of pieces. The note apparently had the resonant frequency of the glass, which broke as the vibrations became large.

Certain electrical circuits have also a resonant frequency. For example, a radio wave will produce a large "response" in a receiver set if its frequency is exactly the same as the natural frequency of electrical circuits in the set.

**Resonance tube. Measurement of velocity of sound in air.** The velocity of sound in air can be measured by means of a closed pipe known as a "resonance tube." A tube $V$ is filled with water, whose level can be adjusted by raising or lowering the tube $W$ (Fig. 277). A sounding tuning fork $F$ is placed over the top of $V$, and as the water level is gradually lowered, a position is reached when a sound of maximum loudness is heard coming from the tube. At this stage, the vibration of the tuning fork prongs sets the air inside $V$ in *resonance*; that is, the frequency of the fork is exactly equal to the natural frequency of the air column.

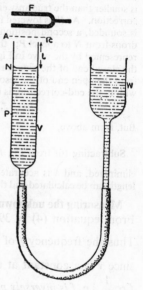

FIG. 277.  Resonance tube

Now we have already seen that when a pipe with one end closed is sounded, an antinode $A$ of a stationary wave exists at the open end and a node $N$ at the closed end. Thus the length, $l (= NA)$, of the air column, which is measured, is $\frac{1}{4}\lambda$, where $\lambda$ is the wavelength of the sound in air. Thus $\lambda = 4l$, and the velocity $v$ of sound is calculated from the relation

$$v = f\lambda = f \times 4l \quad . \quad \quad . \quad \quad . \quad \quad . \quad (4)$$

where $f$ is the known frequency of the tuning fork.

In an experiment to determine the velocity $v$, $l$ was found to be 26·5 cm. when $f$ was 312 cycles per sec. Thus $v = f \times 4l = 312 \times 4 \times 26·5 = 33,080$ cm. per sec., the temperature being 12° C.

**The end-correction.** In practice, the top end of the pipe is not exactly the position of the antinode $A$ of the stationary wave. The layer of air vibrating through the maximum amplitude, the antinode, is a small distance $c$ *outside* the top of the tube. This distance is known as the *end-correction* to the tube (see Fig. 277). Thus $NA = \dfrac{\lambda}{4} = l + c$, and hence $\lambda = 4(l + c)$. Thus $v = f\lambda = f \times 4(l + c)$, so that the velocity calculated above from $v = f \times 4l$ is smaller than the true answer. In practice it is possible to eliminate the end-correction. As $W$ is lowered from the position shown in Fig. 277 and the fork is sounded, a second position of resonance is obtained when the water level drops from $N$ to a level $P$; the stationary wave in the air between $AP$ is then represented by the wave in Fig. 274 (*b*) if we imagine the latter to be inverted, the closed end of the pipe corresponding to $P$ in Fig. 277. The length $l_1$ say, from the open end of the resonance tube to $P$ is now measured, and together with $c$, the end-correction, is equal to $\frac{3}{4}\lambda$.

$$\therefore \quad l_1 + c = \tfrac{3}{4}\lambda \qquad . \qquad . \qquad . \qquad . \qquad (5)$$

But, from above,
$$l + c = \frac{\lambda}{4} \qquad . \qquad . \qquad . \qquad . \qquad (6)$$

Subtracting (6) from (5), $l_1 - l = \dfrac{\lambda}{2}$. The end-correction has thus been eliminated, and $\lambda$ is accurately equal to $2(l_1 - l)$. Consequently the wavelength can be calculated and the velocity of the sound, $v, = f\lambda = f \times 2(l_1 - l)$.

**Measuring the unknown frequency of a fork by the resonance tube.** From equation (4) p. 399, we obtain the relation $v = f\lambda = 4fl$. Thus the frequency $f$ of the tuning fork is given by $f = \dfrac{v}{4l}$; and since $v$ is a constant at a given temperature, it can be said that $f \propto \dfrac{1}{l}$, i.e. *f is inversely proportional to the length of the air column in resonance*.

Suppose that a fork of frequency 256 cycles per sec. produces resonance when the length of the air column is 24·2 cm.; and that a tuning fork of unknown frequency $f$ produces resonance when the length of the air column is 16·0 cm. Then, since $f \propto \dfrac{1}{l}$,

$$\frac{f}{256} = \frac{24 \cdot 2}{16 \cdot 0}.$$

Thus $f = 387$ cycles per sec.

So long as a tuning fork of known frequency is provided, the resonance tube enables the unknown frequency of other forks to be roughly but quickly determined.

**The gramophone.** At a British Association meeting in 1859 it was demonstrated that the wave-forms of sound can be faithfully reproduced. A bristle, attached to a parchment, pressed against the lamp-blackened surface of a cylinder which could be rotated. When a sound was spoken in front of the parchment the latter vibrated, and a trace of the sound wave was marked by the bristle on the cylinder's surface as it was rotated; just as a trace was obtained on the blackened glass plate shown in Fig. 257, p. 371, as it dropped past the vibrating bristle.

Nearly 20 years later the great American inventor EDISON succeeded in reproducing the sound from the wave-form produced. He used a rigid point instead of a bristle and covered the cylinder with tin-foil, so that a continuous mark of varying depth was made in the foil when the point vibrated and the cylinder was turned. On running the point over the trace again, the sound was produced. Edison called his apparatus the *phonograph*, and cylindrical records were used on it.

In 1887 another American, BERLINER, succeeded in making a record of the sound on a flat disc, instead of a cylinder. Essentially, the method consisted of allowing a vertical needle to vibrate sideways while a horizontal lamp-blacked glass disc rotated, thus cutting grooves of a wavy shape into the lamp-black. By etching the traces into the glass and running a needle round them, the sound was produced, and Berliner called his apparatus a *gramophone*. The invention of sound recording and the gramophone thus followed from the method of showing wave-forms devised in 1859; and when we think of the pleasure of hearing the records made by great musicians, we realize once again how the pioneer work of the physicist in the laboratory has contributed to the cultural life of the community.

**The gramophone record.** Since 1925, records have been made by receiving the sound in a microphone, amplifying the resulting varying current by means of valves, and passing the current round the coils of an electromagnet which operates a magnetic stylus or pen. The pen moves over a horizontal disc of wax which revolves steadily round an axis through its centre, and wavy grooves (more than 130 of them to the inch) are obtained on the disc as the pen spirals towards the centre. The recorded wax is then sprinkled

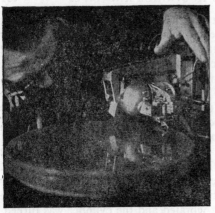

(*a*) Beginning to make a record; the electrically-operated stylus is placed on a revolving wax disc.

(*b*) Removing copper negative from the original wax recording.

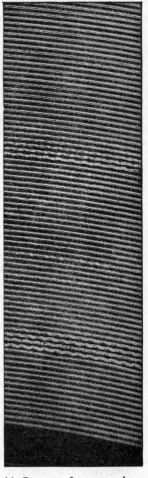

(*c*) Grooves of a gramophone record, highly magnified.

FIG. 278. GRAMOPHONE RECORD

with a very fine metallic powder, and placed in an electroplating bath. After about six hours a "negative" shell of copper is obtained which has the wavy grooves on it, and this is then stripped from the wax. In order to make a record a warm plastic material is pressed on to the copper shell by means of a hydraulic press exerting a pressure of about 100 tons weight per square inch; a double-sided record is made by compressing the plastic material between two copper shells. When the plastic material cools to room temperature it becomes hard and black, and after shaping and polishing it assumes the appearance of the familiar gramophone record. The sound is reproduced by a needle moving round the record and affecting a diaphragm in the "sound box" to which the needle is attached.

---
## Summary
---

1. Stationary waves set up along strings, and along the air in pipes, are due to two waves of the same frequency and amplitude travelling in opposite directions.

2. The "nodes" of a stationary wave are the points permanently at rest; the "antinodes" are the points midway between the nodes, and have maximum amplitude of vibration.

3. **The distance between a node and the next antinode is $\lambda/4$, where $\lambda$ is the wavelength; the distance between successive antinodes (or nodes) is $\lambda/2$.**

4. The fundamental frequency, $f$, of a string is given by $f = \dfrac{1}{2l}\sqrt{\dfrac{T}{m}}$, where $l$ is the length in centimetres, $T$ is the tension in dynes, $m$ is the mass per unit length of the string. This formula can be verified by the sonometer.

5. The fundamental frequency of a closed pipe $= v/4l$, where $l$ is the length and $v$ is the velocity of sound in air; the fundamental frequency of an open pipe of length $l = v/2l$.

6. The velocity of sound in air, and the unknown frequency of a tuning-fork, can be measured by a resonance tube experiment.

## WORKED EXAMPLES

1. *State the relation between velocity, frequency and wavelength of sound waves and calculate the wavelength of the note given by a tuning fork of frequency 384 cycles per sec. when the velocity of sound is 340 metres per sec. at room temperature. How could you determine this wavelength by an experiment with a resonance tube?* (*L.*)

The required relation is $v = f\lambda$. In this expression $v$ is in *centimetres* per sec. when $\lambda$ is in centimetres and $f$ in cycles per sec.; $v$ is in *metres* per sec. when $\lambda$ is in metres and $f$ in cycles per sec.

$$\therefore \quad \lambda \text{ (in metres)} = \frac{v \text{ (metres per sec.)}}{f}$$

$$= \frac{340}{384} = 0 \cdot 885$$

2. *A sonometer wire 1 metre long emits a note of frequency 300 per sec. when under a tension of 2 Kgm. weight. Calculate (a) the length at which it will have a frequency of 600 per sec., the tension being unaltered, (b) the frequency when the tension is increased to 8 Kgm. weight, the length remaining 1 metre.* (*N.*)

(*a*) Under a constant tension, the frequency $f$ of a sonometer wire is inversely proportional to the length, i.e. $f \propto \dfrac{1}{l}$. The frequency of 600 per sec. thus corresponds to *half* the length of the frequency of 300 per sec. Thus length $= \frac{1}{2}$ metre.

(*b*) For a constant length, the frequency of a sonometer wire is proportional to the square root of the tension, i.e. $f \propto \sqrt{T}$. Consequently, the new frequency, $f$, is given by

$$\frac{f}{300} = \frac{\sqrt{8}}{\sqrt{2}} = \sqrt{\frac{8}{2}} = \sqrt{4}$$

$$\therefore \quad \frac{f}{300} = 2, \text{ i.e. } f = 600 \text{ per sec.}$$

3. *Write a short description of the state of the air through which sound waves are travelling. A long vertical glass tube is filled with water, which is then slowly run out whilst a vibrating tuning fork is held over the open end of the tube. It is found that maxima of sound are obtained when the air column measures 13·5 in. and 41·5 in. respectively. Calculate the frequency of the fork.* (*Velocity of sound in air* = 1120 *ft. per sec.*) (*N.*)

In the first position of maximum sound, the distance from the water to the open end of the tube is $\dfrac{\lambda}{4}$, where $\lambda$ is the wavelength (see p. 400).

$$\therefore \quad \frac{\lambda}{4} = 13 \cdot 5 + c \quad . \quad \quad . \quad \quad . \quad \quad . \quad (1)$$

where $c$ is the small distance known as the "end-correction" of the tube.

In the second position of maximum sound, the distance from the water to the open end of the tube is $\dfrac{3\lambda}{4}$ (p. 400).

$$\therefore \quad \frac{3\lambda}{4} = 41\cdot5 + c \quad . \quad . \quad . \quad . \quad . \quad (2)$$

Subtracting (1) from (2),

$$\therefore \quad \frac{3\lambda}{4} - \frac{\lambda}{4} = 41\cdot5 - 13\cdot5$$

$$\therefore \quad \frac{\lambda}{2} = 28, \text{ i.e. } \lambda = 56 \text{ inches} = 4\tfrac{2}{3} \text{ ft.}$$

$$\therefore \quad \text{frequency}, f = \frac{v}{\lambda} = \frac{1120}{4\tfrac{2}{3}} = 240 \text{ cycles per sec.}$$

## EXERCISES ON CHAPTER XXVI

**1.** Draw a diagram of a stationary wave, and mark on it the positions of the antinodes. Define the terms *antinode* and *node*.

**2.** How is a stationary wave formed ? What is the distance in terms of wavelength between (i) two consecutive nodes, (ii) a consecutive node and antinode, (iii) two consecutive antinodes, (iv) two alternate nodes ?

**3.** Draw a diagram of the simplest stationary wave formed in a closed and open pipe when they are sounded. If the length of the pipe is 25 cm. in each case, calculate the wavelength of the notes and the frequencies of the fundamentals. (Velocity of sound = 33,600 cm. per sec.)

**4.** A taut string fixed at both ends is plucked in the middle. Where are the nodes and antinodes of the simplest stationary wave in the string ? If the length of the string is 60 cm., calculate the wavelength and the frequency of the note obtained. (Velocity of sound in string = 72,000 cm. per sec.)

**5.** How does the frequency of the note emitted by a vibrating wire depend on (*a*) the length, (*b*) the tension ? Describe an experiment to verify the relation between frequency and length. A wire 1 metre long gives the same note as a tuning fork of 288 cycles per sec. What are the frequencies of the notes emitted by the two segments of the wire when a bridge is placed at a point 60 cm. along the wire from one end ? (*L.*)

**6.** Explain carefully what is meant by *resonance*. Describe how you would arrange a state of resonance between a column of air and a given tuning fork, and explain how you would use the arrangement to enable

14*

you to determine the velocity of sound in the air, the frequency of the fork being known.  (*N*.)

**7.** Describe how you would arrange for a sonometer wire to have the same fundamental frequency as a given tuning fork.  With such an arrangement, a sounding tuning fork of double the frequency placed on the sonometer board causes the wire to vibrate.  Explain this, and describe the motion of the wire.  (*N*.)

**8.** An organ pipe of effective length 25 cm. is stopped at one end. Calculate the wavelength and frequency of the note emitted by it when blown with air.  What difference, if any, would you expect to find if carbon dioxide were substituted for air ?  The velocity of sound in air is 340 metres per sec. and the density of carbon dioxide is 1·44 times that of air.  (*C*.)

**9.** What factors determine the frequency of vibration of a stretched string ?  Two wires of the same material have lengths in the ratio 2 : 3. If their diameters are the same, what must be the ratio of their tensions for the shorter wire to give a note an octave higher than the longer ?  (*C*.)

**10.** Being provided with a resonance tube, a tall gas-jar full of water and a tuning fork of frequency 512 cycles per sec., how would you determine the frequency of an unmarked tuning fork ?  A resonance box is to be made for use with a tuning fork of frequency 439 cycles per sec. when the velocity of sound in air is 341 metres per sec.  What must be approximately the shortest length of the box if it is closed at one end ?  (*L*.)

**11.** Define *frequency*.  What are the factors which control the frequency of the note emitted by a sonometer wire when this is plucked or bowed ?  Describe and explain a method of comparing the frequencies of two tuning forks.  (*L*.)

**12.** A tuning fork of frequency 330 per sec. is held vibrating above the open mouth of a tube about one metre deep while water is slowly sent into the tube through an inlet at the bottom.  A loud note is heard when the length of the air column above the water is 74 cm., and as more water is introduced the sound dies away, returning at full strength when the length of the air column is reduced to 24 cm.  Explain why this happens and calculate the velocity of sound in air from these observations.  (*N*.)

**13.** How does the frequency of transverse vibration of a stretched string depend on (*a*) the tension, (*b*) the length, (*c*) the density of the material, of the string ?  Describe a sonometer, and show how you would use it to verify the relation between frequency and length under constant tension. (*L.*)

**14.** Define *frequency, wave-length* for a sound wave. State the equation relating these quantities with the velocity of the wave. Describe how the resonance tube can be used to measure the velocity of sound. In an experiment with a resonance tube, closed at one end, the two shortest lengths for which resonance was obtained were (*a*) 15 cm., (*b*) 49 cm. If the frequency of the note was 500, calculate the velocity of sound at the temperature of the experiment. (*C.*)

**15.** Describe how you would determine the velocity of sound in air. The velocity of sound in dry air at 0° C. is 332 metres per sec. How is this velocity affected by (*a*) rise of temperature, (*b*) change of pressure ? Calculate, neglecting end corrections, the length of an open organ which will have a frequency of 500 when blown with dry air at 0° C. (*N.*)

**16.** Describe an experiment to demonstrate that the prongs of a tuning fork vibrate when it emits sound and explain the increased loudness of the sound when the stem of the vibrating fork is held in contact with a table.

A tuning fork has the number 256 stamped on it. What does this signify ? A sonometer wire is adjusted to be in unison with this fork. State exactly two different ways in which the wire could be readjusted so as to be in unison with a fork marked 384. (*L.*)

**17.** Describe a sonometer. How may it be used to compare the frequencies of two notes ?

Two cog wheels, *A* and *B*, in a piece of machinery, revolve at 15 and 8 revolutions per second respectively. *A* has 28 teeth, and a card held against it emits a note in unison with 16·8 cm. of the sonometer wire. When the card is transferred to *B*, the length of the wire has to be changed to 25·2 cm. for unison. Calculate (*a*) the frequency of the card when touching *A*, (*b*) its frequency when touching *B*, (*c*) the number of teeth on *B*. (*L.*)

# ANSWERS TO NUMERICAL EXERCISES

## CHAPTER XXV (p. 384)

**12.** 340, 68 cm.; 200, 50 cycles per sec.

**16.** 2240 ft.

**18.** $3\frac{3}{8}$ ft.

**19.** 275 cycles per sec.

## CHAPTER XXVI (p. 405)

**3.** 336, 672 cycles per sec.

**4.** 120 cm., 600 cycles per sec.

**5.** 480, 720 cycles per sec.

**8.** 100 cm., 340 cycles per sec.

**9.** 16 : 9.

**10.** 19·4 cm.

**12.** 33,000 cm. per sec.

**14.** 34,000 cm. per sec.

**15.** 0·332 metre.

**16.** Length reduced $\frac{2}{3}$ times, or tension increased $\frac{9}{4}$ times.

**17.** (a) 420, (b) 280 c.p.s., (c) 35.

# ELECTRICAL ENERGY

# CHAPTER XXVII

## ELECTROSTATICS

MORE than two thousand years ago it was known that a stick of amber or glass, after being rubbed with silk, had the power of attracting light bodies, such as pieces of paper or wood. In this condition the stick is electrified, or "charged" with electricity. The same phenomenon can be produced by rubbing an ebonite fountain pen on one's sleeve, or combing one's hair briskly with an ebonite comb.

In the seventeenth century it was observed that a charged rod, after attracting to itself a piece of cork suspended at the end of a thread, then began to repel it. As the cork had been electrified by contact with the rod, it was concluded that *similarly charged bodies repel each other*.

**Positive and negative electricity.** Further experiments showed that an ebonite rod $R$ rubbed with fur repelled a similarly electrified suspended rod $X$ (Fig. 279 (*a*) ), but that a glass rod $G$ rubbed

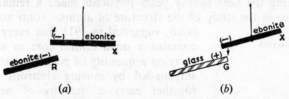

(*a*)        (*b*)

FIG. 279. Law of repulsion and attraction

with silk *attracted* $X$ (Fig. 279 (*b*) ). Electrified rods thus either repelled or attracted $X$, and charged rods could thus be separated into two classes: (1) Those charged similarly to an ebonite rod rubbed with fur; (2) those charged similarly to a glass rod rubbed with silk.

BENJAMIN FRANKLIN, a prominent American scientist of the

eighteenth century, showed that one kind of electricity could be made to neutralize or reduce the effect of the other, and he gave the names of *positive* (+ ve) electricity and *negative* (− ve) electricity to them to indicate their opposite character. The electricity on an ebonite rod rubbed with fur is called "negative," and that on a glass rod rubbed with silk is called "positive" (see Fig. 279).

We have already seen that two ebonite rods rubbed with fur, each carrying negative electricity, repel one another, and that a glass rod rubbed with silk, carrying positive electricity, attracts an ebonite rod rubbed with fur, carrying negative electricity. Similarly, two glass rods rubbed with silk, each carrying positive electricity, repel one another. All these experimental results can be summarized by the statement:

**Like charges repel. Unlike charges attract.**

This is a "fundamental law" of electrostatics.

**Electrons and protons.** In 1897 SIR J. J. THOMSON discovered the existence of a tiny particle carrying a quantity of *negative* electricity, and having a mass about 1/2000th of the mass of the hydrogen atom (p. 623). This particle was called an **electron**. Until the time of its discovery, it was believed that the lightest particle was a hydrogen atom.

During the next twenty years physicists made a remarkable advance in the study of the structure of atoms. LORD RUTHER-

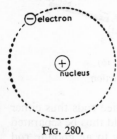

FORD, suggested in 1911 that every atom contains a small central core or **nucleus** carrying a quantity of positive electricity, surrounded by moving electrons which together carry a quantity of negative electricity equal to that on the nucleus. As the positive and negative kinds of electricity are opposite in character, the net amount of electricity in a normal atom is zero. The nucleus and the surrounding electrons take up a tiny fraction of the space occupied by the atom. Electrons are much smaller than the nucleus of their atom, and move about the stationary nucleus

FIG. 280.
The hydrogen atom

under the attraction of its positive charge, just as the planets move round the sun.

Fig. 280 illustrates diagrammatically the structure of a hydrogen atom, with one electron revolving round the nucleus. Helium, the next lightest element, has two electrons moving about its nucleus· Nitrogen has 7 electrons, and uranium, one of the extremely heavy metals, has 92.

**How friction electrifies.** When an ebonite rod is rubbed with fur, the friction results in the transfer to the ebonite rod of some of the loosely attached electrons from the atoms constituting the fur. Since the rod was originally uncharged, it becomes negatively charged, leaving the fur with an *equal* positive charge. Thus, if the fur loses a total of 1000 electrons, (i.e. 1000 units of negative electricity) it is left with a surplus of 1000 units of positive electricity.

When a glass rod is rubbed with silk, electrons are transferred from the rod to the silk. The silk therefore becomes negatively charged, while the rod is left with an equal quantity of positive electricity.

It should be noted that rubbing an ebonite or glass rod does not *create* electricity. The electricity is already present in the atoms of the materials; the rubbing merely results in a transfer of electrons from one substance to another.

**Conductors and insulators.** In 1729 STEPHEN GRAY noticed that metallic substances could not be charged by friction if they were held in the hand. They could be charged, however, if they were held by an ebonite or glass support. Gray concluded that the electricity on a metal rod could move along it and escape through the human body, and he therefore distinguished between substances which easily conduct electricity, called **conductors,** and others which do not easily conduct electricity, known as **insulators** (see p. 450).

All metals, the human body, the earth, impure water, and salt solutions are examples of conductors; glass, ebonite, paper, wood, and mica are normally insulators. Any electricity obtained on the latter substances is stationary or *static*, as the early scientists knew, and "electrostatics" was the branch of electricity first studied extensively.

Modern theories of the atom suggest that some of the electrons in a metal atom are loosely bound to their nucleus, and wander about indiscriminately in the spaces between the atoms; by comparison with those which remain within the sphere of influence of their nucleus, these electrons are known as "free" electrons. No such "free" electrons are found in insulators, all of whose electrons are firmly bound to the nucleus. According to modern ideas, then, *the movement of "electricity" along a metal is simply the movement in one direction of the "free" electrons* in the metal.

**The gold-leaf electroscope.** The gold-leaf electroscope was one of the earliest instruments for testing electric charges, and was designed about 1778. It was used extensively by MICHAEL FARADAY in the 19th century in many of his important researches in electrostatics, and, in a rather more sensitive form, is still in use to-day.

In its simplest form, it consists of a metal rod, $A$, with a gold leaf $L$ attached to its lower end and a metal disc $C$ to its upper end (Fig. 281 ($a$) ). The rod passes through a block $S$ of sulphur, which insulates it from a metal case $D$ containing the leaf. The disc $C$ is outside the case. When the electroscope is in normal use, the case is connected to the earth.

If a negatively charged rod $R$ touches the cap $C$ (Fig. 281 ($a$) ), some of the electrons in the metal system comprising $C$, $A$, $L$ are

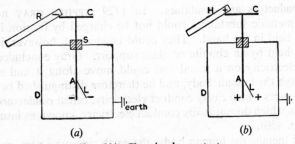

FIG. 281. Charging by contact

repelled to $A$ and to $L$, since like charges repel. The leaf $L$ is then repelled by the similar charge on $A$, and becomes inclined to the latter at an angle which depends on the magnitude of the charges. The electroscope is now said to be "charged" by contact.

If a glass rod $H$ rubbed with silk touches the cap $C$ of an uncharged electroscope (Fig. 281 (b) ), the positive charge on the rod
attracts electrons in $C$, $A$, $L$ to the cap, leaving a surplus positive
charge on $A$ and $L$. The gold leaf now diverges, since like charges
repel. Thus, in general, the electroscope has a charge of the same
kind as that on the rod when the latter touches it.

The nature of a charge can be tested by bringing it near to the
cap of an electroscope having a known charge. If the leaf is
negatively charged and diverges further still, the unknown charge
$A$ is a negative one (Fig. 282 (a) ). If a positively charged leaf
diverges further, the unknown charge is positive. An ebonite rod

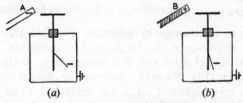

FIG. 282. Testing charges

rubbed with fur causes the leaf of a negatively charged electroscope to diverge further when brought near to the cap; but the
fur itself, held by an insulator, causes the divergence of the same
leaf to diminish. This shows that the ebonite rod has a negative
charge and suggests that the fur has a positive charge. When the
charged rod and the charged fur are held *together* near the charged
electroscope, the position of the leaf remains the same. This shows
that the amounts of positive and negative electricity on the rod and
fur are equal.

A positively charged rod causes a negatively charged electroscope to decrease in divergence (Fig. 282 (b) ). But as an uncharged rod would have the same effect on an electroscope, the
only sure test for a charge is an *increased* divergence, as already
explained.

**Induced charges.** Consider a negatively charged ebonite rod $R$
near to two uncharged metal spheres $A$, $B$ suspended by silk
insulating threads, and suppose $A$, $B$ are in contact with each

other (Fig. 283). Under the influence of the negative electricity on $R$ electrons are repelled from $A$ to $B$, leaving $A$ with a surplus

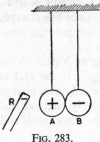

FIG. 283.
Induced charges

positive charge. Thus $A$ and $B$ have equal positive and negative charges. While in the neighbourhood of $R$, the two charges can be separated by means of the insulating threads, and their magnitudes and nature compared by a charged gold-leaf electroscope, as explained before.

The two charges on $A$ and $B$ are obtained without $R$ touching the spheres and they are hence known as *induced charges*. The phenomenon is known as **induction**.

**Charging an electroscope by induction.** Fig. 284 illustrates how a gold-leaf electroscope can be charged by induction. Suppose that a negatively charged rod $R$ is brought near to the cap $C$ of an electroscope (Fig. 284 (*a*) ). Since unlike charges attract, electrons in the metal system $C$, $A$, $L$ are repelled to $A$ and $L$, leaving a positive charge on $C$. The leaf thus diverges. When $C$ is touched the electrons pass through the body to earth and the leaf collapses; but the positive charge remains on $C$ (Fig. 284 (*b*) ). When the finger is removed, the positive charge remains at the same place (Fig. 284 (*c*) ) but the leaf remains uncharged. However, when the

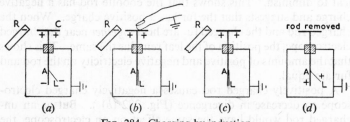

FIG. 284. Charging by induction

rod is removed some of the positive charge spreads over $A$ and $L$, and the leaf therefore diverges (Fig. 284 (*d*) ). The electroscope has thus become charged without being touched by the rod $R$; i.e. it has acquired an induced charge. On bringing a positively charged glass rod near to the cap, the divergence of the leaf increases, showing that the induced charge is positive, i.e. *opposite* to the negative inducing charge, the charge on *R*.

It should be noted that if the rod is taken away *before* the finger is removed in Fig. 284 (*b*), electrons pass through the body from the earth and neutralize the positive charge. In this case, no charge appears on the leaf, which does not diverge.

**Faraday's ice-pail.** Faraday investigated the phenomenon of induction by using a conductor in the form of an ice-pail or can. He placed the can *C* on top of a gold-leaf electroscope, and lowered an insulated charged sphere *A* deeply inside it, without touching *C* (Fig. 285 (*a*) ). The leaf diverged. When *A* was taken out of the can, the leaf collapsed entirely. *A* was now lowered again deeply into the can, whereupon the leaf diverged, and when

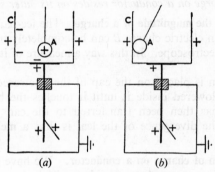

FIG. 285. Ice-pail experiment

the inside of the can was touched by *A* (Fig. 285 (*b*) ) *no change was observed in the divergence of the leaf.* Further, on withdrawing *A* and testing it, *no charge could be detected on A.*

When the sphere is first placed in the can (Fig. 285 (*a*) ), an induced negative charge appears on the inside of *C*, and an induced positive charge is spread over the outside of *C* and the cap and leaf of the electroscope. Since the leaf closes completely when *A* is withdrawn, the induced negative and positive charges must have been equal.

When *A* is placed inside *C* and then touches the can, the

divergence of the leaf is unaltered, and no charge remains on *A*. Thus the negative induced charge on the inside of *C* has completely neutralized the positive charge on *A*. Hence the induced positive charge on the outside of *C* and on the electroscope is equal to the inducing charge *A*. Moreover, as *A* has no charge on it after touching *C*, the positive induced charge on *C* must exist entirely on its *outside*; otherwise *A* would have had some positive charge by contact with *C*.

The ice-pail experiment shows that—

(1) *In the process of induction, the induced positive and negative charges are equal in magnitude.*

(2) *The induced and inducing charges are equal when the inducing charge is completely surrounded by the body on which the charge is induced.*

(3) *The charge on a conductor resides on its outer surface.*

**Measuring the magnitude of a charge.** The ice-pail experiment shows how an electric charge *B* can be completely transferred to a gold-leaf electroscope. In this way a measure of its charge can be obtained.

A deep can is placed on the cap of the electroscope, and the charge *B* is lowered inside it until it touches the bottom. *All* the charge has then been transferred to the can and electroscope, and the divergence of the leaf is thus a measure of the charge on *B*.

**Distribution of charge on a conductor.** We have already seen that the charge on a conductor resides on its *surface*, and this can be explained by the repulsion of like charges, which move outwards from the middle of a conductor until the surface is reached. Faraday demonstrated this property of a charged conductor by charging a butterfly net, made of conducting material, and testing the interior and exterior for charges. The outside was charged, but no charge could be detected on the inside. He then turned the same net inside out by pulling a silk thread at the corner of the net, and found that the charge now appeared on the *new* outside, and that there was no charge on the new interior of the net.

The distribution of a charge on a charged conductor can be tested by means of a *proof-plane*, which consists of a small metal disc $M$, at the end of a long insulating handle $H$, made of ebonite or glass (Fig. 286 (a) ).

Suppose that $X$ is a positively charged conductor of a pear-shape, so that the curvature of its surface varies from place to

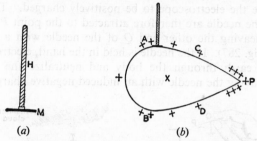

FIG. 286. (a) Proof plane (b) Variation of charge

place (Fig. 286 (b) ). To find the charge round $A$, the proof-plane is placed on $X$ so that the whole of the disc $M$ makes contact with the conductor round this point, and the charged disc is then removed and placed inside a can on the top of a gold-leaf electroscope. By touching the inside, the whole of the charge on $M$ is transferred to the can and electroscope, and the divergence of the leaf is noted. The experiment is repeated with the whole of the disc $M$ of the proof-plane touching different parts of the conductor $X$, and in this way the variation of the *charge per unit area*, or *surface-density of charge*, of $X$ can be roughly measured.

It is found that the pointed end $P$ of the conductor has the greatest surface-density of charge; that the parts, $A$, $B$, of the surface, which have a much smaller curvature than $P$, have a much smaller surface-density; and that the almost plane portions, $C$, $D$, of the conductor have a very small surface-density. In general, then, *that part of the surface of a conductor which has the greatest curvature has the greatest density of charge*. A charged metal sphere has a constant surface-density, as the curvature is the same all over.

**Action at point of a conductor.** If the point $P$ of a needle is held near to the cap of a charged electroscope, the divergence of the leaf slowly diminishes. An action has thus taken place which has resulted in the *discharge* of the electroscope.

The explanation of this phenomenon has practical application in the lightning conductor and the electrostatic generator described later (p. 423).

Suppose the electroscope to be positively charged. The electrons in the needle are therefore attracted to the point $P$ near to the cap, leaving the other end $Q$ of the needle with a positive charge (Fig. 287). If the needle is held in the hand, electrons pass from the earth through the body and neutralize the positive charge, leaving the needle with an induced negative charge.

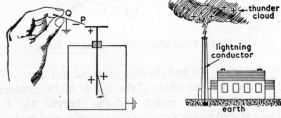

FIG. 287. Action of points    FIG. 288. Lightning conductor

As shown before, the charge on a conductor concentrates at those parts of its surface having the greatest curvature. In particular, the concentration of electricity is very great at *points* on a conductor. The air molecules in contact with the point of the needle thus gain some charge, and are repelled from the needle by the similar charge left on the point. Other molecules of air then become charged by contact and move away from the point, with the result that a stream of negative electricity, carried by the molecules of the air, moves towards the cap of the positively charged electroscope. The net charge on the electroscope then diminishes, and the divergence of the leaf is gradually reduced to zero.

**The lightning conductor.** In a thunderstorm in 1749, Benjamin Franklin at the risk of his life flew a kite with a metal point at the top and a metal key at the lower end, using a silk (insulating) thread. For a short time there was no effect. Then suddenly, as a result of the rain, the silk thread became completely conducting, and Franklin obtained sparks when he brought his knuckles near the key. In this way he proved for the first time that lightning is an electric spark on a grand scale.

As a protection for buildings, Franklin designed the *lightning conductor*. This is a long vertical rod, pointed at the upper end, and connected to earth (Fig. 288). If a thundercloud, containing particles with a positive charge is in the region above the conductor the latter becomes negatively charged by induction. A stream of negative electricity then flows upwards from the point, and the discharge of the cloud takes place more slowly, and with less violence, than if the conductor were absent.

**Electrostatic generators.** In the nineteenth century, machines were invented to produce a large quantity of electricity by electrostatic methods. LORD KELVIN designed several "generators" of static electricity, but their development came to a halt as a result of the greater efficiency of the dynamo, which provides a large and continuous flow of electricity. In 1930, however, VAN DE GRAAFF, an American physicist, designed a new type of electrostatic generator, following an idea gleaned from an account of Kelvin's work on the subject (see p. 423). The principles of three electrostatic generators are given below.

*The electrophorus.* The first, and simplest, of these instruments was the "electrophorus," or "charge carrier," which consisted of an ebonite block *A*, and a metal plate *B* with an insulating

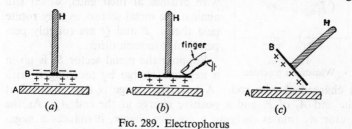

FIG. 289. Electrophorus

handle *H*. The block *A* is rubbed with a piece of fur, and becomes negatively charged (Fig. 289 (*a*) ). *B* is then placed on it and momentarily touched with the finger (Fig. 289 (*b*) ). Finally *B* is removed by means of $H$ (Fig. 289 ($c$) ). On testing, $B$ is found to possess a positive charge, which is opposite to that on $A$.

When $B$ is placed on $A$, the two make very little contact as their surfaces are not perfectly plane, and hence electrons in $B$ are repelled to the upper surface, leaving a positive charge on $B$ near $A$. The electrons pass through the body to earth when the finger is placed on $B$ (Fig. 289 ($b$) ), and the positive charge spreads over $B$ when it is lifted by the handle $H$ (Fig. 289 ($c$) ). Thus $B$ has acquired a charge by induction, leaving the negative charge on $A$ unaltered. The charge on $B$ can be transferred by contact to another conductor $Y$, and $B$ can then be brought back to $A$ and recharged in the same way. In this manner a large charge can be built up on $Y$ from a small charge. It should be noted that in using an electrophorus the finger must be removed *before* the disc is taken away.

***The Wimshurst machine*** was designed in the nineteenth century, and acts like an electrophorus which automatically transfers

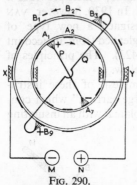

FIG. 290.
Wimshurst machine

charges continuously to two conductors. It consists essentially of two large coaxial glass plates, one behind the other, which have metal sectors $A_1$, $A_2$ . . . , $B_1$, $B_2$ . . . , shown diagrammatically in Fig. 290, spaced equally all round their circumference. Fixed metal conductors, $P$, $Q$, have wire brushes at their ends, which rub against the metal sectors as they rotate past them. $P$ and $Q$ are roughly perpendicular to each other.

Suppose the metal sector $B_1$ is given a negative charge by touching it with a charged ebonite rod. A negative charge is then induced at the end $A_7$ of $P$, and a positive charge at the end $A_1$. As the sector $A_1$ rotates carrying a positive charge, it induces a negative charge on the metal sector then in the position $B_3$ and a positive charge on the metal sector then in the position $B_9$. By following the induced charges on the metal sectors of the two glass plates, which are made to rotate in opposite directions, it can be seen that negative charges are carried towards the metal pointed "comb" $X$, and positive charges are carried towards the other "comb" $Y$. At $X$, the negative charges induce a positive charge on the points and a negative charge on the sphere at $M$. The positive charges stream from the points of $X$ (as explained on p. 420), and neutralize the charge on the metal sectors, leaving the negative charge on $M$. Similarly, by means of the comb $Y$, the sphere at $N$ obtains an induced positive charge. The charges at $M$ and $N$ both increase until the insulation of the air breaks down, when a spark is obtained between them.

**Van de Graaff electrostatic generator.** Electrostatic generators were regarded only as scientific curiosities in the first part of the twentieth century. About 1929, however, Van de Graaff designed a new form of high voltage electrostatic generator, and one was used to maintain the whole of the display lighting at the Chicago Fair Exhibition in 1931.

The principal features of the generator are shown in Fig. 291. The high voltage terminal is a smooth metal hemisphere $M$, which can be seen clearly in Fig. 292, at the top of a tall insulating column. A continuous silk belt $G$, $G$ moves up and down by means of a motor driving the pulleys $P_1$, $P_2$, and pointed conductors $B$, $A$ are placed opposite the belt at the top and bottom respectively. $B$ is joined to the hemisphere $M$; $A$ is joined to the positive pole of a large battery $L$ such as 10,000 volts, the other pole being earthed.

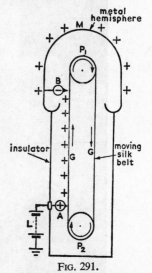

FIG. 291.
Van de Graaff generator

The conductor $A$ obtains a positive charge, and being pointed, it loses its charge quickly. In so doing, it "sprays" the belt facing it with positive charge. The belt rises, carrying the positive charge upward. When the charge arrives opposite $B$, it induces a negative charge on $B$ and an equal positive charge on the

FIG. 292. VAN DE GRAAFF GENERATOR

hemisphere $M$. The negative charge on $B$, a pointed conductor is immediately sprayed on to the belt in front of it, and neutralizes the positive charge on the belt. The belt thus moves round $P_1$ uncharged. When it passes $P_2$ and reaches $A$, the action is repeated. Thus the positive charge on the hemisphere $M$ is increased continuously, and when the sphere is large and the insulation is high, a very high voltage, such as several millions of volts, can be built up on it.

Van de Graaff generators are used nowadays in the study of atomic structure and nuclear energy. They have been built for this purpose at Harwell, London, Cambridge and Manchester.

────────Summary────────

1. **Like electrical charges repel, unlike charges attract.**

2. An ebonite rod rubbed with fur acquires a negative charge; a glass rod rubbed with silk acquires a positive charge. An electron, the lightest particle known, and a constituent of all atoms, carries a negative charge.

3. Metals, the human body, and the earth are examples of conductors; glass, ebonite, porcelain, and plastics are normally insulators.

4. The sign of a charge can be found by means of a charged gold-leaf electroscope; an increased divergence of the leaf indicates a charge similar to that on the electroscope.

5. **An induced charge can be obtained on an electroscope by (i) bringing a charge near, (ii) touching the electroscope cap, (iii) removing the finger, (iv) taking the charge away.**

6. A charged pear-shaped conductor has the greatest concentration of charge round the point.

## EXERCISES ON CHAPTER XXVII

**1.** How could you obtain a charge (quantity of electricity) on (i) a glass rod, (ii) a brass rod? Describe how you would test the sign of the charge.

**2.** State the sign of the charge on an *electron*, and the latter's approximate weight. What is the difference between the electrons in a metal and in an insulator?

**3.** State which of the following are insulators and which are conductors of electricity: *Mica, brass, air, silver, copper, water, china,* the *earth, bakelite.* What happens to the charge on a brass rod when it is held in the hand and rubbed?

**4.** Describe a *gold-leaf electroscope.* How is it used to test (i) a negative charge, (ii) a positive charge?

**5.** Describe the *electrophorus.* How is it used to build up a large charge on a conductor?

**6.** Describe how a gold-leaf electroscope is charged by induction. Draw diagrams of the various stages of the process, showing the appearance of the leaf.

**7.** A pear-shaped conductor is charged. Which part of its surface has the greatest charge per unit area? Explain the principle of the lightning conductor.

**8.** A *Van de Graaff generator* was built at Cambridge in 1947 for atomic research. Describe the generator, and explain its action *briefly.*

**9.** Describe a gold-leaf electroscope with the aid of a labelled diagram. Explain how it would be used (*a*) to show that equal and opposite electric charges are produced by friction, (*b*) to find the nature of the charge produced on an insulated brass rod when rubbed with catskin. (*L.*)

**10.** If an electroscope is charged and left, the leaves gradually collapse. Give *two* possible reasons for this. (*N.*)

**11.** Describe the electrophorus and explain how it works. When the charged metal plate is brought near to an earthed conductor a spark passes. Explain this. Whence is derived the energy manifested by the spark? (*N.*)

**12.** How would you demonstrate that equal and opposite charges are produced by electrostatic induction? How would you transfer the whole of the charge on a small insulated spherical conductor to a large hollow conductor? Give reasons for your answer. (*C.W.B.*)

**13.** Describe a gold-leaf electroscope and explain how you would use it to test the insulating properties of various materials. If you were provided with an ebonite rod and fur, how would you give the electroscope (a) a positive charge, (b) a negative charge ? (L.)

**14.** Give a description of a gold-leaf electroscope. An insulated metal plate is connected electrically to the cap of an electroscope. State how you would charge the system positively by induction; by means of diagrams show the distribution of the charge at various stages of the process. State and explain how the distribution of the charge is altered when an earthed metal plate is brought near the first plate. (N.)

**15.** Three equal uncharged insulated metal spheres, A, B, and C, have their centres in line and only B and C are in contact. A is given a positive charge. Draw a diagram of the resulting lines of force, and describe how you would show experimentally (a) that electrostatic induction has occurred, and (b) that B and C have a uniform potential which is positive. (N.)

**16.** Give a labelled diagram to show the essential parts of a gold-leaf electroscope. Explain how this instrument is used (a) to show that there is no charge on the inside of a hollow charged conductor, (b) to investigate the distribution of charge over a pear-shaped conductor. What would be the result in case (b) ? (L.)

**17.** Describe the gold-leaf electroscope. How would you find out whether the charge on a charged gold-leaf electroscope was positive or negative ? Explain the method of charging an electroscope (a) positively, (b) negatively, using an ebonite rod and a piece of fur. (O. & C.)

**18.** Give an account of *two* of the following: (a) Faraday's ice-pail experiment. (b) The screening of a body from the influence of electrostatic forces by a sheet of metal or gauze. (c) The induction of magnetism in a bar of soft iron. (O. & C.)

**19.** Explain the following: (a) When a charged ebonite rod is brought near to an uncharged gold-leaf electroscope, the leaves of the electroscope diverge. (b) A gold-leaf electroscope can be charged positively by means of a negatively charged ebonite rod. (c) Lightning conductors are sharply pointed. (O. & C.)

# CHAPTER XXVIII

## ELECTRIC FORCE AND POTENTIAL. CONDENSERS

EXPERIMENTS performed by COULOMB in 1780, and by CAVENDISH somewhat earlier, showed that two electrical charges exert a force which varies *inversely as the square of the distance* between them. This is known as the *inverse-square law* of electrostatics. Thus, if the distance between two given charges is doubled, the force between them is one-quarter of the original value, and if the distance is reduced to one-third of its original distance, the force is increased nine times as much. In the usual mathematical notation,

$$F \propto \frac{1}{d^2},$$

where $F$ is the force between two given charges a distance $d$ apart.

The **electrostatic unit** (*e.s.u.*) **of charge** is defined as *that charge which repels a similar charge 1 cm. away in air with a force of 1 dyne.* Thus a charge of 3 e.s.u. placed 1 cm. away in air from a similar charge of 1 e.s.u. repels it with a force of 3 dynes; and a charge of 3 e.s.u. placed 1 cm. away in air from a charge of 4 e.s.u. repels it with a force of 3 × 4, or 12 dynes. In general, it can be seen from the definition of the unit charge that the force between two charges at a given distance apart is proportional to the *product* of the charges.

From this relation and the inverse-square law, it follows that the force $F$ between two charges of $Q_1$, $Q_2$ e.s.u. at a distance $d$ cm. apart in air is given by

$$F = \frac{Q_1 Q_2}{d^2} \text{ dynes.}$$

428

As an example, we may calculate the force between a positive charge of 50 e.s.u. and a negative charge of 40 e.s.u. if they are 10 cm. apart in air.

$$F = \frac{Q_1 Q_2}{d^2} = \frac{50 \times 40}{10^2} = 20 \text{ dynes.}$$

This force is one of attraction, since unlike charges attract.

It should be noted that the electrostatic unit (e.s.u.) of charge is very small. The *practical unit* of charge is the **coulomb**, and the relation between the two is that 1 coulomb = $3 \times 10^9$ e.s.u. An electron carries a tiny quantity of negative electricity numerically equal to $\frac{4 \cdot 8}{10^{10}}$ e.s.u., which is $\frac{1 \cdot 6}{10^{19}}$ coulomb.

**Electric field strength or intensity.** A valve in a transmitting or receiving set contains electrons moving under the influence of stationary charges on metals contained inside the valve. These static charges create round them an *electric field*, a region where an electric force is experienced, and the electrons move in a manner depending on the *strength* or *intensity* of the field in the space between the charged metals.

The intensity of an electric field at a point is measured by the force in dynes exerted on a unit charge placed at that point. Thus if the intensity at a point is 50 units, a force of 50 dynes acts on an electrostatic unit of charge placed at that point. The following example illustrates how the intensity at a point in an electric field can be calculated.

Suppose that a unit positive charge is placed at a point midway between a positive charge of 100 e.s.u. and a positive charge of 60 e.s.u., which are 20 cm. apart.

The force $F_1$, acting on the unit charge and due to the larger charge of 100 e.s.u. is given by $F_1 = \frac{100 \times 1}{10^2}$ dynes, since the two charges are 10 cm. apart. Thus $F_1 = 1$ dyne.

The force $F_2$ on the unit charge due to the smaller charge of 60 e.s.u. is given by $F_2 = \frac{60 \times 1}{10^2} = 0 \cdot 6$ dyne.

Now $F_1$ and $F_2$ act in opposite directions; hence their resultant $= F_1 - F_2$ $= 1 - 0 \cdot 6 = 0 \cdot 4$ dyne. Consequently the intensity at the point midway between the charges is 0·4 unit.

**Electric fields and lines of force.** From the examples just given, it can be seen that the behaviour of an electron passing through

15

an electric field depends on the direction and magnitude of the field's intensity at all points in the path of the electron. By general agreement among scientists, the direction of an electric field at a point is defined as the direction in which a *positive* charge is urged if placed at that point, so that the directions of the electric field at $X$, $Y$ in Fig. 293 (*a*) are as shown by the arrows.

A striking method of "mapping out" an electric field was first conceived by Faraday. He considered that the medium surrounding a positive charge, for example, was in a state of electric stress, even though the charge was in a vacuum, so that the medium was different when the charge was completely removed from its neighbourhood. Faraday imagined *electric lines of force*, or, simply, lines of force, in the electric field, which showed the direction of the electric force at the different points, and the appearance of a few of the lines in some typical fields is illustrated in Fig. 293. The idea is analogous to magnetic lines of force (p. 536).

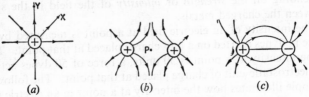

FIG. 293. Lines of electric force

In Fig. 293 (*a*), the lines of force due to an isolated positive spherical charge spread out in straight lines radiating from the centre of the sphere, the arrows indicating the direction in which a small positive charge would move if placed in that field. The lines of force between two small positive charges appear to move away from each other, and none pass through a point $P$ between them, showing that no electric force is experienced at that point (Fig. 293 (*b*) ). $P$ is known as a **neutral point** in the electric field. The lines of force between two unlike charges, however, pass from one charge to the other; the lines can be imagined as "invisible threads" which tend to pull the positive to the negative charge.

Fig. 294 (*a*) illustrates the lines of force in the electric field between a metal plate $A$ positively charged and a metal plate $B$

negatively charged. The lines are parallel between the plates except at the edges, illustrating an electric field whose direction and intensity are constant at all points except near its boundaries.

**Screening.** Fig. 294 (*b*) illustrates a positively charged rod *R* brought near to an uncharged electroscope *D*, with a metal gauze held in the hand between *R* and *D*. The leaf does not diverge, which shows that there is no electric field due to *R* in the region of the electroscope; *D* is said to be *screened* from the influence of *R*

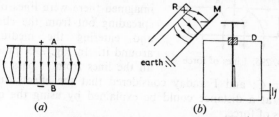

(*a*)                                        (*b*)

FIG. 294. (*a*) Parallel plates, (*b*) screening

by the earthed metal gauze *M*. Some of the valves in radio sets have a metal paint, which is earthed, on the glass envelope, to shield the pieces of metal inside the valve from the electrical influence of neighbouring apparatus, such as another valve or a coil of wire. Electrical instruments in high-voltage laboratories are tested in earthed "cages," so that the instruments may remain unaffected by electrical disturbances outside the "cage."

To return to Fig. 294 (*b*), it can be seen that the earthed metal *M* acquires a negative charge by induction on the side facing the charged rod *R*. Lines of force are then obtained in the region between *R* and *M*; but no lines of force exist round the electroscope, showing that there is no electric force in this region.

**The ice-pail experiment and lines of force.** A further point in connection with the ice-pail experiment (p. 417) can now be noted. Faraday considered that the medium surrounding an electric charge, even though it was a vacuum, played some part when this charge attracted or repelled other charges in its neighbourhood. By using a deep hollow can, *C*, and holding the positively charged sphere *A* well into the interior (Fig. 295), he obtained a conductor (the can) which almost surrounded the charge $A$. In these circumstances, the positive induced charge on the can and electroscope is equal to the positive inducing charge $A$, even though the latter makes no contact with the can (see

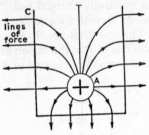

FIG. 295. Lines of force

p. 417). It thus appears that a charge exactly equal to that on $A$ is able to appear on other conductors in its neighbourhood without touching them, and Faraday imagined there were lines of force spreading out from the charge $A$ and entering the medium all around it. In Fig. 295, practically all the lines from $A$ pass through the can $C$; and Faraday considered that the effects due to a charge at a distance could be explained by using the concept of lines of force.

## ELECTRIC POTENTIAL

**Potential energy in a field of force.** In the chapter on Energy and Work (p. 31), we discussed the energy possessed by an object held at a point above the earth's surface. This energy is called "potential energy," and the amount $W$ is given by $W = wh$ ft. lb. wt., where the $w$ is the weight of the object in lb. wt. and $h$ is its height above the earth in feet (p. 42).

Since the effect of gravity is due to the earth, we speak of the earth's "gravitational field"; objects placed at points in the earth's field possess potential energy, as already explained. In this chapter we have discussed "electric fields," and the forces acting on electric charges placed in these fields; and by analogy with the gravitational field, electric charges placed at points in an electric field possess electrical potential energy, which is usually reckoned in "ergs" or "joules" (p. 38).

Points in an electric field are thus said to have an "electric potential," or, briefly, a "potential"; the subject of potential is of vital importance in electricity because it concerns *energy*.

**Movement of electricity in an electric field. The volt.** An object in the earth's gravitational field falls if it is allowed to drop; thus it moves naturally from one point to another where its potential energy is less. In the case of an electric field we have already referred to a movement of electrons along metals, or through the human body, and *an electron (negative electricity) moves naturally from one point in an electric field to another at a higher potential*, if a conducting path is provided. If it could move, a positive charge would go from one point to another of lower potential, the opposite way to an electron. If a mass of 10 lb. moves through a distance of 3 ft. in the earth's gravitational field, the energy change is 30 ft. lb. wt. In the same way, an energy change occurs if an electron moves between two points at a difference of electrical potential.

At this stage we have to define the practical unit of **potential difference** (*p.d.*) in electricity; it is the **volt**, named after a prominent pioneer of electricity (p. 448). *The volt is defined as the p.d. between two points in an electric field such that 1 joule of energy is gained or expended when 1 coulomb of quantity of electricity moves between the points.* Thus, if 5 coulombs pass between two points of a wire maintained at a p.d. of 1 volt by a battery (see below), 5 joules of energy are released. If 5 coulombs pass between two points of a wire at a p.d. of 4 volts, 20 joules of energy are released. It can thus be seen that if $Q$ coulombs of electricity flow along a wire joining two points whose p.d. is $V$ volts, the energy change W is given by

$$W = QV \text{ joules.}$$

A **battery** is a device which maintains its two terminals, or poles, at a difference of potential as a result of a transformation of chemical energy to electrical energy. A lead-acid accumulator, used in some receiver sets, has a p.d. of about 2 volts at its terminals. A **dynamo** is another device which maintains two points at a difference of potential (p. 603). The p.d. at the mains in our homes has a value in most districts of about two-hundred volts. The p.d. between *a charged ebonite rod* and the human body is normally a few thousand volts; a spark passes from the rod to the knuckles of the hand when the latter is brought near.

**The electrostatic unit (e.s.u.) of potential.** The volt is the practical unit of potential difference because it concerns the practical unit of quantity of electricity, the coulomb (p. 429). Now on p. 428 we defined the electrostatic unit (e.s.u.) of quantity of electricity; linked with this charge is the *electrostatic unit (e.s.u.) of potential*. The latter is defined as *the potential difference between two points such that* 1 *erg of energy is expended, or released, when* 1 *e.s.u. of electricty moves from one point to another*. From this definition, it follows that the energy change $W$, when $Q$ e.s.u. of charge moves between two points at a p.d. of $V$ e.s.u., is given by $W = QV$ ergs (compare p. 433).

The e.s.u. of potential is related to the volt as follows: 1 volt $= \frac{1}{300}$ e.s.u. of potential.

### Earth potential; positive and negative potentials.

It can be seen that the idea of "potential" in electricity is analogous to "temperature" in heat. In the latter case the temperatures of two bodies decide which way heat flows from one to the other when they are joined; in the former case the potential values of the two points decide which way electricity flows between the points when they are connected by a wire. Now just as a "zero" is required in temperature measurement, which is chosen as the temperature of melting ice in the centigrade scale, so a choice of a practical zero of potential has to be made.

The obvious property of the chosen "zero" is that its electrical potential should remain constant. Now the earth (a conductor) is so huge, that its electrical potential is always constant, i.e. its potential is practically unaffected by any electricity it may gain or lose from day to day. Accordingly, the potential of the earth is reckoned as the practical zero, and points at a higher potential than the earth have a positive value of potential, while points at a lower potential than the earth have a negative value of potential.

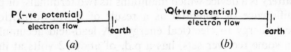

FIG. 296. Negative and positive potential

Suppose that a battery terminal $P$ is at a negative potential relative to the earth. If a wire is connected from $P$ to the earth, electrons flow in the wire from $P$ to the earth as long as a p.d. exists (Fig. 296 (a) ). When a point $Q$ at a positive potential is connected by a wire to the earth, electrons flow from the earth to $Q$ as long as a p.d. exists (Fig. 296 (b) ). After a time, if there is no agency maintaining the points at their respective potentials, the potentials of $P$ and of $Q$ become zero, the same as that of the earth, and at this stage no more electricity flows between either point and the earth. In an analogous way, heat flows between two points as long as they have a temperature difference, but no heat flows if the temperature of the two points is the same. Fig. 297 illustrates the conventional way of drawing the terminals, $A$, $B$, of a battery, where $B$ is the terminal at a lower potential than $A$ (see p. 452). If a wire is connected to the terminals, electrons flow from $B$ to $A$. $A$ is usually coloured red, while $B$ is usually coloured dark blue or black.

FIG. 297.
Battery

**The gold-leaf electroscope measures p.d.** We can now return to a special feature of the gold-leaf electroscope which had to be omitted until the topics of potential and potential difference had been discussed.

Fig. 298 (*a*) illustrates a negatively charged rod $R$ placed near to the cap $C$ of a gold-leaf electroscope whose case is earthed, causing the leaf to diverge (p. 414). $C$ has a negative potential, since it is insulated and in the neighbourhood of $R$, and as the case is earthed, there is a p.d. between the cap and the case. If the case is insulated and the cap is earthed, the leaf diverges when the charged rod is brought near to the case (Fig. 298 (*b*) ). The case

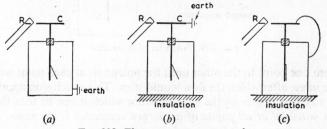

FIG. 298. Electroscope measures p.d.

has now a negative potential and $C$ is at earth (zero) potential, so that a p.d. exists between the cap and the case. However, if the case is insulated, and the cap and case are connected by a metal wire, the leaf does not diverge when the charged rod is brought near (Fig. 298 (c) ). Even when the rod now touches the case or cap and gives up some of its charge to the electroscope, the leaf does not diverge. In this special case there is no p.d. between the cap and the case as they are connected.

It follows from these experiments that *the leaf of an electroscope diverges only when there is a potential difference between the cap and case*; further experiments show that *the divergence is a measure of the p.d. between the cap and the case.*

**Potential of surface of charged conductor. Measuring potential.** Consider a positively charged pear-shaped conductor, *A*, which, as seen on p. 419, has a high concentration of charge at the point *P*, and a low concentration at *Q* and other parts of its surface, where the curvature is small (Fig. 299). If the potential at *Q* were different from that at *P*, electrons would flow along the conducting surface

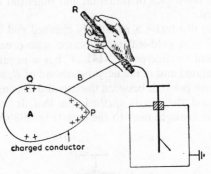

FIG. 299. Potential of conductor

from one point to the other until the potentials at each point were the *same*, after which the flow would stop. This conclusion applies to any two points on the surface, from which it can be seen that *the potential at all points of a charged conductor is the same, no matter what shape it has.*

Fig. 299 illustrates a simple experiment to verify this theoretical result. Part of a metal wire *B* is wound round an insulating ebonite rod *R*, and one end of the wire is connected to the cap of a gold-leaf electroscope whose case is earthed. The other end of the wire is moved all over the surface of the charged conductor *A*, and the divergence of the leaf is carefully noted. *The divergence remains unaltered.* Thus the potential of $A$ is the same at all points.

**Potential due to charges.** Consider a small insulated metal plate $A$, with a positive charge of 10 units (Fig. 300) ). If a positive charge $X$ approaches it in the direction $XMA$, a repulsive force is exerted on $X$ continuously, and *work* will have to be done to bring the latter to the point $M$. Thus $M$ has a "potential" due to the

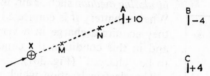

Fig. 300. Potential due to charge

positive charge of 10 units on $A$. More work will have to be done to bring $X$ to $N$, a point nearer to $A$, so that the potential at $N$ is greater than at $M$. In this way, by imagining $X$ to *reach* $A$, it can be understood that $A$ has a potential due to its own charge; suppose its potential is $+ 50$ units.

Consider an insulated metal plate $B$ with a negative charge of 4 units (Fig. 300) ), and suppose all other charges are absent. If a positive charge $Y$ (not shown) approaches $B$, an attractive force is exerted on $Y$, and hence various amounts of work are done by the charge $B$ as $Y$ reaches different points. By comparison with the case of the positive charge on $A$, it can be understood that $B$ has a negative potential due to its charge at all points surrounding it. At the surface of $A$, therefore, the net potential $= + 50 - 20$ $= + 30$ units, where $- 20$ units is the potential at $A$ due to the negative charge of $B$. The potential of $A$ is thus *lowered* if a negative charge is in its neighbourhood. If an insulated plate $C$ with a positive charge of 4 units is further away from $A$ than $B$, the potential due to it at $A$, may perhaps be $+ 12$ units. Thus the net potential of $A$ due to its own charge, the negative charge on $B$, and the positive charge on $C = + 50 - 20 + 12 = 42$ units. On the whole, then, the potential of $A$ is less than if $B$ and $C$ were both absent. We shall see later how this result is used in the theory of condensers (p. 440).

15*

## CONDENSERS

**Charging a condenser.** A *condenser* is an apparatus for storing an electric charge (i.e. a quantity of electricity). It is an essential component in all radio transmitters and receivers, besides being used in the Post Office telephone and telegraph service.

The simplest practical form of a condenser consists of two

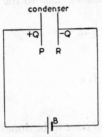

FIG. 301.
Condenser principle

parallel metal plates, *P*, *R*, *separated by an insulating medium* such as air, mica, or paper. When a battery *B* is connected to the plates they acquire a charge in a very short time, and in this condition the condenser is said to be "charged" (Fig. 301).

The electrical action which occurs is due to the fact that a battery is a device whose terminals are always maintained at a difference of potential; it thus acts like an "electric pump," driving electrons round any metal circuit connected to it (p. 451). Electrons thus begin to move in the direction *PBR* when the battery *B* is connected and hence *R* gains a negative charge and *P* an *equal* positive charge. Since like charges repel, the electrons already on *R* begin to oppose the movement of the electrons in the wires, and the flow of electrons quickly ceases. The two plates, *P* and *R*, then have equal but opposite charges of magnitude *Q* units, for example, and the charge stored in the condenser is said to be *Q* units, the charge on *either* plate.

**Capacitance, *C*, of a condenser.** Suppose that a condenser, such as the pair of plates *P*, *R*, has a charge of 100 e.s.u. when a battery of p.d. 300 volts is connected to it. If two batteries are used instead of one, so that a p.d. of 600 volts is applied to the condenser, more electrons than before are driven to the plate *R* (Fig. 301), since, to use an analogy, the "electric pump" is now twice as powerful. Experiment shows that the condenser now has a charge of 200 e.s.u.; if the p.d. is increased to 900 volts by using three batteries, the charge increases to 300 e.s.u. Thus the ratio $\frac{\text{charge}}{\text{p.d.}}$ is equal to $\frac{1}{3}$ in each of the three cases. In general, it can be seen that the ratio $\dfrac{\text{charge}}{\text{p.d.}}$ is always a *constant* for a given condenser, and this ratio is called the *capacitance* of the condenser.

The capacitance of the condenser, $X$, formed by the plates $P$, $R$ is thus $\frac{1}{3}$ unit. If another condenser, $Y$, has a charge of 400 e.s.u. when a p.d. of 200 volts is applied to it, its capacitance $= \dfrac{\text{charge}}{\text{p.d.}}$ $= \frac{400}{200} = 2$ units. Thus $Y$ has a capacitance six times as large as that of $X$; which means that $Y$ can store six times the charge that $X$ can store when the same p.d. is applied.

**Formula for capacitance. Units.** If $C$ is the capacitance of a condenser and $Q$ is the charge on it when a p.d. $V$ is applied, then, by definition,

$$C = \frac{Q}{V}.$$

A unit of capacitance is the **farad** (named after Faraday) which is defined as the capacitance of a condenser which has a charge of 1 coulomb (p. 429) when a p.d. of 1 volt is applied to it. In practice a condenser of capacitance 1 farad (1$F$) would have plates of an impossibly large size; practical condensers, such as those used in receiver sets, for example, have capacitances of the order of millionths of a farad (see p. 444). The **microfarad** ($\mu F$) is a millionth of a farad.

**Discharging a condenser.** If a wire $A$ is connected to the two plates $P$, $R$ of a charged condenser, a spark passes as the connection is made, and no charge remains on either plate after a very short time (Fig. 302). Suppose the plate $R$ was originally negatively charged and that $P$ had a positive charge (see Fig. 301). When the wire $A$ is connected to the plates, electrons from $R$ flow to $P$ and completely neutralize the positive charge there; the charges on $R$ and $P$, it will be remembered, were originally equal (p. 438). Thus $R$ and $P$ have no charges left on them, and the condenser is now said to be *discharged*.

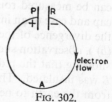

FIG. 302.
Discharge of condenser

**The effect of an earthed plate near a charged plate.** In the early days of electrostatics it was easier to give a plate a charge by means of a charged rod, and *then* to study its p.d. with another plate by an electroscope, than to charge the two plates with a battery. In those days batteries had not been fully developed, and their p.d.'s were uncertain.

If a metal plate *A* is given a positive charge *Q* by touching it with a charged rod, *A* has then a positive potential due to its own charge. When an insulated uncharged metal plate, *B*, is brought near to *A* (Fig. 303 (*a*)), an induced negative charge *Q'* appears on the side of *B* facing *A*, and an equal positive charge appears on the

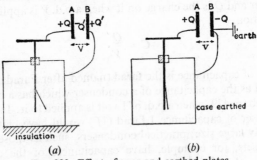

FIG. 303. Effect of near and earthed plates

other side of *B*. Since the negative charge is nearer *A* than the equal positive charge, the potential of *A* is *less* than it would be if *B* were absent. The p.d., *V*, between the two insulated plates *A*, *B* can be measured roughly by connecting them respectively to the cap and case of an insulated electroscope, and noting the extent of the divergence of the leaf. If the plate *B* is now earthed (Fig. 303 (*b*) ), observation shows that the divergence of the leaf *diminishes*, showing that the p.d., *v*, between *A* and *B* is now less than when *B* was insulated. This is explained by the fact that electrons flow from the earth to neutralize the positive charge on *B* completely, leaving only the negative charge on the latter. Since the positive charge on *B* is now absent, the potential of *A* is lowered further still compared with Fig. 303 (*a*).

The capacitance of the condenser formed by *A*, *B* in Fig. 303 (*a*) is given by $C = \dfrac{Q}{V}$, where $V$ is the p.d. between the plates and $Q$ is the charge on $A$. Suppose $Q$ is 400 e.s.u. and $V$ is 500 volts; then $C = \frac{400}{500} = \frac{4}{5}$ unit. The p.d. when $B$ is earthed is less than 500 volts; suppose it becomes 200 volts. Then the capacitance between the plates alters to $\frac{400}{200}$, or 2 units, as the charge on $A$ is unaltered. Consequently, it can be seen that *the capacitance of A, B is increased by earthing the plate B when A has a given charge.*

**Factors affecting capacitance of condenser.**

(1) *The distance between the plates.* If the plate $A$ of the condenser $A, B$ is given a charge $Q$, and $B$ is earthed and connected to the case of an electroscope (Fig. 303 (*b*) ), experiment shows that the divergence of the leaf diminishes when $B$ is moved nearer to $A$. Thus the p.d. between $A, B$ is diminished, which is explained by the fact that the potential of $A$ is further reduced the nearer the induced negative charge on $B$ gets to it. The capacitance $C$ of the condenser $= \dfrac{Q}{V}$; and since $V$ is diminished and $Q$ remains constant, $C$ is *increased.* Thus the

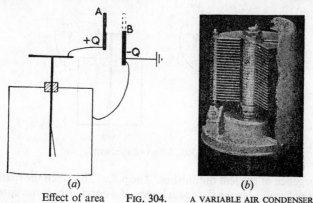

(a)    Effect of area    FIG. 304.    A VARIABLE AIR CONDENSER

capacitance of a condenser increases when the distance between the plates decreases.

(2) *The common area and size of the plates.* If the earthed plate $B$ is lowered slowly, keeping the distance between the plates

constant (Fig. 304 (a)), experiment shows that the divergence of the leaf increases. Thus the p.d. between the plates is increased, and since $C = \dfrac{Q}{V}$ and $Q$ has been constant, the capacitance is reduced.

The experiment illustrates that the smaller the common area between the plates of a condenser, the smaller is its capacitance.

From the above experiment, it follows that a *variable condenser* can be made by having one set of metal plates moving between, but insulated from, another fixed set of plates (Fig. 304 (b)) in such a way that the common area is varied. This type of condenser has extensive application in radio receiver sets for "tuning" purposes. When a different station is required, a knob on the panel of the radio set is rotated and turns the movable set of plates, thus altering the capacitance in a circuit inside the set known as a "tuning circuit."

(3) *The medium between the plates.* If a charged condenser, with air between its plates, is connected to an electroscope (Fig. 305 (a)) and glass is interposed between the plates (Fig. 305 (b)), the

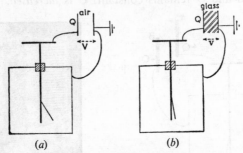

Fig. 305. Effect of medium

divergence of the leaf diminishes. The p.d., $V$, between the plates is thus reduced. Hence the capacitance of a condenser is *increased* if glass is the medium between the plates instead of air. A similar experiment shows that mica, paper, and other insulators also increase the capacitance when used instead of air. The name *dielectric* is given to the medium between the plates of a condenser, and the *dielectric constant* of the medium is defined as the ratio of the capacitance when the medium is used to the capacitance when air is used. The dielectric constant of air is thus 1; the dielectric constant of glass is about 2, so that the capacitance of a given plate condenser is doubled when glass is used instead of air. Paper and mica have dielectric constants of about 2 and 5 respectively, and are commonly used in modern condensers (p. 444).

Note carefully that the capacitance of a condenser is increased (1) when the distance between the plates is *decreased*, (2) when the common area between the plates is increased, (3) when the dielectric constant of the medium between the plates is increased.

**The Leyden jar** was one of the earliest condensers made, and consisted of a glass jar partly lined on the inside and outside with metal foil A, B (Fig. 306). The "plates" of the condenser are A and B, while the dielectric between the plates is glass. A metal knob, K, was connected to the inside metal foil by means of a chain D, which hung loosely, and the jar was given a charge by connecting the terminal of a Wimshurst machine to K. A violent electrical shock is obtained if K is touched after the jar is charged, as the large charge stored passes through the body to the earth. The Leyden jar was used as a scientific toy in the eighteenth century when this effect was first discovered. The sense of humour

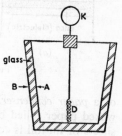

FIG. 306. Leyden jar

enjoyed by the Abbé Nollet, a French scientist, may perhaps be gleaned from the fact that he lined up hundreds of monks, connected them together by wires, and discharged a Leyden jar through them; the monks are reputed to have all jumped into the air together as a result of the shock.

**Modern condensers.** When a radio receiver, a transmitter, or a television set is made, many condensers are required of different capacitances. Constant research is still being made to provide high-grade condensers, so that the listener can enjoy the programmes without the irritation arising from distorted or weak sounds.

We have already mentioned the *variable condenser*, which has air as its dielectric (p. 441). The maximum value of its capacitance is 0·0005 $\mu F$. Another type of condenser used in a radio set is a *mica condenser*, so called because mica is used as the dielectric between the plates. Fig. 307 (*a*) illustrates two sets of metal plates of a mica condenser, each connected to a terminal, $T$; the capacitances of these fixed condensers depend on their sizes, of course, and range from very small capacitances such as 0·0001 $\mu F$ to larger capacitances such as 1 $\mu F$.

Paraffin-waxed paper is a much cheaper dielectric than mica, and it has a further advantage in saving space because, unlike mica, it can be rolled. Fig. 307 (*b*) illustrates the essential features

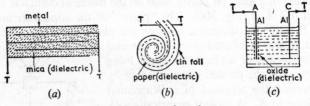

Fig. 307. Types of condensers

of a *paper condenser*, which consists of two sheets of paraffin-waxed paper rolled between two sheets of tin-foil. Each of the sheets of tin-foil is connected to a terminal, $T$, of the condenser.

In designing power circuits for converting A.C. to D.C. power (see p. 621), and in some parts of a radio circuit, a condenser of very high capacitance, such as 16 $\mu F$, is required. Such a condenser would also have to be economical in size to be of practical use, and the problem of its design was solved by passing a current between two aluminium plates immersed in ammonium borate solution (Fig. 307 (*c*) ). A *very* thin film of aluminium oxide was formed on the anode plate by electrolytic action, and the oxide constituted a very thin dielectric (insulating medium) between the anode and the solution when the action stopped. Now it was pointed out on p. 441 that the capacitance of a condenser was greater if the distance between its two plates was smaller; the very thin dielectric consequently makes the capacitance between the anode plate and the solution exceptionally high. A condenser of several hundred microfarads can be made by this method. The *electrolytic condenser*, as it is called, has the anode plate as one terminal; the cathode plate can be used as the other, because the solution has a low resistance.

## Summary

1. The electrostatic unit (e.s.u.) of charge is that charge which repels a similar charge 1 cm. away in air with a force of 1 dyne. The force between two small charges $Q_1$, $Q_2$ e.s.u. in air $= Q_1Q_2/d^2$ dynes, where $d$ is their distance apart in centimetres.

2. Electrons move from one point to another point at a higher potential when they are joined by a wire. Earth potential is chosen as "zero" potential.

3. **The volt is the p.d. between two points if 1 joule of energy is needed to move 1 coulomb of charge between them.** The energy change, $W$, when $Q$ coulombs move between two points at a p.d. of $V$ volts is $QV$ joules.

4. **The capacitance, $C$, of a condenser in farads $= Q/V$,** where $Q$ is the charge in coulombs, and $V$ is the p.d. in volts.

5. The capacitance of a condenser increases (i) the smaller the distance between the plates, (ii) the greater the common area of the plates, (iii) the higher the dielectric constant, $K$, of the medium between the plates.

## EXERCISES ON CHAPTER XXVIII

**1.** A small positive charge of 10 e.s.u. is 5 cm. away in air from a small positive charge of 8 e.s.u. Calculate the force between the charges. What is the force if the charges are separated by a distance of 10 cm. ?

**2.** Draw the electric lines of force between (i) two small positive charges, (ii) a small positive and negative charge, (iii) two parallel-plane plates with unlike charges, (iv) a small positive charge and an earthed plate near it.

**3.** A small charge of 24 e.s.u. repels a similar charge with a force of 36 dynes. Calculate the distance apart of the charges.

**4.** A point $X$ is a short distance from a positive charge. What is meant by the statement that "the electric potential at $X$ is $+ 8$ e.s.u." ? Why is the potential of the earth usually chosen as the "zero" of potential ? Describe the effect on the electrons in a wire joining $X$ to the earth.

**5.** What is a *condenser* ? Describe how you would make (i) a simple condenser, (ii) a practical condenser.

**6.** Explain why the potential of a positively charged insulated metal plate $A$ is lowered when another insulated metal plate $B$ is placed near it. Why does the potential difference between $A$, $B$ decrease further when $B$ is earthed ?

**7.** A condenser has a charge of 20 micro-coulombs when a battery of 10 volts is connected to it. What is the capacitance of the condenser ? What is the capacitance of a condenser gaining a charge of 32 micro-coulombs when a 4-volt battery is applied ?

**8.** What is the *dielectric* of a condenser ? Describe an experiment to show the effect of the dielectric on the capacitance of a condenser.

**9.** Show by diagrams the effect of bringing an uncharged insulated metal plate near a charged insulated metal plate. Describe what happens when (*a*) the first plate is earthed, (*b*) a slab of paraffin wax is then inserted between the plates. How may these results be demonstrated ? (*L.*)

**10.** State how the capacity of a condenser depends only on (*a*) the area of its plates, (*b*) the distance between its plates. How would you increase the capacity of a parallel plate condenser without moving the plates ? (*N.*)

**11.** Point out the essential features of an electrostatic condenser. State the factors which determine the capacity of a condenser, and describe and explain an experiment to demonstrate *one* of them. (*C.W.B.*)

**12.** What do you understand by (*a*) the potential, (*b*) the capacity, of a conductor ? How is each affected (if at all) by the charge on the conductor ? An insulated metal plate is connected to the cap of a gold-leaf electroscope. State and explain what happens when (*a*) an earthed metal plate is placed parallel to and about 5 cm. from the charged plate; (*b*) the earthed plate is then approached nearer the charged plate; (*c*) a thin sheet of ebonite is then placed between the plates. (*L.*)

**13.** Describe experiments, *one* in each case, to illustrate how the capacity of a condenser depends on (*a*) the area of the plates, (*b*) the distance between the plates, (c) the material between the plates.  A positively-charged insulated metal sphere is brought up to the knob of an uncharged Leyden jar and finally makes contact with the knob; the outer coating of the Leyden jar is earthed.  Describe, with the aid of diagrams, the changes which occur in potential and in the distribution of electric charges.  (N.)

**14.** What factors determine the capacity of a condenser ?  Describe an experiment to illustrate the effect of *one* of them.

Describe *one* type of variable condenser and point out how the factors you have mentioned determine its capacity.  (L.)

# CHAPTER XXIX

## PRINCIPLES OF CURRENT ELECTRICITY

IF two points at different electric potentials are connected by a conductor such as a metal wire, electron flow takes place along it in one direction (p. 434). This flow of electricity, known as an *electric current*, continues as long as there is a potential difference between the two points.

The study of current electricity began in 1790, when GALVANI noticed the twitching of a frog's leg as it lay between two metals. He mentioned the observation to his friend, VOLTA, who made many experiments on the phenomenon, which he recognized as an electrical one. He succeeded in making the first *battery*, a device which can maintain two points at a difference of potential for a considerable time. This opened up an entirely new advance in the study of electricity, and enabled scientists at the beginning of the nineteenth century to begin investigations into the continuous flow of electricity through metals and liquids. These led to the invention of the electric lamp, the electric cooker, the telephone, the dynamo, and made possible the manufacture of certain metals such as aluminium, to mention only a few results.

**The effects of electric current.** When an electric torch is switched on, the ends of the wire filament inside the torch bulb are connected to the battery in the torch, and an electric current flows continuously in the filament (Fig. 308 (*a*) ). Since the wire glows, we conclude that *an electric current has a heating effect*.

If a battery is connected to two copper plates, *A*, *C*, dipping into dilute copper sulphate solution, which is a liquid conductor of electricity (Fig. 308 (*b*) ), a current flows through the liquid. When the plate *C* is removed from the solution after about half an hour, a deposit of fresh copper is observed on that part of it which has been submerged in the liquid. It thus appears than *an electric current has a chemical effect*.

If a coil, *C*, of insulated copper wire (known as a **solenoid**) is wound round a piece of unmagnetized iron, *A*, and a battery is connected to the ends of the coil, the iron attracts steel pins, *X*,

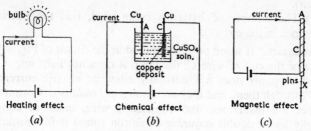

| Heating effect | Chemical effect | Magnetic effect |
| (a) | (b) | (c) |

FIG. 308. Effects of current

to itself (Fig. 308 (*c*)). When the battery is disconnected from the coil, so that the current ceases, the pins fall to the ground. Thus an electric current causes an unmagnetized piece of iron to become a magnet, i.e. *an electric current has a magnetic effect.*

**Factors affecting current strength.** Suppose *A* is a small lamp, lit up by direct connection to a battery, *B*. If a coil of wire, *C*, is now placed in the circuit (Fig. 309), the lamp glows less brightly than before. As more wire is introduced into the circuit, the light

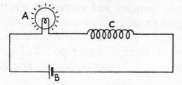

FIG. 309. Current strength

becomes still duller; we conclude that the introduction of more wire into the circuit reduces the "strength" of the current flowing. The "resistance" of the circuit, or its "opposition" to the flow of electric current, thus appears to have increased when more wire is added.

If more batteries are added to *B* (Fig. 309), the lamp *A* burns more brightly. The current "strength" in the circuit is therefore increased by the use of more batteries.

From these simple experiments, it appears that the "strength" of the current in a circuit depends upon two factors—

(1) The amount of wire in the circuit (i.e. the resistance of the circuit).

(2) The number of batteries being used, (i.e. the potential difference in the circuit).

**Insulators.** If wood or paper is used in the circuit of Fig. 309 in place of the coil of wire $C$, the lamp $A$ does not light up. Thus wood or paper does not normally allow an electric current to flow through them, and they are known as *insulators*. Cotton and rubber are insulators, and connecting wire, made of copper, usually has a double covering of cotton round it for insulation purposes. Flex like that used for connecting radio sets and electric heaters to the mains, contains copper wire surrounded by rubber and cotton. Mica and bakelite are insulators, and porcelain supports are used on pylons to insulate cables carrying electricity from power stations (see Fig. 312). Porcelain insulators may also be observed in the London Underground Railway, supporting rails carrying electric current, as well as at the top of telegraph poles.

Normally, air is an insulator. When lightning occurs, however, the insulation of the air is broken down, and a momentary electric current flows (p. 420).

**Relation between quantity and current of electricity.** When a battery is connected to the ends of a wire, a steady electric current flows along the metal. The steady flow of electricity is analogous to the steady flow of water along a pipe $AB$ (Fig. 310), and in the latter case we speak of a "water current." The water current can

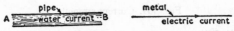

FIG. 310. Currents

be measured by collecting the water flowing out of the end of the tube in 10 sec., for example, and weighing the water. If the mass is 50 gm., the water current can be classified as $\frac{50}{10}$ or 5 gm. per sec. rate of flow; if the water current increases and 90 gm. of water are collected in 10 sec., the current is 9 gm. per sec. rate of flow.

In an analogous way, a steady electric current in a metal can be expressed in terms of the quantity, $Q$, of electricity flowing past a point of the metal in a time $t$. The current $I$ is the *quantity of electricity per second* passing the point, and thus

$$I = \frac{Q}{t} \quad . \quad . \quad . \quad . \quad (1)$$

In order to facilitate the interpretation of scientific results all over the world, scientists have reached agreement about the letters, or symbols, to be used for most electrical quantities. The same is true for the names of the units of many electrical quantities, and these have usually been chosen in honour of scientists who have made outstanding discoveries in the subject. The practical unit of current is the **ampere** (*amp.*), named after a famous French scientist (p. 566); in the formula for $I$ given above, the latter is in amperes when $Q$ is in coulombs (p. 429), and $t$ is in seconds. Thus 1 ampere can be defined as 1 coulomb per second rate of flow of electricity.

**Conventional direction of electric current.** Soon after the effects of the electric current began to to be studied, it became necessary to refer to the current direction. From their studies in electrostatics scientists knew there were two kinds of electricity, positive and negative, and it was agreed to call the direction of an electric current the direction of movement of *positive* electricity. This convention has persisted, and to-day circuit diagrams have arrows which denote the movement of positive electricity.

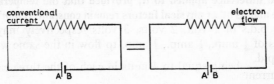

FIG. 311. Conventional current and electron flow

However, it was discovered early in the twentieth century that "free" electrons (p. 414) are the carriers of the electric current in metals, and these particles carry *negative* electricity. The electrons thus drift in a direction opposite to that of the conventional current (Fig. 311). There should be no difficulty about the matter if it is realized that we can refer *either* to the conventional current direction *or* to the electron flow, whichever suits us better at the time. In order to avoid confusion, the adjective "conventional" will precede "current" in the former case. In Fig. 311, the (red) terminal *A* of the battery is at a higher potential than the (black) terminal *B*, so that the conventional direction of the current is from *A* to *B* in the metal joining the terminals. Actually, electrons flow in the metal from *B* to *A*.

**Ohm's Law. Resistance.** The practical unit of potential difference is the **volt**, named after Volta (p. 448), and the symbol used is *V*. Special pains should be taken at the outset to distinguish between the meanings of "potential difference" and "current." For example, a potential difference must exist between two points before a current will flow along a metal joining them. To make an analogy, we can compare potential difference to the difference in water pressure between two places which is necessary to make water flow along a pipe connecting them (Fig. 310).

If a wire is connected to a varying number of batteries, the current in the wire increases as the number of batteries is increased. This is not surprising, because, to return to the water analogy, water flows faster or slower through a given pipe according as the pressure difference between the ends is increased or decreased. OHM discovered in 1826 that

**the electric current in a given wire is directly proportional to the potential difference applied to it, provided that the temperature of the wire and its other physical factors remain constant.** (*Ohm's Law*).

Thus p.d.s of 1 volt, 2 volts, 3 volts respectively may cause currents of $\frac{1}{4}$ amp., $\frac{1}{2}$ amp., $\frac{3}{4}$ amp. to flow in the same wire, the ratio $\dfrac{\text{p.d.}}{\text{current}}$ being equal to 4 units in each of the three cases.

If another wire has p.d.s of 1 volt, 2 volts, 3 volts respectively to its ends, the current flowing may be $\frac{1}{2}$ amp., 1 amp., $1\frac{1}{2}$ amps. in the three cases. The ratio $\dfrac{\text{p.d.}}{\text{current}}$ is again constant, but the ratio has now a value of 2 units. Now this wire has a current of $\frac{1}{2}$ amp. flowing when a p.d. of 1 volt is connected to it, whereas the previous wire had a current of $\frac{1}{4}$ amp. flowing in it when the same p.d. was connected to it. It appears, then, that the latter wire has a greater electricity resistance to the flow of electricity; the magnitude of the **resistance** of a conductor is *defined* by the ratio

$$\frac{\text{p.d. applied to its ends}}{\text{current flowing}}.$$

The symbol for resistance is $R$, and its unit is the **ohm.**

FIG. 312. HASTINGS MAIN GRID SUB-STATION
Note how the cables are supported on porcelain insulators

**Fundamental formulae.** LORD KELVIN, one of the greatest scientists of the nineteenth century, once said: "When you can measure what you are speaking about and express it in numbers, you know something about it, and when you cannot measure it, when you cannot express it in numbers, your knowledge is of a meagre and unsatisfactory kind." He was, of course, speaking about subjects such as engineering, chemistry, and physics. Electrical and radio engineers, for example, have need of formulae which concern the magnitudes of current, p.d., and resistance in electrical circuits.

Since, as we have seen, resistance $= \dfrac{\text{p.d.}}{\text{current}}$, it follows that

$$R = \frac{V}{I} \qquad . \qquad . \qquad . \qquad . \quad (2)$$

and $R$ is in ohms when $V$ is in volts and $I$ is in amperes. Suppose that a p.d. of 8 volts ($V$) is applied to a wire and causes a current of 2 amps. ($I$) to flow; then the resistance of the wire, $R$, equals $\dfrac{V}{I} = \dfrac{8}{2} = 4$ ohms. If a current of $\frac{1}{4}$ amp. flows when a p.d. of 3 volts is connected to a wire, its resistance, $R$, equals $\dfrac{V}{I} = \dfrac{3}{\frac{1}{4}} = 12$ ohms.

From (2) above, it follows that:

$$V = IR \qquad . \qquad . \qquad . \qquad (3)$$

and

$$I = \frac{V}{R} \qquad . \qquad . \qquad . \qquad (4)$$

If any of these formulae is known, the other two can be found, but the reader is advised to memorize each of the three. In particular, the formula for p.d., $V$, is the only one given by a *product* relation, i.e. $I \times R = V$.

*Other units.* In dynamos and motors, the electrical engineer deals with amperes of current and sometimes with thousands of volts of p.d. Accordingly, a larger unit of p.d. is used in these cases, and is the **kilo-volt**, which represents 1000 volts. Again, the radio engineer frequently deals with currents much smaller than one ampere, as the currents through valves are of the order of thousandths of an amp. The **milli-ampere** is defined as $\frac{1}{1000}$ of an ampere. Radio sets contain resistances very much larger than an ohm. The **Kilohm** is 1000 ohms, and the **megohm** is one million ohms. A small unit of resistance is the **microhm**, one-millionth of an ohm, which is used in discussing the specific resistances of materials (p. 518).

**Connecting wire and resistance wire.** In general, metals differ considerably in their resistance to an electric current. The pure metal with the least resistance (the best conductor) is silver, but copper has a very low resistance and is much cheaper, so that this is the metal commonly used in industry as *connecting wire* in electrical circuits.

Research has shown that alloys generally have higher resistances than pure metals, and that the resistance of certain alloys varies very slightly with temperature change. Consequently, when a resistance coil is needed, it is made from one of the special alloys designed for the purpose. Manganin, an alloy of copper, manganese, and nickel, and eureka (or constantan) an alloy of copper and nickel, are the trade names of two metals used for *resistance wire*. These have resistances about 30 and 60 times respectively, as high as that of copper. Fig. 313 (*a*) shows a resistance wire *P* joined to a resistance wire *Q* by a connecting wire *cd*; the connecting wire is represented by a straight line in a diagram and a resistance wire by a "zig-zag" line. If *P* is 10 ohms and *Q* is 5 ohms, for example, and the resistance of the connecting wires *ab*, *cd*, *ef* is about 0·01 ohm, it can be seen that the resistance of

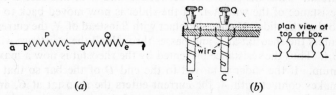

(*a*)          (*b*)

FIG. 313. Connecting and resistance wire

the connecting wire in a circuit can be neglected. This is usually the case.

**Resistance boxes and rheostats.** *Resistance boxes* are those which contain coils of wire of known resistance. The wire is wound on a wooden bobbin such as *B*, *C* (Fig. 313 (*b*) ), and the ends are connected to blocks of brass on the top of the box, across an opening which can be fitted with a metal plug such as *P* or *Q*. When the box is used in an electrical circuit, and the plug *P* is taken out, the current flows through this coil and across the metal plug *Q*. Note that a plug in the box must be taken *out* when the corresponding resistance is required.

*Rheostat* is the name given to a variable resistance whose value can easily be adjusted but need not be accurately known. Fig. 314 (*a*) illustrates one type of rheostat, consisting of resistance wire joined between two terminals *A*, *B*. The bar *DC* is made of a low-resistance metal such as brass, and *M* is a "slider" which makes contact with different points of the resistance wire by moving it along *DC*. When a variable resistance is connected in a circuit, the current flows into the rheostat at *C*, along the low-resistance bar to the slider, and then through *XA* back to the

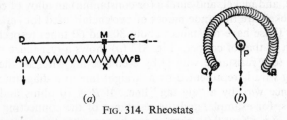

(*a*)            (*b*)

FIG. 314. Rheostats

circuit. Thus the resistance between *C* and *A* in this case is the resistance of the wire *XA*. If the slider is now moved back to *C* so that the slider *M* makes contact with *B* instead of *X*, the current enters the rheostat at *C* and passes through all the resistance wire *BA*. Thus the resistance presented by the rheostat is now a maximum. If the slider is moved to the end *D* of the bar so that it makes contact with *A*, the current enters the rheostat at *C*, and then flows along *CD* and passes out of the rheostat through *A*. The rheostat resistance is zero.

Fig. 314 (*b*) illustrates another type of rheostat in common use. The wire is wound on a circular frame between the terminals *Q*, *R*, and a knob at *P* moves a slider round the wire when it is rotated. This rotary type of rheostat is used in radio receiver sets, as a volume control.

**The ammeter.** An *ammeter* is an instrument for measuring an electric current, and the principles on which it acts are discussed later (p. 587). For the present, all we require to know is that it consists of a wire which can rotate between the poles of a permanent magnet. The coil is connected to the terminals of the instrument, and an attached pointer rotates when a current flows through the coil. Fig. 315 (*a*) illustrates an ammeter, *A*, which measures the current flowing in a resistance $R$ joined to the terminals of a battery $B$. The current which flows in $R$ must also flow in $A$, if $A$ is to be effective, i.e. *an ammeter must always be placed directly, or "in series," in the circuit*. The ammeter $A_1$ in Fig. 315 (*a*) is also correctly positioned for measuring the current in $R$, as it is in series in the circuit, and would register the same current as $A$. Fig. 315 (*b*) illustrates an ammeter $A$ measuring the current in

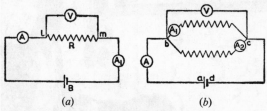

FIG. 315. Positions of ammeter and voltmeter

the wire $ab$, and ammeters $A_1$, $A_2$ measuring the currents in the resistance wires shown. It was stated above that a coil of wire is connected between the terminals of an ammeter, and the resistance of the coil is the resistance of the ammeter. If we do not wish the resistance of the circuit to be seriously altered by the introduction of the ammeter into it, the resistance of the latter should be very small by comparison with the rest of the circuit; hence, in general, *an ammeter is a low-resistance instrument*.

**The voltmeter.** The *voltmeter* is an instrument for measuring potential differences; its principle of action is similar to that of an ammeter, though it is different in detail (p. 587). Since it has to measure the p.d. between two points, its terminals must be connected to those two points. For example, Fig. 315 (*a*) illustrates a voltmeter, $V$, measuring the p.d. between the ends, $l$, $m$, of the resistance $R$. Fig. 315 (*b*) illustrates a voltmeter, $V$, measuring the p.d. between the ends, $b$, $c$, of another circuit. In distinction from an ammeter, it can be seen that *a voltmeter is always connected "outside," or "in parallel" with, a circuit*.

Like the ammeter, a voltmeter has a coil inside connected to the terminals, and the coil rotates when a current flows in the voltmeter. Now a voltmeter is connected in parallel with a circuit,

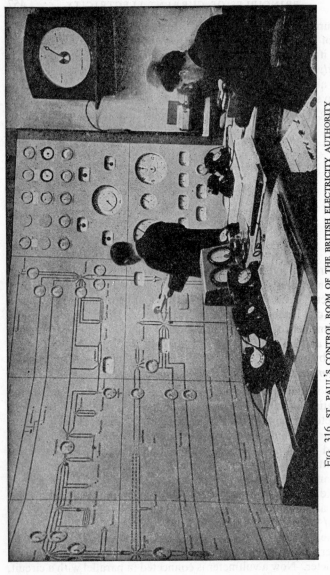

FIG. 316. ST. PAUL'S CONTROL ROOM OF THE BRITISH ELECTRICITY AUTHORITY

The instruments show the power being used at any instant throughout the London and South-Eastern region. The large meter, top right, indicates the frequency of the generators, usually 50 cycles per second.

as illustrated in Fig. 315 (*a*), and hence some current will always be diverted from the main circuit and flow through the voltmeter. In order to make the least disturbance to the circuit the current through the voltmeter must be as small as possible, and this is arranged in practice by making the voltmeter resistance as high as possible. In general, *a voltmeter is a high-resistance instrument*.

**Measurement of resistance.** The resistance of a wire can be measured simply by connecting a battery to it, and measuring the current in the wire and the p.d. across it. Thus the resistance, $R$, of the wire in Fig. 315 (*a*) can be found from the circuit arrangement shown. Suppose the current, $I$, on the ammeter is 0·5 amp. and the p.d., $V$, on the voltmeter is 1·5 volts; then, since $R = \dfrac{V}{I}$ (p. 454), $R = \dfrac{1\cdot5}{0\cdot5} = 3$ ohms.

**Resistances in series.** In wiring electrical circuits such as transmitting and receiving radio sets, resistances have often to be placed in series. Let us consider a resistance $P$ connected in series with a resistance $Q$ and a battery $B$ (Fig. 317). A current flows in $P$ and in $Q$, and no matter where an ammeter is placed in the circuit,

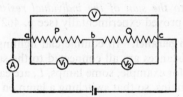

Fig. 317. Resistances in series

at *a*, *b*, or *c*, the current registered is always the same. The current is thus the same at all parts of a series circuit, and in particular, the current flowing in $P$ equals the current flowing in $Q$.

The potential difference, $V_1$, between the ends, *a*, *b*, of the resistance $P$ and the p.d., $V_2$, between the ends, *b*, *c*, of the resistance $Q$ can be measured by connecting voltmeters across each wire, as shown; $V_1$ may then be 0·5 volt, and $V_2$ may be 1·0 volt. If the p.d., $V$, between the ends *a*, *c*, of both wires is measured, experiment shows that it is 1·5 volts, so that $V = V_1 + V_2$. If the p.d.

across $P$ and $Q$ is 0·9 volt and 0·3 volt respectively, the potential difference at $a$, $c$, across the two wires is 1·2 volts. This relation between $V$, $V_1$, and $V_2$ is explained by the fact that the work done in moving the charge on an electron, for example, from $a$ to $c$ equals the work done in moving the charge from $a$ to $b$ + the work done in moving it from $b$ to $c$; so that the potential difference between $a$, $c$ is always the sum of the potential differences between $a$, $b$ and $b$, $c$ (see p. 433).

**Combined resistance.** Suppose that the potential difference, $V_1$, across the resistance $P$ is 0·6 volt, and the current, $I$, in it is 0·3 amp.; then the resistance $P = \dfrac{V_1}{I} = \dfrac{0·6}{0·3} = 2$ ohms. Suppose that the potential difference, $V_2$, across the resistance $Q$ is 0·9 volt. Then, as the current in it is also 0·3 amp., the resistance $Q = \dfrac{0·9}{0·3}$ $= 3$ ohms. Now the combined or total resistance, $R$, of $P$ and $Q$ in series $= \dfrac{\text{potential difference across } a, \ c}{I} = (0·6 + 0·9)/0·3 =$ 5 ohms; thus $R = P + Q$. It can be seen from the calculation that, in general, *the combined resistance of any number of wires in series is equal to the sum of the individual resistances*, and this relation can be proved experimentally (see p. 462).

**Resistances in parallel.** In commercial lighting and other circuits, many resistances are all connected to the same two points. In Fig. 318 (*a*), for example, some lamps, $L$, are each connected to the p.d. at the mains, so that switching a lamp on or off makes no effect to the p.d. applied to other lamps in the same building. These lamps are said to be in *parallel* across the mains. Fig. 318 (*b*) illustrates two resistances $P$, $Q$ in parallel across the points $a$, $b$ connected to a battery. Thus, unlike the case of resistances in series, *the p.d. is the same across conductors in parallel*.

If there were a "water circuit" with pipes in the place of the wires $da$, $acb$, and $anb$ in Fig. 318 (*b*), the current of water flowing along $da$ would divert into two parts at $a$ and flow along $acb$ and $anb$ respectively. At $b$ the two currents would unite, and the current along a pipe $bm$ would be the same in magnitude as that along $da$. In the same way, *the electric current $I$ in the wire $da$ in*

*Fig. 318 (b) is equal to the sum of the currents, $I_1$ and $I_2$, in the wires acb and anb respectively.* Thus the current, $I$, in *da* is 3 amps. if the current $I_1$ is 1 amp. and the current $I_2$ is 2 amps.

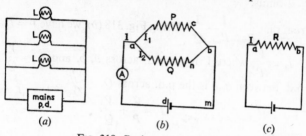

FIG. 318. Resistances in parallel

**Combined resistance.** Suppose the p.d., $V$, across the resistances $P$, $Q$ is 12 volts when the three currents have the values just given. Then $P = \dfrac{V}{I_1} = \dfrac{12}{1} = 12$ ohms, $Q = \dfrac{V}{I_2} = \dfrac{12}{2} = 6$ ohms. The "combined resistance," $R$, of $P$ and $Q$ in parallel is the resistance of that wire which can replace $P$ and $Q$ without disturbing the current $I$ flowing along $d$, $a$, assuming the p.d. between the two points is unaltered. In this case the current $I$ is also the current along the resistance $R$ (Fig. 318 (c) ). Hence, since $I = 3$ amps. and $V = 12$ volts, $R = \dfrac{V}{I} = 4$ ohms.

In general, as proved on p. 462, the combined resistance $R$ of two resistances $P$, $Q$ in *parallel* is given by

$$\frac{1}{R} = \frac{1}{P} + \frac{1}{Q}.$$

Thus if $P = 12$ ohms, $Q = 6$ ohms, the resistance $R$ is calculated from $\dfrac{1}{R} = \dfrac{1}{12} + \dfrac{1}{6}$.

$$\therefore \quad \frac{1}{R} = \frac{3}{12}$$

$$\therefore \quad R = \frac{12}{3} = 4.$$

Thus $R$ is 4 ohms. Similarly, the combined resistance of 4, 5, and 20-ohm wires in parallel is given by $\frac{1}{R} = \frac{1}{4} + \frac{1}{5} + \frac{1}{20}$, from which $R = 2$ ohms.

**Proof that $\frac{1}{R} = \frac{1}{P} + \frac{1}{Q}$.** Using Fig. 318 (b), we have $I = I_1 + I_2$.

Now $I_1 = \frac{V}{P}$, where $V$ is the p.d. across $a$, $b$, and $I_2 = \frac{V}{Q}$, since the p.d. between $a$, $b$ is the p.d. across $Q$.

$$I = \frac{V}{P} + \frac{V}{Q}.$$

But $I = \frac{V}{R}$ (Fig. 318 (c)).

$$\therefore \frac{V}{R} = \frac{V}{P} + \frac{V}{Q}.$$

Dividing throughout by $V$,

$$\therefore \frac{1}{R} = \frac{1}{P} + \frac{1}{Q}.$$

**Verification of combined resistance formula.** *Resistances in series.* The combined resistance of two resistances $P$, $Q$ in series can be measured by connecting an accumulator, an ammeter $A$,

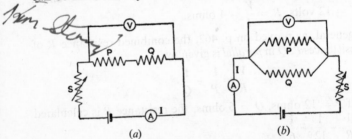

FIG. 319. Resistances in series and parallel

and a rheostat $S$ to them, and joining a high-resistance voltmeter $V$ across both resistances (Fig. 319 (a)). The rheostat is adjusted until suitable readings are obtained on $V$ and $A$, and the combined resistance $R$ is then given by $R = \dfrac{V}{I}$, where $V$ is the voltmeter reading and $A$ is the ammeter reading.

The magnitude of $Q$ is obtained by disconnecting $P$ from the circuit, then re-forming the circuit, and noting the p.d. ($V$) across $Q$ and the current ($I$) in it. From Ohm's Law, $Q = \dfrac{V}{I}$. The resistance of $P$ is measured in the same way, and the combined resistance $R$ will be found equal to the sum of the resistances $P$, $Q$.

**Resistances in parallel.** The combined resistance, $R$, of two resistances $P$, $Q$ in parallel can be measured by the circuit arrangement shown in Fig. 319 (b). $R$ is then given by $\dfrac{V}{I}$, where $V$ is the p.d. measured on the high-resistance voltmeter used, and $I$ is the current measured by the ammeter $A$. The two separate resistances, $P$ and $Q$, are measured in the same way as described above, and it will be found that the measured value $R$ is very closely equal to the value calculated from the relation $\dfrac{1}{R} = \dfrac{1}{P} + \dfrac{1}{Q}$, which is the formula for resistances in parallel.

**Converting a milliammeter to an ammeter.** A *milliammeter* is an instrument measuring currents of the order of milliamps,. i.e. thousandths of an ampere (p. 454); a *galvanometer* is an instrument usually capable of detecting and measuring smaller currents than a milliammeter (p. 588). Both of these instruments can easily be adapted to measure much larger currents by connecting a resistance of a certain magnitude *in parallel* with each; and as an illustration of the principles concerned, we shall consider how a milliammeter is converted into an ammeter.

Suppose that a milliammeter has a 10-ohm resistance coil and is constructed to read 0–5 milliamps. If the milliammeter is to be converted to read 0–2 amps., a resistance $R$ must be placed in parallel with the instrument so that a current of 5 milliamps. flows in the latter when a current of 2 amps. flows in the outside circuit (Fig. 320). If this is the case, the maximum reading on the milliammeter scale corresponds to a current of 2 amps. outside the instrument.

The resistance $R$ and the 10-ohm resistance of the instrument are in parallel across the terminals $a$, $b$. Since 2 amps. is the current outside the circuit, and 5 milliamps. flows through the instrument, it follows that a current $I$ flows through $R$ which is equal to the difference of the two currents. Now 5 milliamps. $= 5/1000$ amp. $= 0.005$ amp.

$$\therefore \quad I = 2 - 0.005 = 1.995 \text{ amps.}$$

To find $R$, we require the p.d., $V$, between its ends, i.e. the p.d. between $a$, $b$. Now between $a$, $b$, there is a resistance of 10 ohms

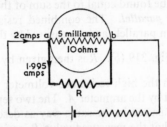

FIG. 320. Conversion of milliammeter to ammeter

in which a current of 5 milliamps. flows (Fig. 320). Thus, since p.d. = current × resistance, from Ohm's Law,

$$V = 10 \times 0.005 = 0.05 \text{ volt} = \text{p.d. between } a, b$$

$$\therefore \quad R = \frac{V}{I} = \frac{0.05}{1.995} = 0.0250 \text{ ohm.}$$

This small resistance is made as accurately as possible by the manufacturers, and is supplied with the milliammeter instrument as a *shunt* for measuring currents from 0–2 amps. Other ranges of current, e.g. 0–4 amps., are obtained by using shunt resistances of other magnitudes, whose values are calculated in exactly the same way as shown above.

**Converting a milliammeter to a voltmeter.** One of the advantages of a moving-coil milliammeter is that it can easily be converted into an instrument measuring potential difference, a fact which may at first occasion surprise.

Suppose we again consider the milliameter of 10 ohms resistance, reading 0–5 milliamps. When it carries a current, $I$, of 5 milliamps., or 0.005 amp., there is a p.d., $V$, across its terminals given by $V = IR$, where $R$ is 10 ohms. Thus $V = 0.005 \times 10$ = 0.05 volt. Consequently, if the instrument is used in the same way as a *voltmeter*, by joining it in parallel with the circuit concerned, a reading of 5 milliamps on the instrument indicates that the p.d. it now measures is 0.05 *volt*, or 50 millivolts. Since $V = IR$, it can be seen that a p.d. always exists across the terminals of a moving-coil instrument when it carries a current.

In order to increase the magnitude of the p.d. which can be measured, a resistance $R$ is added *in series* with the instrument. The new terminals of the voltmeter formed are then $a$, $c$, instead of $a$, $b$ (Fig. 321). Suppose the instrument is to be adapted to

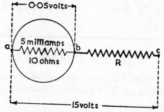

FIG. 321. Conversion of milliammeter to voltmeter

measure 0–15 volts. Then $R$ must be chosen so that a p.d. of 15 volts applied to $a$, $c$ causes a maximum deflection in the instrument, *i.e.* a current of 5 milliamps. then flows in the coil of the instrument.

Since resistance $= \dfrac{\text{p.d.}}{\text{current}}$ (p. 454), the resistance between $a$, $c = \dfrac{15 \text{ volts}}{0.005 \text{ amp.}}$, the current of 5 milliamps. flowing along $a$, $c$. Hence the resistance between $a$, $c = 3000$ ohms.

But the instrument resistance $= 10$ ohms.

$\therefore$ series resistance, $R = 3000 - 10 = 2990$ ohms.

Thus the milliammeter reads 0–15 volts when 2990 ohms is added in series with it. If other ranges of p.d. are required, e.g. 0–150 volts, other resistances are added in series whose magnitudes are calculated in exactly the same way as described above. The higher the maximum potential difference to be measured, the higher the series resistance must be.

─Summary─

1. **Ohm's law states: The ratio p.d./current for a given conductor is constant, provided the physical conditions of the conductor, such as its temperature, remain analtered.**

2. Fundamental formulae are: $I = V/R$; $V = IR$; $R = V/I$, where $I$ is in amperes, $V$ is in volts, $R$ is in ohms.

3. The ammeter (or milliammeter) measures current; it has a **low** resistance, and is placed **in series** with the circuit. A voltmeter measures potential difference; it has a **high** resistance, and is placed **in parallel** with the circuit.

4. The combined resistance, $R$, of two resistances, $P$, $Q$, in series is given by $R = P + Q$; in parallel, it is given by $1/R = 1/P + 1/Q$.

## WORKED EXAMPLES

1. *A resistance of 10 ohms is in series with one of 20 ohms, and the p.d. across both of them is 60 volts (Fig. 322). Calculate the current in the 10-ohm wire and the p.d. across each wire.*

The total resistance between $A$ and $B$ is $(10 + 20)$, or 30 ohms. Since

FIG. 322

$I = \dfrac{V}{R}$, from Ohm's law, the current in each wire is given by

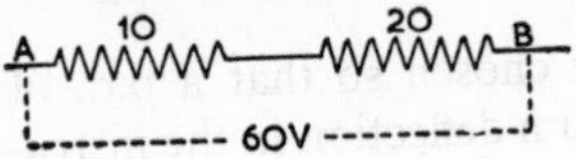

$$I = \tfrac{60}{30} = 2 \text{ amps.} \quad . \quad . \quad (1)$$

The p.d., $V$, across the 10-ohm wire $= IR = 2 \times 10 = 20$ volts . (2)

The p.d. across the 20-ohm wire $= IR = 2 \times 20 = 40$ volts . (3)

*Note.* It is wrong to state that the current in the 10-ohm wire is $\tfrac{60}{10}$ amps., because the p.d. between the ends of this resistance is not 60 volts; this is the p.d. across *both* wires.

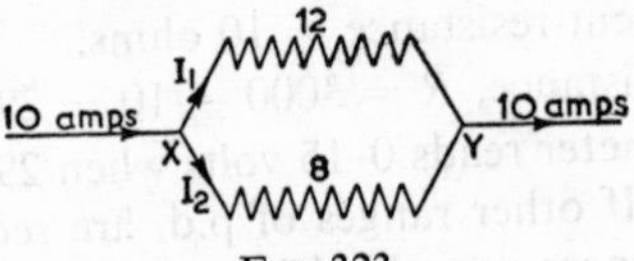

FIG. 323

2. *A resistance of 12 ohms is in parallel with one of 8 ohms, and a current of 10 amps. flows outside the parallel combination in the main part of the circuit (Fig. 323). Find the p.d. between X, Y and the currents flowing in the two wires.*

If the two wires were replaced in the circuit by their equivalent resistance, R, the current of 10 amps. would flow through $R$. Consequently, the p.d. $V$ between $X$, $Y$ is given by $V = IR$, where $I$ is 10 amps, and $R$ is the combined resistance of 12 and 8 ohms.

Now $\dfrac{1}{R} = \dfrac{1}{12} + \dfrac{1}{8} = \dfrac{5}{24}$. Thus $R = \dfrac{24}{5}$ ohms.

$$\therefore \quad \text{p.d. between } X, Y = V = IR = 10 \times \frac{24}{5} = 48 \text{ volts} \qquad . \quad . \quad (1)$$

The current, $I_1$, in the 12-ohm wire is given by $I_1 = \dfrac{V}{R}$, where $V$ is the p.d. across the wire and $R = 12$. But $V = 48$ volts.

$$\therefore \quad I_1 = \frac{48}{12} = 4 \text{ amps.} \qquad . \qquad . \qquad . \qquad (2)$$

Similarly, the current $I_2$ in the 8-ohm wire $= \dfrac{V}{R} = \dfrac{48}{8} = 6$ amps., since the p.d. across this wire is also 48 volts.

*Note.* In this circuit, part of the current of 10 amps. flows along the 12-ohm wire and the remainder along the 8-ohm wire. As the p.d., $V$, is the same across each wire and $I = \dfrac{V}{R}$, it follows that $I \propto \dfrac{1}{R}$. Thus the current divides at $X$, for example, in the *inverse* ratio of the two branch resistances, i.e. $I_1 : I_2 = 8 : 12$. Since there are $(8 + 12)$ "parts" altogether, the actual current in the 12-ohm wire is given by $I_1 = \frac{8}{20} \times 10$ amps. $= 4$ amps., and the actual current in the 8-ohm wire is given by $I_2 = \frac{12}{20} \times 10$ amps. $= 6$ amps. This is a quick method of calculating the current in either of the two branches of a parallel combination of *two* wires.

3. *A circuit consists of a 12-ohm resistance in series with a parallel combination of 10- and 40-ohm wires, and a p.d. of 40 volts is connected across the whole circuit* (Fig. 324). *Calculate the currents in each of the three wires.*

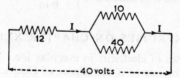

Fig. 324.

Since 40 volts is the p.d. across the whole circuit, the current $I$ flowing in the main part of the circuit is given by $I = \dfrac{V}{R}$, where $V = 40$ volts and $R$ is the *total* resistance of the circuit.

The combined resistance, $S$, of the parallel combination is found first. This is given by

$$\frac{1}{S} = \frac{1}{10} + \frac{1}{40} = \frac{5}{40}, \text{ i.e. } S = \frac{40}{5} = 8 \text{ ohms}$$

$$\therefore \quad \text{total resistance, } R = 12 + 8 = 20 \text{ ohms}$$

$$\therefore \quad I = \frac{V}{R} = \frac{40}{20} = 2 \text{ amps.}$$

and this is the current flowing in the 12-ohm wire.

The current in the 10- and 40-ohm wires can be found by the first method given in example 2. As an exercise, however, the method in the "Note" of example 2 will be given. In this method, we utilize the fact that the current divides in the inverse ratio of the resistances. Hence the current in the 10-ohm wire is given by $\frac{40}{40 + 10} \times 2$ amps. or 1·6 amps., and the current in the 40-ohm wire is given by $\frac{10}{40 + 10} \times 2$ amps. or 0·4 amp. The sum of the currents = 1·6 + 0·4 = 2 amps. which is the current in the main part of the circuit.

4. *Describe the construction of a moving-coil galvanometer of the pivoted coil type. Explain why such a galvanometer cannot be used to measure alternating current. An ammeter and a resistance of 1090 ohms were connected in series with 110-volt D.C. mains. The ammeter reading was 0·10 amp. What was its resistance? A voltmeter was connected across the terminals of the 1090-ohms resistance. What voltage did it read?* (C.)

Since the ammeter reading is 0·10 amp., the p.d. across the 1090-ohms resistance is given by $V = IR = 0.10 \times 1090 = 109$ volts. This is the reading on the voltmeter placed across the wire. But the ammeter is in series with the resistance.

Hence the p.d. across the ammeter = 110 − 109 = 1 volt.

$$\therefore \quad \text{resistance } R \text{ of ammeter} = \frac{V}{I} = \frac{1}{0.10} = 10 \text{ ohms.}$$

## EXERCISES ON CHAPTER XXIX

1. Name six uses of electricity in everyday life. What social conveniences would be missed if the science of Electricity had not been developed?

2. A 20 volt battery is connected to a resistance of (i) 100 ohms, (ii) ½ ohm, (iii) 2 ohms. Calculate the current in each case.

3. A current of 3 amps. is flowing in (i) a 20 ohm, (ii) a 2 ohm wire. Calculate the potential difference across each wire.

4. What is an *ammeter* and a *voltmeter*? Draw a circuit sketch

showing how you would use them for measuring the current in a lamp and the potential difference across it.

**5.** What is the combined resistance of the following arrangements: (i) 6- and 4-ohm wires in parallel, (ii) 5, 20, and 10-ohm wires in series, (iii) 5, 20, 10-ohm wires in parallel, (iv) a 5-ohm wire in series with a parallel arrangement of 60 and 30 ohms ?

**6.** A potential difference of 120 volts is connected across a 6- and 4 ohm wire arranged (i) in series, (ii) in parallel. Calculate the current in the wires in each case and the potential difference across each.

**7.** Are lamps connected in parallel or in series in a lighting circuit ? Give a reason for your answer. Three similar electric lamps are arranged in a lighting circuit. The mains is 200 volts, and each lamp has a filament resistance of 250 ohms. Find the current through the mains.

**8.** A current of 15 amps. is flowing towards the junction of (i) a 6- and a 3-ohm wire in parallel, (ii) three wires of 1, 4, and 5 ohms in parallel. What is the current in each wire, and the potential difference across each in the two cases ?

**9.** Draw a sketch showing how a 0–10 milliamp instrument of 5-ohm resistance can be converted into (i) an ammeter reading 0–1 amp., (ii) a voltmeter reading 0–3 volts. Calculate the resistance required in each case.

**10.** State *Ohm's law*. Deduce an expression for the equivalent resistance of two resistances $R_1$ and $R_2$ in parallel and describe how you would check the result by experiment. (*L.*)

**11.** A voltmeter gives a full-scale deflection when the potential difference between its terminals is 1 volt. Its resistance is 400 ohms. How would you convert it (i) to give full-scale deflection with a p.d. of 10 volts, (ii) to work as an ammeter reading up to 250 milliamps. ? (*C.*)

**12.** A resistance of 4 ohms is in series with a parallel arrangement of 12 and 6 ohms, and a p.d. of 32 volts is connected across the whole arrangement. Find the current in each of the wires.

**13.** A wire of 7 ohms is in series with a parallel arrangement of 4- and 12-ohm wires. If the current in the 7-ohm wire is 3 amp., find (i) the current in the 4-ohm wire, (ii) the p.d. across the whole circuit. (*O. & C.*)

16*

# CHAPTER XXX

# ELECTRICAL ENERGY. HEATING EFFECT
## OF CURRENT

THE heating effect of an electric current serves a most useful purpose in the home and in industry. It causes the electric lamp
filament to light up, the electric cooker to function, and the electric
fire to radiate; it also heats the filaments of the valves in radio sets.

The heating effect of an electric current is obtained by a transformation of electrical energy to heat energy. It is therefore
necessary to consider first how much electrical energy is used in a
circuit when a current flows in it.

**How electrical energy is calculated.** In the section on electrostatics, it was explained that an energy change occurs when electricity moves from one point to another at a different potential,
just as an energy change occurs when a mass moves from one
point above the earth to another at a different height (see p. 433).
The volt, the practical unit of p.d., is defined as the p.d. between
two points such that 1 joule of energy is liberated when 1 coulomb
of quantity of electricity moves from one point to another; from
this definition it follows that 2 joules of energy are liberated when
1 coulomb moves between two points at a p.d. of 2 volts, and that
20 joules of energy are liberated by the movement of 10 coulombs
of electricity between two points at a p.d. of 2 volts. It can thus be
seen that the energy $W$ liberated when $Q$ coulombs move between
two points at a p.d. of $V$ volts is given by

$$W = QV \text{ joules} \qquad \cdot \qquad \cdot \qquad \cdot \qquad (1)$$

If a continuous current of $I$ amps. flows in a wire for $t$ seconds,
the quantity of electricity, $Q$, which has moved across a section of
the wire in that time is given by $Q = It$ coulombs (p. 451). Thus
the energy liberated, from (1), is also given by

$$W = IVt \text{ joules} \qquad \cdot \qquad \cdot \qquad \cdot \qquad (2)$$

and this is a more useful formula than (1) for calculating the energy liberated in electrical circuits, since current ($I$) can be measured more easily than quantity ($Q$) of electricity.

Suppose that a battery is connected to an *electric motor*, as represented in Fig. 325 (*a*). The electrical energy liberated inside the motor is $IVt$ joules, from (2), where $I$ is the current in amps. flowing, $V$ is the p.d. between the terminals in volts, and $t$ is the time in seconds. Now a motor is a device which converts electrical

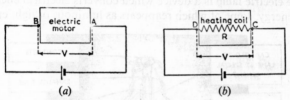

FIG. 325. Conversion of electrical energy

energy to mechanical energy of rotation, and most of the $IVt$ joules is hence converted into mechanical energy. A radio *transmitter* converts some of the electrical energy from a battery into electrical energy in the form of radio waves, which travels out from the transmitter, and the rest of the energy is converted into heat. When, however, a battery is connected to an ordinary coil of wire (Fig. 325 (*b*) ), *all* the electrical energy of $IVt$ joules is converted to heat energy in the wire. This occurs in the coil inside an electric iron or electric radiator. In this special case, therefore, the heat energy produced is $IVt$ joules, and is due to the resistance to motion in the wire encountered by the moving electrons.

**Electrical power.** The *power* of a machine is the rate at which it delivers energy, i.e. power $= \dfrac{\text{energy delivered}}{\text{time taken}}$ (see p. 43). The electrical power $P$ delivered to the motor or to the coil in Fig. 325 is thus given by $P = \dfrac{IVt \text{ joules}}{t \text{ secs.}}$, from which $P = IV$ joules per sec. The watt is a unit of power equal to 1 joule per sec. rate of working (see p. 43), and hence the power in an electrical circuit is given by

$$P = IV \text{ watts.}$$

This relation is sometimes remembered by the formula—

*watts = amps. × volts.*

An electric fire which carries a current of 2·5 amps. when the p.d. of the mains is 200 volts has thus a power rating of 2·5 × 200 or 500 watts, i.e. half a kilowatt (p. 474). The total power of all the transmitters of the B.B.C. is about 1,200,000 watts, or 1200 kilowatts.

**The electric lamp** is a device which converts electrical energy to heat energy, some of which reappears as luminous or light energy.

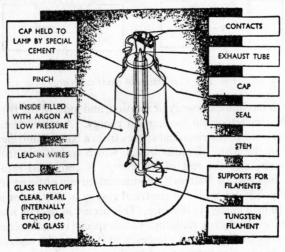

FIG. 326. Electric filament lamp

The great majority of present-day electric lamps contain a tungsten filament or wire, with a small amount of an inert gas, such as argon, to prevent oxidation and evaporation of the metal (Fig. 326). The lamps are connected *in parallel* across the mains (Fig. 318 (*a*) ).

The electrical power consumed by the lamp filament is given by $P = IV$ watts, from above, where $I$ is the current in amps. flowing in the lamp and $V$ the p.d. in volts across it. If the resistance of the lamp filament is $R$ ohms, then $V = IR$ from the Ohm's law relation on p. 454; substituting for $V$ in the expression for power, we have

$$P = IV = I \times IR = I^2R \text{ watts.}$$

Further, $I = V/R$, from Ohm's law, and substituting for $I$ in the expression $P = IV$, we have

$$P = \frac{V}{R} \times V = \frac{V^2}{R} \text{ watts.}$$

Any one of three expressions can therefore be used for calculating electrical power in the case of a resistance wire; they are respectively $\boldsymbol{P = IV, \ P = I^2R,}$ or $\boldsymbol{P = V^2/R.}$

As an illustration of the application of the formula, suppose that a 100-watt lamp is used normally on a 200-volt mains. The normal current, $I$, flowing through the filament is thus given, from $P = IV$, by

$$100 = I \times 200.$$

Consequently, $I = \frac{1}{2}$ amp. The *resistance*, $R$, of the filament is given by $R = \dfrac{V}{I}$, from Ohm's law, and hence $R = 200/\frac{1}{2} = 400$ ohms.

**Fluorescent lamps.** Towards the end of the nineteenth century, scientists began to study how electricity was carried through *gases* (p. 622). A long glass tube $T$, containing a little gas, was used, and a high p.d. $V$ of several thousand volts was connected to two pieces of metal, $A$, $K$, at each end of the tube (Fig. 327). It was then observed that the gas glowed, and in certain circumstances a band of light stretched across the tube from $A$ to $K$. This is an example of the conversion of mechanical energy to light energy, as fast-moving electrical particles (electrons and ions) in the gas collide with the gas molecules and in so doing cause them to emit light.

Since about 1930, scientists have been investigating the possibility of using the glowing gas as the basic part of a new form of lamp, known as a *fluorescent lamp*. Mercury vapour produces invisible ultra-violet rays (p. 353) when it glows; most of the light coming from the lamp is obtained from fluorescent powders coating the inside of the tube, which are exposed to the ultra-violet light. Different coloured light is obtained by using different powders. There is no glare from such lamps, and little shadow. They have the advantage of being more efficient than the filament type of lamp in converting electrical energy to light energy. Further, the cost of running them is much less than the filament

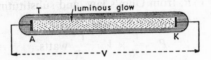

FIG. 327. Luminous gas

type as the power they absorb is much less. Fluorescent lamps are now used for lighting streets, offices, and factories, and are being installed in mines (see Fig. 328).

FIG. 328. FLUORESCENT STREET LIGHTING IN REGENT STREET, LONDON

**Commercial unit of electrical energy.** In electrical engineering a larger unit of power than the watt is frequently used. This is the *kilowatt*, which is 1000 watts, so that 1500 watts is 1·5 kilowatts.

From the definition of power (p. 43), it follows that energy = power × time. In industry a unit of energy known as the *watt-hour* is used, and is defined as the energy consumed when 1 watt of power is delivered continuously for 1 hour. The number of watt-hours consumed by a 60-watt lamp burning continuously for 30 hours is hence $60 \times 30$, or 1800 watt-hours. A larger unit of energy, used by commercial power companies, is the *Board of Trade unit*, which is defined as the *kilowatt-hour*, or 1000 watt-hours. Thus 1800 watt-hours of energy = 1·8 Board of Trade units. The latter are the "units" of electrical energy for which house-holders pay the electric power companies. The cost of 1·8 Board of Trade units at 4d. a unit is 7·2d., that is, about ¼d. an hour for a 60-watt lamp.

Suppose that a factory contains 30 100-watt and 50 60-watt lamps, which are lit continuously for 15 hours a week at a cost of 3d. a (Board of Trade) unit. The total number of watt-hours of energy used = $30 \times 100 \times 15 + 50 \times 60 \times 15 = 90,000$; hence the number of kilowatt-hours = $\dfrac{90,000}{1000} = 90$.

$$\therefore \quad \text{cost} = 90 \times 3\text{d.} = \text{£1 2s. 6d.}$$

**Heating effect of an electric current.** In 1843 Joule (see p. 210) discovered the laws relating to the heat produced in a wire by the flow of an electric current. *Joule's laws of heating* state that—

*The heat developed in a wire is proportional to (i) the time ( for a given resistance and current), (ii) the square of the current ( for a given resistance and time), (iii) the resistance of the wire ( for a given current and time).*

These laws are easily derived from our previous work, since it has been shown that the heat energy, $W$, in a wire is given by $W = IVt$ joules (p. 470). Now $V = IR$, from Ohm's law, where $R$ is the resistance of the wire; hence $W = IVt = I \times IR \times t = I^2Rt$ joules.

Since 4·2 joules of mechanical energy are equivalent to 1 calorie of heat energy (p. 211), the heat developed in a resistance of $R$ ohms in a time $t$ seconds is given by $\dfrac{I^2Rt}{4\cdot2}$ calories, where $I$ is the current in amps.

Suppose that a current of 2 amps. is passed into a wire of resistance 50 ohms for 1 minute, and that the wire is totally immersed inside 80 gm. of water contained in a calorimeter of water equivalent 10 gm. Then, if the rise in temperature is $x°$ C., the heat obtained by the water and calorimeter

$$= (80 + 10)x \text{ calories.}$$

But the heat supplied $= \dfrac{I^2 Rt}{4 \cdot 2}$ calories $= \dfrac{2^2 \times 50 \times 60}{4 \cdot 2}$ calories.

$$\therefore \quad 90x = \dfrac{2^2 \times 50 \times 60}{4 \cdot 2}$$

$$\therefore \quad x = \dfrac{2^2 \times 50 \times 60}{90 \times 4 \cdot 2} = 31 \cdot 7° \text{ C.}$$

### Verification of Joule's laws.

(1) *Heat proportional to time and to square of current.* Fig. 329 (*a*) illustrates how Joule's laws can be verified. A wire is immersed inside a calorimeter $C$ containing water, and a battery, an ammeter $A$, and a rheostat $S$ are connected in series with the wire.

To show that the heat is proportional to the time, a constant current is passed into the wire and the temperature rise is noted by the thermometer $T$ after 2, 4, 6, 8, and 10 minutes. It will be found that the temperature rises obtained are in the ratio 1 : 2 : 3 : 4 : 5;

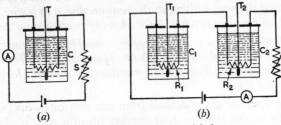

FIG. 329. Verification of Joule's laws

and since the heat produced $= (W + m) \times$ temp. rise, where $W$ is the water equivalent of the calorimeter and $m$ is the mass of water, the heats produced are also in the ratio 1 : 2 : 3 : 4 : 5. Thus the heat produced $\propto$ the time.

To show that the heat produced is proportional to $I^2$, the square of the current, different currents are passed into the calorimeter for the *same* period of time. This is done by altering the rheostat $S$. If the currents are 1, 1·5, 2, 2·5 amps., the temperature rises obtained will be found to be proportional to $1^2$, $1 \cdot 5^2$, $2^2$, $2 \cdot 5^2$, i.e. in the ratio $1 : 2\frac{1}{4} : 4 : 6\frac{1}{4}$. Thus, as explained above, the heat produced $\propto I^2$.

(2) *Heat proportional to the resistance for a given current.*
Two wires of different resistances $R_1$, $R_2$ are connected in
series with a battery, a rheostat, and an ammeter (Fig. 329 (b) ).
The respective wires are totally immersed in calorimeters filled
with water. A current is passed into the wires for a certain time
and the respective temperature rises are noted on the thermometers $T_1$, $T_2$.

The heats produced in the two calorimeters are $(W_1 + m_1)x_1$
and $(W_2 + m_2)x_2$ calories respectively, where $W_1$, $W_2$ are the
respective water equivalents, $m_1$, $m_2$ are the respective masses of
water, and $x_1$, $x_2$ are the respective rises of temperature. It will
be found that

$$\frac{(W_1 + m_1)x_1}{(W_2 + m_2)x_2} = \frac{R_1}{R_2},$$

which shows that the heat produced is proportional to the
resistance.

**The fuse.** The heating effect of an electric current is utilized to
safeguard circuits against excessive currents. The device known as
a *fuse* consists simply of a short piece of fine wire made of a
tin-lead alloy, for example, which has a low melting-point. Fuses
are usually fitted across two brass terminals in porcelain or bakelite holders, and connected in the circuit. When an excessive
current flows, the fuse melts and the circuit becomes broken.

Fuse wire is rated as "5-amp.," "10-amp.," "15-amp.," and so
on, which means that the wire will melt if the current exceeds the
stipulated value. Suppose that a lighting circuit in a house has
fuses of 5 amps., and the mains is 200 volts. A $\frac{1}{2}$-kilowatt
electric fire connected in the circuit has then a current $I$ flowing
in it given by $500 = I \times 200$, since power $P = IV$ (p. 471); thus
$I = 2\cdot5$ amps. The electric fire can hence be safely used, as the
current is much below the value of the fusing current. If a
2-kilowatt electric fire were connected across a 200-volt mains,
the current flowing would be given by $2000 = I \times 200$, so that $I$
would be 10 amps. in this case. A 2-kilowatt fire cannot therefore be used in the lighting circuit.

**The hot-wire ammeter.** The heating effect of a current was
applied to the construction of a measuring instrument known as a

"hot-wire" ammeter. The essential details of the instrument are shown in Fig. 330. *PQ* is a length of resistance wire connected to fixed terminals at *P*, *Q*, and the current to be measured flows through the wire, which then becomes hot. The length of the wire now increases, so that it begins to sag slightly; the point *M* on the wire is thus lowered. A silk thread passing round a grooved wheel *A* has one end connected to *T*, a point on a wire *MN*, and the other end to a spring *S*. As the wire sags, *A* turns, and a pointer *X* connected to the axle moves over a scale which is calibrated in amps. by passing known currents into *PQ*. The heating effect is proportional to the *square* of the current, so that the scale is not a uniform one; *i.e.* equal steps along the divisions of the scale do not represent equal increases in current (see p. 588).

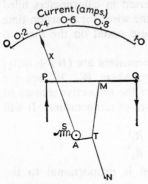

FIG. 330. Hot-wire ammeter

**High-tension (H.T.) transmission.** Since power *P* = *IV*, a given amount of power can be delivered either at high current (*I*) and low voltage (*V*), or at low current and high voltage; for example, a power of 2000 watts is obtained if the p.d. (voltage) is 1000 volts and the current supplied is 2 amps., or if the p.d. is 100 volts and the current supplied is 20 amps. If the current flows along wires of fixed resistance, the power converted to heat is proportional to the square of the current, since *P* = *I*²*R* (p. 473). Consequently, less power is wasted in the wires as heat if the given amount of power is delivered at *low current* and *high voltage*. This is the way commercial power is distributed throughout the country (see p. 606). The distribution is known as "high tension (H.T.)" transmission, because the voltage is high.

**Electric radiant heaters.** The heating effect of a current is utilized commercially to warm buildings. A nickel-copper-resistance wire is embedded in a large insulating panel, which radiates heat when a current flows in the wire, and the panels are suspended horizontally from ceilings or placed vertically on walls.  The indoor temperature is then maintained at an even and comfortable temperature.

**Determination of Joule's mechanical equivalent of heat by electrical method.**  Since electrical energy is expressed in the same units as mechanical energy (joules or ergs), the heating effect of a current has been used to determine the value of the mechanical equivalent of heat, $J$ (*i.e.* the number of ergs or joules of mechanical energy which give rise to 1 calorie of heat, p. 212).  The value of $J$ obtained by this method is much more accurate than that obtained by allowing weights to fall (see p. 211), as electrical instruments have been developed to a high pitch of perfection.

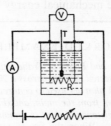

FIG. 331. Determination of $J$

Fig. 331 illustrates the circuit required.  A current is passed through a coil $R$ immersed in some water, and the rise in temperature $\theta$ after a time $t$ is noted on a thermometer $T$.  A voltmeter is used to determine the p.d. $V$ across $R$, and the steady current $I$ in the coil is noted from an ammeter $A$.

Suppose $m$ is the mass of water used, and $W$ is the water equivalent of the calorimeter.  The heat developed by the current is then $(m + W)\theta$ calories.  The energy which provides this heat is given by $IVt$ joules (p. 470), and consequently the number of joules of mechanical energy providing 1 calorie of heat is $\dfrac{IVt}{(m + W)\theta}$ joules.

Thus $J = \dfrac{IVt}{(m + W)\theta}$ joules per calorie, or $J = \dfrac{IVt}{(m + W)\theta} \times 10^7$ ergs per calorie, as 1 joule $= 10^7$ ergs (p. 38).  Since $I$, $V$, $t$, $m$, $W$, and $\theta$ can all be measured, the magnitude of $J$ can be calculated.

---

─────────────Summary─────────────

1. **Energy** $(W) = IVt$ **joules,** where $I$ is in amps., $V$ is in volts, $t$ is in seconds.

2. **Power** $(P) = IV$ **watts,** where $I$ is in amps., $V$ is in volts. A Board of Trade unit = 1 kilowatt-hour = 1 kilowatt × 1 hour.

3. **Joule's laws of heating state: The heat produced is proportional to the square of the current, to the resistance, and to the time. The heat** $= I^2Rt/4\cdot2$ **calories** $= V^2t/4\cdot2R$ **calories** $= IVt/4\cdot2$ **calories.**

4. Joule's mechanical equivalent of heat can be measured by passing a current through a coil totally immersed in water in a calorimeter; the mechanical energy = $IVt$ joules.

---

## WORKED EXAMPLES

1. *Define the terms watt and Board of Trade unit. An electric fire takes a power* 1 *kilowatt when connected directly across a* 200 *volt main. If long leads having a total resistance of* 1 *ohm are used to connect the fire to a* 200-*volt main, calculate* (a) *the power taken from the main, and* (b) *the power used in the fire itself.* (*Assume that the resistance of the fire remains constant.*) (*N.*)

1 kilowatt = 1000 watts. When the fire is connected across the mains, ($V$ = 200 volts), the current $I$ in amps. is given by

$$\text{power} = IV$$

i.e. 1000 = $I$ × 200. Hence $I$ = 5 amps.

$$\therefore \text{ resistance, } R, \text{ of fire} = \frac{V}{I} = \frac{200}{5} = 40 \text{ ohms.}$$

(a) When the leads are used, the *total* resistance in the circuit alters from 40 to 41 ohms. Hence new current, $I = \dfrac{V}{R} = \dfrac{200}{41}$ amps.

$$\therefore \text{ power taken from mains} = IV = \frac{200}{41} \times 200 = 975\cdot6 \text{ watts.}$$

(b) Power used in fire = $I^2R$, see p. 473, where $I$ is the current in the fire and $R$ is its resistance.

$$\therefore \text{ power} = \left(\frac{200}{41}\right)^2 \times 40 = 951\cdot8 \text{ watts.}$$

2. *Describe an experiment to show the relation between the rate of production of heat in a coil of wire and the potential difference applied to the ends of the coil. Two lamps, connected separately to a* 100-*volt supply, are found to take*

*60 watts and 75 watts respectively. Determine the resistance of each lamp. If the two lamps are now connected in series to a 200-volt supply, find (a) the total watts taken by the two lamps, (b) the cost of using the two lamps for 60 hours if the cost of electricity is 4d. per kilowatt-hour.* (C.W.B.)

Since
$$\text{power, } P = \frac{V^2}{R} \text{ watts,}$$

where $V$ is the p.d. in volts and $R$ is the resistance of the lamp in ohms, it follows that, for one lamp of resistance $R_1$,

$$60 = \frac{100^2}{R_1}, \text{ i.e. } R_1 = \frac{100^2}{60} = 166\tfrac{2}{3} \text{ ohms.}$$

For the other lamp, $75 = \dfrac{100^2}{R_2}$, where $R_2$ is the resistance of this lamp

$$\therefore R_2 = \frac{100^2}{75}, \text{ i.e. } R_2 = 133\tfrac{1}{3} \text{ ohms.}$$

(a) When the lamps are in series across a 200-volt supply, the current flowing in them is given by $I = \dfrac{200}{R}$, where $R$ is the combined resistance.

But
$$R = R_1 + R_2 = 166\tfrac{2}{3} + 133\tfrac{1}{3} = 300 \text{ ohms}$$

$$\therefore I = \frac{200}{300} = \tfrac{2}{3} \text{ amp.}$$

$$\therefore \text{ Power in first lamp} = I^2 R_1 = (\tfrac{2}{3})^2 \times 166\tfrac{2}{3} \text{ watts}$$
$$\text{and power in second lamp} = I^2 R_2 = (\tfrac{2}{3})^2 \times 133\tfrac{1}{3} \text{ watts}$$
$$\therefore \text{ total power} = (\tfrac{2}{3})^2 \times 166\tfrac{2}{3} + (\tfrac{2}{3})^2 \times 133\tfrac{1}{3} = 133\tfrac{1}{3} \text{ watts}$$

(b) The number of kilowatt-hours used $= \dfrac{133\tfrac{1}{3} \times 60}{1000}$

$$\therefore \text{ cost} = \frac{133\tfrac{1}{3} \times 60}{1000} \times 4d. = 2s. \ 8d.$$

## EXERCISES ON CHAPTER XXX

**1.** Name four applications of the heating effect of an electric current. What factors govern the amount of heat produced?

**2.** Name two units of energy used in electricity. What is a *joule*? A motor runs under a potential difference of 20 volts, and takes a current of 2 amps. from the supply. What energy is used up in (i) 100 sec., (ii) ½ hour? What is the *power* in each case?

**3.** What current is normally taken by a 120-watt, 240-volt lamp? If the Board of Trade unit is 2½d., calculate the cost of running it 12 hours a day for a week.

**4.** Describe the interior of an electric filament lamp.

**5.** State *Joule's laws of electrical heating*. Describe briefly how you would verify the law relating to the current, drawing a circuit sketch and making a list of imaginary results.

**6.** Two lamps of filament resistance 500 ohms and 200 ohms respectively are connected to a 200-volt mains supply. Which would glow brighter ? Give reasons.

**7.** A resistance of 40 ohms and one of 60 ohms are arranged in series across a 200 volt battery. Find the heat in calories produced in each wire in $\frac{1}{2}$ minute. ($J = 4\cdot2$ joules per calorie.)

**8.** Calculate the filament resistance of a 80-watt, 120-volt lamp when it is used.

**9.** Describe an experiment to show how the rate at which heat is produced in a given conductor by the passage of an electric current depends on the magnitude of the current. (*L.*)

**10.** Name *two* requirements in a wire to be used as a fuse in a house circuit. (*N.*)

**11.** A fully-charged 6-volt accumulator can give a current of 2 amps. for 30 hours. Calculate, in joules, the energy given out during its discharge. (*N.*)

**12.** Describe the essential parts of a modern electric lamp of the filament type, and indicate on a diagram how the current is passed via the lampholder to the filament. What is meant by the statement on a lamp "230 volts, 100 watts." Calculate, for such a lamp, (*a*) the current taken, (*b*) the resistance of the filament, (*c*) the cost of using the lamp for 100 hours at $2\frac{1}{2}d.$ per kilowatt-hour. (*C.W.B.*)

**13.** "The *power* of an electric lamp is measured in *watts*." Explain this statement with particular reference to the words in *italics*. Deduce that the power supplied to a constant resistance is proportional to the square of the p.d. between its ends, and describe how you would verify this relation for a given length of wire. (*N.*)

**14.** A lamp is rated at 230 volts–60 watts. What does this rating indicate ? What current does the lamp take and what is its resistance ? Find the cost of using this lamp for 500 hours if the cost of electrical energy is $3d.$ per unit. (*L.*)

**15.** Define the *watt* and the *joule*. If an electric kettle contains a 1000 watt heating unit, what current does it take from 230-volt mains ? How long will the kettle take to raise 1000 gm. of water at 15° C. to the boiling point, if 90 per cent of the heat produced is used in raising the temperature of the water? How much would this cost if the charge is 1*d*. for a kilowatt-hour? (*C.*)

**16.** What is meant by *power* in an electric circuit? How does the power depend on the potential difference between the ends of a wire of fixed resistance? An electric immersion heater is placed in a calorimeter containing 192 gm. of water and is connected to 210-volt mains. The temperature rises 10° C. in one minute. Find the resistance of the heater, assuming the water equivalent of calorimeter and heater to be 18 gm. If the mains voltage drops to 200, what is the new rise of temperature per minute? (*L.*)

**17.** Give an account of a method you would use to determine Joule's equivalent by a method of electrical heating. An aluminium kettle weighing 2 Kgm. holds 2000 c.c. of water and consumes electric power at the rate of 2 kilowatts. If 40 per cent of the heat supplied is wasted, find the time taken to bring the kettle of water to boiling point from an initial temperature of 20° C. (Specific heat of aluminium, 0·2; Joule's equivalent 4·2 joules per cal.) (*N.*)

**18.** Define the terms *ampere, volt, ohm, watt*. Assuming that the electrical energy costs 2*d*. per unit (kilowatt-hour), calculate the cost: (*a*) Of running for 2 hours a carpet-sweeper fitted with a ⅙ h.p. motor; (*b*) of raising a litre of water from room temperature (15° C.) to boiling point, assuming that the electric kettle has an efficiency of 80 per cent. (1 h.p. = 746 watts; *J* = 4·2 joules per calorie.) (*O. & C.*)

**19.** How is the heating effect of a current passing through a constant resistance related to the strength of the current? How would you demonstrate experimentally the correctness of your relationship? Two exactly similar heating resistances are to be used connected across a mains supply to heat some water. Is more heat obtained per minute if they are connected in series or if they are connected in parallel? What is the ratio of the rates of obtaining heat for the two modes of connexion? (*O. & C.*)

**20.** Describe an experiment to show that the rate of production of heat in a wire by a current is directly proportional to the square of the current.

An electric iron is rated at 350 watts when used on a 230-volt supply. Find: (*a*) the current flowing in the heating coil, (*b*) the heat developed in the iron per hour, (*c*) the cost of using the iron for 4 hr., given that the cost of a kilowatt-hour is ⅞ penny. (*J* = 4·2 joules per calorie.) (*N.*)

# CHAPTER XXXI

## ELECTROLYSIS

SO far we have discussed the flow of electricity through metals, or solid conductors. In 1834 FARADAY began to investigate the behaviour of liquid conductors, such as acids and salt solutions; the study of the flow of electricity through liquids is called **electrolysis**, and the liquids are called **electrolytes.**

Some liquids do not conduct an electric current; but when an acid, a base, or a salt is dissolved in water, the solution is usually a good conductor. The two materials which lead the current into and out of the liquid are known as the **electrodes.** The electrode, C, by which the conventional current leaves the solution is known as the **cathode**, while the current enters the solution by the electrode, A, known as the **anode** (Fig. 332). The whole arrangement containing the electrodes and electrolyte is known as a **voltameter.**

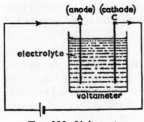

FIG. 332. Voltameter

**Typical examples of electrolysis.** There are many examples of electrolysis, but to understand the principles concerned we need only consider two typical cases.

*Electrolysis of copper sulphate solution.* (1) *With copper electrodes.* If an electric current is passed through a dilute solution of copper sulphate with copper electrodes, weighing shows that some copper is deposited on the *cathode* plate after a time, and that the anode plate has then lost an equal amount of copper (Fig. 333 (*a*) ). The cathode plate is thus heavier, and the anode plate is lighter, at the end of the experiment. On testing the density of the copper sulphate solution it will be found to be unaffected by the electrolysis which has taken place.

(2) *With platinum electrodes.* The materials of the electrodes play an important part in determining the products liberated in electrolysis. If an electric current is passed through dilute copper

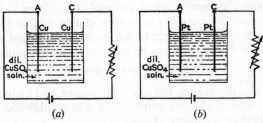

(a)                              (b)

FIG. 333. Copper voltameter

sulphate solution with platinum electrodes (Fig. 333 (*b*) ), copper is again deposited on the cathode; but oxygen gas is now liberated at the anode. Further, the density of the copper sulphate solution is different from what it was at the beginning of the experiment.

**Arrhenius' theory of electrolytic dissociation.** Until electrolysis was studied no one had thought that a beaker of dilute sulphuric acid, copper sulphate solution, or salt solution contained anything but molecules of these substances in solution mixed with molecules of water. In 1887 ARRHENIUS suggested that the molecules of electrolytes were "dissociated" into particles *each carrying a charge*, which he called **ions**, and all the experimental evidence available agreed with this theory, which was therefore accepted. A molecule of sulphuric acid ($H_2SO_4$) in solution dissociates into 2 hydrogen ions (made up of 2 atoms of hydrogen, H, each carrying a positive charge numerically equal to $e$, the charge on an electron), and a sulphate ion (made up of 1 atom of sulphur and 4 atoms of oxygen, $SO_4$, carrying a negative charge equal to double that on an electron). The total charge on the two ions is thus zero, and hence no electrical shocks are experienced on dipping one's finger into dilute sulphuric acid solution. A hydrogen ion is represented by $H^+$ since it consists of an atom of hydrogen with a positive charge, and a sulphate ion is represented by $SO_4^{--}$ since it carries a negative charge equal to the total positive charge on the two hydrogen ions. The dissociation of 1 molecule of sulphuric acid is thus represented by the equation

$$H_2SO_4 = 2H^+ + SO_4^{--}.$$

Similarly, a copper sulphate molecule dissociates in solution into a copper ion carrying a positive charge double that on a hydrogen ion, and a sulphate ion carrying an equal negative charge. The dissociation can be represented by the equation

$$CuSO_4 = Cu^{++} + SO_4^{--}.$$

Water also dissociates, but to a slight degree, and hydrogen and hydroxyl ions ($OH^-$) are obtained in dilute sulphuric acid or copper sulphate solutions. Thus—

$$H_2O = H^+ + OH^-.$$

The ions in an electrolyte are moving about haphazardly in the solution and making frequent collisions; on the whole, there is no general movement of the ions in any one direction. When, however, a battery is connected to the electrodes, a potential difference is set up across the electrolyte, and, as we have seen with electrons in the case of metals, the ions with a negative charge begin to drift across the liquid towards the anode, $A$ (the higher potential plate). At the same time, the ions with a positive charge drift in the opposite direction towards the cathode, $C$. If the battery is disconnected the drift of electricity between the plates ceases, and the ions once again have a completely random motion.

Electrons are the particles which carry the electric current through a metal, and are a very small fraction of the total weight of the atoms. In contrast, the ions which carry the current through a liquid conductor are particles massive by comparison with electrons because they are made up of the atoms of the electrolyte.

**Explanation of electrolysis of copper sulphate solution,** We are now in a position to explain the electrolysis of copper sulphate solution.

(1) *With copper electrodes.* The copper sulphate solution contains positive copper ions ($Cu^{++}$), positive hydrogen ions ($H^+$)

negative hydroxyl ions ($OH^-$), and negative sulphate ions ($SO_4^{--}$). When the battery is connected to the electrodes, the positive ions drift towards the cathode. *Now some ions are more easily discharged than others*; in particular, the copper ions are more easily discharged than the hydrogen ions, and hence copper is deposited at the cathode. The discharge can be represented by—

$$Cu^{++} + 2e = Cu.$$

While copper is being deposited at the cathode, the copper atoms of the anode electrode go into solution as ions. This is a

FIG. 334 (*a*). PURE COPPER PRODUCTION

The tank house of an African copper mine, showing the vats and a raised cathode plate

process which occurs more easily than the discharge of hydroxyl and sulphate ions, which move to the cathode while the current flows, and it can be represented by—

$$Cu - 2e = Cu^{++}$$

Consequently copper is lost from the anode, an equal amount is deposited on the cathode, and the density of the copper sulphate solution is unaltered.

(2) *With platinum electrodes.* When platinum electrodes are used instead of copper electrodes, copper is again deposited on the cathode, as explained above. At the anode, however, hydroxyl ions ($OH^-$) are discharged, a process which takes place more easily

than the discharge of the sulphate ions ($SO_4^{--}$), or the solution of platinum atoms from the anode as ions. Oxygen is formed at the anode from the discharge of the hydroxyl ions, a process which can be represented by—

$$4OH^- - 4e = 2H_2O + O_2.$$

Consequently, oxygen is obtained at the anode, copper is deposited on the cathode, and the density of the copper sulphate solution decreases.

FIG. 334 (*b*). AUTOMATIC SILVER-PLATING PLANT
showing forks as cathodes

It can now be seen that the nature of the materials of the electrodes, as well as the nature of the electrolyte, affect the final products obtained in electrolysis.

**Electrolysis of water.** If a battery is connected to platinum electrodes in a solution of dilute sulphuric acid, hydrogen and oxygen, in the ratio 2 : 1 by volume, are obtained at the cathode and anode respectively. Fig. 335.

The current in the solution is carried by the positive and negative ions present. The positive hydrogen ions ($H^+$) drift towards the cathode, where they are discharged—

$$2H^+ + 2e = H_2.$$

The negative hydroxyl ions (OH⁻) and sulphate ions ($SO_4^{--}$) drift towards the anode, where the hydroxyl ions are discharged and oxygen is formed—

$$4OH^- - 4e = 2H_2O + O_2.$$

It can be seen that the amount of sulphuric acid in solution

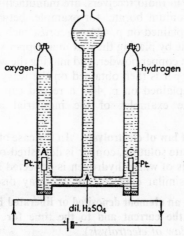

oxygen → ←hydrogen

dil. H₂SO₄

FIG. 335. Electrolysis of water

remains constant as the current flows, and the water in the solution is decomposed into hydrogen and oxygen.

In 1959 the *Hydrox* fuel cell was announced, invented by F.T. BACON. Basically, this cell creates electrical energy from the combination of hydrogen and oxygen in the presence of a catalyst, which is the reverse process to the electrolysis of water.

**Industrial applications of electrolysis.** In order to present a bright appearance and to prevent rusting, steel articles are often *chromium-plated*. The handlebars of a bicycle, for example, are made the cathode of a voltameter containing chromic acid, and a current is passed through the solution until a suitable deposit of chromium is obtained on the frame.

*Silver plating* of spoons is effected by making them the cathode of a voltameter containing potassium silver cyanide, and passing a current through the solution. (See Fig. 334 (*b*). )

Certain metals, such as *sodium* and *aluminium*, are extracted by electrolysis. In Scotland, for example, the British Aluminium Company obtains pure aluminium by electrolysis. Pure *copper*, which must be used in the manufacture of cables, is produced commercially by electrolysis (Fig. 334 (*a*)). *Electrolytic condensers*, used extensively in radio receivers, are manufactured by the electrolysis of ammonium borate, for example, between aluminium electrodes, as explained on p. 444. *Records*, such as gramophone records, are made by placing the wax in a copper voltameter after sprinkling it with copper powder, and making the wax the cathode. A "shell" of copper is then obtained on the wavy grooves of the wax, and, as explained on p. 403, a record can then be made. These are some examples of the industrial applications of electrolysis.

**Faraday's first law of electrolysis.** In the case of the electroysis of copper sulphate solution, copper is deposited on the cathode; in the electrolysis of water, hydrogen is collected at the cathode. By performing similar experiments, Faraday discovered that—

**The weight of an element deposited or liberated in electrolysis is proportional to the current and to the time for which it flows.** (*Faraday's first law of electrolysis*).

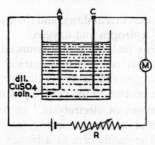

FIG. 336. Faraday's first law

Fig. 336 illustrates how the law can be verified. A circuit is made up of a copper sulphate solution with copper electrodes, known as a copper voltameter, a battery, a rheostat *R*, and an ammeter *M* for measuring the current. A steady current is then passed into the voltameter for 10, 20, 30, 40, and 50 minutes, and at the end of the respective times the cathode plate $C$ is taken out, dried with a spirit flame, and then weighed, so as to determine the weight of copper deposited. It will then be found that the weights are in the ratio of $1:2:3:4:5$; showing that $w \propto t$ for a given current, where $w$ is the weight deposited and $t$ is the time.

To verify that the weight deposited is proportional to the current for a given time, different steady currents are passed into the voltameter each time for 30 minutes, for example; the current is altered by using the rheostat each time. If currents of 0·5, 1·0, 1·5, 2·0 amps. are used, the respective weights of copper deposited will be found to be in the ratio $1:2:3:4$. Thus $w \propto I$ for a given time, where $w$ is the weight deposited and $I$ is the current.

**Electrochemical equivalent.** From Faraday's experiments, it became clear that the same weight of copper, for example, was always deposited by the same current in the same time, no matter which compound of copper was electrolysed, and the name *electrochemical equivalent* was given to the *weight of an element deposited when 1 ampere flowed for 1 second*. Since 1 coulomb of electricity passes when a current of 1 ampere flows for 1 second (see p. 451), the electrochemical equivalent (e.c.e.) of copper is stated as 0·00033 gm. per coulomb; the e.c.e. of silver is 0·001118 gm. per coulomb; the e.c.e. of hydrogen is 0·0000105 gm. per coulomb.

It follows from the e.c.e. value that 3 amperes flowing for 20 seconds deposit a weight of copper equal to $0·00033 \times 3 \times 20$ gm. The weight $w$ of an element deposited by a current $I$ in a time $t$ is hence given by

$$w = zIt,$$

where $z$ is the e.c.e. of the element. In this formula $w$ is in gm. when $z$ is in gm. per coulomb, $I$ is in amperes, and $t$ is in *seconds*.

The e.c.e., $z$, of an element is easily found by weighing the amount of it deposited by a known current in a known time, and the apparatus of Fig. 336 can be used for this purpose.

Suppose, in an experiment, that 0·591 gm. of copper is deposited by a current of 1 amp. in 30 minutes. Then $w = 0.591$, $I = 1$, $t = 30 \times 60$. Substituting in the formula $w = zIt$,

$$\therefore \quad 0.591 = z \times 1 \times 30 \times 60$$

$$\therefore \quad z = \frac{0.591}{1800} = 0.00033 \text{ gm. per coulomb.}$$

**Measurement of current.** Silver is an element which can be obtained in a high degree of purity, and many careful and elaborate experiments have been performed to find its electro-chemical equivalent. The result is 0.001118 gm. per coulomb, and this can be utilized for measuring a current in a circuit by including in it a silver voltameter, which contains a silver salt, e.g. a nitrate. The current causes silver to be deposited on the cathode, and as an illustration, suppose 2.173 gm. are deposited in 20 minutes.

Then $t = 20 \times 60$ sec., $w = 2.173$ gm., $z = 0.001118$ gm. per coulomb. Substituting in the formula $w = zIt$,

$$\therefore \quad 2.173 = 0.001118 \times I \times 1200$$

$$\therefore \quad I = \frac{2.173}{0.001118 \times 1200} = 1.62 \text{ amps.}$$

An ammeter in a laboratory can be checked by arranging it in series with a silver voltameter, and calculating the *true* current as shown above. If the ammeter reading is 1.5 amps. and the true current is 1.56 amps., then 0.06 amp. is the correction to be added to the ammeter reading when the pointer registers 1.5 amps. It is interesting to note that years ago scientists agreed internationally to define one ampere as that current which deposits a weight of 0.001118 gm. of silver in 1 second.

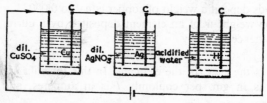

Fig. 337. Faraday's second law

**Faraday's second law of electrolysis.** When the same current is passed through electrolytes of copper sulphate solution, silver nitrate solution, and acidified water, observation shows that different weights of copper, silver, oxygen, and hydrogen are deposited at the end of a given time (Fig. 337). As Faraday showed, the respective weights have a ratio of $31\cdot5 : 108 : 8 : 1$, and this is also the ratio of the chemical equivalents of the elements.

*Faraday's second law of electrolysis states that—*

**When the same current passes through different electrolytes for the same time, the weight of the element deposited is proportional to its chemical equivalent.**

As an illustration, suppose in Fig. 337 that a weight of 3 gm. of copper is deposited at the end of a certain time when a current flows. The chemical equivalent of copper is $31\cdot5$, and the chemical equivalent of oxygen is 8. Since the current and the time is the same for the copper and water voltameters, the weight $w$ of oxygen deposited is given by

$$\frac{w}{3 \text{ gm.}} = \frac{8}{31\cdot5},$$

from Faraday's second law. Hence $w = \dfrac{8}{31\cdot5} \times 3 = 0\cdot76$ gm. Since the chemical equivalent of silver is 108, the weight, $w$, of silver deposited in the same time is given by $\dfrac{w}{3 \text{ gm.}} = \dfrac{108}{31\cdot5}$, from which $w = 10\cdot3$ gm.; and the weight, $w$, of hydrogen deposited in the same time is given by $\dfrac{w}{3 \text{ gm.}} = \dfrac{1}{31\cdot5}$, from which $w = 0\cdot095$ gm.

**Explanation of Faraday's laws.** At the beginning of the chapter it was pointed out that the carriers of the electric current through an electrolyte were the ions, the charged particles, inside it. Now these ions are the actual material of the electrolyte; for example, the copper ions and the sulphate ions together form the copper sulphate in a solution of this compound. Consequently the current through the electrolyte results in the transport of the material in it to the electrodes; and, the greater the current flowing, the greater is the deposit on the cathode. Again, the longer the same current flows, the greater is the deposit on the cathode, so that Faraday's first law is completely explained. Furthermore, the weights of different elements reaching the electrodes in voltameters in series (Fig. 337) are proportional to the weights of the ions in the particular electrolyte, so that the weights are proportional to the combining or *equivalent* weights of the elements in their compounds. Thus Faraday's second law can be explained by the ionic theory.

17

---
**Summary**
---

1. Electrolytes are liquids which conduct an electric current; they contain charged particles known as "ions" which carry the current through the liquid.

2. When copper sulphate solution is electrolysed with copper electrodes, copper is deposited at the cathode, an equal weight of copper is lost from the anode, and the density of the solution is unaltered; with platinum electrodes, copper is deposited at the cathode, oxygen is liberated at the anode, and the density of the solution decreases.

3. **Faraday's laws of electrolysis state:** (a) **The weight of an element deposited or liberated in electrolysis is proportional to the current and to the time.** (b) **When the same current passes through different electrolytes for the same time, the weight of the element deposited is proportional to its chemical equivalent.**

4. The "electrochemical equivalent" (z) of an element is the weight of it deposited or liberated in electrolysis by 1 ampere in 1 second. **The weight, $w$, of an element deposited in $t$ seconds by a current of $I$ amp. $= zIt$.**

## WORKED EXAMPLES

1. *State Faraday's laws of electrolysis. Explain the method used in either copper-plating or silver plating, and state the action which occurs. Give a labelled diagram of a suitable circuit, indicating the direction of the current and also the plate on which the metal is deposited. How long would it take to give a coating of copper, 0·01 cm. thick, to a metal article of surface area 27·0 sq. cm., using a current of 0·5 amp.? Take the density of copper to be 8·8 gm. per c.c. and its electro-chemical equivalent to be 0·00033 gm. per coulomb. (C.W.B.)*

The volume of copper deposited $= 0·01 \times 27·0 = 0·27$ c.c.

$$\therefore \quad \text{mass of copper} = \text{volume} \times \text{density} = 0·27 \times 8·8 \text{ gm.}$$

But
$$zIt = w \text{ (p. 491)}$$

$$\therefore \quad 0·00033 \times 0·5 \times t = 0·27 \times 8·8$$

$$\therefore \quad t = \frac{0·27 \times 8·8}{0·00033 \times 0·5} = 14,400 \text{ sec.} = 4 \text{ hours.}$$

2. *Describe and explain how you would use a copper voltameter to find the error at the 1-amp. mark of an ammeter which is known to be inaccurate.*

*A current of 2 amps. is passed through a copper voltameter. Given that the copper is deposited evenly on an electrode of area 66 sq. cm., find the thickness of the layer deposited after the current has been flowing for 30 min. (Electrochemical equivalent of copper = 0·00033 gm. per coulomb. Density of copper 9·0 gm. per c.c.) (N.)*

The mass, $w$, of copper deposited is given by

$$w = zIt$$
$$= 0·00033 \times 2 \times (30 \times 60) = 1·188 \text{ gm.}$$

Since volume $= \dfrac{\text{mass}}{\text{density}}$, the volume of copper deposited $= \dfrac{1·188}{9·0}$ c.c. $= 0·132$ c.c.

$$\therefore \quad \text{thickness} = \frac{\text{volume}}{\text{area}} = \frac{0·132}{66} \text{ cm.} = 0·002 \text{ cm.}$$

3. *State Faraday's laws of electrolysis. Describe and explain as fully as you can what is observed when an electric current is passed through a solution of common salt, using carbon electrodes. A Daniell cell gives a current of 0·2 amp. for 45 minutes. Calculate the change in weight of both electrodes. (E.C.E. of copper = 0·00033 gm. per coulomb. Equivalent weight of copper = 31·8.; of zinc =. 32·6). (N.)*

When a current flows, $H^+$ and $Na^+$ ions move to the cathode. $OH^-$ and $Cl^-$ ions move to the anode. Hydrogen gas is obtained at the cathode, chlorine gas at the anode and sodium hydroxide is formed in solution.

The weight, $w$, of copper deposited $= zIt = 0·00033 \times 0·2 \times 45 \times 60$
$$= 0·1782 \text{ gm.}$$

By Faraday's *second* law,

$$\frac{\text{wt. of zinc lost}}{\text{wt. of copper}} = \frac{\text{equivalent wt. of zinc}}{\text{equivalent wt. of copper}}$$

$$\therefore \quad \frac{\text{wt. of zinc}}{0·1782} = \frac{32·6}{31·8}$$

$$\therefore \quad \text{wt. of zinc} = \frac{32·6}{31·8} \times 0·1782 = 0·184 \text{ gm.}$$

## EXERCISES ON CHAPTER XXXI

**1.** What is meant by *electrolysis, electrolyte, cathode, anode*? Name three applications of electrolysis.

**2.** What are the particles which carry a current through (i) copper sulphate solution, (ii) sodium chloride solution, (iii) sulphuric acid solution, (iv) copper wire, (v) platinum wire, (vi) silver nitrate solution? (*Note.* Remember water is present in each solution.)

**3.** What is obtained when a current is passed through copper sulphate solution (i) with copper electrodes, (ii) with platinum electrodes ? Explain how the products are formed.

**4.** What is meant by the statement that *the electrochemical equivalent* of silver is 0·001118 *gm. per coulomb* ? Calculate the weight of silver deposited in $\frac{1}{2}$ hr. when a current of 2 amps. is passed through a silver solution.

**5.** 0·6 gm. of copper is deposited when a current of 1·2 amps. is passed through a copper solution for 25 mins. Calculate the e.c.e. of copper.

**6.** Describe how you would copper-plate a piece of nickel by electrolysis. A metal disc, 6 cm. in diameter and 1 mm. thick, is to be coated with a film of copper 0·01 mm. thick by electrolysis. What time will be required if a current of 2 amps. is employed ? (The electrochemical equivalent of copper is 0·000334 gm. per coulomb; the density of copper is 8·8 gm. per c.c.) (*L.*)

**7.** State briefly what is meant by an electrolyte. Underline the electrolytes in the following list: paraffin, common salt solution, mercury, molten copper, hydrochloric acid. (*N.*)

**8.** Describe and explain what occurs when an electric current is passed through the following aqueous solutions: (*a*) dilute sulphuric acid with platinum electrodes, (*b*) copper sulphate with platinum electrodes, (*c*) copper sulphate with copper electrodes.

What can be deduced about the masses of the elements liberated in the first solution if the same current passes through each solution for the same time and 0·26 gm. of copper is deposited from the last solution ? (Chemical equivalent of copper relative to hydrogen = 31·5; of oxygen = 8.) (*N.*)

**9.** Describe, in detail, how you would determine the value of the electro-chemical equivalent of copper. (*L.*)

**10.** State Faraday's laws of electrolysis. Explain the method used in *either* copper-plating *or* silver-plating, and state the action which occurs. Give a labelled diagram of a suitable circuit, indicating the direction of the current and also the plate on which the metal is deposited. How long would it take to give a coating of copper, 0·01 cm. thick, to a metal article of surface area 27·0 sq. cm., using a current of 0·5 amp. ? Take the density of copper to be 8·8 gm. per c.c. and its electrochemical equivalent to be 0·00033 gm. per coulomb. (*C.W.B.*)

**11.** One ampere flowing for one second in a silver voltameter liberates 0·001118 gm. of silver, the equivalent (or equivalent weight) of silver being 108. Calculate the mass of hydrogen liberated in a water volt-ameter when a current of 5·4 amps. flows for 20 minutes. (*N.*)

**12.** State the laws of electrolysis and explain the meaning of the terms: *cathode, anode, electro-chemical equivalent.* A steady current is passed through a copper voltameter in series with an ammeter. With the ammeter reading 0·4 ampere, the mass of copper deposited in half-an-hour is 0·25 gm. What is the error in the ammeter reading? [Electro-chemical equivalent of copper = 0·00033 gm. per coulomb.] (*L.*)

**13.** An ammeter, a battery, and a copper voltameter were connected in series. The ammeter recorded a steady current of 1 ampere. After 30 minutes, the weight of copper deposited on the cathode was found to be 0·588 gm. What was the error in the ammeter reading: (Electro-chemical equivalent of copper = 0·00033 gm. per coulomb.) (*C.*)

**14.** State Faraday's laws of electrolysis and explain what is meant by *electro-chemical equivalent.* Describe, with the aid of a diagram, a copper voltameter and explain how it works. A copper voltameter and a 20 ohm coil are connected in series and an electric current passed through them. Find the weight of copper deposited in the voltameter in 15 minutes when the potential difference across the terminals of the 20 ohm coil is maintained at 10 volts. (E.C.E. of copper = 0·00033 gm. per coulomb.) (*O. & C.*)

**15.** Define the terms *coulomb, ampere, kilowatt-hour.* The potential difference between the plates of a copper voltameter is 8 volts and 3·3 gm. of copper are deposited. Calculate (*a*) the quantity of electricity (in coulombs) that has passed through the voltameter, (*b*) the energy consumption in kilowatt-hours. (Electro-chemical equivalent of copper = 0·00033 gm. per coulomb.) (*O. & C.*)

**16.** State Faraday's laws of electrolysis and describe an experiment to illustrate *one* of them.

Calculate the volume of mercury liberated in 40 min. by a current of 2 amps. passing through a solution of mercury salt. (Electro-chemical equivalent of hydrogen = 0·00001044 gm. coulomb$^{-1}$, chemical equiva-lent of mercury = 200·6, density of mercury = 13·6 gm. cm.$^{-3}$.) (*L.*)

# CHAPTER XXXII

# PRIMARY CELLS AND ACCUMULATORS

MILLIONS of *batteries* are made annually for use in torches, bicycle lamps, and some radio receiver sets. *Accumulators* (p. 504) are used extensively as a source of electric current for lighting and for power in certain vehicles and radio receiver sets. These all depend upon the discovery of VOLTA (p. 448), who in the eighteenth century made the first electric battery. His "pile" consisted of a series of copper and zinc plates with cloths of brine

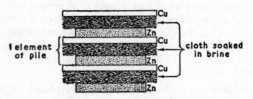

FIG. 338. Volta's pile

between alternate pairs of plates (Fig. 338), and he found that a current could be obtained by connecting the ends of the pile to a wire. The pile was shown to the Royal Society in London, where it excited considerable interest, and the efficiency of the **voltaic cell** was soon increased. Essentially, the cell is a device which converts chemical energy to electrical energy.

**The simple cell.** One of the earliest cells to be constructed, known as a *simple cell*, consisted of a copper and a zinc plate immersed in dilute sulphuric acid (Fig. 339). By means of a voltmeter, experiment shows that the copper plate or pole is at a higher potential than the zinc plate; scientists believe that an electrical action takes place at the surfaces of the two plates and the acid which results in this electrical p.d. To denote that copper is at a higher potential than the zinc, the former is known as the "positive pole" (+) and the latter as the "negative pole" (−) of the cell. The reader should note carefully that the + and − have no algebraic significance.

When the two poles are joined by a wire, electrons flow along it from the zinc to the copper plate (Fig. 339). Since the electric

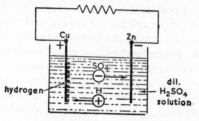

FIG. 339. The simple cell

current is continuous in the circuit the negative sulphate ions in the acid flow towards the zinc plate, and hence the positive hydrogen ions flow towards the copper plate. Thus hydrogen is given off continuously at the copper plate, while zinc sulphate is formed at the zinc plate and goes into solution. As the current flows, therefore, some zinc is used up, and the net chemical change can be represented by

$$Zn + H_2SO_4 = ZnSO_4 + H_2.$$

**The electro-motive force (e.m.f.) of a cell.** Since the copper and zinc plates are connected by dilute sulphuric acid, which is a conductor, it might be expected that the potentials of the copper and zinc would become equal. A voltmeter shows that there is a p.d., $V$, between the plates so long as there is acid between them, so that there must be some agency *inside* the cell which acts in opposition to this p.d.

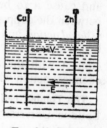

FIG. 340. E.m.f.

This agency is termed the *electro-motive force (e.m.f.)* of the cell, and *it is equal numerically to the p.d. between the plates when no wire is connected to them.*

The e.m.f. of a cell drives the current round any circuit connected to the terminals and also through the cell itself, and its value in volts depends only on the nature of the two metals of the cell and the liquid used. It does not depend on the size of the cell. The e.m.f. in Fig. 340 is represented by $E$, and its direction is the direction of movement of *positive* electricity.

**Polarization. Local action.** After a short time the current in a wire joined to a simple cell almost ceases, and experiments show that this is primarily due to the *hydrogen* produced at the copper plate (Fig. 339). Hydrogen and zinc are dissimilar elements, and they produce an e.m.f. in the acid which is *opposite* to that set up between the copper and zinc plates, with the result that the net e.m.f. inside the cell soon becomes very small. This phenomenon is termed *polarization*. As soon as its harmful effect was discovered, efforts were made to get rid of the hydrogen by continually "scrubbing" the copper plate where it was produced, but better methods of eliminating it were soon forthcoming, as we shall see.

Besides the effect of polarization, the current maintained by a simple cell drops to a low value because the layer of hydrogen gas on the copper plate acts as an insulator, and causes a fairly large increase in the resistance between the plates of the cell. Another disadvantage of the cell arises from the impurities, such as iron, present in commercial zinc. These form tiny cells with the zinc and cause it to be used up even when no circuit is completed between the copper and zinc plates. This phenomenon is known as *local action*, and is eliminated by rubbing the zinc with mercury, which covers the impurities and prevents them from making contact with the zinc and acid. With pure zinc, there is no chemical action when no circuit is completed between the plates.

**The Daniell cell.** JOHN DANIELL, a professor of chemistry at London University, designed a cell in which the hydrogen produced at the positive pole was eliminated by chemical means. The *Daniell cell* consists of a copper plate in the form of a cylindrical vessel $M$, and a zinc rod in a porous pot $P$ containing dilute sulphuric acid. The pot $P$ is placed inside copper sulphate solution contained in the vessel $M$ (Fig. 341).

When the copper and zinc are connected by a wire, the positive hydrogen ions of the sulphuric acid drift towards the copper vessel (Fig. 339). However, copper sulphate solution is present outside $P$, and an action takes place which results in *copper* being deposited on $M$. The chemical action can be represented by

$$H_2 + CuSO_4 = H_2SO_4 + Cu.$$

Copper, and not hydrogen, is thus deposited on the copper vessel, and hence polarization is overcome. The copper sulphate solution is known as the **depolarizer** in the cell, and the strength of the solution is maintained by crystals of copper sulphate placed on a gauze just above the liquid.

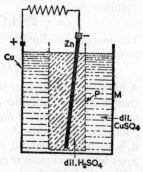

FIG. 341. Daniell cell

The negative sulphate ions of the sulphuric acid drift towards the zinc pole (see Fig. 341), and zinc sulphate is formed. Some zinc is therefore used up when the cell is used.

The e.m.f. of the Daniell cell is originally about 1·1 volts, and the cell can be used in the laboratory when a small current is required for a long period. The resistance of the liquids between the plates is called the **internal resistance** of the cell, and depends on the amount and nature of the chemicals used, as well as on the length of time the cell has been in use. The internal resistance of a Daniell cell is initially several ohms.

The Daniell cell is not used extensively, mainly because the sulphuric acid and copper sulphate solution have a tendency to mix through the porous pot, so that the cell cannot be left set up for use.

**The Leclanché cell. Wet type.** The wet *Leclanché cell* consists of (i) a zinc negative pole, $B$, (ii) a carbon positive pole, $C$, (iii) a depolarizer of manganese dioxide ($MnO_2$) powder (mixed with powdered carbon, which is a good conductor, to lower the internal resistance of the cell), (iv) a solution of sal ammoniac, or ammonium chloride ($NH_4Cl$). The zinc rod is inside the sal ammoniac solution, which is contained in a glass vessel $G$, and the depolarizer is packed round the carbon rod in a porous pot $P$ (Fig. 342).

17*

When the carbon and zinc are joined by a wire, electrons flow in the wire from the zinc to the carbon. *Inside* the cell, therefore, the negative chlorine ions of the ammonium chloride solution move towards the zinc, and form zinc chloride which goes into solution. The positive ammonium ($NH_4^+$) ions drift towards the carbon pole through the porous pot, and hydrogen is formed at this pole. The net chemical action can be represented by the equation—

$$Zn + 2NH_4Cl = ZnCl_2 + 2NH_3 + H_2.$$

The manganese dioxide then attacks the hydrogen, oxidizing it to water—

$$H_2 + 2MnO_2 = H_2O + Mn_2O_3.$$

A harmless substance is thus produced at the carbon pole by the chemical action, but *after a short time the hydrogen is produced too fast for the manganese dioxide to cope with it*, and the cell then polarizes. This is the defect of the Leclanché cell. Nevertheless, it is used in household bell circuits, for example, and in other cases where a current is required only intermittently.

The e.m.f. of a Leclanché cell is originally about 1·5 volts, and the internal resistance is originally of the order of 10 ohms or more.

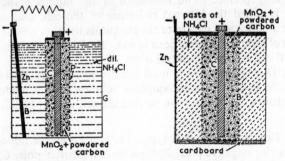

FIG. 342. Leclanché cell      FIG. 343. Dry cell

**The dry (Leclanché) cell.** The Leclanché cell described above is inconvenient to carry about owing to the solution of ammonium chloride used. This disadvantage is overcome by making a paste of ammonium chloride, flour, and gum, and Fig. 343 illustrates the comparatively "dry" cell manufactured commercially for use in torches and as "high tension" and "grid bias" batteries. In a high tension battery of 120 volts there are 80 cells connected together to assist each other, i.e. in *series* (p. 511), each having an e.m.f. of about 1·5 volts.

The "dry" cell has a zinc container with the ammonium chloride paste inside it. The carbon rod and manganese dioxide with

FIG. 344. BATTERY ROOM AT THE BIRMINGHAM TOLL EXCHANGE

powdered carbon are contained in a muslin bag, *B*, separated from the zinc by cardboard (Fig. 343). The cell has a lower internal resistance than the "wet" type owing to the closer spacing of the electrodes, but the e.m.f. is exactly the same as the "wet" cell as this depends only on the nature of the chemicals used (p. 500). When the battery is used in a torch, the positive (carbon) pole is in electrical contact with the lower end of the bulb, to which one end of the filament is connected. The other end of the filament is connected to the metal casing of the bulb. The negative (zinc) pole of the battery is in contact with the bottom of the metal torch case, and when the switch is pressed, contact is made between the case and the metal casing of the bulb, thus causing the filament to light up. The depolarizing action in the dry cell is better than in the "wet" cell; a cycle-lamp, for example, will burn for two hours with a dry battery supplying the current.

**Accumulators.** The Leclanché and Daniell cells are known as primary cells; when current is drawn from them, their chemicals are consumed, and eventually they have to be thrown away. Cells which can be re-charged by passing a current through them from an outside supply are called *secondary cells*, or *accumulators*. These cells are convenient because they can be re-charged many times; they have also the advantages of low internal resistance and negligible polarization. Thus they can be used to supply a large current for a long time. Accumulators are used for providing the electric lighting on ships, on buses, and in motor-cars, for driving the motors on light trolleys used at railway stations, and for providing the steady current in telephone cables.

**The lead-acid accumulator.** There are a number of different types of accumulators, but the most common one manufactured is the *lead-acid* type, which has a positive pole of lead peroxide, a negative pole of lead, and a solution of dilute sulphuric acid (Fig. 345 (a) ). The chemical materials are on grid plates connected respectively to a red (positive) and a black (negative) terminal at the top of a glass container.

Initially, the e.m.f. of a lead-acid accumulator is about 2 volts and the specific gravity of the acid is about 1·25. When a circuit is connected between the terminals, the accumulator is said to be **discharging**, and electrons flow from the negative to the positive pole in the circuit. Inside the accumulator, therefore, negative sulphate ions of the acid move towards the lead and the positive hydrogen ions drift towards the lead peroxide, and a chemical action, represented below, results in a little lead sulphate being formed at *both* plates.

Positive pole: $PbO_2 + H_2 + H_2SO_4 = PbSO_4 + 2H_2O$

Negative pole: $Pb + SO_4 = PbSO_4$

Some of the sulphuric acid is thus used up, and observation shows that its specific gravity diminishes while a current flows. At first the e.m.f. remains almost constant at 2 volts, but as discharge continues it begins to fall. When the e.m.f. has dropped to 1·8 volts, the cell should be re-charged; the specific gravity at this point is about 1·17. If the cell is not re-charged at this point, the lead sulphate on the plates becomes hard, and the re-charging process, which we are about to describe, becomes difficult if not impossible. The specific gravity of the acid is a sensitive indication of the condition of an accumulator, and this is measured by a hydrometer (p. 114).

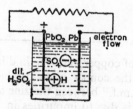

FIG. 345 (*a*). Discharging of accumulator

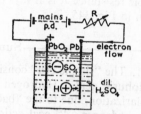

FIG. 345 (*b*). Charging of accumulator

**Charging an accumulator.** To restore the materials of a discharged accumulator to their original condition, a current is passed through the accumulator in the *opposite* direction to that obtained by using it. This is usually done by connecting the mains to the accumulator. Fig. 345 (*b*) illustrates the *charging* process, as it is called, and the electron flow in the wires is now opposite to that in Fig. 345 (*a*). The negative sulphate ions in the acid therefore drift towards the lead peroxide pole, while the positive hydrogen ions drift towards the lead plate. A chemical action takes place at the plates which results respectively in the production of lead peroxide and lead from the lead sulphate on the plates.

At positive pole: $PbSO_4 + SO_4 + 2H_2O = PbO_2 + 2H_2SO_4$

At negative pole: $PbSO_4 + 2H = Pb + H_2SO_4$

At the same time it can be seen that the specific gravity of the acid increases, and the charging should be stopped when the specific gravity reaches the value of about 1·25, which is again determined by using a hydrometer.

**The internal resistance** of an accumulator is very low, e.g. $\frac{1}{100}$ ohm, and if by accident the terminals are both touched by a copper wire, an exceptionally large current flows through the accumulator. Thus, if the e.m.f. is 2 volts the current is given roughly by $2/\frac{1}{100}$, or 200 amps., from Ohm's law; a large amount of heat is generated inside the cell, which results in the plates becoming buckled. On account of the low internal resistance a suitable resistance $R$ is always provided in the charging circuit, as shown in Fig. 345 (*b*).

──────Summary──────

1. The "simple cell" consists of copper and zinc in dilute sulphuric acid. Hydrogen at the copper plate produces polarization, due to the "back e.m.f." between the zinc and the hydrogen. "Local action" is due to impurities in the zinc, and is overcome by rubbing the zinc with mercury.

2. The Daniell cell has a positive copper pole, a negative zinc pole inside a porous pot containing dilute sulphuric acid, and a depolarizer of copper sulphate solution. The cell has an e.m.f. of 1·1 volts and an internal resistance of several ohms when first made.

3. The Leclanché cell has a positive carbon rod surrounded by manganese dioxide and powdered carbon in a porous pot, and a negative zinc rod in sal ammoniac solution. The "dry" cell is a Leclanché cell containing a paste of sal ammoniac instead of solution.

4. The lead-acid accumulator has a positive lead peroxide plate, a negative lead plate, and dilute sulphuric acid of specific gravity about 1·25 when freshly made. When discharged the specific gravity decreases to about 1·17, and a little lead sulphate forms at both plates. The accumulator is charged by passing a current through it from the positive to the negative pole, with a resistance in series.

## EXERCISES ON CHAPTER XXXII

**1.** Name four uses of batteries in everyday life.

**2.** Describe the disadvantages of a *simple cell*. What happens in the cell when the terminals are connected by a wire?

**3.** Describe a Leclanché cell, and draw a diagram of one. What are its advantages and disadvantages?

**4.** A small current is required for an hour. Describe the cell you would use, and explain its action.

**5.** What is the *internal resistance* of a cell? State what you know about the internal resistances of a Leclanché cell and an accumulator. Draw a labelled diagram of an accumulator.

**6.** Describe a *dry cell*, and explain its advantages.

**7.** Explain the precautions to be observed in maintaining an accumulator in good condition.

**8.** Describe the action of a simple voltaic cell. State and explain the defects of this type of cell and show what steps have been taken to overcome these defects in *either* the Leclanché *or* the Daniell cell. How far have they been successful? (*L.*)

**9.** What are the chief merits and demerits of any two cells in common use in a school laboratory? How may the e.m.f.s of the cells be compared, no voltmeter being available? (*L.*) [See p. 512 for second part.]

**10.** Describe the contents of an *accumulator*. What happens to the acid when it is charged and discharged? Draw a circuit showing how you would charge two accumulators.

**11.** Draw a labelled diagram to show the structure of a dry cell such as is used in electric torches. (*N.*)

**12.** Describe, by drawing a labelled diagram, the detailed construction of one form of voltaic cell. [Chemical actions are *not* required.] State two disadvantages which it has compared with a lead accumulator. A battery consisting of 12 accumulators is to be charged from 200 volt D.C. mains. Draw a diagram of the circuit required. If, during charge, each cell has an E.M.F. of 2·5 volts, and the charging current is specified as 3 amperes, what extra resistance will be needed, that of the battery itself being 1 ohm? If electrical energy costs 1*d.* per unit, what will it cost to charge the battery for ten hours? (*C.*)

**13.** Explain the terms: *electrolyte, local action.* Describe, with the aid of a diagram, the construction of a Daniell cell, and give an account of the chemical changes that take place when the cell is in use. State the main factors which decide the magnitude of the internal resistance of a voltaic cell. (*O. & C.*)

**14.** What is meant by *local action* and *polarization* in connection with a simple cell? How are they minimized in the Leclanché cell? Draw a diagram of this cell in its dry form. (*L.*)

**15.** What are the defects of the simple cell? Describe the disadvantages caused by each defect and explain how the disadvantages are prevented in *one* type of practical cell.

A wire of resistance 4·0 ohms is joined to the terminals of a battery formed by connecting 2 cells, each of e.m.f. 1·5 volts and internal resistance 1·0 ohm, (*a*) in series, (*b*) in parallel. Find the strength of the current flowing in the wire in each case. (*N.*)

**16.** (*a*) State the defects of the simple voltaic cell, and describe fully how these defects are minimized in a Daniell cell.

(*b*) Describe, with the aid of a diagram, *either* a soft-iron ammeter *or* a hot-wire ammeter. Explain the action of the instrument you choose, and explain why it is able to measure alternating current. (*L.*)

# CHAPTER XXXIII

## E.M.F.: THE POTENTIOMETER
## MEASUREMENT OF RESISTANCE

THE electro-motive force (e.m.f.) of a cell is measured by the p.d. at its terminals when a circuit is not connected to it (p.499). The cell, which is not now maintaining a current, is said to be on **open circuit**, and a voltmeter $D$ connected to the terminals $a$, $b$ gives a very accurate reading of the e.m.f. (Fig. 346 ($a$) ). It should be carefully noted that a moving-coil voltmeter requires a current in it to make it function (p. 587), but if its resistance is very high

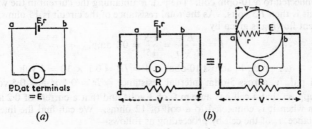

FIG. 346. E.m.f. and p.d.

the current is extremely small, and the cell can be considered to be almost on "open circuit."

When a circuit of resistance $R$ is connected to the terminals, $a$, $b$, of the cell, a **closed circuit** is formed. A current $I$ then flows outside and inside the cell, i.e. through its internal resistance $r$ as well as through the external resistance $R$. *The p.d. which maintains the current in the whole circuit is the e.m.f.*, $E$; just as, to use an analogy, the pressure in a pump must be sufficient to drive the water in a complete circuit through itself as well as through the external water circuit. Now the whole circuit has a resistance of $(R + r)$, since the internal resistance is in series with the external resistance; this can be seen from Fig. 346 (*b*), which shows the e.m.f. and internal resistance separated diagrammatically inside the cell. Thus, from Ohm's law,

$$\text{current} = \frac{\text{p.d.}}{\text{resistance}}$$

$$\therefore \quad I = \frac{E}{R + r} \quad . \quad . \quad . \quad . \quad (1)$$

From (1), it follows that $E = I(R + r) = IR + Ir$. Now from Ohm's law, $IR = V$, the p.d. across $R$; and $Ir = v$, the p.d. across $r$. Thus $E = V + v$, which shows that the e.m.f. $E$ of a cell is always greater than the p.d. $V$ across the external resistance, unless the internal resistance is zero. In the lead-acid accumulator (see p. 506), the p.d. $v$ across the internal resistance is very small, and the p.d. $V$ is then practically equal to the e.m.f., $E$.

Suppose a cell has an e.m.f. of 1·5 volts and internal resistance 5 ohms, and is connected to a 10-ohm coil. The p.d. maintaining the current in the whole circuit is then 1·5 volts. As the total resistance of the circuit is 15 ohms, the current flowing is given by

$$I = \frac{E}{R + r} = \frac{1·5}{15} = 0·1 \text{ amp.}$$

$$\therefore \quad \text{p.d., } V, \text{ across 10-ohm coil} = IR = 0·1 \times 10 = 1 \text{ volt}$$

and p.d., $v$, across 5-ohm internal resistance $= Ir = 0·1 \times 5 = 0·5$ volt.

Suppose that a cell has an e.m.f. of 3 volts and that a current of 0·2 amp. flows when it is connected to a coil of 12 ohms. We can find the internal resistance, $r$, of the cell by proceeding as follows—

The p.d., $V$, across 12-ohm coil $= IR = 0·2 \times 12 = 2·4$ volts

$$\therefore \quad \text{p.d., } v, \text{ across } r = 3 \text{ volts} - 2·4 \text{ volts} = 0·6 \text{ volt}$$

$$\therefore \quad r = \frac{v}{I} = \frac{0·6}{0·2} = 3 \text{ ohms,}$$

as the current in the internal resistance is the same as that in the external resistance of 12 ohms.

**Terminal potential difference.** Since $V$ is the p.d. across the external resistance $R$, it follows that the current in the circuit can be calculated from the formula

$$I = \frac{V}{R},$$

when $V$ and $R$ are known. Now if a voltmeter is left connected to the terminals when an external resistance is joined to the cell, the reading on the voltmeter drops from the e.m.f. value to a reading *which is the p.d., V, across the external resistance.* Thus, in Fig. 346 (*b*), the voltmeter *D* reads the p.d. across *R*. The same reading would be obtained if the voltmeter were disconnected from *a*, *b* and attached directly across *R* at *c*, *d*. It has become common to refer to the p.d. *V* as the *terminal p.d.* when a circuit is connected to a cell, and it should be kept in mind that it is the p.d. across the resistance *R* external to the cell and is given by $V = IR$.

**Arrangement of cells.** *Series arrangement.* As was mentioned on p. 449, the current in a circuit can be increased by increasing the number of cells. Fig. 347 (*a*) illustrates three identical cells in series; if they are arranged to assist one another, their total e.m.f. is 3*E*, where *E* is the e.m.f. of each cell. The total internal

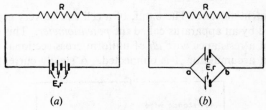

(*a*)                                    (*b*)

Fig. 347. Series and parallel cells

resistances of the cells is 3*r*, where *r* is the internal resistance of each cell. From Ohm's law, it follows that the current in the circuit is given by $I = \dfrac{3E}{R + 3r}$, where *R* is the resistance connected to the cells.

*Parallel arrangement.* In charging accumulators, it is sometimes useful to arrange a number of them in parallel. Fig. 347 (*b*) illustrates a resistance *R* connected to two *identical* cells in parallel. Now the e.m.f. of a cell depends only on the nature of the chemicals used, and is independent of its size. Since the two positive poles are connected to *a*, and the two negative poles to *b*, we can imagine that a cell is obtained between *a*, *b* whose size is the sum of the sizes of the two cells. Consequently, from above, the total e.m.f. of the two identical cells is *E*, the e.m.f. of one of them.

This value, $E$, is the total e.m.f. if any number of similar cells are connected between $a$, $b$.

The internal resistances, $r$, of the cells are in parallel across $a$, $b$. The total internal resistance, $S$, is thus given by $\frac{1}{S} = \frac{1}{r} + \frac{1}{r}$, from the usual formula. Hence $\frac{1}{S} = \frac{2}{r}$, from which $S = \frac{r}{2}$. By similar reasoning, the total internal resistance across $a$, $b$ is $\frac{r}{5}$ if five similar cells are used. Thus, although the total e.m.f. is $E$ in Fig. 347 (b), the total internal resistance is $\frac{r}{2}$, and hence the current flowing in $R$ is given by $I = \dfrac{E}{R + \dfrac{r}{2}}$ (see also example 1, p. 521).

**The potentiometer.** The e.m.f. of a cell can be very accurately measured by an apparatus called the *potentiometer*. This consists simply of a resistance wire $ab$ of uniform cross-sectional area, to which an accumulator $A$ is connected. A steady current $I$ then

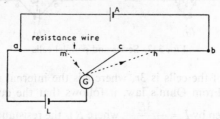

FIG. 348. Potentiometer principle

flows along the wire, and since $V = IR$, the p.d. across a portion of the wire is proportional to the resistance $R$ of that portion. Now $R$ is proportional to the length $l$ of the wire considered; *hence the p.d. across the length of the wire is proportional to l.* Thus, if the p.d. across the length $am$ of the wire is 0·8 volt, the p.d. across $an$ is 1·6 volts if $an$ is twice the length of $am$.

If the e.m.f. of a Leclanché cell, $L$, is required, for example, its positive pole is joined to the potentiometer wire at the same terminal $a$ as the positive pole of the accumulator (Fig. 348). The negative pole of $L$ is joined through a sensitive galvanometer $G$ to a "tapping key," which can be placed on the potentiometer wire to make contact with any selected point on it. At one point $c$ between $m$ and $n$, *no current flows in $G$*. The p.d. between $a$, $c$ is now exactly equal to the e.m.f. of the cell, as the latter does not maintain a current in this special case (see p. 509), and the length of $ac$ is measured. Suppose it is 74·2 cm.

The Leclanché cell $L$ is now disconnected from $a$ and the galvanometer, and a cell whose e.m.f. is accurately known to one part in ten thousand, known as a **standard cell**, is connected in place of $L$. The experiment is then repeated, and the new point $x$ is obtained on the wire where no current flows in $G$. Suppose that the distance $ax$ is 51·2 cm.

Then, since the p.d. along the potentiometer wire is proportional to the length (see p. 512),

$$\frac{\text{e.m.f. of } L}{\text{e.m.f. of standard cell}} = \frac{74 \cdot 2}{51 \cdot 2}.$$

Thus, if the e.m.f. of the standard cell, obtained from tables, is 1·0186 volts,

$$\frac{\text{e.m.f. of } L}{1 \cdot 0186 \text{ volts}} = \frac{74 \cdot 2}{51 \cdot 2}.$$

$$\therefore \text{ e.m.f. of } L = \frac{74 \cdot 2}{51 \cdot 2} \times 1 \cdot 0186 = 1 \cdot 48 \text{ volts}$$

The p.d. between $a$ and a point $m$ to the left of $c$ is less than the p.d. between $a$ and $c$ (Fig. 348). Consequently the p.d. between $a$, $m$ is less than the e.m.f. of the cell $L$, and a current will therefore flow in the direction $LamL$ through the galvanometer $G$ when contact is made with $m$. The p.d. between $a$ and a point $n$ to the right of $c$ is *greater* than the e.m.f. of $L$, so that a current flows in the direction $aLna$ through $G$ when contact is made with $n$. It can thus be seen that opposite deflections occur in $G$ when contact is made with points on opposite sides of the balance-point, enabling the balance-point to be determined quickly.

**Advantages of a potentiometer.** A potentiometer is more accurate for measuring e.m.f. than a moving-coil voltmeter, as the cell maintains no current when the potentiometer readings are taken.

The moving-coil voltmeter, on the other hand, requires a small current in it to function (p. 587), so that the p.d. at the terminals of the cell is a little less than its e.m.f.; the voltmeter readings may also be in error owing to a faulty zero.

## RESISTANCE AND ITS MEASUREMENT

**The Wheatstone bridge circuit.** The resistance of a wire can be measured within a few per cent. of its value by using a voltmeter and an ammeter, as explained on p. 459. A much more accurate method is attributed to CHARLES WHEATSTONE, the first professor of Physics at King's College, London, who utilized a circuit now famous as the *Wheatstone bridge*.

The bridge is illustrated in Fig. 349, and consists of four resistances, $P$, $Q$, $R$, $S$, joined to form a complete circuit, with a battery

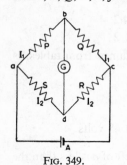

FIG. 349.
Wheatstone bridge

$A$ across one pair, $a$, $c$, of opposite junctions, and a sensitive galvanometer $G$ across the other junctions $b$, $d$. A current then usually flows in $G$, but if the resistances are suitably adjusted the current becomes zero, and a "balance" is said to be obtained. In this case, as we shall now show, there is a simple relation between $P$, $Q$, $R$, $S$.

Suppose $I_1$ is the current along $ab$ when the bridge is balanced. Then, since no current is diverted at $b$ through the galvanometer $G$, the current along $bc$ is also $I_1$. If the current along $ad$ is $I_2$, the current along $dc$ is also $I_2$ as no current is diverted at $d$. Further, since no current flows along $bd$, the potential at $b$ must be equal to that at $d$, and hence

$$\text{p.d. between } a, b = \text{p.d. between } a, d \quad . \quad . \quad (2)$$

and

$$\text{p.d. between } b, c = \text{p.d. between } d, c \quad . \quad . \quad (3)$$

Since potential difference = current × resistance (Ohm's law), it follows from (2) and (3) that

$$I_1 P = I_2 S \quad . \quad . \quad . \quad . \quad (4)$$

and

$$I_1 Q = I_2 R \quad . \quad . \quad . \quad . \quad (5)$$

Dividing (4) by (5) to eliminate $I_1$ and $I_2$, we obtain

$$\frac{P}{Q} = \frac{S}{R}.$$

This is the relation between $P$, $Q$, $R$, $S$ when the bridge is balanced.

**The metre bridge.** One of the earliest applications of the Wheatstone bridge circuit was the *metre bridge*. This consisted of a 100 cm. length of resistance wire soldered to terminals at $a$, $c$, and strips of brass, $am$, $np$, $qc$, are connected so that gaps are left between $m$, $n$, and $p$, $q$.

The unknown resistance, $P$, to be measured is connected in one gap, and a known resistance $Q$, of the same order as $P$, is connected in the other gap (Fig. 350). A battery $A$ is connected to $a$, $c$, and a galvanometer $G$ is connected to $b$, the junction between $P$ and $Q$,

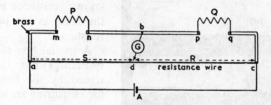

Fig. 350. Metre (slide-wire) bridge

and to a tapping key which can make contact with any point on the wire $ac$. When the key is at some point $d$, no current flows in $G$, and a balance is then obtained; the lengths $ad$, $dc$ are now measured.

Fig. 350 illustrates four resistances forming a complete circuit; they are $P$, $Q$, the resistance $R$ of the wire $dc$, and the resistance $S$ of the wire $da$. As $G$ is connected across two opposite junctions, and a battery $A$ is connected across the other pair of junctions, Fig. 350 is a Wheatstone bridge circuit. Hence, from the relation above,

$$\frac{P}{Q} = \frac{S}{R}.$$

But
$$\frac{S}{R} = \frac{\text{length of } ad}{\text{length of } cd}$$

$$\therefore \quad \frac{P}{Q} = \frac{\text{length of } ad}{\text{length of } cd}.$$

Suppose $Q = 10$ ohms, $ad = 45\cdot2$ cm., and $cd = 54\cdot8$ cm. The unknown resistance $P$ is then given by

$$\frac{P}{10 \text{ ohms}} = \frac{45\cdot2}{54\cdot8}$$

$$\therefore \quad P = \frac{45\cdot2}{54\cdot8} \times 10 = 8\cdot2 \text{ ohms.}$$

The metre bridge thus enables an unknown resistance to be measured. It is most accurate when the balance-point is near the middle of the wire, which occurs when the known resistance is nearly the same as the unknown resistance.

FIG. 351 (a). ELECTRIC STRAIN GAUGE
(lower one is in section)

A special box of known resistances has been designed by the Post Office for measuring an unknown resistance rapidly and accurately by the Wheatstone bridge principle. The interested reader should consult other books for the action of the *Post Office box.*

**The electric strain gauge.** By means of the Wheatstone bridge circuit, the changes of the resistance of a wire can be measured to a high degree of accuracy. In recent years a new method of examining the strains in aircraft surfaces and the surfaces of roads has been developed, which utilizes the change of resistance of a wire. The apparatus is called an *electric strain gauge*, and consists simply of a fine eureka or nichrome wire of several hundred ohms resistance, which is firmly attached by a special glue to the surface of the aeroplane under test, for example. As the surface is strained, the metal wire is extended or compressed, and the wire undergoes a change of resistance bearing a definite relationship to the change of its dimensions, which in turn is related to the strain of the surface. By employing various wires attached to different parts of

FIG. 351 (*b*). ELECTRIC STRAIN GAUGES IN USE FOR
TESTING AIRCRAFT

the surface, the strains at many parts of the surface can be measured simultaneously when the aircraft is tested (Fig. 351 (*a*), (*b*) ).

**Resistivity, or specific resistance.** Experiments with the metre bridge show that the resistance of a piece of wire of given material, such as copper, is proportional to the length of the wire and inversely proportional to its area of cross-section. If we know the resistance of 1 cm. length of it when the area of cross-section is 1 sq. cm., we can deduce the resistance of any length of the wire, whatever its cross-sectional area may be. The resistance of a specimen 1 cm. long and 1 sq. cm. cross-sectional area is called the *resistivity*, or *specific resistance*, of the material of the specimen.

As an illustration, the resistivity of 1 cm. length of manganin wire of 1 sq. cm. cross-section is 0·000044 ohm (approx.). If we have 120 cm. of manganin wire of 1 sq. cm. cross-section, its resistance = 0·000044 × 120 ohm, since the resistance is *proportional* to the length. Suppose the manganin wire has an area of cross-section of $\frac{1}{10}$ sq. cm., i.e. its area is 10 times as small as before. Now just as the resistance to the flow of water along a pipe increases when the pipe is narrower, so the resistance to the flow of electricity along a wire increases when its cross-sectional area decreases. The resistance of the manganin wire increases ten times as much since the resistance is inversely proportional to the area, and the resistance $R$ of 120 cm. length is hence given by

$$R = 0.000044 \times 120 \times 10 \text{ ohm} = 0.0528 \text{ ohm}.$$

We are now in a position to obtain a useful formula for the resistance of a wire used by manufacturers of standard coil resistances and of shunts. The resistance, $R$, is given by

$$R = \frac{\rho l}{a}, \qquad . \qquad . \qquad . \qquad . \qquad (6)$$

where $\rho$ is the resistivity in "ohms-cm." (the unit when ohms and cm. are used), $l$ is the length of wire in centimetres, and $a$ is the cross-sectional area in sq. cm. The formula in (6) shows that the resistance is proportional to the length $l$, and *inversely* proportional to the area $a$, i.e. the greater the value of $a$, the smaller is the resistance.

Tables of the resistivity of different materials of wire have been compiled for use by manufacturers, and reels of wire of different diameter (gauge) are made. Copper has the low resistivity of 1·72 microhms-cm. (1 microhm is 1 millionth of an ohm), and nichrome has a resistivity of about 100 microhms-cm. (or 0·0001 ohm-cm.). Suppose that a coil of 10 ohms is to be made from a reel of manganin wire of diameter 0·4 mm., the resistivity of manganin being 44 microhms-cm., or 0·000044 ohm-cm. Since the area, $a$, of the circular cross-section $= \pi r^2$, where $r$ is the radius of the circle, and $r = 0.02$ cm., it follows that $a = \pi \times 0.02^2$ sq. cm. The length, $l$, of wire required is given by $R = \dfrac{\rho l}{a}$, and since $R = 10$ ohms and $\rho = 0.000044$ ohm-cm., we have

$$10 = \frac{0.000044 \times l}{\pi \times 0.02^2} = \frac{0.000044 \times l}{3.14 \times 0.0004}$$

$$\therefore \quad l = \frac{10 \times 3.14 \times 0.0004}{0.000044} = 285.4 \text{ cm.}$$

The resistivity, $\rho$, of a wire can be measured by measuring the resistance $R$ of a length, $l$, of it on a metre bridge (p. 515), and measuring the diameter of the wire by a micrometer screw gauge. The area of cross-section, $a$, can then be calculated, as shown in the example above; and the resistivity, $\rho$, can be calculated from the formula $R = \dfrac{\rho l}{a}$, as it is the only unknown in the expression.

**Measurement of resistance by substitution. The ohm-meter.** A moving-coil instrument is capable of measuring *resistance*, as well

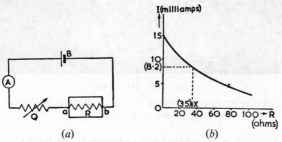

FIG. 352. Resistance by substitution

as current and potential difference, and the principle of the method is illustrated in Fig. 352 (a).

A battery is connected to a milliammeter $A$, a rheostat $Q$, and a box of known resistances, $R$. Before any readings on $A$ are taken $R$ is made zero, and the rheostat $Q$ is adjusted until the maximum deflection is obtained on $A$. Suppose the instrument reads 0–15 milliamps., for example; then the current $I$ is made 15 milliamps. when $R = 0$. The resistance $R$ in the box is then increased, and each time the smaller value of current and the corresponding value of $R$ are noted. In this way a graph can be drawn up of the current $I$ indicated on $A$ and the known resistance $R$, as shown in Fig. 352 (*b*).

Suppose an unknown resistance $S$ requires to be measured. The box $R$ is then disconnected from the terminals *a*, *b* in the circuit

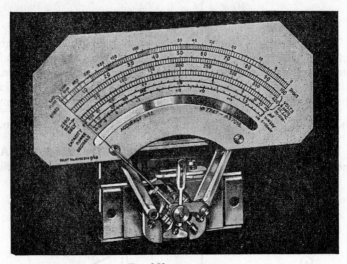

FIG. 353. AVO-METER

Showing moving-coil system and scale calibrated in amperes, volts and ohms

(Fig. 352 (*a*) ), and keeping the rheostat $Q$ unaltered, $S$ is connected to *a*, *b*. The reading on the instrument $A$ is noted; suppose it is 8·2 milliamps. From the graph (Fig. 352 (*b*) ), the resistance $OX$ in ohms is obtained corresponding to this current; and if $OX = 35$ ohms, for example, the unknown resistance $S$ is 35 ohms.

**The AVO-meter.** The above method of measuring resistance is used in commercial instruments such as the *AVO-meter*, which has scales calibrated in amperes, volts, and ohms, and is basically a milliammeter. (Fig. 353). A battery and a rheostat are incorporated at the back of the meter for measuring an unknown resistance, in accordance with Fig. 352 (*a*). Different ranges of current are measured by using a switch which incorporates various *shunt* resistances (p. 463), and different ranges of potential difference are measured by using another switch which incorporates various *series resistances* (p. 464).

───Summary───

1. The electromotive force (e.m.f.), $E$, of a cell is the p.d. of the cell on open circuit. **The current flowing, $I = E/(R + r)$** where $R$ is the external resistance and $r$ is the internal resistance.

2. When cells are in series, the total e.m.f. is the sum of the separate e.m.f.s. When *identical* cells are in parallel, the total e.m.f. is the e.m.f. of one cell.

3. The potentiometer is used for measuring accurately the e.m.f. of a cell. It consists of a uniform straight wire connected to an accumulator. The e.m.f. of the cell is "balanced" against the p.d. along the wire.

4. The Wheatstone bridge is used for measuring resistance accurately. It consists of four resistances, $P$, $Q$, $R$, $S$, forming a closed circuit, with a battery across one pair of junctions and a galvanometer across the other pair. At a balance, $P/Q = S/R$. The metre bridge is a form of Wheatstone bridge.

5. **The resistivity (specific resistance), $\rho$, of a material is the resistance of a cube of unit length of the material.** $R = \dfrac{\rho l}{a}$,

where $\rho$ is in ohm-cm., $l$ is the length of wire in centimetres, the area $a$ ($\pi r^2$) of cross-section is in sq. cm., and $R$ is in ohms.

## WORKED EXAMPLES

1. *Three cells, each of e.m.f. 2 volts and resistance 1 ohm are joined in parallel and the combination is connected to a coil of resistance 5 ohms. Find the current through the coil and the current through each cell.* (N.)

The combined e.m.f. of the cells is 2 volts, the e.m.f. of one cell. The combined internal resistance, $S$, is given, from the parallel formula, by

$$\frac{1}{S} = \frac{1}{1} + \frac{1}{1} + \frac{1}{1} = \frac{3}{1}, \text{ i.e. } S = \frac{1}{3} \text{ ohm.}$$

Thus the total resistance of the circuit $= 5 + \frac{1}{3} = 5\frac{1}{3}$ ohms, and hence the current *in the coil* is given by $I = \dfrac{\text{e.m.f.}}{\text{total resistance}} = \dfrac{2}{5\frac{1}{3}}$

$$\therefore \quad I = \frac{6}{16} = \frac{3}{8} \text{ amp.}$$

The current through each cell is one-third of the current in the coil. Hence the current $= \frac{1}{3} \times \frac{3}{8} = \frac{1}{8}$ amp.

2. *Draw a fully-labelled diagram to show the essential features of any one type of voltmeter. Explain the working of the instrument. A cell of e.m.f. 1 volt is connected in series with a resistance of 8 ohms and an ammeter, and the ammeter registers 0·1 amp. What will be the reading of the ammeter if an additional resistance of 6 ohms is joined in parallel with the 8 ohm resistance?* (N.)

The total resistance of the circuit, i.e. that of the ammeter together with the 8-ohm resistance and the internal resistance of the battery, is given by

$$\frac{\text{e.m.f.}}{\text{current}}, \text{ from Ohm's law for the complete circuit.}$$

$$\therefore \quad \text{total resistance} = \frac{1}{0·1} = 10 \text{ ohms}$$

$$\therefore \quad S = 10 - 8 = 2 \text{ ohms,}$$

where $S$ is the resistance of the circuit other than the 8 ohms.

When the 6 ohms is joined in parallel with the 8 ohms resistance, its combined resistance, $R$, is given by

$$\frac{1}{R} = \frac{1}{6} + \frac{1}{8} = \frac{4 + 3}{24} = \frac{7}{24}$$

$$\therefore \quad R = \frac{24}{7} = 3\frac{3}{7} \text{ ohms}$$

$$\therefore \quad \text{total resistance of circuit} = 3\frac{3}{7} + S = 3\frac{3}{7} + 2 = 5\frac{3}{7} \text{ ohms}$$

$$\therefore \quad \text{new current, } I = \frac{\text{e.m.f.}}{\text{total resistance}} = \frac{1}{5\frac{3}{7}} = \frac{7}{38} \text{ amps.}$$

$$\therefore \quad I = 0·18 \text{ amp.}$$

3. *Deduce from first principles the value of the effective resistance R of three resistances $r_1$, $r_2$, $r_3$, connected in parallel. Two accumulators, each of e.m.f. E volts and internal resistance 0·2 ohm, are joined in series, and then connected in a series circuit with a fixed resistance of 3 ohms, a combination of 1 ohm and 1·5 ohms in parallel, a resistance X, and an ammeter of negligible resistance. When the circuit is completed a voltmeter joined to the ends of X reads 2·0 volts, while the ammeter reads 0·5 ampere. Give a circuit diagram, and calculate the value of (a) the resistance X, (b) the e.m.f. E, (c) the reading on a voltmeter joined across the two accumulators.* (C.W.B.)

(*a*) The circuit arrangement is shown in Fig. 354. The current in $X$ is 0·5 amp., the reading on the ammeter $A$, and the p.d. across $X$ is 2·0 volts. The resistance of $X$ is thus given by

$$R = \frac{V}{I} = \frac{2·0}{0·5} = 4 \text{ ohms.}$$

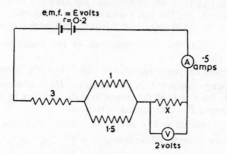

FIG. 354. Calculation

(*b*) The combined resistance of 1 ohm and 1·5 ohms in parallel is given by

$$\frac{1}{R} = \frac{1}{1} + \frac{1}{1·5} = \frac{1}{1} + \frac{2}{3} = \frac{5}{3}, \text{ i.e. } R = \frac{3}{5} \text{ ohm.}$$

Since the two cells are in series, their combined internal resistance $= 0·2 + 0·2 = 0·4 = \frac{2}{5}$ ohm. Thus the total resistance of the whole circuit is given by

$$R = \tfrac{2}{5} + 3 + \tfrac{3}{5} + X = 8 \text{ ohms, as } X = 4 \text{ ohms.}$$

But the current, $I = \dfrac{2E}{8}$,

since the total e.m.f. of the two cells $= E + E = 2E$. Also, $I = 0·5$ amps.

$$\therefore \quad 0·5 = \frac{2E}{8}$$
$$\therefore \quad 4 = 2E$$
$$\therefore \quad E = 2 \text{ volts}$$

(*c*) A voltmeter joined across the outer terminals of the two cells reads the *p.d. across the circuit joined to them.* The resistance of the circuit

$$= 3 + \tfrac{3}{5} + 4 = 7\tfrac{3}{5} \text{ ohms,}$$

and the current in it is 0·5 amps.

$$\therefore \quad V = IR = 0·5 \times 7\tfrac{3}{5} = 3·8 \text{ volts.}$$

4. *Describe, by drawing a labelled diagram, the detailed construction of one simple form of voltaic cell. (Chemical actions are* NOT *required.) State two disadvantages which it has compared with a lead accumulator. A battery consisting of 12 accumulators is to be charged from 200 volt D.C. mains. Draw a*

*diagram of the circuit required. If, during charge, each cell has an e.m.f. of* 2·5 *volts, and the charging current is specified as* 3 *amperes what extra resistance will be needed, that of the battery itself being* 1 *ohm? If electrical energy costs* 1*d. per unit, what will it cost to charge the battery for ten hours?* (C.)

(i) The charging circuit is that shown in Fig. 345 (*b*). The 12 accumulators have a total e.m.f. of $12 \times 2·5$, or 30 volts. But the e.m.f. of the accumulators acts *in opposition* to the 200 volt mains e.m.f. in the circuit. Hence the net e.m.f. = $200 - 30 = 170$ volts, and this is the p.d. which drives the current in the circuit. Thus the total resistance of the circuit $= \dfrac{\text{p.d.}}{\text{current}} = \dfrac{170}{3}$ $= 56\frac{2}{3}$ ohms.

$$\therefore \quad \text{extra resistance required} = 56\frac{2}{3} - 1 = 55\frac{2}{3} \text{ ohms.}$$

(ii) The energy dissipated in the circuit $= IEt$ watt-hours, where $I$ is in amps., $E$ is the e.m.f. of the mains in volts, and $t$ is in hours (p. 470). Since $I = 3$ amps., $E = 200$ volts, $t = 10$ hours,

$$\text{energy} = 3 \times 200 \times 10 = 6000 \text{ watt-hours}$$

$$\therefore \quad \text{number of units (kilowatt-hours)} = 6$$

$$\therefore \quad \text{cost} = 6\text{d.}$$

5. *Explain what is meant by the resistance of a conductor. Describe the Wheatstone bridge method of measuring resistances. Failing a suitable length of wire, a metre bridge wire is improvised using two* 50 *cm. lengths, AB and BC, oined in series. The left-hand portion, AB, has a resistance of* 0·36 *ohm and the right-hand portion, BC,* 0·22 *ohm. Where is the balance point when two equal resistances are placed in the gaps of the metre bridge?* (N.)

Since *equal* resistances are in the gaps, the resistances of the lengths of wire from $A$ and $C$ respectively must be equal. See "ratio relation", p. 516. The balance point, D, is thus on the portion $AB$ of the wire, and the resistances from $A$ to $D$ equal the resistance from $B$ to $D$.

$$\therefore \quad \text{resistance from } A \text{ to } D = \tfrac{1}{2} \text{ of } (0·36 + 0·22) \text{ ohm}$$

$$= 0·29 \text{ ohm}$$

$$\therefore \quad \text{length of } AD = \frac{0·29}{0·36} \times 50 \text{ cm.} = 40·3 \text{ cm.}$$

$$\therefore \quad \text{balance-point is } 40·3 \text{ cm. from } A.$$

6. *Define ampere, volt, and ohm. What length of wire of diameter* 2 *mm. and resistivity (specific resistance)* 0·000049 *ohm-cm. is necessary to make a resistance of* 5 *ohms? Describe briefly how the value can be roughly checked by experiment.* (L.)

Since $R = \dfrac{\rho l}{a}$, we have $l = \dfrac{Ra}{\rho}$. Now $a = \pi r^2 = 3·14 \times 0·1^2$, since $r = 1$ mm. $= 0·1$ cm.

$$\therefore \quad l = \frac{5 \times 3·14 \times 0·01}{0·000049} = 3204 \text{ cm.}$$

The value can be checked by using a metre bridge to measure the resistance of 3204 cm. of the wire. It should be 5 ohms. Details of the experiment are given on p. 515.

## EXERCISES ON CHAPTER XXXIII

**1.** A cell has an e.m.f. of 1·5 volts and an internal resistance of 2 ohms. Calculate the current flowing when it is connected to resistances of (i) 8 ohms, (ii) 13 ohms, (iii) 38 ohms .

**2.** A cell of internal resistance 4 ohms is connected to a resistance of 20 ohms. The current flowing is then 0·2 amp. Calculate the p.d. across the internal and external resistance, and the e.m.f. of the cell.

**3.** A resistance of 10 ohms is joined to a cell of internal resistance 5 ohms. Find the e.m.f. of the cell if the p.d. across the terminals of the cell is 2 volts.

**4.** A battery of e.m.f. 20 volts and internal resistance 2 ohms is joined to a parallel arrangement of two wires of 6 and 3 ohms. Calculate the current in the cell and in each wire.

**5.** How would you compare the e.m.f.s of a Daniell cell and a Leclanché cell if a voltmeter is not available ? (*L.*)

**6.** The resistance of a piece of wire of length 90 cm. and diameter 0·03 cm. is 5·6 ohms. Find the specific resistance (resistivity) of the material of which the wire is made. (Take $\pi = 22/7$.) (*N.*)

**7.** Draw a labelled diagram of a Wheatstone bridge circuit suitable for measuring the resistance of a length of wire. Give details of the observations you would make in using the bridge to determine the resistivity (specific resistance) of the material of the wire, and show how you would make the necessary calculation.

Two wires, *A* and *B*, of the same material are of equal resistance, but the diameter of *A* is twice that of *B*. What is the ratio of the lengths of *A* and *B* ? (*C.W.B.*)

**8.** Describe how you would compare the e.m.f.s of a flash lamp battery (about 3 volts) and a Leclanché cell (about 1½ volts) by means of a potentiometer. Give the theory of the method employed, and point out any details of experimental procedure which ensure accuracy. (*N.*)

**9.** State *Ohm's law* and show how it leads to a definition of resistance. Describe an experiment to find the resistance of a coil of wire. What length of manganin wire of cross section 0·00089 sq. cm. is required to

18

make a resistance of 5 ohms ? (Specific resistance of manganin = $4.43 \times 10^{-5}$ ohm cm.) (*L.*)

**10.** Give diagrams to show how you would connect three cells (i) in series, (ii) in parallel. If each cell has an e.m.f. $E$ volts and an internal resistance $b$ ohms, point out in each case the values of the e.m.f. and the internal resistance of the battery of cells so formed. Using thick leads, a cell of internal resistance 1 ohm is connected in series with a variable resistance and an ammeter of negligible resistance. When the variable resistance is adjusted to a certain value $R$ ohms, the reading on the ammeter is 0·5 amp. On increasing the value of $R$ by 4 ohms, the ammeter reading is observed to fall to 0·25 amp. Determine the value of $R$ and the e.m.f. of the cell. (*C.W.B.*)

**11.** Give a labelled diagram of the apparatus you would use to determine the resistance of a piece of wire. The circuit should be shown connected up. (*N.*)

**12.** Describe an experiment to compare the e.m.f. of a dry cell with that of a Daniell cell; an accumulator is available if required. Summarize the advantages and disadvantages of these three cells, stating any values you know for their e.m.f.s and internal resistances. (*L.*)

**13.** The Wheatstone Bridge and the potentiometer employ "null" methods. Explain what is meant by this statement, and suggest reasons why such methods are used in physics. Describe how you would use a potentiometer to compare accurately the e.m.f.s of two cells. (*C.*)

**14.** Given a supply of resistance wire how would you determine the value of its resistance per metre length ? 100 cm. of nichrome wire, area of cross-section 0·50 sq. mm. has a resistance of 2·20 ohms. Calculate (*a*) the resistivity (specific resistance) of nichrome, (*b*) the length of the wire which, connected in parallel with the 100 cm. length, will give a resistance of 2·00 ohms. (*N.*)

**15.** How would you compare the electromotive forces of two cells given *either* an ammeter and some resistances *or* a potentiometer and its accessories ? Give the theory of the method you describe. A potentiometer whose wire has a length of 200 cm. and a resistance of 10 ohms is used to compare the electromotive forces of two cells $A$ and $B$. It is found that cell $A$, whose e.m.f. is 1·1 volts, gives a balance point 121 cm. from the common end of the wire and that cell $B$ gives a balance point 165 cm. from the same end. Calculate (*a*) the e.m.f. of $B$, (*b*) the potential difference between the ends of the potentiometer wire, (*c*) the current through the wire. (*N.*)

**16.** Distinguish between *resistance* and *specific resistance*. Describe a method of measuring the specific resistance in the form of a wire, and mention any check measurements you would make to improve the accuracy of the result. Explain why the alloys manganin and constantan are largely used in making standard resistance coils. (*O. & C.*)

**17.** State the relation between the electric *current* flowing through a wire and the *potential difference* across its ends, and define *resistance* and *resistivity*. A voltaic cell of e.m.f. 1·5 volts is connected in series with an ammeter of resistance 0·2 ohm and a resistance $R$. The reading of the ammeter is $\frac{1}{3}$ amp., and the p.d. between the terminals of the cell is 1·3 volts. Calculate the value of $R$ and the internal resistance of the cell. (*O. & C.*)

**18.** Explain the principle of a slide-wire potentiometer. Describe how you would use it to compare the e.m.f.'s of two cells.

One of the cells needed 30 cm. of the slide wire to balance its e.m.f. but only 20 cm. to balance its potential difference when a resistance of 4 ohms was connected across its terminals. What was the internal resistance of the cell? (*N.*)

**19.** Draw a diagram of the circuit known as a Wheatstone bridge and state the condition for its balance.

If you were provided with a reel of wire, an accurate 1 ohm coil, a metre-wire bridge and other necessary apparatus, describe and explain how you would construct a coil of resistance 2 ohms. What would be the resistance of these two coils when joined in parallel? (*L.*)

**20.** State Ohm's law and describe an experiment to determine the resistance of a conductor by the direct application of this law.

Explain what is meant by the internal resistance of a primary cell, and state *three* factors which affect its value in different cells.

When a battery is connected to a conductor of resistance 300 ohms the potential difference between the terminals of the battery is 15 volts. On disconnecting the conductor, the potential difference rises to 16 volts. Calculate the internal resistance of the battery. (*L.*)

# CHAPTER XXXIV

## MAGNETISM

MODERN civilization relies to a great extent on efficient *magnets*, some permanent and others temporary. They play an important part in the production of electrical power and in long distance communications. The earpiece of a telephone, for example, and the microphones used in broadcasting stations, depend on magnets; the loudspeaker in a receiving set has a powerful magnet without which it could not function. The dynamos at power stations which produce electrical power, and the electric motors which drive trains and buses, also contain magnets.

**The poles of a magnet; magnetic length.** Many years ago it was known that a mineral called a *lodestone* always pointed roughly north and south when it was suspended, and it was used for navigation across country and sea (the name "lodestone" was given because it acted as a "leading" stone). It was also found that iron and steel were attracted to the ends of the lodestone, which was a natural magnet. Little progress, however, was made in the study of the subject until 1603, when DR. GILBERT became interested in the phenomenon. He was a physician to Queen Elizabeth, and his extensive researches, culminating in a learned volume called *De Magnete*, have led him to be regarded as the founder of magnetism.

Observation shows that iron filings are attracted mainly towards the ends of a magnet, and the regions where the attracting power is greatest are known as the two **poles** of the magnet. An important property of a magnet is that it always comes to rest pointing approximately in a north-south direction, and the end of the magnet pointing northwards is said to have a north ($N$) magnetic polarity, while that end pointing southwards is said to have a south ($S$) magnetic polarity.

528

The distance between the poles of a magnet is known as the *magnetic length* of the magnet. It is impossible to define the exact position of the poles, and the magnetic length is therefore never known exactly. The magnetic length is certainly less than the

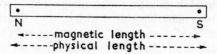

FIG. 355. Magnetic and physical lengths

physical length of the magnet, as the poles are never located at the very ends (Fig. 355).

**Fundamental law of magnetism.** If the *N*-pole of a magnet is brought near to the *N*-pole of a compass-needle, repulsion is observed to take place. If the *S*-pole of the magnet is brought near to the same pole of the compass-needle attraction takes place, and further experiment shows that a *S*-pole repels a *S*-pole. These observations are summed up by a fundamental law in magnetism which states that—

**Like poles repel; unlike poles attract.**

**Magnetic and non-magnetic substances.** A magnetic substance is one which can be magnetized. Copper, wood, and glass cannot be made into magnets, and are termed "non-magnetic." Iron, nickel, cobalt, and certain alloys of these metals are capable of being made into strong magnets, and are magnetic substances. Researches show that the magnetic property of such metals is partly bound up with their special crystalline structure and partly with the structure of their atoms. *Steel* is made by adding a small percentage of carbon to pure iron, and is generally much harder than iron physically. Correspondingly, the magnetic properties of steel form a complete contrast to that of iron; steel is more difficult to magnetize than iron, but loses its magnetism much less easily than iron. Bar magnets should therefore be made of steel, not iron.

A vast amount of research has been expended in finding alloys which can be made into powerful magnets. Special alloys such as *mumetal* have been developed for the electromagnet and the transformer, in which temporary magnets are used. *Alni* and *ticonal* are alloys of nickel and cobalt which are used for making powerful permanent magnets.

**The magnetic filter and separator.** The attraction of a magnet for iron is utilized in industry for cleaning oil and separating iron from non-ferrous materials mixed with it. Fig. 356 illustrates the essential features of a *magnetic filter system*, which is used for extracting small iron particles contaminating oil or any other liquid. $M$ is a powerful magnet made of ticonal, and $C$ is an iron filter cage which is magnetized by $M$. The dirty oil is poured into the filter at $A$, and the iron particles collect between the spaces of the iron cage under the powerful attractive force on them. The clean oil issues through $B$, the bottom of the filter cage.

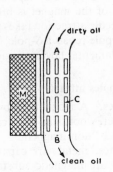

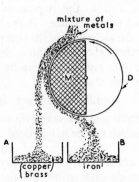

FIG. 356. Magnetic filter     FIG. 357. Magnetic separator

Fig. 357 illustrates the principle of a *magnetic separator*, used in metal works for separating iron from copper, brass, and other non-ferrous solids. It consists of a drum $D$, revolving about an axis through its middle, with a powerful permanent magnet, represented by $M$, in one half of it. The mixture of metal dust is poured through a funnel (not shown) above the drum, and the particles are carried round slowly. The non-ferrous materials, such as copper and brass, are not attracted by the magnet $M$, and drop into a container $A$ placed as shown. The iron particles, however, are carried round the drum towards the bottom under the attraction of *M*, and eventually fall into a container *B* placed below the drum. In this way, ferrous and non-ferrous particles can be separated quickly and efficiently.

**Methods of making magnets.** (1) *Single touch.* A simple method of making a steel knitting needle, or a piece of clock spring, into a magnet consists of laying the specimen, *X*, on a table, and stroking it repeatedly in *one* direction with a pole of a magnet *Y* (Fig. 358 (*a*) ). This is known as the method of *single touch*. *Y* must be raised clear of *X* each time the end of the latter is reached, as the magnetism induced in *X* in one movement of *Y* would be annulled on the opposite movement if *Y* always moved along the surface of *X*. Experiment shows that the end of *X* last touched has an *opposite* polarity to the stroking pole.

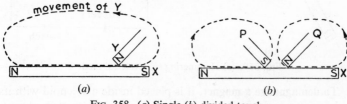

FIG. 358. (*a*) Single (*b*) divided touch

(2) *Divided touch.* Fig. 358 (*b*) illustrates the method of magnetizing by *divided touch*, which makes a strong magnet more quickly than the method of single touch. Beginning at the middle of the specimen *X*, two magnets, *P*, *Q*, are used to magnetize one half of *X*; the procedure for each half is the same as in the method of single touch. The end of *X* last touched is again opposite in polarity to the magnetizing pole, so that the poles of *P* and *Q* used must be opposite in polarity to make *X* a normal magnet.

(3) *Hammering in the earth's field.* In the seventeenth century, a weak magnet was made by hammering one end of the specimen *X*, which pointed north; the best position to hold *X* is about 70° inclined to the horizontal (Fig. 359 (*a*) ). Instead of a magnet, such as *Y* in Fig. 358 (*a*), the magnetic influence of the *earth* is utilized in this case (see p. 555), and in this country the lower end of the specimen is found to possess a north polarity.

(4) *Electrical method.* The best method of making a magnet is to utilize the effect of an electric current flowing in a long coil of insulated wire, known as a **solenoid**. This is the method employed in industry for making magnets. The specimen $X$ is placed inside the solenoid, the current is switched on for a moment, and then switched off (Fig. 359 (*b*) ). When $X$ is withdrawn and tested it is found to be a magnet with poles as shown in the figure.

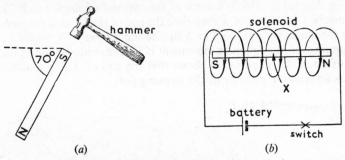

FIG. 359. (*a*) Hammering (*b*) electrical method

**To demagnetize a magnet,** it is placed inside a solenoid with its axis pointing east-west and which carries an alternating current. The magnet is then withdrawn slowly a long distance away from the solenoid, and is now demagnetized.

**How magnetized, unmagnetized, and non-magnetic substances can be distinguished.** Suppose we are told that one of three specimens of metal, $X, Y, Z$, is a magnet, another is unmagnetized iron, and the third is non-magnetic like brass or copper. If iron filings are not available to help in distinguishing the three specimens, they can each be suspended horizontally in turn by a thread. The specimen, $Y$ say, which always comes to rest pointing in the same direction (approximately north and south) is the magnet. $Y$ can now be taken and placed close to $X$ and $Z$ when they are suspended in turn. The specimen which is attracted to $Y$, $Z$ say, is the un-magnetized iron, so that $X$ is non-magnetic. If the attraction is too small, the magnet $Y$ can be used to stroke each of the remaining specimens in turn by the method of single touch (p. 531). The specimens are now suspended horizontally, and the one which comes to rest always pointing in the same direction is the magnetic specimen. The other is the non-magnetic specimen.

If iron filings are available, the same problem is solved by seeing which of the specimens attracts large numbers of filings at both ends; this is the magnet. The unmagnetized iron and brass can then be distinguished as explained above.

**Induced magnetism.** Experiment shows that a magnet $X$ can pick up a number of iron pins, each hanging from the other as shown by $a$, $b$, $c$, in Fig. 360 ($a$). If the lower end of $c$ is now tested by a compass needle it will be found to be a north ($n$) pole, and $c$ has therefore become a magnet. If $c$ is removed and $b$ is tested in

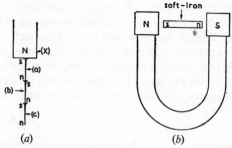

FIG. 360. Induced magnetism

the same way, the latter will be found to be a magnet; $a$, too, is found to be a magnet. Since the magnet $X$ has not stroked any of the pins, as in magnetizing by single touch, for example, the magnetism acquired by $a$, $b$, $c$ is known as *induced magnetism*. If, in Fig. 360 ($a$), the top pin $a$ is removed from $X$, the pins $b$ and $c$ fall to the ground, showing that the magnetism in the pins is temporary and due to the influence of $X$.

Induced magnetism always occurs in soft iron when it is in the neighbourhood of a magnet or in any other magnetic field. Thus a piece of soft iron between the poles $N$, $S$ of a curved magnet becomes magnetized by **induction**, as the phenomenon is called, with unlike poles facing $N$, $S$ respectively (Fig. 360 ($b$) ). It is much harder to obtain induced magnetism in steel than in soft iron since, in general, steel is harder to magnetize than iron (p. 529).

18*

**The molecular theory of magnetism.** The search for a theory to explain all the observed phenomena in magnetism has occupied the attention of many famous scientists. Even to-day the problem cannot be said to have been fully solved. It is, however, usually agreed that the magnetism in the specimen is due to the *electrons* surrounding the nucleus of the atom, which carry a quantity of electricity and move and rotate in the atom (see p. 631). These moving electrons constitute tiny circulating electric currents; on this theory it appears that magnetism is an "offspring" of electric currents.

In 1898, however, EWING developed a *molecular theory* of magnetism (first proposed by WEBER about 1850) which explains many of the common phenomena of magnetism. The theory stated that each molecule of a magnetic substance is itself a tiny magnet, even though the substance is unmagnetized. In this case the molecules are arranged in groups having a *closed chain* formation. Fig. 361 (*a*), which is exaggerated and must not be taken too literally, illustrates the idea of a closed chain; similar groups of molecular magnets exist throughout the material. These closed

FIG. 361 (*a*). Unmagnetized specimen (closed chains)

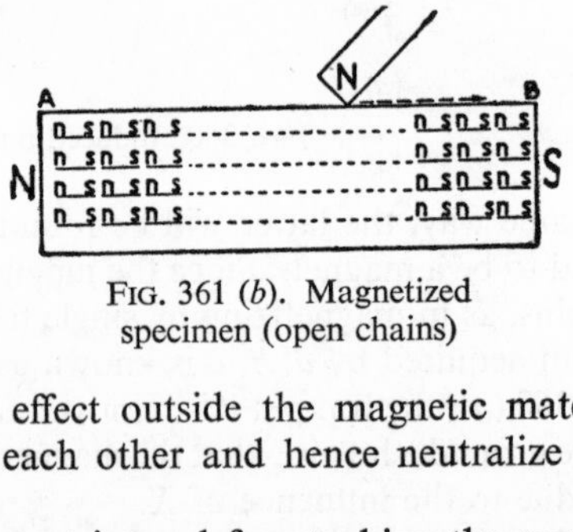

FIG. 361 (*b*). Magnetized specimen (open chains)

chains have no magnetic effect outside the magnetic material, as the poles are close to each other and hence neutralize each other's effect.

When the pole of a magnet is used for stroking the material (Fig. 361 (*b*) ), the molecules move round under the magnetic influence with the south poles, *s*, facing the *N* stroking pole as it moves towards *B*. The closed chains then become *open chains*, as illustrated in the figure; and as the *s* poles at *B* have no north poles to neutralize them, unlike the case of closed chains (Fig. 361 (*a*) ), the end *B* acts like a south magnetic pole, *S*. Similarly, the "free" north poles *n* of the molecules at the end *A* makes this end act like a north pole, *N*. Experiment shows that the strength of a magnet cannot be increased beyond a certain limit, when the magnet is said to be **saturated**; on the molecular theory, a magnet is saturated when all the closed chains inside it have become open chains.

## Experimental evidence for the molecular theory of magnetism.

Ewing proceeded to test his theory by placing a group of pivoted magnets in the form of a closed chain and bringing a magnet towards them, when the magnets opened into the form of an open chain. The following simple experiments are also evidence for the molecular theory.

*Each molecule of a magnet is itself a magnet.* If a knitting needle is magnetized and then broken into two halves, experiment shows that each half is a magnet (Fig. 362 (*a*) ). If each half is now broken again, four magnets are obtained. No matter how small the broken parts may be, each of them is a magnet; which suggests that the smallest particle of a magnet, the molecule, is itself a magnet, as the molecular theory states.

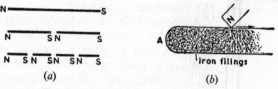

(*a*)                                          (*b*)

FIG. 362. Evidence for molecular theory

*Magnetizing iron filings in a test tube.* A test tube is filled with iron filings and laid on the table. The end, A, of the tube has then no magnetism, since it attracts each end of a compass needle on testing, but when the tube is stroked by the *N*-pole of a magnet (Fig. 362 (*b*) ) as in magnetizing by single touch, the filings at the top of the test tube can be seen to be aligning themselves horizontally. The end *A* is now found to exhibit a north polarity, on testing with a compass. This experiment suggests that the molecules in an unmagnetized specimen align themselves into "open" chains when the specimen is magnetized.

*Closed and open chains, using compass needles.* A group of compass needles can be arranged to form a closed chain of small magnets, as illustrated in Fig. 363 (*a*). When a magnet is brought near them, the needles turn round and form an open chain (Fig. 363 (*b*) ). When the magnet is taken away, the needles return to the closed chain formation. This experiment demonstrates that a closed chain of magnets can be made into an open chain by a magnet.

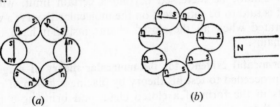

Fig. 363. Closed and open chains

**Keepers of magnets.** If magnets are stored in a box in a haphazard manner, their open chains tend to become partly closed owing to (i) the effect of other magnets and (ii) the vibration experienced by buildings from outside and inside traffic. To prevent

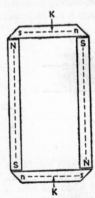

Fig. 364. Keepers

demagnetization, magnets are stored in boxes in pairs, with opposite poles facing each other, and with pieces of soft iron, known as *keepers* (K), placed across both ends.

An open chain of molecular magnets will return to a closed chain formation if it is disturbed, and in general, a closed chain of magnets is much more stable than an open chain. When the soft-iron keepers, K, are placed across the two bar magnets, they become magnetized by induction with poles as shown in Fig. 364. The molecules of the two bar magnets and of the two keepers now form a *closed chain* throughout the whole of the four pieces of metal, and when stored in this manner, the magnets retain their magnetism for a very long time. The keepers are made of soft-iron and not of steel because the former metal is easier to magnetize by induction.

**Magnetic fields, and lines of force.** If a compass needle is placed at a point *a* near to the north pole, *N*, of a bar magnet, experiment shows that the needle settles in a definite direction (Fig. 365). At other points near the magnet, such as *b*, *c*, *d*, the needle settles in other directions. Now we refer to an oil "field," or a gold "field," when these substances are discovered in a region; in the same way, we refer to the *magnetic field* round a bar magnet, because a magnetic force is experienced there. A compass needle settles in the direction of the magnetic force, and if it is moved from one end of a magnet to the other, a continuous curve can be drawn through all the positions of the ends of the compass needle. For obvious reasons this curve is known as a *line of magnetic force*, or more simply as a "line of force."

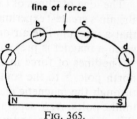

FIG. 365.
Line of magnetic force

By using a compass needle, a magnetic field can be "mapped out" by lines of force covering the whole region. A vivid demonstration of the field, however, can be obtained by sprinkling iron filings on a sheet of paper in the field, and then tapping the paper

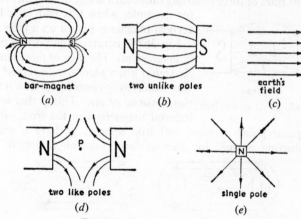

FIG. 366. Some magnetic fields

lightly. The filings become tiny magnets by the process of induction (p. 533), and tapping the paper gives them the necessary freedom to settle in the direction of the magnetic field, like miniature compass needles. Fig 366 illustrates a few of the lines of force in some typical magnetic fields; it should be noted that the lines occur in space, so that the sketches represent a horizontal section of them.

The idea of "lines of force" is mainly due to FARADAY, perhaps Britain's greatest experimental scientist (see p. 595). He considered that a medium like air or a vacuum is in a condition of "strain" when a magnet is present.

The lines of force in the case of a bar magnet pass from the north pole $N$ to the south pole $S$ in the air, and are continuous through the magnetic material. The lines of force between two unlike poles, $N$, $S$, separated in air pass from one pole to the other (Fig. 366 ($b$)); but the lines between two like poles, $N$, $N$, proceed away from each other, and none pass through some place $P$, which is known as a *neutral* point in the field on this account (Fig. 366 ($d$)). Fig. 366 ($c$) illustrates the lines of force due to the earth's field locally; they are straight and parallel, and point in the direction of the magnetic north and south. Fig. 366 ($e$) shows straight lines of force radiating out from a single north pole, $N$, as, for example, when this end of a long vertical magnet is placed on a table.

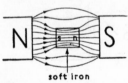

soft iron

FIG. 367. Effect of soft iron

Fig. 367 illustrates an important point in magnetism. *The lines of force passing through a given place are increased perhaps hundreds of times if soft iron is used there instead of air.* This is due to the induced magnetism in the iron. From a popular point of view, soft iron appears to have a greater "conducting power" for lines of force than air (compare Fig.

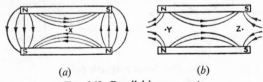

($a$)       ($b$)

FIG. 368. Parallel bar magnets

366 ($b$)), and this property of iron is employed in transformers, motors, and dynamos.

Fig. 368 ($a$), ($b$) illustrate respectively the lines of force between two parallel bar magnets when their unlike and like poles are opposite each other. Neutral points occur at $X$ and at $Y$, $Z$.

**The direction of a magnetic field at a given point** is defined as the direction along which a *north* pole would tend to move if it were placed at that point. The arrows on the lines of force in Fig. 368 hence indicate the direction of movement of a north pole. In practice, as shown in Fig. 365, the direction of the field at a point is obtained by placing a compass needle there, and it can be seen that the compass needle points along the tangent to the line of force where it is placed. A *line of force* can thus be accurately defined as a *line such that the tangent to it at any point is the direction of the magnetic field there.* Lines of force cannot cross each other. If they did, two tangents to the lines could be drawn at the point of intersection; this is impossible, as there cannot be two different directions of the magnetic field at the same point in it.

FIG. 369. A SHIP'S COMPASS

**Ship's magnetic compass.** Although it has been displaced in large ships by another form of compass which is gyroscopic, the *magnetic compass* is still used extensively in ships for navigation. Fig. 369 illustrates one form of compass, which consists of a few magnets fixed side by side, underneath a compass needle contained in a bowl. The card is pivoted at its centre by means of a metal point resting on the bottom of the bowl, which is non-magnetic, and the bowl is supported freely by an axle (not shown) passing through an outer metal bowl. The latter is supported by an axle passing at right angles to the first axle through an outer ring, a method of mounting known as *gimbals*. In view of the mounting, the card takes up a horizontal position if the bowl rocks about either axle, so that the gimbals counteract the pitching and rolling of the ship.

The iron and steel in a ship causes a deviation of the compass-needle from its true direction. This deviation is found by experiment before the ship starts on its journey, and the correction to be applied to the reading on the card is thus known. A ship's compass or *binnacle* is shown in Fig. 379.

---

## Summary

1. Like magnetic poles repel; unlike poles attract.

2. Magnets can be made (i) by single touch, (ii) by double touch, (iii) by electrical method, (iv) by hammering in the earth's field.

3. Soft iron is more easily magnetized than steel, and loses its magnetism more easily.

4. According to the molecular theory, a magnetized specimen has "open chains" of molecular magnets and an unmagnetized specimen has "closed chains." Keepers are pieces of soft iron used in storing magnets; they help to form closed chains with the magnets.

5. Lines of force indicate the appearance of a magnetic field; they are obtained by means of a compass needle, or by using iron filings. **A line of force is a line such that the tangent to it at any point indicates the direction of the magnetic field there.**

---

### EXERCISES ON CHAPTER XXXIV

1. Name three practical applications of magnets. How are magnets made in industry?

2. Explain how you would distinguish between three substances *A*, *B*, *C*, painted the same colour, if one was a magnet, another was

an unmagnetized piece of iron, and the third was a piece of brass. No other piece of apparatus except thread and a stand were available.

**3.** Describe three completely different methods of making a magnet. Draw diagrams showing the polarity of the magnet in each case.

**4.** Draw the lines of force, and mark their direction, round (i) an isolated south pole, (ii) a bar magnet, (iii) two equal parallel bar magnets with like poles facing each other.

**5.** Using the molecular theory of magnetism, explain why a steel pin is attracted to a magnet. Why can several pins be picked up, one under the other, and why do most of the pins fall off on taking away the magnet?

**6.** Describe three experiments which support the molecular theory of magnetism.

**7.** Describe a method of magnetizing a bar of steel without using a magnet and give a diagram showing the polarity which would be obtained. What are the chief differences in the magnetic behaviour of soft iron and steel? Describe and explain *two* cases in which the magnetic properties of soft iron are used. (*L.*)

**8.** How would you demagnetize a steel rod? (*N.*)

**9.** What is meant by a magnetic substance? Underline those of the following substances which are magnetic: ebonite, iron, chromium, nickel. (*N.*)

**10.** A small bar of soft iron is placed lengthwise along the magnetic meridian on a horizontal table. How would you plot the lines of force in the neighbourhood of the iron? Give a diagram of the result to be expected. Describe briefly how this result is applied (*a*) in the storing of bar magnets, (*b*) to a ship's compass. (*L.*)

**11.** Explain what is meant by the *molecular theory of magnetism* and how it accounts for the difference between a non-magnetic substance, an unmagnetized piece of magnetic substance, and a magnet. Describe an experiment in favour of the theory. How does the molecular theory of magnetism account for (*a*) the fact that the poles of a magnet are not at its ends, (*b*) the loss of magnetism of a magnet when made red hot? (*C.W.B.*)

**12.** An unmagnetized soft iron rod is held in a vertical position and its upper end is struck a few blows with a hammer. State and explain what happens to the magnetic state of the bar. The *N*-pole of a long vertical magnet rests on a table. Draw diagrams showing the distribution of the lines of force in the plane of the table round the pole (*a*) if no external field is present, (*b*) if the earth's field is acting. (*L.*)

13. Describe a method of magnetizing a steel knitting needle *XY* so that the end *X* shall have a north-seeking polarity. Give a diagram and state reasons for the procedure adopted. How would you prove the poles to be equal in strength?

State the magnetic properties desirable in (*a*) the magnet of a moving-coil ammeter, (*b*) a transformer core. Name the type of material to be used in each case. (*N.*)

14. How do you account for the following facts: (*a*) Iron becomes magnetized when placed in a coil carrying direct current. (*b*) Bar magnets lose their magnetism when heated strongly. (*c*) Steel makes better permanent magnets than soft iron. (*d*) Keepers help to prevent magnets losing their magnetism? (*C.*)

15. How could the lines of force due to a magnet be traced out? Give a diagram showing their arrangement in the case of a bar magnet. Describe and explain how your diagram would be modified (*a*) if the magnet lies in the earth's field with its south-seeking pole pointing north, (*b*) if a ring of soft iron is near one pole. (*L.*)

16. Explain the terms *magnetic induction, magnetic screening.* Describe an experiment to show that a bar of soft iron can be strongly magnetized by magnetic induction but that the induced magnetism disappears when the inducing magnet is removed. How would you expect a bar of steel to behave in similar circumstances? Explain the effects observed. (*O. & C.*)

17. Describe how an iron bar *AB* may be magnetized so that the end *B* becomes a north pole, (*a*) using a bar magnet, (*b*) using an electric current, (*c*) using the magnetic field of the earth. Describe briefly experiments to illustrate *three* properties which *AB* possesses after being magnetized which it did not possess before. (*L.*)

18. (*a*) Define *magnetic pole strength, magnetic field strength.* Describe a method for determining the positions of the poles of a bar magnet.

(*b*) Give an account of the molecular theory of magnetism and show how it can account for magnetization by (i) stroking, (ii) induction, and (iii) an electric current. How are differences of magnetic porperties of iron and steel explained on this theory? (*L.*)

## CHAPTER XXXV

# MAGNETIC MEASUREMENTS
# EARTH'S MAGNETISM

WHEN one pole of a long magnet $Y$ is placed at different distances from one pole of another long magnet $X$, experiment shows that the force $F$ between the poles increases as the distance $d$ between them decreases. Fig. 370 illustrates an apparatus, known as HIBBERT's *balance*, which can be used to investigate how $F$ varies with $d$. The $n$ pole of the magnet $Y$ is placed opposite the $n$ pole of the magnet $X$, which can tilt about its mid-point $O$, and the moment of the force $F$ is counter-balanced

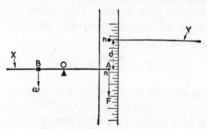

FIG. 370. Hibbert's balance

by a small weight $w$ on the other side of $O$. Taking moments about $O$, $F \times AO = w \times OB$; thus $F = \dfrac{w \times OB}{AO}$, and hence $F$ can be calculated if $w$, $AO$, $OB$ are known. By altering the distance $d$ and finding the corresponding value of $F$, it can be shown that—

**the force between two poles varies inversely as the square of their distance apart.**

This is known as the *inverse-square law* of magnetism; it means that the force is only one-quarter as much when the distance between the same two poles is doubled, and nine times as great

543

when the distance between the poles is diminished to one-third. Mathematically, the force $F \propto \dfrac{1}{d^2}$, where $d$ is the distance.

**Unit pole-strength. Law of force between poles.** Magnets differ widely in their strength; some are very strong, others may be weak. It has been found necessary to define a *unit* of pole-strength. The *unit pole-strength*, or more briefly, the *unit pole*, is defined as *the strength of that pole which repels a similar pole with a force of 1 dyne when placed 1 cm. away in air.*

It follows from the definition that a pole of strength 5 units repels a similar pole of 1 unit at a distance 1 cm. away in air with a force of 5 dynes, and hence, that a pole of strength 5 units repels a similar pole of 4 units with a force of $5 \times 4$ dynes. Thus the force between two poles at a given distance apart is proportional to the *product* of the pole-strengths. Taking this fact in conjunction with the experimental inverse-square law, we arrive at the *law of force* between two poles. This states that the force $F$ between two poles of strengths $m_1$ and $m_2$ units separated a distance $d$ cm. in air is given by

$$F = \frac{m_1 m_2}{d^2} \text{ dynes} \qquad . \qquad . \qquad . \qquad . \quad (1)$$

**Magnetic field strength or intensity.** The magnetic field inside a moving-coil measuring instrument is provided by a powerful magnet, so that the field inside is a strong one (p. 587). On the other hand, the earth has a magnetic field which is a very weak one by comparison. Since a magnetic field is a region where a magnetic force is experienced, it is natural to define the *strength or intensity* of a magnetic field at a point in it as *the force in dynes exerted on a unit pole* placed at that point. The unit of magnetic field strength is the **oersted**, named after the discoverer of the magnetic effect of the electric current (p. 566).

The field strength inside a moving-coil instrument (p. 587) may thus be 1000 oersteds, for example; a unit pole would then experience a force of 1000 dynes when placed in the field. The earth's field strength in a horizontal plane, its horizontal component $H_o$, is about 0·2 oersted (p. 556), so that a unit pole in this field experiences a force of 0·2 dyne. The force on a pole of

10 units in the earth's field is thus $10 \times 0.2$ dynes, and it can hence be seen that the force $F$ exerted on a pole of strength $m$ units in a field of $H$ oersteds is given by

$$F = Hm \text{ dynes} \qquad . \qquad . \qquad . \qquad . \qquad (2)$$

**The pole-strengths of a magnet are equal.** If a magnet is placed gently on a piece of cork floating in a trough of water, observation shows that the magnet *turns round* and finally points in the direction of the magnetic north and south. The magnet does not move across the water to one side or the other of the trough, showing there is no resultant force acting on it.

Since the earth's magnetic field intensity operates equally on each pole of the magnet, it follows that *the pole-strength of the N-pole of a magnet must equal the pole-strength of its S-pole.* This is true for all magnets.

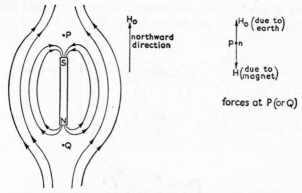

FIG. 371. Bar-magnet with S-pole pointing northwards

**Neutral points in a magnetic field. N-pole pointing southwards.** If a bar magnet is placed with its north pole pointing southwards and iron filings are sprinkled round it, the latter assume an appearance similar to that shown in Fig. 371 (*a*). Very few filings are found round a point $P$, and a compass needle placed at this point settles in no fixed direction; from which it follows that there is no magnetic force at $P$, and this point is therefore known as a *neutral point* in the magnetic field. There is another neutral point, $Q$, on the other side of the magnet.

To explain the production of the neutral points, consider a unit north pole, $n$, placed at the point $P$. The force acting on it due to the earth's field is $H_o$ dynes, where $H_o$ is the intensity of the earth's magnetic field (0·2 oersted, approx.), and this force acts northwards as this is the direction to which the north pole of a compass-needle points (Fig. 371 (b)). The net force on the unit $n$ pole due to the magnet acts southwards, since the $S$-pole of the magnet is nearer to it than the $N$-pole. Suppose this net force to be $H$ dynes. As we proceed northwards from the $S$-pole, a point will be reached where $H$ has the value of $H_o$. At this place, which is the neutral point $P$, the force due to the earth cancels out exactly the force due to the magnet, and consequently there is no resultant magnetic field.

**North pole pointing northwards.** When a bar magnet is placed with its $N$-pole pointing northwards, the magnetic field has an appearance similar to that shown in Fig. 372 (a). Neutral points,

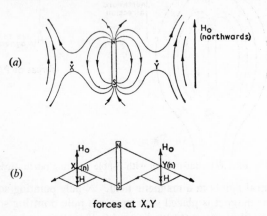

Fig. 372. Bar magnet with $N$-pole pointing northwards

$X$, $Y$, are found at equal distances on either side of the magnet on a line bisecting the latter perpendicularly, and, as we shall see, they are due to the earth's magnetic field neutralizing completely the magnet's field at these two points.

To explain why a neutral point appears at $X$, consider a unit north pole ($n$) placed at this point (Fig. 372 ($b$) ). The repulsive force on the latter due to the $N$-pole of the magnet acts along $NX$ produced, and the attractive force on it due to the $S$-pole acts along $XS$. These two forces are equal in magnitude, as the distance $NX =$ the distance $SX$, and their resultant $H$ can be found by the parallelogram of forces method. $H$ acts southwards, as shown in Fig. 372 ($b$), and represents the force in dynes due to the magnet acting on the unit $n$ pole at $X$. The force on the latter due to the earth's magnetic field acts northwards, and is equal to $H_o$ dynes, where $H_o$ is the intensity of the earth's field. The latter force remains constant as we move to the left of the magnet along the line $YX$, and at the point $X$ the force due to the magnet becomes exactly equal and opposite to the force due to the earth. The resultant force at $X$, and the total intensity there, is thus zero. A similar explanation accounts for the presence of a neutral point at $Y$.

**How field strength (or intensity) can be measured.** The strength of a magnetic field can be measured by means of a **deflection magnetometer**, which consists essentially of a small magnet $ns$

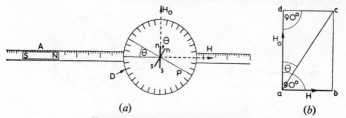

(a)          (b)

FIG. 373. Deflection magnetometer

pivoted at the middle of a circular scale $D$ graduated in degrees. A light aluminium pointer, $P$, is rigidly attached to $ns$, and any deflection of $ns$ is measured by the movement of $P$. When a magnet $A$ is placed at right angles to $ns$ (i.e. east and west), thus creating a field of strength $H$ at $ns$, the latter is deflected and comes to rest inclined at an angle $\theta$ to its original direction (Fig. 373 ($a$) ). $ns$ settles along the resultant of $H$ and $H_o$.

Fig. 373 ($b$) shows the field strengths acting on the needle in its deflected position. They are: (1) The earth's horizontal component, $H_o$, (2) the strength $H$ of the magnetic field at the needle $ns$ due to the specimen $A$. Now since the magnet was placed east and west, the field $H$ it sets up is perpendicular to $H_o$. By the parallelogram of forces, the resultant magnetic field at $ns$ is in the direction of the diagonal $ac$ of the rectangle $abcd$, where $ab$ represents $H$ and $ad$ represents $H_o$. Since $adcb$ is a rectangle,

$$\tan \theta = \frac{cd}{ad} = \frac{H}{H_o}$$

$$\therefore \quad H = H_o \tan \theta \qquad . \qquad . \qquad . \qquad (3)$$

$$\therefore \quad H \propto \tan \theta, \qquad . \qquad . \qquad . \qquad (4)$$

since $H_o$ is constant at a given place on the earth.

Thus, from equation (3), the intensity $H$ of the magnetic field at $ns$ can be calculated by observing $\theta$ and finding the value of $H_o$ at the locality concerned from tables. Further, from equation (4), the intensity $H$ is proportional to $\tan \theta$ at a given locality. If the specimen $A$ is kept fixed and magnetized more strongly, the deflection $\theta$ will be observed to increase, since $H$ increases; at some stage, however, the deflection $\theta$ remains constant, showing that the specimen has become saturated (p. 535).

(a)　　　　　　　　　　(b)

FIG. 374. Verification of inverse-square law

**Verification of the inverse-square law.** The deflection magnetometer can be used to verify the inverse-square law of magnetism, which states that the force between two given poles varies inversely as the square of the distance between them (p. 543).

A long thin magnetized rod, having two small steel spheres at its ends, is used. This magnet is known as a *ball-ended Robison magnet* (Fig. 374 (a) ), and the positions of its poles can be taken as the centres of the spheres at its ends. The magnet is positioned with its north pole $N$ at some distance $d$ from the needle $P$ of a magnetometer, so that (i) its south pole is vertically above $P$, (ii) the line joining $N$ to $P$ is east and west. The deflection $\theta$ of the needle and the distance $d$ are then noted. The experiment is repeated with varying values of $d$, and each time the corresponding deflection $\theta$ is observed. Fig. 374 (b) illustrates a plan view of the arrangement, the south pole (not shown) being vertically above $P$.

The south pole has no influence on the needle $P$ in a horizontal plane, since it is vertically above $P$. Thus the intensity $H$ in a horizontal plane at $P$ due to the magnet is given by $H = \dfrac{m}{d^2}$, where $m$ is the pole-strength of $N$, *assuming the inverse-square law is true.* But $H = H_o \tan \theta$, where $H_o$ is the earth's horizontal component (see p. 548). Thus

$\dfrac{m}{d^2} = H_o \tan \theta$, and hence

$$\frac{1}{d^2} = \frac{H_o}{m} \tan \theta.$$

Since $\dfrac{H_o}{m}$ is a constant, it follows that

$$\frac{1}{d^2} \propto \tan \theta,$$

if the inverse-square law is true.

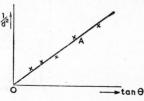

FIG. 375. Graph of $1/d^2$ v. $\tan \theta$

Experiment shows that a straight line graph $A$ passing through the origin is obtained when the values of $\dfrac{1}{d^2}$ are plotted against $\tan \theta$ (Fig. 375). Thus $\dfrac{1}{d^2} \propto \tan \theta$, and the inverse-square law must be true.

**The moment of a magnet.** If a magnet is suspended so as to move in a horizontal plane in the earth's field, experiment shows that the magnet oscillates for some time about the magnetic north and south direction before coming to rest in that direction. It thus appears that, when a magnet is inclined to a magnetic field, the forces acting on it have a turning-effect or *moment*.

To explain how these forces arise, consider a magnet held at right angles to a uniform magnetic field of intensity 10 oersteds. The latter means that the force on a unit pole in the field is 10 dynes (p. 544); hence the forces on the poles of the magnet $NS$, of strength $m$ units, say, are $10m$ dynes respectively (Fig. 376). These two equal forces act in opposite directions, since a north pole is urged in the direction of the field, and a south pole is urged in the opposite direction. It can be seen that the forces tend to *rotate* the magnet in an anti-clockwise direction, and they are an example of two forces which, taken together, are known in mechanics as a **couple**.

In an earlier chapter the *moment* or turning-effect of a force was considered, and it was calculated by multiplying the force by its perpendicular distance from the point about which the moment was taken (p. 70). In a similar way, the moment of a couple is defined as the product of *one* of the forces and the perpendicular distance between them. Thus, since the forces of 10*m* dynes are

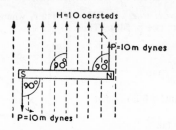

Fig. 376. Moment of couple on magnet

each perpendicular to the magnet *NS* (Fig. 376), the moment of the couple is given by 10*m* × *b*, where *b* is the magnetic length of the magnet. Thus the moment = 10*mb* units.

If the same magnet is placed at right angles to a field of 50 oersteds, the moment of the couple tending to rotate it is given by 50 *mb* units, by similar reasoning to the above. It thus appears that the quantity "*mb*" or "pole-strength × magnetic length," known as the *magnetic moment*, enters into formulae for the moments of the forces acting on a magnet in a magnetic field. If the magnet is placed at right angles to a uniform field of intensity 1 oersted, the moment of the couple acting on it is 1*mb*, or *mb* units, from above. Thus the magnetic moment *M* of a magnet can be defined as *the moment of the couple acting on the magnet when it is held at right angles to a field of 1 oersted.* Hence

*magnetic moment, M = pole-strength × magnetic length.*

**The importance of magnetic moment** lies in the fact that it is a property of a magnet which has a physical significance, and it can be accurately measured by experiment. For example, there is a formula connecting the magnetic moment (*M*) of a magnet and its period of swing in a magnetic field, and thus *M* can be calculated from an experiment of this nature. Further, as shown later, a deflection magnetometer can be used to measure and compare the moments of two magnets.

On the other hand, the positions of the poles of a magnet cannot be defined; each pole occupies some unknown volume inside the magnet. Hence the magnetic length, the distance between the two poles can never be ascertained. Moreover, the pole-strength of a magnet has little physical significance, as it is defined in terms of a unit pole-strength, which is an abstract quantity (p. 544). Thus the "moment" of a magnet is one of its important physical properties.

**Intensity along the axis of a magnet.** Consider a point $P$ on the axis of a magnet $NS$, situated at a distance $d$ from the *mid-point* of $NS$ (Fig. 377). Then the distance $PN = d - l$, where $l$ is *half* the length of the magnet, and the distance $PS = d + l$.

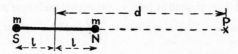

FIG. 377. Intensity along axis

Suppose the pole-strength of the magnet is $m$ c.g.s. units, and in order to find the intensity at $P$, imagine a north pole at $P$ of 1 c.g.s. unit (p. 544). The force on the unit pole due to the $N$-pole $= \dfrac{m \times 1}{(d - l)^2}$, from the law of force between two poles. The force on the unit pole due to the $S$-pole is $\dfrac{m \times 1}{(d + l)^2}$, but this force acts in the *opposite direction* to the force due to the $N$-pole, because the $S$-pole attracts the unit north pole and the $N$-pole repels it.

$$\therefore \quad \text{resultant force} = \text{intensity} = \frac{m}{(d - l)^2} - \frac{m}{(d + l)^2}$$

$$= \frac{m[(d + l)^2 - (d - l)^2]}{(d - l)^2(d + l)^2}$$

$$= \frac{m[4ld]}{(d^2 - l^2)^2}.$$

But the moment of the magnet, $M$

$$= m \times \text{magnetic length}$$
$$= m \times 2l, \text{ since } SN = 2l$$

$$\therefore \quad \text{intensity, } H = \frac{2Md}{(d^2 - l^2)^2},$$

substituting $2lm = M$ in the numerator of the fraction.

*Short magnet.* The above is a general formula for the intensity due to a magnet of any length. Suppose that a short bar magnet is used, and the distance $d$ of the point $P$ from the magnet is made large compared with $l$, the half-length of the magnet.

For example, suppose $l = 4$ cm. and $d = 40$ cm. The intensity $H$ at the point considered is then given by $H = \dfrac{2M \times 40}{(40^2 - 4^2)^2} = \dfrac{80M}{(1600 - 16)^2}$. But as 16 is a hundred times as small as 1600, we can neglect the former compared with the latter number. Thus $H$ is given by $\dfrac{80M}{1600^2}$, which simplifies to $\dfrac{M}{32,000}$.

To obtain a formula for the approximate value of $H$, consider the general relation $H = \dfrac{2Md}{(d^2 - l^2)^2}$. If $l$ is made small compared with $d$, we can neglect the value of $l^2$ compared with the value of $d^2$; the denominator is then $(d^2)^2$, or $d^4$. Thus $H = \dfrac{2Md}{d^4} = \dfrac{2M}{d^3}$.

The intensity at a point on the axis of a short bar magnet when the point is a long way from the magnet is hence given by

$$H = \frac{2M}{d^3}.$$

**Comparison of moments of short bar magnets by deflection magnetometer method.** The deflection magnetometer can be used to compare the moments of two short bar magnets. The magnetometer is set up first, and one of the short magnets, $A$, is placed in an east-west position, with its mid-point a long distance $d$ from the small magnet $a$ compared with its half-length (Fig. 378). The deflection $\theta$ of the pointer attached to $a$ is then measured.

Since the intensity $H$ at $a$ due to the magnet $A$ is at right angles to the earth's horizontal component, $H_o$ (see Fig. 373), it follows from p. 548 that

$$H = H_o \tan \theta$$

But

$$H = \frac{2M}{d^3}$$

$$\therefore \quad \frac{2M}{d^3} = H_o \tan \theta$$

$$\therefore \quad M = \frac{d^3}{2} H_o \tan \theta \qquad . \qquad . \qquad . \quad (1)$$

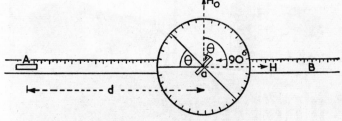

FIG. 378. Comparison of moments

Suppose the mid-point of the second short bar magnet $B$ is placed at the same distance $d$ from the small magnet $a$ when $A$ is taken away. If $M_1$ is the moment of $B$ and $\theta_1$ is the new deflection, it follows from similar reasoning to before that

$$M_1 = \frac{d^3}{2} H_o \tan \theta_1 \qquad . \qquad . \qquad . \quad (2)$$

Dividing equation (1) by (2), we have

$$\frac{M}{M_1} = \frac{\tan \theta}{\tan \theta_1}.$$

Thus knowing $\theta$ and $\theta_1$, the ratio $\dfrac{M}{M_1}$ can be calculated.

**Comparison of moments by null (zero deflection) method.** The moments of two magnets can also be compared by placing them on *opposite* sides of a magnetometer in an east-west position; for example, at $A$ and $B$ in Fig. 378. If similar poles of the magnets face each other, their intensities at the needle will act in opposite directions, and if one of the magnets is moved, a position will be found when the deflection of the needle is zero. In this case *the intensity $H_1$ at the needle due to one magnet is exactly equal to the*

FIG. 379. COMPASS AND BINNACLE ON BRIDGE OF SHIP

The two iron spheres are used to correct for induced magnetism in the ship due to the earth's field. A radar scanner is near the funnel.

*intensity $H_2$ at the needle due to the other magnet.* But intensity $= \dfrac{2M}{d^3}$ (p. 552).

$$\therefore \quad \frac{2M_1}{d_1{}^3} = \frac{2M_2}{d_2{}^3},$$

where $M_1$, $M_2$ are the respective moments of the magnets, and $d_1$, $d_2$ are the measured distances from the needle to the respective mid-points of the short bar magnet

$$\therefore \quad \frac{M_1}{M_2} = \frac{d_1{}^3}{d_2{}^3}.$$

## THE EARTH'S MAGNETISM

Centuries ago it was known that a pivoted magnetic needle came to rest pointing approximately north and south, and the magnetic compass needle is still used for navigation purposes. It follows, from the fundamental law of magnetism (p. 529), that the earth acts like a magnet, with a *south magnetic* pole A located near the geographic north pole a of the globe and a *north magnetic*

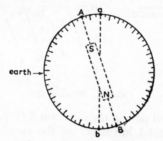

Fig. 380. Earth as a magnet

pole B near the geographic south pole b. Scientists have still not solved the mystery of the origin of the earth's magnetism, but some of the common observations can be explained by imagining a fictitious magnet at the middle of the earth (Fig. 380).

**The angle of dip.** If a non-magnetic uniform piece of metal, such as brass, is suspended at its centre of gravity M, the brass remains horizontally in equilibrium (Fig. 381 (a) ). If however, a magnetic needle is suspended at its centre of gravity G so that it is free to take up any position, it "dips" at an angle of about 70° to the horizontal in this country, with its north pole pointing downwards (Fig. 381 (b) ). Now, in general, a magnetic needle points in the direction of the total (resultant) intensity of the magnetic field in which it is situated. It therefore follows that the *resultant intensity R* of the earth's magnetic field acts at an angle of about 70° to the horizontal in this country.

So far we have studied the effect of the earth's field in a horizontal plane only, using a compass needle. An ordinary compass needle is pivoted so that it can only move in a horizontal plane.

We now see that the earth's field acts not only in a north-south direction, but also downwards; the compass needle does not show the latter effect because it cannot move in a vertical plane. The horizontal force on a unit pole in the earth's field, which we have denoted by $H_0$, is called the "horizontal component" of the field.

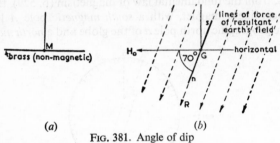

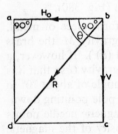

(a)                              (b)

FIG. 381. Angle of dip

The force on a unit pole in the direction $R$ (Fig. 381) is called the "resultant" or "total intensity" of the earth's magnetic field. Thus, $H_0$ is related to $R$ by $R \cos 70° = H_0$. Likewise, $R$ has a *vertical component*, $V$, given by $R \cos 20°$, but this has not concerned us previously in the book because the compass needles hitherto used were only at liberty to move in a horizontal plane.

A magnetic needle which can settle in the direction of the resultant intensity of the earth's field is said to be "freely suspended" in the field, and the *angle of dip* at a place on the earth's surface is the angle made with the horizontal by a freely-suspended magnetic needle there. Otherwise defined, the angle of dip is the angle made with the horizontal by the resultant intensity of the earth's field at the place concerned.

FIG. 382. Resultant and components

Fig. 382 illustrates the magnitudes of $H_0$, $V$, and their resultant $R$. Since $H_0$ and $V$ are at right angles, $abcd$ is a rectangle. The angle of dip, $\theta$, is thus given by

$$\tan \theta = \frac{ad}{ab} = \frac{V}{H_0},$$

so that $V$ can be found if the angle of dip and the earth's horizontal component $H_o$ are measured. It will be noted that, since $\theta$ is about 70° in this country, the length of $ad$ is greater than $ab$ in Fig. 382, and hence $V$ is greater than $H_o$ in magnitude.

Suppose that at a certain place it is found that the magnitude of the horizontal component of the earth's magnetic field is 0·21 oersted and that of the vertical component is 0·35 oersted. The value of the angle of dip ($\theta$) can be calculated as follows—

Tan $\theta = \dfrac{V}{H_o}$, where $V$ and $H_o$ are the vertical and horizontal components.

$$\therefore \quad \tan \theta = \frac{0·35}{0·21} = \frac{5}{3} = 1·66$$

$$\therefore \quad \theta = 59°$$

**Variation of angle of dip over the earth's surface.** If a freely suspended magnetic needle is taken to various points on the earth's surface, the needle "dips" at various angles to the horizontal. Near the geographic north and south poles, for example, the needle dips vertically, and the angle of dip is then 90°. Near the equator the freely suspended needle is horizontal, and the angle of dip is then zero. In this country the angle of dip is about 70°, as stated previously.

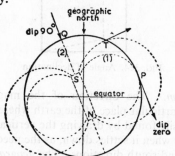

FIG. 383. Variation of dip

Fig. 383 illustrates how a fictitious magnet at the middle of the earth can account for the variation of the angle of dip. The lines of force due to the magnet are shown by dotted lines, and the tangent to the line of force (1) at $P$ on the earth is the direction of

19

the magnetic field there (p. 539). Now at this place the tangent is parallel to the earth's surface, since the latter is the tangent to the circle representing the globe. It follows that the resultant intensity of the earth's field is horizontal at *P*, and the angle of dip is therefore zero. At *Q*, near the geographic north pole, a tangent to the line of force is vertical, and hence the angle of dip is 90° here. This region contains the magnetic north pole of the earth, and in recent years an aeroplane equipped with a specially designed magnetic dip-needle has located this pole about 900 miles from the geographic north pole. At *T* on the globe, the tangent to the line of force is less than 90° to the earth's surface, and hence the angle of dip is less than 90°.

**Magnetic and geographic meridians. Angle of declination.** A vertical plane containing the earth's total intensity is known as a

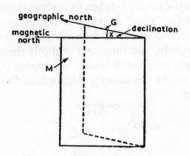

FIG. 384. Angle of declination

*magnetic meridian.* The direction of the magnetic meridian depends on the particular place of the earth where such a plane is drawn, and is easily found by drawing the vertical plane through a compass needle when it settles down, as the needle points in the magnetic north and south directions. The *geographic meridian* is the vertical plane passing through the geographic north and south poles, and at a given place on the earth this meridian *G* makes an angle *x* with the magnetic meridian *M* (Fig. 384). Seamen require to know the angle *x* to find their way about the globe, as a "correction" can then be applied to the magnetic compass needle direction to obtain the true (geographic) north. The angle *x* between the magnetic and geographic meridians is known as the *angle of declination*, or the *variation*. For navigation by magnetic compass, maps are prepared showing **isogonic lines**, i.e. lines connecting places having the same declination.

As the poles of a magnet are not at the ends (p. 529), the magnetic north and south direction is not the line joining the ends of a magnetic needle. The direction can be found by attaching two pins to points $P$, $Q$ on opposite sides of a magnet, and suspending the latter in a stirrup so that it can oscillate in a horizontal plane just above a sheet of paper (Fig. 385 (*a*)). When it comes to rest, the positions of $P$, $Q$ are noted on the paper. The magnet is then turned over and allowed to come to rest again, and the new

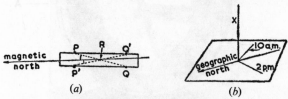

*(a)*          *(b)*

FIG. 385. Magnetic axis and declination

positions $P'$, $Q'$ of the two pins are again noted on the paper. Now the line joining the two pins is a fixed line in the magnet, and hence it always makes the same angle with the magnetic north and south when the magnet comes to rest. Thus $PQ$ and $P'Q'$ are inclined at equal angles to the magnetic north and south, and hence the latter direction is given by the *bisector* of the angle $PRP'$. The magnetic meridian, of course, is the vertical plane passing through the bisector.

The geographic north and south direction is the direction of the shadow of a vertical pole $X$ at noon, on a day when the sun is shining. Alternatively, the direction of the shadow is noted two hours before noon, for example, and two hours after noon, and the geographic north and south is that direction corresponding to the bisector of the angle between the two shadows (Fig. 385 (*b*)). This method was first given by Dr. Gilbert (p. 528). Knowing the magnetic north and south direction, the angle of declination is easily obtained.

**Measurement of the angle of dip.** At meteorological stations all over the world, measurements of the angle of dip are made daily. A *dip circle* is an instrument capable of measuring the angle of dip directly, and is illustrated in Fig. 386. It consists of a vertical circle $A$ graduated in degrees, which can rotate about a vertical axis passing through the centre of a horizontal circle $B$, also

FIG. 386. DIP CIRCLE

graduated in degrees. A magnetic needle $M$ is placed at the middle of $A$ so that it can rotate about a horizontal axis through its centre of gravity.

To measure the angle of dip, the vertical circle $A$ is turned until the needle is vertical. In this case the earth's horizontal component $H_o$, which acts in the magnetic meridian, has no effect in the plane of $A$, and as the needle is then only affected by the vertical component, it sets vertically. Thus, when the needle is vertical, the magnetic meridian is perpendicular to the plane of $A$. The reading on the lower graduated circle $B$ is noted, and the plane of $A$ is turned through an angle of 90° so that the magnetic needle $M$ now lies in the magnetic meridian. The inclination of $M$ to the horizontal is read on $A$, and this reading is the angle of dip.

---

## Summary

1. The force $F$ between two poles of $m_1$, $m_2$ c.g.s. units in air is given by $F = m_1 m_2/d^2$ **dynes**, where $d$ is the distance apart in centimetres. A pole of unit strength is one which repels a similar pole 1 cm. away in air with a force of one dyne.

2. The intensity of a magnetic field is measured by the force in dynes exerted on a unit pole placed in the field. Its value, $H$, is measured by the deflection magnetometer, and $H \propto \tan \theta$, where $\theta$ is the angle of deflection, provided that the field of intensity $H$ is at right angles to the earth's field.

3. When a bar magnet is situated in the earth's field with its **south pole pointing northwards,** two neutral points are obtained on the **axis** of the magnet. When the north pole of the magnet points northwards, two neutral points are obtained on the perpendicular bisector (equator) of the magnet.

4. **The moment of a magnet, $M =$ pole-strength $\times$ magnetic length.**

5. The *angle of dip* is the angle with the horizontal made by the resultant earth's magnetic field. The *declination* is the angle between the magnetic and geographic meridians at the place concerned. If $\theta$ is the angle of dip, and $R, V, H_0$ are the resultant, vertical component, and horizontal component respectively of the earth's field, then

$$\tan \theta = V/H_0, \quad R^2 = V^2 + H_0{}^2, \quad H_0/R = \cos \theta.$$

## WORKED EXAMPLES

1. *Briefly describe a magnetometer. If you were provided with a long bar-magnet how would you (a) find the position of a pole of the magnet, (b) use it with the magnetometer to verify the inverse-square law for a magnetic pole? Two magnets, each of magnetic length 8 cm. and of pole-strength 40 c.g.s. units,*

*are placed so that they lie along the same straight line with like poles 4 cm. apart. Calculate the resultant force of repulsion between the magnets. (C.W.B.)*

Suppose the magnets are A, B, arranged as shown in Fig. 387. To find the force of repulsion between them, consider the force exerted by each of the poles of A on those of B.

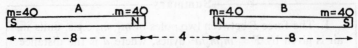

FIG. 387. Force between magnets

*North pole of A.* The force $F_1$ between this pole and the north pole of B is given by $\frac{m_1 m_2}{d^2}$, or $\frac{40 \times 40}{4^2}$ dynes. Hence $F_1 = 100$ dynes.

The force $F_2$ between the north pole of A and the south pole of B is given by $F_2 = \frac{40 \times 40}{12^2} = 11\frac{1}{9}$ dynes. But this force is one of attraction, whereas the force $F_1$ is one of repulsion. Hence resultant force $= F_1 - F_2 = 100 - 11\frac{1}{9} = 88\frac{8}{9}$ dynes.

*South pole of A.* The force $F$ between this pole and the north pole of B is given by $F = \frac{40 \times 40}{12^2} = 11\frac{1}{9}$ dynes; the force between the south pole of A and the south pole of B is given by $\frac{40 \times 40}{20^2} = 4$ dynes. The resultant of these forces is $(11\frac{1}{9} - 4)$, or $5\frac{1}{9}$ dynes, and is a force of *attraction*.

∴ force of repulsion of A on B $= 88\frac{8}{9} - 5\frac{1}{9} = 83\frac{7}{9}$ dynes.

2. *Define magnetic pole-strength and magnetic field strength. A bar-magnet 10 cm. long and of pole-strength 40 units is placed horizontally in an east-west position with its north pole pointing east. Find, from first principles, the field strength at a point P 10 cm. east of the north pole on the prolongation of the axis. If the horizontal component of the earth's field is 0·2 oersted, what is the resultant magnetic field at P ? (L.)*

To find the field strength at P, suppose that a *unit north* pole n is placed at P. The force $F_1$ on n due to the north pole N of the magnet is given by $F_1 = \frac{m_1 m_2}{d^2} = \frac{40 \times 1}{10^2}$ dynes, since $d = NP = 10$ cm. Thus, $F_1 = \frac{40}{100} = 0·4$ dyne. Similarly, the force $F_2$ due to the south pole on n is given by $F_2 = \frac{40 \times 1}{20^2}$, since $d = SP = 20$ cm. Thus $F_2 = \frac{40}{400} = 0·1$ dyne. *But the forces $F_1$, $F_2$ have opposite directions*, since N repels the n pole and S attracts it.

∴ resultant field strength, H, at P $= 0·4 - 0·1 = 0·3$ oersted and acts in the direction NP (Fig. 388).

It can now be seen that the horizontal component, 0·2 oersted, which is in a north-south direction, is *at right-angles* to the field of intensity 0·3 oersted. By the parallelogram of forces, the resultant intensity $R$ is hence given by

$$R = \sqrt{0·3^2 + 0·2^2} = \sqrt{0·13} = 0·36 \text{ oersted.}$$

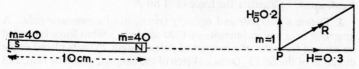

FIG. 388. Calculation on intensity

Further, the direction of $R$ is at an angle $\theta$ to the direction $H$ given by $\tan \theta = \dfrac{H_o}{H}$.

$$\therefore \quad \tan \theta = \frac{0·2}{0·3} = 0·66$$

$$\therefore \quad \theta = 33·5°.$$

3. *Describe the deflection magnetometer and explain how you would use it to investigate the way in which the magnetic intensity at a point on the axis of a short bar magnet varies with the distance of this point from the centre of the magnet. A short bar magnet A is placed 20 cm. from the needle of a magnetometer used in the ordinary way, and causes a deflection; a second short magnet B is brought up on the opposite side of the needle, A still remaining in place. When B is at a distance of 25 cm. from the needle the deflection is reduced to zero. Compare the magnetic moments of A and B.* (N.)

The intensity at the needle due to $A$ is given by $\dfrac{2M}{d^3}$, or $\dfrac{2M}{20^3}$, where $M$ is the moment of $A$. Similarly, the intensity at the needle due to $B$ is given by $\dfrac{2M_1}{25^3}$, where $M_1$ is the moment of $B$. Since the deflection of the needle is zero, the resultant intensity at the needle must be zero. Consequently the two intensities at the needle must be exactly equal and act in opposite directions.

$$\therefore \quad \frac{2M}{20^3} = \frac{2M_1}{25^3}$$

$$\therefore \quad \frac{M}{M_1} = \frac{20^3}{25^3} = \frac{64}{125} = 0·51.$$

The ratio of the moments is thus equal to 0·51.

## EXERCISES ON CHAPTER XXXV

**1.** Calculate the force between (i) two north poles of 10 and 20 units placed 5 cm. apart, (ii) a north pole of 35 units and a south pole of 25 units placed 10 cm. apart, in air.

**2.** A magnet *A* 15 cm. long has a pole-strength of 16 units. The axis of the magnet is placed in line with another magnet *B*, 20 cm. long and pole-strength 8 units, such that the two north poles face each other 5 cm. apart. Calculate the force of *A* on *B*.

**3.** Define a *unit pole* and *intensity* (*strength*) *of a magnetic field*. A magnetic field has an intensity of 1000 oersteds. What force is exerted on each of the poles of a magnet of pole-strength 25 units placed in the direction of the field? Draw a sketch of the magnet, magnetic field, and the two forces.

**4.** What magnetic polarity has the earth's north and south geographic poles? Give a reason for your answer.

**5.** Define *magnetic meridian, geographic meridian, declination* (or *variation*). Why is the variation required in navigation with magnetic compasses?

**6.** What reason have you for believing that the strength of the vertical component of the earth's magnetic field in England is greater than that of the horizontal component? (*N.*)

**7.** Write a detailed description of a method of comparing the moments of two magnets.

Two short bar magnets are placed with their magnetic axes perpendicular to the sides *AB* and *BC* of a horizontal square *ABCD*, of which the diagonal *BD* lies in the magnetic meridian. The centres of the magnets are at the centres of the sides. Show that their moments are equal if the centre of the square is a neutral point. Draw a diagram of the arrangement with the polarity of the magnets marked. (*N.*)

**8.** What is meant by *magnetic field-strength, magnetic line of force*? Give a diagram of the lines of magnetic force round a bar magnet placed on a table with its *N*-pole pointing south. If the magnet is 10 cm. long and has a pole-strength of 50 c.g.s. units, what is the magnetic field strength at a point 10 cm. from the *N*-pole and on the prolongation of the axis of the magnet? ($H_o = 0·18$ oersted.) (*L.*)

**9.** Define magnetic *declination* (*variation*) and magnetic *dip*. If you were provided with an unmagnetized steel knitting needle, how would you use it to demonstrate magnetic dip? Give details of your method of magnetizing the needle. Explain, giving reasons, how you would magnetize a piece of mild steel as strongly as possible by induction in the earth's magnetic field. (*C.W.B.*)

**10.** Describe the deflection magnetometer and explain how you would use it to investigate the way in which the magnetic intensity at a point on the axis of a short bar magnet varies with the distance of this point from the centre of the magnet. A short bar magnet $A$ is placed 20 cm. from the needle of a magnetometer used in the ordinary way, and causes a deflection; a second short magnet $B$ is brought up on the opposite side of the needle, $A$ still remaining in place. When $B$ is at a distance of 25 cm. from the needle the deflection is reduced to zero. Compare the magnetic moments $A$ and $B$. (N.)

**11.** Describe the structure of *either* a ship's compass *or* an aeroplane compass, and give an explanation of the chief features of its design. Explain the use, to a navigator, of isogonic lines on a map. (N.)

**12.** Define *magnetic meridian, declination, dip.* Describe the construction of a dip-circle and explain how you would use it: (a) To find the magnetic meridian: (b) to measure the angle of dip. (O. & C.)

**13.** Define the terms *pole strength, strength of a magnetic field.* Give an account of a method of comparing the pole strengths of two very long bar magnets.

You are provided with a steel rod marked at one end. Explain clearly how you would magnetize the rod strongly so that it had a north pole at the marked end. (C.)

**14.** Describe experiments, *one* in each case, to show that (a) the direction of the earth's magnetic field (in England) is inclined downwards; (b) a soft iron rod can be magnetized by induction in the earth's magnetic field; (c) iron is more intensely magnetized than steel by the action of the same magnetic field. (N.)

**15.** How would you determine the angle of dip at a point on the earth's surface? If the strength of the horizontal component of the earth's field is 0·18 oersted and the angle of dip is 60°, what is the magnitude of the vertical component of the earth's field? Show diagrammatically the general form of the earth's magnetic field in a section through the earth's magnetic poles. (O. & C.)

**16.** Describe and explain the magnetization acquired by a soft iron rod when placed vertically and hammered.

Sketch the magnetic field in a horizontal plane through the $N$-pole of a very long vertical bar magnet and calculate its pole strength if the neutral point is formed at a distance of 12 cm. from the pole ($H=0·18$ oersted). (L.)

# CHAPTER XXXVI

## MAGNETIC EFFECT OF CURRENT

IN previous chapters the chemical effect (electrolysis) and the heating effect of an electric current were discussed. The most important effect of an electric current, however, is its *magnetic effect*, which was discovered by OERSTED in 1820 while he was lecturing to students. He noticed that a magnetic needle was slightly disturbed whenever an electric current was passed along a wire. Up to the time of Oersted's observation, scientists had failed to connect the phenomena of magnetism and electricity, but a brilliant series of investigations by AMPÈRE, a French scientist, shortly after Oersted's observation was published, showed conclusively that *a current in a wire always gives rise to a magnetic field round it*. As we shall see later, the importance of the magnetic effect of a current lies in the fact that it gives rise to mechanical forces. Among numerous industrial applications of the magnetic effect are the telephone, the motor and the dynamo.

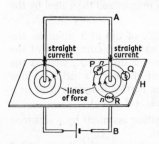

FIG. 389. Straight conductor

The appearance of the magnetic field round a current-carrying conductor depends on the *shape* of that conductor, and we will now discuss four cases of practical importance.

**The straight conductor.** A cable laid below the sea to carry telegraph messages from this country to America, for example, consists of a straight conductor which carries an electric current when a message is sent. The overhead cables which supply electric current to trolley-buses are mainly straight conductors.

Fig. 389 shows a simple experiment to illustrate the appearance of the magnetic field round a straight vertical wire *AB*. The wire passes through a horizontal sheet of cardboard, *H*, and compass needles such as *P*, *Q*, *R* are arranged round *AB*. Before the current is passed the needles all point northwards; but when the current flows they are observed to move, and point somewhat as shown in the figure. By sprinkling iron filings on the cardboard and using a very strong current, it is possible to show that the lines of force round the straight wire are *concentric circles* whose centre lies on the wire. The same result is obtained on all other horizontal planes passing through the wire.

**The circular coil.** In the early days of electricity an electric current was measured by passing it through a circular coil, and observing the deflection of a magnetic needle at its centre (see "Tangent Galvanometer," p. 575). Fig. 390 (*a*) illustrates the appearance of the magnetic field in a plane passing at right angles through the middle of the coil. The lines of force are circular near the wire, but they become straight and parallel at *M*, the middle of the coil.

Fig. 390 (*b*) illustrates a section of a circular coil of wire joined to the terminals of copper and zinc plates contained in a small

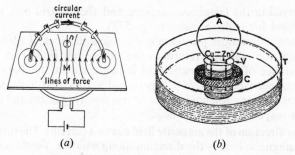

*(a)*                    *(b)*

FIG. 390. Magnetic effect of circular coil

vessel of dilute sulphuric acid. The vessel *V* is floated on water in a trough *T* by means of cork *C*, and a current flows in the coil as a result of the action of the simple cell (p. 498). When the south pole of a magnet is brought near to the face *A* of the coil, the whole arrangement drifts across the water away from the magnet. If the magnet is turned round and the north pole is presented to the face

*A*, the arrangement is drawn across the water towards the magnet. This simple experiment shows that the current in the coil has a magnetic effect which causes the face *A* to behave like a south magnetic pole.

**The solenoid.** The solenoid is a long coil usually containing a large number of close turns of insulated wire (Fig. 391). It is

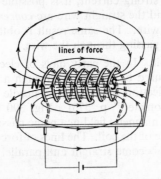

FIG. 391. Magnetic effect of Solenoid

employed in the telephone earpiece and the electric bell, and is also used for making magnets (p. 532).

When a current is passed into the solenoid, lines of force pass through the coil and return to the other end, as shown. The appearance of the magnetic field is hence similar to that of the bar magnet (see Fig. 366); experiment shows that a suspended current-carrying solenoid comes to rest pointing north and south, like a suspended magnetic needle.

**The direction of the magnetic field due to a current.** The direction of a magnetic field is the direction along which a *N*-pole tends to move if placed in that field, and hence this is the direction of the lines of force in the field (p. 539). A general rule for the magnetic field direction due to a current-carrying conductor was given by Clerk Maxwell, and is known as *Maxwell's corkscrew rule.* It states: *If a right-handed corkscrew is turned so as to move in the direction of the (conventional) current in a conductor, the direction of rotation of the corkscrew is the direction of the lines of force.*

Maxwell's corkscrew rule is easiest to apply in the case of the straight conductor; Fig. 392 (a) shows the direction of a line of force round a straight wire A carrying a current in the conventional direction. A magnetic needle below A is hence deflected with its north pole into the paper, while the north pole of a needle above A is deflected out of the paper towards the reader. The direction of the lines of force in Fig. 389 should now be verified.

In the case of the circular conductor, such as the circular coil or solenoid, Maxwell's corkscrew rule can be applied if the corkscrew is rotated in the direction of the current in the coil; the direction

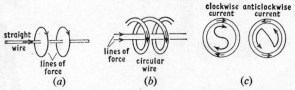

FIG. 392. Rules for magnetic effect of current

of movement of the corkscrew is then the direction of the lines of force. Fig. 392 (b) illustrates the directions of the conventional current and lines of force in part of a coil B.

Other rules than Maxwell's have been put forward for the direction of the magnetic field, especially for the case of the circular conductor. A *clock rule* states: *If the current flows in a clockwise direction, the end of the coil viewed appears to act like the S-pole of a magnet; if the direction is anti-clockwise, that end acts like a magnetic N-pole.* Fig. 392 (c) illustrates one way of remembering the rule; arrows at the ends of the letters S and N follow each other round in a clockwise and anti-clockwise direction respectively. Yet another rule for the direction of the magnetic field, a *clenched fist rule,* states: *If a coil is held with the right hand so that the fingers are curled round it in the same direction as the current, the outstretched thumb points in the direction of the magnetic field.* By applying one of the above rules, it will be seen that the current in a solenoid makes the ends of the coil appear to act like a south and north magnetic pole respectively. The lines of force are thus similar to those obtained with a bar magnet (see Fig. 366 (a) ).

**The electromagnet.** An electric current can be used for making a temporary magnet known as an *electromagnet*, which has extensive use in industry. Fig. 393 shows an electromagnet used for lifting heavy masses of iron. Electromagnets are also used in

FIG. 393. ELECTROMAGNET LIFTING PIG IRON

scrap iron yards for separating iron and steel from non-magnetic materials like copper and brass.

The electromagnet consists essentially of a soft iron core with insulated copper wire wound round it. In the case of a horse-shoe type *A* (Fig. 394), the wire must be wound in opposite directions on the two limbs to obtain a magnet with *N*- and *S*-poles at the free ends. This follows from the clock rule for determining polarity since the current flows in opposite directions round each limb. When the current is switched off, the magnetism in $A$ diminishes quickly to a low value. This would not be the case if steel were used instead of soft iron (see p. 529).

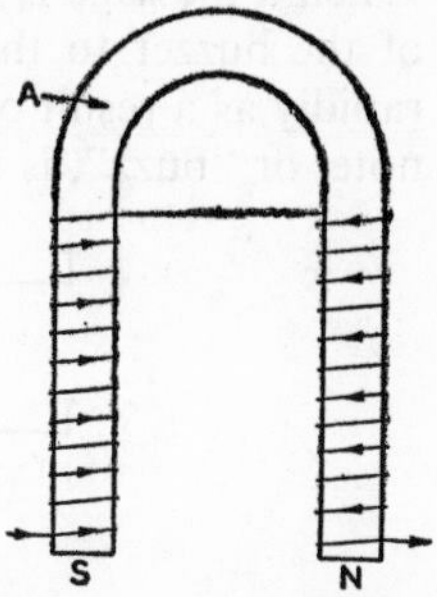

**The electric bell.** The essential features of an electric bell circuit are illustrated in Fig. 395. The Leclanché battery $A$ is connected to a bell-push $B$, and an electromagnet $M$, and the circuit is completed through the point $N$ of a screw and a spring $S$ attached to a piece of soft iron $T$. A clapper, $C$, is attached to $T$, and is initially a short distance away from a gong $G$.

FIG. 394. Electromagnet

When the bell-push $B$ is depressed, the circuit is completed. The electromagnet acts and attracts the soft iron $T$, and $C$ strikes the gong, as illustrated by the dotted lines in Fig. 395. Since $T$ leaves the point $N$, the circuit is now broken. Thus the electromagnet ceases to hold the iron, and $T$

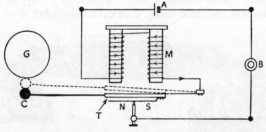

FIG. 395. Electric bell

falls back to $N$; whereupon, as the circuit is now made, the whole action described begins again. The gong is struck continually at a rate which depends partly on the elasticity of the spring $S$ and partly on the degree of contact between the screw-point and $T$.

**The buzzer.** The *buzzer* is an apparatus producing an audible note, and is used as the calling apparatus in telegraph and telephone installations. It consists of an electromagnet $M$, with a flexible soft iron armature *A* arranged just above the poles (Fig. 396). One end of *A* is connected to the end of one of the coils *C*, and contact is made with the middle of *A* by an adjustable screw *S*. When a message is received, a current flows from one terminal *T* of the buzzer to the other terminal, and the armature *A* vibrates rapidly as a result of the make-and-break action. A high-pitched note, or "buzz", is thus produced.

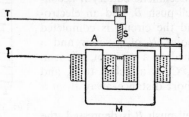

FIG. 396. Buzzer

**The telephone and telephone ear-piece.** After years of continued efforts, ALEXANDER GRAHAM BELL designed the first telephone in 1876. Nowadays we are so used to the existence of the telephone that we are apt to overlook Bell's service to society in enabling people to communicate easily with each other; but if we think of

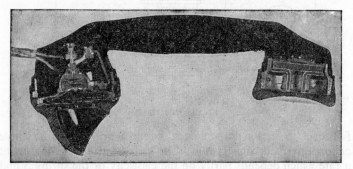

FIG. 397. G.P.O. MICROPHONE-TELEPHONE EARPIECE

the ease with which we can to-day get in touch with the doctor, we obtain some idea of the practical value of Bell's researches.

A telephone system requires essentially two devices: (1) a device at one end to make an electric current varying in the same way as the air pressure due to the sound, known as a *microphone* (see p. 574); (2) a device at the other end to convert the varying current into varying air pressure, so that the sound is reproduced (the *telephone earpiece*). Fig. 397 shows a modern hand-microphone telephone in section. The telephone earpiece is further illustrated in Fig. 398. A permanent magnet $X$ is bolted to pieces of soft iron curved at one end, and coils of insulated wire, represented by $B$, $D$ are placed round the other ends of the soft iron. A soft iron circular plate $A$ is near to the latter, and when the earpiece is used, $A$ is close to the ear.

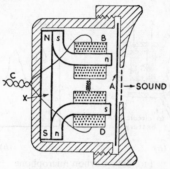

Fig. 398. Telephone earpiece

The presence of the magnet $X$ results in the ends of the curved soft iron pieces near $A$ becoming a north ($n$) and a south pole ($s$) respectively. When a person speaks into the microphone at the other end of the telephone line the sound energy is converted into electrical energy (see p. 574), and electric currents, *varying in magnitude at the frequency of the sound*, travel along and pass through $C$ into the coils $B$, $D$, which are connected together. Since the current varies, the strength of the magnetic effect due to the current also varies. Thus the total attractive force on the soft iron plate $A$ now varies, and $A$ begins to vibrate to and fro. This movement gives rise to sound waves in the air, and the notes produced are a close replica of those spoken into the microphone.

**The carbon microphone.** A *microphone* is an apparatus which converts sound energy to electrical energy, used in modern tele-phones and broadcasting. There are a number of microphones working on different principles, but we shall deal only with the *carbon microphone*, which was first designed in 1878 by HUGHES.

This microphone consists essentially of two carbon blocks $X$, $Y$, with carbon particles between them (Fig. 399 (*a*) ). A battery $B$ is connected to $X$, $Y$ so that a steady current, $I$, flows in the circuit. The value of $I$ is given by $\dfrac{E}{R}$, from Ohm's law, where $E$ is the battery e.m.f. and $R$ is the total resistance of the carbon between

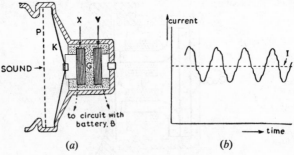

FIG. 399. Carbon microphone

$X$, $Y$, the connecting wire in the circuit, and the internal resistance of the battery. Fig. 399 (*b*) shows the steady current $I$ as a dotted straight line.

$K$ is a thin conical diaphragm; $P$ is a perforated plate protecting it from damage. When a person speaks into the microphone, sound waves impinge on $K$ and make it vibrate. $K$ thus moves very slightly to and fro in accordance with the varying pressure of the sound wave. When it moves to the right the carbon particles are compressed; when $K$ moves to the left the particles are loosened. Since the contact between the particles is worse in the latter case, the resistance between $X$, $Y$ is increased, and, conversely, the resistance between $X$, $Y$ is decreased when the particles are compressed. From Ohm's law, it follows that a *varying* current flows in the circuit whose magnitude varies at the same frequency as the pressure variation of the sound waves. This is illustrated in Fig. 399 (*b*). Sound energy is thus converted to electrical energy in the microphone circuit, and the varying electric currents travel along to a telephone earpiece.

**The moving-iron ammeter.** Soon after the magnetic effect of a current was discovered, scientists devised instruments for detecting and measuring a current which utilized this effect. One type of commercial instrument, known as a *moving-iron ammeter*, is illustrated in Fig. 400. It consists essentially of a solenoid, *S*, connected to the terminals *TT* of the instrument, with a fixed piece of soft iron *F* and a movable soft iron *M* inside it. When a

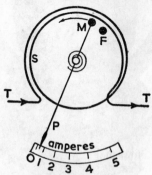

FIG. 400. Moving-iron ammeter

current flows in the solenoid, *F* and *M* become magnetized with like poles facing each other, and *M* is therefore repelled. An attached pointer *P* is then deflected. Since the repulsion depends on the strength of the poles of *M* and *F*, and this in turn depends on the strength of the current, the deflection of *P* is a measure of the current. The deflection may be controlled by a spring and the scale is graduated in amperes. As the force of repulsion is not proportional to the current in *S*, equal steps of current do not correspond to equal intervals on the scale, and the latter is thus known as a *non-uniform* scale.

**The tangent galvanometer.** The tangent galvanometer was used in the early days of electricity for measuring an electric current, but it is no longer used commercially. It consists essentially of a vertical circular coil of wire with a large number of turns, and a small magnetic needle suspended at the middle of the coil. The magnet has a light aluminium pointer rigidly attached at right angles to it, and the deflection of the magnet is noted by the movement of the pointer over a horizontal scale graduated in degrees (Fig. 401).

FIG. 401. Tangent galvanometer

Before the instrument is used, the coil, which can turn about a vertical axis, is rotated so that its plane is in the magnetic meridian. In this position the plane of the coil points in the same direction as the small magnet $NS$, which is the direction of $H_o$, the intensity of the earth's horizontal magnetic field. When a current is now passed into the coil a new magnetic field, of intensity $H$, is set up at the middle, the direction of $H$ being *perpendicular* to the plane of the coil (Fig. 402). (This direction is illustrated in Fig. 390 (*a*) by the lines of force at $M$.)

The magnetic needle now swings round through an angle $\theta$ from the direction of $H_o$, its original direction, and since it is in equilibrium in this position, the *resultant* of $H$ and $H_o$ acts at angle $\theta$ to the direction of $H_o$. Since $H$ and $H_o$ are perpendicular, the parallelogram of forces which represents them is a rectangle $abcd$ (Fig. 402). Consequently $\dfrac{bc}{ab} = \tan \theta$; but $bc$, $ab$ represent $H$, $H_0$ respectively. Hence $\dfrac{H}{H_0} = \tan \theta$.

$$\therefore \quad H = H_0 \tan \theta \quad . \quad . \quad . \quad . \quad (1)$$

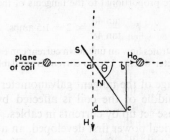

FIG. 402. Theory

Now the magnetic intensity $H$ due to the current depends on its magnitude, $I$ amps., the radius, $r$ cm., of the coil, and the number of the turns, $n$, in the coil; and a calculation beyond the scope of this book shows that

$$H = \frac{2\pi n I}{10r} \quad . \quad . \quad . \quad (2)$$

Substituting for $H$ in (1),

$$\therefore \quad \frac{2\pi n I}{10r} = H_0 \tan \theta$$

$$\therefore \quad I = \frac{10 r H_0}{2\pi n} \tan \theta.$$

Consequently if $r$, $H_0$, $n$ are known, and the deflection $\theta$ is observed, the current $I$ can be calculated. Further, the alarming-looking equation for $I$ can be simplified by using a single letter $k$ for $\dfrac{10 r H_0}{2\pi n}$, so that we can now write

$$I = k \tan \theta \quad . \quad . \quad . \quad (3)$$

This quantity $k$ is a constant number if the same galvanometer is always at the same place, since $H_0$, $n$, and $r$ are then all constant. From (3), therefore, it follows that $I \propto \tan \theta$.

Suppose that a current of 2 amps. produces a deflection of 48°, and an unknown current $I$ produces a deflection of 40° in the same instrument.

Then
$$\frac{I}{2 \text{ amps.}} = \frac{\tan 40°}{\tan 48°},$$

since the currents are proportional to the tangents of their respective angles of deflection.

$$\therefore \quad I = \frac{\tan 40°}{\tan 48°} \times 2 = 1·5 \text{ amps.}$$

This calculation illustrates how an unknown current can easily be measured by a tangent galvanometer when a known current is available.

The disadvantage of the tangent galvanometer is that the small magnet at the middle of the coil is affected by stray magnetic fields, such as those set up by currents in cables. Years ago, when commercial electrical power first developed, an attempt was made to "screen" the instrument from stray magnetic fields by housing it in a box of iron, but this was found unsatisfactory. The moving-coil instrument (p. 587) has now completely displaced the tangent galvanometer as a commercial instrument for measuring current.

**The astatic galvanometer.** The magnetic needle of the tangent galvanometer is subject to control by the earth's field when the needle is deflected. In order to minimize the effect of the earth's field, and hence to have greater sensitivity, a galvanometer was designed which had a needle consisting of two small magnets of about equal strength pointing in opposite directions. The two magnets are together equivalent to a weak magnet, so that the earth's magnetic field has a small control over the deflection of the needle. Fig. 403 illustrates the needle of an *astatic galvanometer*, as this type is called. A coil $C$ is wound round the lower of the two magnets, and when a current is passed into it, the magnetic field created causes the lower magnet to be deflected. Both magnets rotate, since they are rigidly connected, and the angle of rotation, which gives a measure of the current, can be measured by the movement of

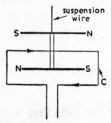

FIG. 403. Astatic needle

the top magnet over a circular graduated scale (not shown) placed just below it.

**The telegraph.** If you wish to send an urgent message to a relative or friend living a long way off, the quickest way is by telegram. This method of communication was devised by scientists who saw that the magnetic effect of a current could be turned to useful account, and in 1837 WHEATSTONE (see p. 514) and COOKE invented the first telegraph system in this country, which can be seen at the Science Museum, South Kensington, London.

The *basic principle* of the early telegraph is shown in Fig. 404. Each station contains a battery, a galvanometer, and a rocking

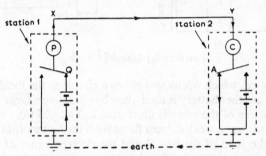

FIG. 404. Principle of early telegraph

key, with a wire $XY$ connecting them. With the keys as shown, a message can be sent from station 1 to station 2 by making and breaking the circuit at $Q$. In the early days, the magnetic needles in the galvanometers $P$, $C$ could be deflected to the left or right according to whether a dot or a dash of the Morse code was intended. If a message is to be sent from station 2 to station 1, the positions of the keys in Fig. 404 are interchanged.

**The sounder and relay.** As the operator became tired quickly when watching the magnetic needles, the *sounder* was invented. This consists of a brass lever $A$ which can rock about $R$ and make contact with a stop at $X$ or $Y$ (Fig. 405 (*a*) ). $B$ is a soft iron armature attached to $A$, and $C$ is part of an electromagnet whose coils carry the current flowing along the telegraph line. When a message is received, the current in the coils makes the electro-magnet function. The lever $A$ makes audible "clicks" as it strikes $Y$ or $X$, and thus enables the operator to hear the Morse code transmission.

When a long telegraph line is used, a small current flows along it because its resistance is fairly large, and the signals along the line are then weak. In order to overcome this disadvantage a

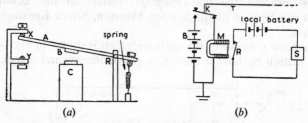

(a)                                    (b)

FIG. 405. (a) Sounder (b) relay

*relay* is used, which opens and closes a circuit at the local station in which another battery is used, thereby obtaining a large current. The principle of the relay is illustrated in Fig. 405 (b). With the key $K$ as shown, a weak current flows from the battery (not shown) at the other end of the line $T$, and the electromagnet $M$ attracts the soft iron armature $R$. This closes the circuit at the local station, which contains a battery and the sounder $S$, and hence the message is relayed to the sounder. Nowadays the telegraph system is a complex but highly efficient circuit. Two wires are used for the to and fro passage of the current, the use of the earth being avoided owing to the presence of underground wires. Many relays are used on long distance lines. The message is now automatically printed on tapes.

**Cables.** The telegraph system transmits messages over land. The cable system which transmits messages under the ocean resulted from the pioneer work of Lord Kelvin, who investigated theoretically the conditions necessary for the efficient flow of electricity along cables. The first cable was successfully laid between Dover and Calais in 1850, and in 1857 attempts were made to lay more than 2000 miles of cables across the Atlantic.

There were many heart-breaking failures. Against the advice of Lord Kelvin, who suggested on theoretical grounds that a p.d. of only a few volts was sufficient to obtain a satisfactory current, a few thousand volts was applied to the long cable, which was ruined. The first cables kept breaking, and ships were sent out to locate the ends of the cables in mid-ocean and to repair the break. A sensitive mirror galvanometer, improved by Kelvin, was used at that time to detect the current which flowed along the cables, and one can well imagine the relief of the engineers at the reception end when the spot of light from the mirror began moving across the scale, indicating that the break was repaired. By 1866 there was an efficient cable link across the Atlantic, and to-day there are about 365,000 miles of cable over the ocean bed, a distance nearly 13 times the circumference of the earth.

The insulation of the copper conductor in the cable must be very high, as sea-water is a conductor. Layers of other materials are used to prevent corrosion and to guard against injury by sea creatures.

————Summary————

1. When a conductor carries an electric current, a magnetic field exists at right angles to the plane of the conductor.

2. Maxwell's corkscrew rule states: **If a right-handed corkscrew is turned so as to move in the direction of the current in a conductor, the direction of rotation of the corkscrew is the direction of the lines of force. In a coil, a clockwise current makes that face act like a south pole; anti-clockwise current makes the face act like a north pole.**

3. The lines of force round a *straight* current-carrying wire are circular. For a flat *circular coil*, they are straight in the middle of the coil in a plane through its centre perpendicular to the coil, but circular near the edges. The lines of force pass through and round a *solenoid*, similarly to the lines round a bar magnet.

4. The current in a tangent galvanometer is proportional to $\tan \theta$, where $\theta$ is the angle of deflection, provided the plane of the coil is set along the magnetic meridian.

## WORKED EXAMPLE

*Give a labelled diagram of a good form of primary cell and explain its action. Point out the chief advantages and disadvantages of the cell given.*

*Two different primary cells are joined in series with a tangent galvanometer and a suitable resistance, and the deflection on the galvanometer is observed to be 60°. On reversing one of the cells, so that it is in opposition to the other cell, the deflection on the galvanometer is observed to fall to 30°. Compare the e.m.f.s of the cells. (C.W.B.)*

Let $E_1$, $E_2$ be the respective e.m.f.s of the two cells, and suppose $R$ is the total resistance of the circuit. When the two cells are arranged so that the two e.m.f.s assist each other, their net e.m.f. is $(E_1 + E_2)$. Consequently the current, $I_1 = \dfrac{E_1 + E_2}{R}$. But $I_1 = k \tan 60°$ (p. 577).

$$\therefore \quad \frac{E_1 + E_2}{R} = k \tan 60° \quad . \quad . \quad . \quad (4)$$

When the two cells are arranged so that their e.m.f.s are in opposition, their net e.m.f. is $(E_1 - E_2)$. The current now flowing is given by $\dfrac{E_1 - E_2}{R}$, as the total resistance remains unaltered when the cells are reversed.

$$\therefore \quad \frac{E_1 - E_2}{R} = k \tan 30° \quad . \quad . \quad . \quad (5)$$

Dividing (4) by (5) to eliminate $R$ and $k$, we obtain

$$\frac{E_1 + E_2}{E_1 - E_2} = \frac{\tan 60°}{\tan 30°}.$$

But $\qquad \dfrac{\tan 60°}{\tan 30°} = 3$, by calculation

$$\therefore \quad \frac{E_1 + E_2}{E_1 - E_2} = 3$$

$$\therefore \quad E_1 + E_2 = 3E_1 - 3E_2$$

$$\therefore \quad 4E_2 = 2E_1$$

$$\therefore \quad \frac{E_2}{E_1} = \frac{1}{2}.$$

## EXERCISES ON CHAPTER XXXVI

**1.** Name four practical applications of the magnetic effect of an electric current.

**2.** Draw sketches of an electromagnet and a telephone earpiece. Explain briefly how each functions.

**3.** Name two rules for finding the directions of the magnetic field round a current-carrying conductor, drawing diagrams to illustrate your answer.

**4.** Explain how the electric bell works. What possible causes would you seek if your bell ceased to function one day?

**5.** Draw diagrams of (i) a tangent galvanometer, (ii) a moving-iron ammeter. Explain *briefly* how each functions, and state if either instrument is commercially used.

**6.** Describe and explain *two* tests which you would make to determine whether an electric current is flowing in a wire. How could *one* of your methods be adapted to estimate the magnitude of the current in the wire? (*L.*)

**7.** How does the strength of the magnetic field at the centre of a circular coil through which a current is passing depend on (*a*) the number of turns, (*b*) the radius of a coil? (*N.*)

**8.** Describe an experiment to demonstrate the nature of the magnetic field associated with a circular coil of wire carrying an electric current. Give a diagram of the field and state a rule relating the directions of the current and the field. Explain, by the aid of diagrams, the use of such a coil to measure an electric current. (*C.W.B.*)

**9.** Describe an experiment to show the magnetic field due to a current flowing in a long straight wire. Give a diagram of the magnetic lines of force and indicate clearly the directions of the field and the current. Describe briefly any practical application of the magnetic field produced by a current and explain its mode of action. (*L.*)

**10.** Describe the construction of a simple form of electromagnet and state three factors on which its lifting power depends. Describe a simple method of ascertaining which of two magnetic materials would be best for use in an electro-magnet, if you were given a small bar-shaped specimen of each. (*N.*)

**11.** Describe the tangent galvanometer and explain how it is used for measuring currents. The coil of a tangent galvanometer has 20 turns, each of radius 10 cm., wound with wire of total resistance 30 ohms. This coil is connected across the terminals of a 2-volt accumulator of negligible internal resistance. Calculate (*a*) the intensity of the magnetic field at the centre of the coil, (*b*) the deflexion of the needle, given that the magnitude of the horizontal component of the earth's field is 0·2 oersted (dynes per unit pole). (*N.*)

**12.** Explain the following effects: (*a*) The reading of a compass placed on the north side of a vertical cable alters when a current flows in the cable. (b) A long spiral of wire hung from one end with its axis vertical, the lower end just dipping into mercury in a dish, oscillates up and down when a cell is joined between the top of the coil and the mercury. (L.)

**13.** State the nature of the magnetic effect when an electric current flows through a wire. Draw diagrams to show the lines of magnetic force for: (a) A straight wire; (b) a circular coil of wire, when carrying an electric current; indicating clearly on the diagrams the corresponding directions of the current and magnetic field. Describe a method of magnetizing a steel knitting needle by an electric current so that a chosen end becomes a north pole. (O. & C.)

**14.** (a) Draw a diagram of an electric bell. Show clearly the path of the current through it and explain its method of working. What cell would be the most convenient to work it? Give your reasons.

(b) Describe the construction and action of a Daniell cell and explain how the faults of a simple cell are avoided. (L.)

**15.** Draw diagrams showing the lines of magnetic force in a horizontal plane associated with current flowing in (a) a long, straight, vertical wire, (b) a solenoid with its axis horizontal. In each diagram show the direction of the current and in (b) state a rule which gives the polarity of the coil. Describe the method you would use to plot the lines of force in either (a) or (b). (N.)

# CHAPTER XXXVII

## FORCE ON A CONDUCTOR
## THE MOTOR PRINCIPLE

SOON after Oersted's discovery of the magnetic effect of a current, Ampère showed in 1821 that, under certain conditions, a force was exerted on a current-carrying conductor situated in a magnetic field. From Ampère's fundamental work in the laboratory, there have evolved the electric motor, the moving-coil ammeter and voltmeter, and the moving-coil loudspeaker, to quote a few examples.

Fig. 406 illustrates a simple experiment to show the force on a current-carrying conductor *ab* in the magnetic field *H* of a horseshoe magnet *M*. It will be noted that the field *H* is arranged *perpendicular* to the wire carrying the current, *I*. The lower end *b* of the conductor just touches the surface of a pool of mercury inside a vessel *P*, and the upper end is suspended loosely at *a*, where it joins a wire from an accumulator *A*. The circuit is completed by dipping another wire from the accumulator into the mercury pool at *e*.

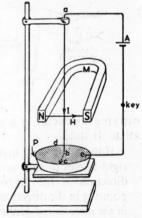

When a current *I* is passed into the wire, the wire swings about *a* and the end *b* skims along the mercury surface in the direction *bc*. As soon as the circuit is broken the wire moves back to its original position at *b*. When the magnet *M* is turned round so that the direction of the magnetic field is opposite to that shown in Fig. 406, the conductor swings in the

FIG. 406. Force on conductor

direction *bd* about *a*, in the opposite direction to *bc*. When the magnet is kept fixed and the current in the wire is reversed, the force on the conductor is reversed.

**Why a force is exerted on a current-carrying conductor in a magnetic field.** When an engine collides with the buffers at a railway station, the buffers exert a force equal and opposite to that of the engine on it. This is an example of the "law of action and reaction" (see p. 31). Now, owing to the magnetic effect of a current, a magnet is deflected when a current-carrying conductor is placed near to it, i.e. a force is exerted on the magnet. It follows, from the law of action and reaction, that an equal and opposite force is exerted *on the conductor*. By using a fixed magnet and a movable current-carrying conductor, as illustrated in Fig. 406, the force on the conductor can be observed.

**Fleming's left-hand rule for direction of force.** FLEMING gave a useful rule for determining the direction of the force on a current-

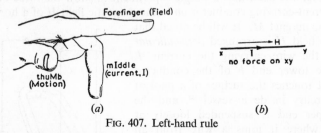

(a)                                        (b)

Fig. 407. Left-hand rule

carrying conductor in a magnetic field, known as the LEFT-HAND RULE. It states—

If the thumb and first two fingers of the left hand are held at right-angles to each other, with the *F*ore-finger pointing in the direction of the magnetic *F*ield (*H*) and the m*I*ddle finger pointing in the direction of the current (*I*), *then the thu*M*b points in the direction of Motion of the conductor, i.e. in the direction of the force acting on it.* (See Fig. 407 (*a*).)

The reader should now verify the direction of the force on the conductor *ab* in Fig. 406.

**The magnitude of the force** depends on the magnitudes of the current $I$, the intensity $H$ of the magnetic field, the length $l$ of the conductor, and its inclination to the magnetic field. Experiment shows that the force is zero when the current is *parallel* to the magnetic field (Fig. 407 (*b*) ), and that the force increases as the inclination to the field becomes bigger until the current is *perpendicular* to the magnetic field. In this position the force $F$ on the conductor is a maximum, and is given by

$$F = IHl/10 \text{ dynes} . \qquad . \qquad . \qquad . \qquad (1)$$

where $I$ is the current in amps., $H$ is the field strength in oersteds, and $l$ is the length of the conductor in centimetres. The formula shows that the greater the current, and the greater the magnetic field intensity, the greater is the force on the conductor.

**The moving-coil instrument.** The most accurate commercial instrument for measuring current and potential difference utilizes the action of the force on a conductor in a magnetic field. The essential features are (i) an insulated rectangular coil *abcd*, (ii) a powerful magnetic field $H$ produced by the curved poles, $N, S$, of

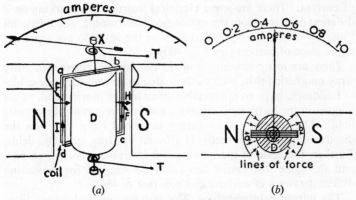

FIG. 408. Moving-coil instrument

a horse-shoe magnet, (iii) a soft iron cylinder $D$ to concentrate the lines of force towards its centre, (iv) two springs to control the rotation of the coil about a spindle moving in jewelled bearings at $X, Y$. Fig. 408 (*a*) illustrates a side view of the apparatus, while Fig. 408 (*b*) is a view from above. When the terminals of the instrument, T, T, are connected in an electrical circuit, a current flows through the coil *via* the springs, each spring being connected to one terminal.

By Fleming's left-hand rule, the forces $F$ on the two vertical sides *ad*, *bc* of the rectangle *abcd* act opposite to each other, while there are no forces on the sides *ab*, *dc* since the currents in them are parallel to the magnetic field $H$ (see Fig. 407 (*b*) ). It can be seen that the two forces $F$ on *ad*, *bc* tend to *rotate* the coil about $XY$ in a clockwise direction, and a pointer $P$ attached to the coil is deflected through an angle depending on the elasticity of the springs, which oppose the motion. The pointer moves over a scale, shown in Fig. 408 (*b*). Since the force on a given conductor in a fixed magnetic field is proportional to the magnitude of the current (p. 587), the rotation of the coil is proportional to the current. Thus a current of 1·5 milliamps, deflects the coil through an angle from its zero position which is three times the deflection caused by a current of 0·5 milliamp. On account of this proportional relation, equal divisions on the scale represent equal steps of current. There are some electrical instruments, working on a different principle from the moving-coil instrument, which have an "uneven" scale, i.e. equal divisions on the scale do not represent equal steps of current (see Fig. 400).

*There are many advantages of the moving-coil instrument.* Any stray magnetic fields, such as those due to current-carrying cables in buildings, have no appreciable effect on the moving coil, as the magnet in the instrument creates a much more powerful magnetic field. The tangent galvanometer, described on p. 575, has the disadvantage that the needle is affected by stray magnetic fields. Further, the scale of the moving-coil instrument is an even one, and the same instrument can easily be adapted for measuring different ranges of current and p.d. (see p. 464).

**The mirror galvanometer.** The moving-coil instrument illustrated in Fig. 408 is a robust instrument, and can easily be carried from place to place. A more sensitive arrangement, capable of measuring fractions of a millionth of an ampere, is made by suspending the coil on a fine phosphor-bronze wire and dispensing with the two springs (Fig. 409). This wire is a more sensitive control over the coil's rotation than the springs, and a small mirror is rigidly attached to it. A light from a lamp is reflected by the mirror on to a scale graduated in millimetres (see Fig. 185), and when the coil rotates, the small mirror turns. The light from the

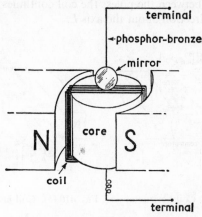

FIG. 409. Mirror galvanometer

mirror is thus reflected in another direction. As explained on p. 274, the angle of rotation of the mirror is obtained from the movement of the light which then takes place along the scale.

**The electric motor.** The electric motor, which developed from Ampère's experiments in the laboratory in 1821, consists essentially of a coil of wire placed in a powerful magnetic field. This is the same apparatus as that used in the moving-coil instrument, except that there are no springs in the motor to hamper the coil's rotation when a current is passed into it.

To explain the principle of the motor, suppose that the coil *abcd* is horizontal and situated between the poles *N*, *S* of a permanent magnet (Fig. 410 (*a*) ). When it carries a current in the direction *dcba*, the latter appears to flow clockwise to an observer from above, and hence, from the clock rule on p. 569, this face of *abcd* acts like a *S*-magnetic pole. This is illustrated in Fig. 410 (*a*). Since like poles repel and unlike poles attract, the coil begins to turn about the horizontal axis *L* in an anti-clockwise direction, as shown by the arrow *X*. *When the coil passes the vertical, the current*

20

*in the coil is reversed by a split-ring commutator A, B, so that it now flows in the direction abcd.* The side of the coil facing the S-pole of the magnet thus acts as a S-magnetic pole, and hence, by the action between the poles, the coil continues to spin round in the same direction about the axis L.

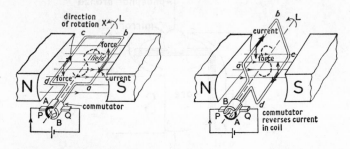

FIG. 410 (*a*). Coil horizontal      FIG. 410 (*b*). Coil just past vertical

The direction of rotation of the coil in Fig. 410 can also be found by applying Fleming's left-hand rule (p. 586) to the current-carrying conductors *ab*, *cd*. It will then be found that the force on *ab* is upwards in Fig. 410 (*a*) and the force on *cd* is downwards, so that the coil spins round in an anti-clockwise direction, as we have already deduced. If Fleming's rule is applied to Fig. 410 (*b*), when the coil has just passed the vertical, it will be found that the forces on *ab*, *cd* have reversed since the current in the coil has reversed, so that the coil continues to move in the same direction.

**The commutator.** The simplest split-ring commutator consists of two separated halves, *A*, *B*, of a copper ring, with brushes, *P*, *Q*, pressing against them as they rotate with the coil. *A*, *B* are each mounted on an insulating disc or shaft and permanently connected to the ends of the coil. When the coil just moves past the vertical, as illustrated in Fig. 410 (*b*), *A* moves over quickly to make contact with the brush *Q*, and *B* with the other brush *P*. By tracing the current from the battery applied to the brushes, it can be seen that the current has now reversed in the coil. *The reversal occurs every time the coil passes the vertical*, thus enabling the coil to keep on rotating in the same anti-clockwise direction about *L*.

A more powerful motor is obtained if the commutator is arranged into a large number of separated equal parts, and the coil is wound into a number of equal parts in different planes. A larger turning effect is also obtained if a greater current and a more powerful magnetic field are employed, since the force on a conductor is proportional to the magnitudes of these two quantities

FIG. 411. ELECTRIC MOTOR OF 3000 H.P.
driving a mill (*in background*) at the works of the British Aluminium Company, Scotland

(see p. 587). An electric motor developing 3000 h.p. is shown in Fig. 411. Further discussion of the motor is outside the scope of this book.

**The moving-coil loudspeaker.** The moving-coil loudspeaker is used in radio receiver sets designed for the home. Like the telephone earpiece, it converts electrical energy to sound energy, but operates on a different principle from the telephone earpiece, because it utilizes the force on a current-carrying conductor in a magnetic field.

The loudspeaker contains a permanent magnet $B$, with circular north ($N$) and south ($S$) poles (Fig. 412 ($a$), ($b$) ). These poles create a **radial magnetic field**, so called because the lines of force between the poles spread out like the radii of a circle (see Fig. 412 ($b$) ). A coil of wire $A$, known as the **speech coil**, is wound round a small cylindrical "former," and is placed between the two circular

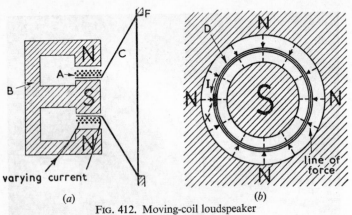

varying current

($a$)

($b$)

FIG. 412. Moving-coil loudspeaker

poles of the magnet. A large paper cone $C$ is rigidly attached to the former, and is loosely connected to a circular board $F$ which surrounds the cone and is known as a **baffle board**.

When a radio receiver set functions, electric currents varying at sound frequencies flow through $A$, the speech coil. Suppose the current $I$ flows in the direction shown in Fig. 412 ($b$) at some instant. Then, applying Fleming's left-hand rule (p. 586) to the small portion $x$ of the speech-coil, whose circular section $D$ is shown, the force on $x$ is upwards towards the reader; this is true for all parts of the speech-coil, as can easily be verified. If the current reverses later, the force on the coil also reverses. Further, the magnitude of the force is proportional to the magnitude of the current (p. 587); so that as the speech current varies, the force on the coil increases and decreases. Since the coil is wound on the former to which the cone $C$ is attached, it follows that *the cone vibrates at the same frequency as the varying current in the coil.*

The vibratory movement of the large mass of air in contact with the cone gives rise to a loud sound, and hence the loudspeaker converts electrical energy to sound energy.

──────Summary──────

1. The force in a current-carrying conductor at right angles to a magnetic field is proportional to the current, the field intensity, and the length of the conductor ($F = IHl/10$ dynes). There is no force on the conductor if it is parallel to the field.

2. Fleming's left-hand rule states: **If the thumb and first two fingers of the left hand are held at right-angles to each other, with the fore-finger pointing in the direction of the field and the middle finger in the current direction, the thumb points in the direction of motion of the conductor.**

3. **The moving-coil instrument has a coil, a permanent magnet, a soft iron core to provide a radial field, and two springs.** The current is proportional to the deflection.

4. The simple electric motor has a coil in a magnetic field. A commutator reverses the current in the coil when it passes the vertical, thereby enabling the coil to keep on spinning.

## EXERCISES ON CHAPTER XXXVII

**1.** A vertical rectangular coil of wire is suspended with its plane parallel to a horizontal magnetic field. The coil carries a current first in a clockwise direction and then in an anti-clockwise direction. Draw a diagram showing the current, magnetic field, and direction of rotation of the coil in each case.

**2.** State *Fleming's left-hand rule* for the force on a current-carrying straight conductor. A straight wire is placed (*i*) at right angles, (*ii*) parallel to a horizontal magnetic field. Draw diagrams illustrating the directions of any force acting on the conductor when it carries a current first in one direction and then in the reverse direction.

**3.** Draw a sketch of a simple motor. How can the speed of rotation of the motor be increased?

**4.** Explain, with diagrams, the action of a split-ring commutator in a d.c. motor.

**5.** Describe, with the help of a diagram, some type of ammeter and explain briefly how it works. (*L.*)

**6.** Describe an experiment to show that a wire, when carrying an electric current and suitably placed in a magnetic field, tends to move. State a rule for the direction of motion in terms of the direction of the field and current. Describe and explain *one* practical application of this phenomenon. (*C.W.B.*)

**7.** Describe a moving-coil galvanometer and explain how it works. What modifications in its construction would be necessary to convert it into a simple d.c. motor ? (*C.*)

**8.** Draw a fully labelled diagram to show the essential features of any one type of voltmeter. Explain the working of the instrument. A cell of e.m.f. 1 volt is connected in series with a resistance of 8 ohms and an ammeter, and the ammeter reads 0·1 amp. What will be the reading of the ammeter if an additional resistance of 6 ohms is joined in parallel with the 8-ohm resistance ? (*N.*)

**9.** Describe the *moving-coil loudspeaker*, and explain its action.

**10.** A rectangular-shaped coil supported between the poles of a horse-shoe magnet is free to move. Draw diagrams, in which the poles of the magnet and directions of motions are marked, to explain what happens when a current is passed through the coil. How may this phenomenon be applied to the construction of (*a*) an ammeter, (*b*) an electric motor ? (*O. & C.*)

**11.** Describe the moving-coil galvanometer (or ammeter), explaining its action and enumerating its merits. An instrument of this type has a resistance of 60 ohms and gives a full-scale deflection for 5 milliamperes. What is the maximum current which may be measured when a resistance of 12 ohms is joined in parallel with the instrument ? (*L.*)

**12.** Give a labelled diagram of a moving-coil milliammeter, and explain the function of each of its component parts. When a moving coil milliammeter is viewed from the front the north pole of the magnet and the zero of the instrument are on the left-hand side. Show by a diagram how the current must flow in the coil in order that the pointer may be deflected in the proper direction, and state reasons for your answer. (*O. & C.*)

# CHAPTER XXXVIII

# ELECTROMAGNETIC INDUCTION
## THE DYNAMO

THE year 1791 saw the birth of one of the greatest of English experimental scientists. MICHAEL FARADAY was the son of poor parents, who sent him to work at a bookshop at the age of 15; there, undeterred by his lack of education, he began to take an interest in the books around him, especially on the scientific side. When he was 20 he attended the lectures at the Royal Institution given by SIR HUMPHRY DAVY, and at the end of the course sent him his notes, beautifully written and illustrated, and asked if there was a vacancy at the Institution. Davy was impressed by the work, and offered him the post of laboratory assistant. From this position he rose to be Davy's personal assistant and, at the age of 33, when his pre-eminence in science was already established, he was made Director of the Royal Institution on Davy's retirement. By his pioneer work on the physical principles of electromagnetism, he led the way to the transformer and to the dynamo, without which commercial electrical lighting and heating could never have been obtained.

**Using magnetism to produce electricity.** It had been known since 1820 that an electric current gave rise to magnetic effects (p. 566). Faraday, like most experimental physicists all over the world at that time, began experiments to obtain the reverse effect, and in 1831 he succeeded in "converting magnetism to electricity."

Fig. 413 (a) illustrates one of Faraday's fundamental experiments. C is a coil connected to a sensitive galvanometer G, so that a small current in the coil can be detected. When the north pole N of a magnet is moved towards C a movement of the needle in G indicates that a small current has flowed in the coil. The more rapidly the magnet is moved, the larger is the current observed in

*G*. However, if the magnet is *kept stationary* close to (or inside) the coil no current is observed (Fig. 413 (*b*) ). When the magnet is moved away from *C* a momentary current again flows in *G*

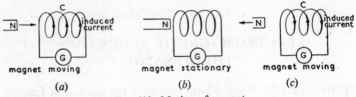

FIG. 413. Motion of magnet

(Fig. 413 (*c*) ), but this time the current is in the opposite direction to that obtained when the magnet is moved towards *C*.

In a previous chapter we have seen that a battery is required to make a current flow in a coil. In the above cases, however, no such agency is present—only a magnet is moved to or from the coil. The name **induced current** is therefore given to the current obtained in this way, and the whole subject concerned is known as

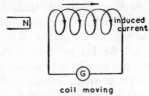

FIG. 414. Motion of coil

**electromagnetic induction** because it concerns both magnetism and electricity. At this stage it is well to note than an induced current is also obtained if the *magnet is kept stationary and the coil is moved*, as illustrated in Fig. 414. This is the case in a dynamo, as we shall see later. In general, an induced current is obtained whenever there is *relative motion* between a conductor (such as a coil) and a magnetic field (such as that produced by a magnet).

We noted above that, while the coil is stationary, the magnitude of the induced current is increased when the speed of the magnet is increased. When the experiment is repeated with soft-iron inside the coil instead of air an induced current many times greater is obtained, and with some specimens of iron the increase may be several hundred times. Increasing the number of turns in the coil also increases the induced current.

**Faraday's law of electromagnetic induction.** A current can only flow in a circuit as a result of an applied electromotive force (e.m.f.). When an induced current is obtained there must be an *induced e.m.f.* in the circuit, and the latter is considered a more fundamental quantity than the current to which it gives rise, since the latter is determined partly by the resistance of the circuit.

As noted in the section on magnetism (p. 538), Faraday considered that the medium in a magnetic field was in a state of stress or strain, and he pictured lines of force in every part of the medium. It was soon evident to Faraday that a *change* had occurred in the number of lines linking, or "threading," a coil when a magnet was moved near to it, for example, in Fig. 413 (*a*). Taking into account the observation of the speed of the magnet and the e.m.f. obtained, Faraday stated that—

**The induced e.m.f. in a circuit is proportional to the rate of change of the number of lines of force linking the circuit.**

Thus,

$$\text{induced e.m.f.} \propto \frac{\text{change in number of lines}}{t},$$

where $t$ is the time in which the change has taken place.

By using the relationship "1 oersted of field strength is equivalent to 1 line per sq. cm. in air," scientists such as electrical engineers are able to deal with *numbers* of lines of force. In Fig. 415 (*a*), (*b*), suppose a magnetic pole is moved from position $A$, where 200 lines link the turns of the coil, to position $B$, where the number of lines passing through the coil increases to 2000. If the magnet is moved from $A$ to $B$ in $\frac{1}{2}$ sec., the rate of change of the lines linking the coil $= (2000 - 200)/\frac{1}{2} = 3600$ lines per sec. If the magnet is now moved quickly away from the coil from position $B$ so that the number of lines through the coil diminish from 2000 to 1600 lines in one-tenth sec., the rate of change of lines through the coil is 4000 lines per sec. Thus the induced

20*

e.m.f. in the latter case is $1\frac{1}{3}$ times greater than in the former case. It should be noted that, the smaller the time in which a given change is made, the greater is the induced e.m.f.; and when the time element is very small, the e.m.f. may reach several thousand volts.

**Lenz' law of electromagnetic induction.** The direction of the induced e.m.f. and current was first expressed in a concise form by LENZ about 1834. His law states—

**The induced electromotive force and current are in such a direction as to oppose the motion producing them.**

In Fig. 413 (a), for example, the movement of the N-pole of the magnet would be opposed by a north pole at the left end of the coil C, since like poles repel. Consequently the current at this end

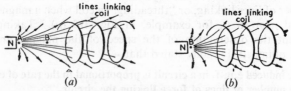

(a)                                (b)

FIG. 415. Change of lines linking coil

of C flows *anti-clockwise* to an observer, as shown, since, from the "clock rule" (p. 569), this makes the end of C act like a north pole. In Fig. 413 (c), the movement of the N-pole away from C would be opposed by a south pole, since unlike poles attract; hence the current in C flows *clockwise* to an observer to the left of it.

Lenz' law follows from the law of conservation of energy, the principle which holds in every branch of science. Suppose that, when a N-pole approached the face A of a coil, the induced current flowed in such a direction that the face A acted like a S-pole as a result of the magnetic effect of the current (Fig. 416). In this case, since unlike poles attract, the magnet would be urged towards the coil with an increasing speed. The induced current would then increase, by Faraday's law, and consequently the magnet's speed would be increased still more. Following this argument it can be seen that the energy of the magnet and the energy produced in the coil by the flow of current would both keep on increasing. Thus, an unlimited amount of energy would be obtained without the expenditure of much energy, which is contrary to the law of conservation of energy.

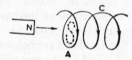

FIG. 416. Direction of current is contrary to law of conservation of energy

The direction of the induced current, therefore, cannot be as shown in Fig. 416. Its actual direction must be such as to *oppose* the motion of the north pole *N*. The induced current consequently makes the face *A* act like a *north* pole, and hence the current flows anti-clockwise to an observer. See Fig. 413 (*a*).

**The primary and secondary coils experiment.** Since a current has a magnetic field round it, a current-carrying coil can be used to produce an induced current in another coil.

Consider a battery connected to a coil *P*, with a key *K* in the circuit, and suppose a coil *S* connected to a galvanometer *G* is placed near to *P* (Fig. 417 (*a*) ). Before any current flows in *P*, no magnetic lines enter *S*. When the key *K* is depressed the current in

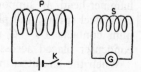

FIG. 417 (*a*). No lines linking *S*

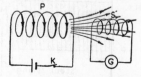

FIG. 417 (*b*). Lines linking *S*

*P* rises to a steady value in a short time *t*, and lines of force now link *S* (Fig. 417 (*b*) ). From Faraday's law, an induced current is therefore obtained in *S* during the time *t*; but *after* this time no induced current flows, since there is then no *change* in the number of lines linking the coil. Thus the induced current is only momentary. When the circuit is broken by means of *K* the current in *P* dies away to zero in a very short time, and a change in the lines linking *S* occurs once more, giving rise to an induced current.

The latter is in the opposite direction to the current obtained when K was depressed, as an opposite change in the lines through S takes place.

If a soft iron core is placed in the coil P or S, or if the number of turns in either coil is increased, an increase in the induced current is observed. Faraday's first discovery of induced currents was obtained by winding two insulated copper coils round an iron ring and connecting a battery and key to one coil and a galvanometer to the other; he found the results which we have just described in connection with the coils P and S. The coil S, in which the induced effects are obtained, is known as the *secondary coil* or, simply, the *secondary*, while the coil P to which the battery is connected is known as the *primary*.

**The induction coil** is an application of electromagnetic induction from which a p.d. of several thousand volts in one direction can be obtained. It was invented by RUHMKORFF about 1836.

Fig. 418 illustrates the essential features of the apparatus. A battery T, of perhaps 4 volts, is connected to a coil P of possibly 50 turns with a key K in the circuit. The coil is made of thick copper wire and contains a bundle of soft iron wires. The circuit

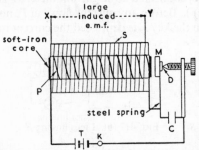

has a make-and-break arrangement, as in the electric bell, provided by a soft iron hammer-head M supported by a steel spring, and platinum contacts D. C is a condenser placed across D.

When the circuit is made by K a current is obtained in the coil P. The iron hammer-head M is then attracted towards P, when the

FIG. 418. Induction coil

circuit is immediately broken at D, and M falls back to renew contact at D. This intermittent to and fro movement of M goes on at a definite frequency, and the current in P therefore rises and falls rapidly. The number of lines of force inside the coil thus keeps changing rapidly, and a condition is therefore obtained in which an induced e.m.f. may be produced. For this purpose a coil $S$, the secondary, is wound over the primary coil. $S$ has a much larger number of turns than $P$, perhaps several thousand, and is therefore made of thinner wire than the primary coil, from which it is insulated.

The p.d. at the secondary terminals $X$, $Y$ is first in one direction and then in the opposite direction, since the change in the secondary lines occurs in opposite directions as $P$ is made and broken. Experiment shows, however, that the p.d. at the break is several thousand volts while the p.d. at the make is a few hundred volts, which is explained by the much shorter time the change takes in the former case. On the average, then, the p.d. at $X$, $Y$ is high and in one direction. The induction coil is used to provide a source of high e.m.f. for car radios, and is employed in the laboratory to operate gas-discharge tubes, which are similar to the neon tubes used for sign lighting.

**Eddy currents.** When the induction coil is functioning, a varying number of lines of force pass through the soft iron inside the coils (Fig. 418), and as iron is a conductor induced currents are obtained in it. These currents are termed *eddy currents*, from the way in which they whirl round, and they produce heat in the iron. If the iron in Fig. 418 were solid the amount of heat produced would be a wastefully high percentage of the power expended by the primary battery and, to diminish the loss of power due to eddy currents, the core is made of a bundle of iron wires which are insulated from each other by the air between them.

**The induction furnace.** In industry, eddy (induction) currents are used to produce high temperatures to melt metals. A type of furnace known as an *induction furnace* is used, as illustrated in Fig. 419. The metal $S$ to be melted is contained in a crucible $A$ lined with

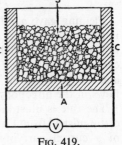

FIG. 419.
Induction furnace

a special heat-resistant material, and coils in the form of a copper tube $C$ are wound round the crucible. An alternating voltage $V$ which varies at the rate of several thousand cycles per second (p. 609), is connected to the coils, and induced (eddy) currents circulate in the metal $S$ as the result of the continuous change of lines of force in it. A continuous supply of heat is then obtained in $S$, and as no heat is lost from $A$, the temperature of the metal rises until it finally melts. In this way very pure steel can be made. The copper tube $C$ also becomes hot by eddy currents circulating in itself, so the tube is cooled by passing water through it while the induction furnace is working.

**The induced e.m.f. in a straight conductor.** In many commercial dynamos, straight copper wires move in a fixed magnetic field. Fig. 420 ($a$) illustrates a straight conductor $AB$ moving vertically downwards across a horizontal magnetic field between the poles, $N$, $S$, of a magnet. As the conductor "cuts" the lines of force an induced current $I$ is obtained in the direction $BA$.

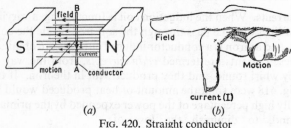

FIG. 420. Straight conductor

Fleming gave a rule for deducing simply the direction of the induced current in a straight conductor: "If the thumb and the first two fingers of the RIGHT hand are held at right angles to each other, with the *F*ore-finger held in the direction of the *F*ield and the thu*M*b in the direction of *M*otion, the induced current ($I$) flows in the direction of the m*I*ddle finger."

The rule is illustrated in Fig. 420 ($b$), and should be carefully distinguished from Fleming's *left*-hand rule which gives the motion of a current-carrying conductor in a magnetic field (p. 586). Further, just as there is no force exerted on the latter conductor when it is *parallel* to the magnetic field, so it should be noted that no induced current would flow along the conductor in Fig. 420 ($a$) if it were to move downwards with its length $AB$ parallel to the magnetic field shown.

**The principle of the dynamo.** We have seen that a battery supplies electric current. The *dynamo* is also an apparatus which supplies electric current, but is an application of electromagnetic induction. Huge dynamos at central power stations provide the main source of electrical power all over the country.

A dynamo consists essentially of a coil rotating at a steady speed in a magnetic field. Fig. 421 illustrates various positions of the

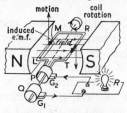

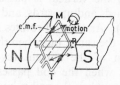

FIG. 421 (*a*). Coil
horizontal—maximum
induced current

FIG. 421 (*b*). Coil
rotating towards
vertical

coil during its rotation, from which it can be realized that a change occurs continuously in the number of lines passing through the coil. Thus, when the latter is horizontal the lines pass over the face of the coil but none pass *into* it (Fig. 421 (*a*) ). When the coil is vertical the number of lines passing through it is a maximum (Fig. 422 (*a*) ).

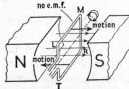

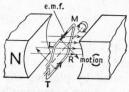

FIG. 422 (*a*). Coil
vertical—no induced
current

FIG. 422 (*b*). Current
reverses after
vertical passed

The direction of the induced e.m.f. and current in the coil may be found by applying Fleming's right-hand rule. Thus, since *LM* is a straight conductor moving vertically upwards in the magnetic field between the poles *N, S* (Fig. 421 (*a*) ), the direction of the induced current is along the direction *LM*. Since *RT* is a straight conductor moving vertically downwards, the induced current in it is in the direction *RT*. Thus, considering the whole coil, a current flows in the direction *LMRT*. As the coil rotates towards the vertical the conductors *LM*, *RT* are still moving respectively up and down in the same direction, and hence the induced current remains in the same direction (Fig. 421 (*b*) ). When the vertical is reached (Fig. 422 (*a*) ) the conductors *LM*, *RT* are moving *parallel* to the magnetic field, and hence at this instant the induced current is *zero*. (See p. 602.)

When the vertical is passed, *LM* and *RT* move respectively downwards and upwards, and the current in the coil *reverses* (Fig. 422 (*b*) ). As the coil rotates further, the direction of the current remains the same, but when the coil again reaches the vertical the direction of the current reverses. In general (i) the current reverses when the coil passes the vertical, (ii) the current is a maximum when the coil is horizontal and zero when it is vertical.

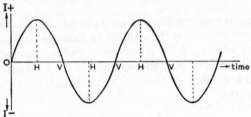

FIG. 423. *H* = *Coil horizontal*; *V* = *Coil vertical*

Fig. 423 illustrates the variation with time of the current *I* obtained in the coil; as is usual in graphs, the reversal of current is indicated by plotting the values below the time-axis. This current is known as the simplest **alternating current (A.C.)**, and the p.d. in the coil is known as an **alternating p.d.** (or, popularly, as an A.C. *voltage*). The electrical supply in most areas of the country is A.C. in character, although some commercial companies supply steady or direct current (D.C.), which is the type of current obtained from an accumulator.

**Obtaining the induced current.** (1) *Alternating current.* The alternating current in the rotating coil can be obtained in a resistance *R* by using two copper rings *P*, *Q* to which each end of the coil is permanently connected (see Fig. 421 (*a*)). These **slip-rings** press against metal brushes $G_1$, $G_2$ as they rotate with the coil, and the resistance $R$, a lamp, is connected to the brushes.

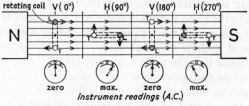

FIG. 424. Alternating current from slip rings

(2) *Direct current*. A dynamo can also supply a current in *one* direction by using a **commutator**. The latter consists of a split copper ring (p. 590); the ends of the coil are connected permanently to each half, $X$, $Y$, of the ring (Fig. 424 (*b*)). As the coil

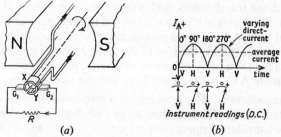

FIG. 425. Commutator action and direct current

rotates to the vertical the half-ring $X$ gradually leaves the brush $G_1$, and $Y$ gradually leaves $G_2$ (Fig. 425 (*a*) ). As the vertical is passed $X$ changes over to make contact with $G_2$, and $Y$ with $G_1$. But the current in the coil reverses as the vertical is passed, hence the current continues to flow *in the resistor R* connected to $G_1$, $G_2$ in the *same* direction. Fig 425 (*b*) shows the variation of the current in $R$ with time. It may be termed a varying direct current, and it has an average value in one direction.

The current obtained, however, is not a very steady one. A very steady direct current is obtained in practice by having many insulated straight conductors equally spaced round a soft iron drum which rotates in a magnetic field. The commutator is divided into a number of equal segments, to which each conductor is connected.

**The grid; use of transformers.** The dynamos in main power stations, like the great station at Battersea, London, are driven by steam turbines, which may require thousands of tons of coal and thousands of gallons of water daily. On this account power stations should be near a river, so that coal may be brought by ships or barges and water be readily available. The e.m.f. of the dynamos can be built up to about 11,000 volts at the main power stations. On account of the advantages of high-tension transmission (p. 478), the e.m.f. is stepped up to 132,000 volts by *transformers*, whose principle of action is explained shortly. This voltage is transmitted to different parts of the country along cables supported by steel pylons (the grid system). Voltages of the order of 30,000 volts are obtained at sub-stations (Fig. 312) by means of step-down transformers, and after passing through local transformers, the electrical energy is conveyed to the consumer at a few hundred volts A.C. This is considered a safer value, though it is still dangerous to life. Because of the greater ease of transforming it from a high to a low voltage and vice-versa, it is more convenient to distribute A.C. voltage than D.C. voltage over an area.

**The transformer.** The principle of a step-down transformer is illustrated in Fig. 426 (*a*). Just as in Faraday's original experiment, two insulated copper coils are wound round a soft iron core,

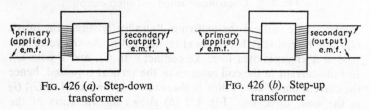

| FIG. 426 (*a*). Step-down transformer | FIG. 426 (*b*). Step-up transformer |

one being the primary coil and the other the secondary coil. The A.C. voltage to be decreased is connected to the primary, which in this case has a larger number of turns than the secondary, and an alternating current therefore flows in the primary. As a result, lines of force pass through the iron core and *link the secondary coil*; since an alternating current changes continuously in magnitude, the number of magnetic lines through the secondary changes

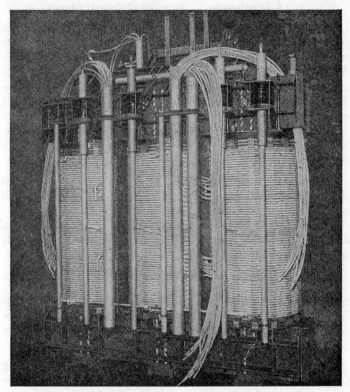

Fig. 427. A 33,000 VOLT TRANSFORMER
in the course of construction, showing the case and windings

continuously. Thus an induced e.m.f. is obtained across the terminals of the secondary coil. The number of lines in the secondary coil are less than in the primary coil at every instant, since the former has less turns, and hence the induced e.m.f. in the secondary is less than that applied to the primary coil.

The e.m.f.s in the primary and secondary are governed by the ratio law:

$$\frac{\text{e.m.f. in secondary}}{\text{e.m.f. in primary}} = \frac{\text{number of secondary turns}}{\text{number of primary turns}}.$$

Thus if the number of turns in the secondary is 50 and that in the primary is 2000, an A.C. voltage of 240 volts applied to the primary is transformed to 6 volts A.C. *The ratio law still holds in the case of a step-up transformer*, but in this case the number of secondary turns is *greater* than the number of primary turns (Fig. 426 (*b*) ). Thus an *increase* in voltage is obtained at the secondary coils when a voltage is applied to the primary.

The transformers needed for power distribution are large. (See Fig. 427.) Small step-up and step-down transformers are used in radio sets.

**Current in the transformer coils.** From the law of conservation of energy it follows that the electrical power in the primary coil must be equal to that in the secondary coil if no power is lost. Since power, $P, = IV$ (p. 471), we have $I_p V_p = I_s V_s$ in this case, where $I_p$, $I_s$ are the primary and secondary alternating currents respectively at an instant, and $V_p$, $V_s$ are the corresponding e.m.f.s.

Thus $\dfrac{I_p}{I_s} = \dfrac{V_s}{V_p}$. As the e.m.f. in the secondary is greater than the e.m.f. in the primary in a step-up transformer, it follows that the secondary current is *smaller* than the primary current in this case. By the same reasoning, the secondary in a step-down transformer has a greater current than the primary.

**Alternating current frequencies.** The simplest type of alternating current (A.C.) has a waveform similar to that obtained from the dynamo discussed on p. 603, and is illustrated in Fig. 428; it is known as a **sinusoidal A.C.** This is a current which increases from zero at an instant $O$ to a maximum value at an instant $A$; it then decreases to zero at an instant $B$ and reverses. It increases to a maximum value in the opposite direction (instant $P$), and finally decreases to zero at an instant $Q$ and reverses again, as shown in Fig. 428. The variation of current now begins all over again, and hence the variation from the instant $O$ to the instant $Q$ is known as a **complete cycle.** From the time $P$ to the time $T$ the current also goes through one complete cycle. The frequency of the A.C. mains in this country is standardized at 50 cycles per second, in

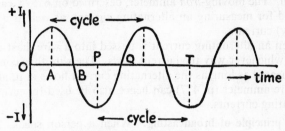

FIG. 428. Alternating current

America the frequency is 60 cycles per second. An A.C. mains receiver set sometimes gives rise to a low hum when it is first switched on, indicating a note of double the mains frequency or 100 cycles per second in this country.

If a person speaks into a microphone, alternating currents of **audio-frequency** are obtained. In speech and music, the range of audio-frequency is from about 20 to 15,000 cycles per second. The electrical circuits of radio transmitters usually generate alternating currents of much higher frequency than 15,000 cycles per second. At a station broadcasting on a wavelength of 300 metres, the current has a frequency of a million cycles (megacycle) per second, and some of the electrical energy is radiated into space in the form of radio waves of this frequency. These high frequencies are termed **radio-frequencies.**

**Effects of alternating current.** If an alternating current is passed through a copper voltameter (p. 484) for some time and the cathode plate is removed, no deposit of copper is observed, unlike the case when a steady p.d. was applied to the copper voltameter. An alternating current causes no electrolysis, because the effect of the current when it flows in one direction is neutralized when the current flows in the opposite direction.

If an alternating p.d. is connected to a moving-coil ammeter, no deflection is obtained in the instrument. The frequent reversals of current are too rapid to be followed by the coil, and the needle is hence not deflected. However, a piece of soft iron is attracted to the end of a coil carrying an alternating current, since attraction occurs whether a south or a north pole is produced at the end of the coil. The moving-*iron* ammeter, described on p. 575, can thus be used for measuring an alternating current, as well as a direct (steady) current.

When an alternating current is passed into a wire, heat is produced whichever way the current flows. The wire thus continues to be heated as long as the alternating current flows in it, and the hot-wire ammeter (p. 477) can hence also be used for measuring alternating currents.

**The principle of broadcasting.** When a person speaks into a microphone at Broadcasting House, alternating currents of audio-(speech) frequency are produced in the transmitter circuit. Now experiment shows that waves of these comparatively low frequencies are unable to travel very far from a transmitter, so that

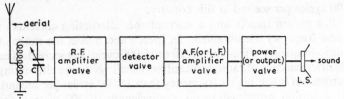

FIG. 429. Principle of receiver

the device is introduced of using a wave of much higher frequency, i.e. *radio*-frequency, to carry the sound or audio-frequencies with it. This wave travels in space, and when it arrives at an aerial, which consists of a metal wire, it induces in the wire an alternating p.d. of the same frequencies as the "mixture" of radio- and audio-frequencies. A large alternating p.d. appears across the variable condenser $C$ when it is "tuned" to the incoming wave, and a valve known as a **radio-frequency (R.F.) amplifier** then amplifies the mixture of radio-frequency (R.F.) and audio- or low-frequency (A.F. or L.F.) alternating p.d.s (Fig. 429). A **detector valve** then separates the A.F. from the R.F. p.d., and passes the A.F. p.d. to an **A.F. (or L.F.) amplifier valve,** which increases the p.d. A **power (or output) valve** then increases the A.F. power as much as possible, and delivers it to the loudspeaker (L.S.), which converts it to sound. Connected between the valves are condensers, coils of wire, and resistances of the order of tens and hundreds of thousands of ohms, and they also play an important part in the action of the receiver set. Radio waves travel in space with the same enormous velocity as light, 186,000 miles per second, so that the interval between a person speaking into a microphone at a B.B.C. studio, and the instant the words are heard from the loudspeaker, is extremely small.

It can be seen that a knowledge of alternating current theory is essential for understanding radio. Further discussion of the subject is outside the scope of this book.

## Summary

1. An induced e.m.f. and current are obtained in a coil in a magnetic field whenever there is a **change in the number of lines of force linking the coil.**

2. **The magnitude of the induced e.m.f. is proportional to the rate of change of the lines of force; the direction is such as to oppose the motion producing the induced e.m.f., which is known as Lenz's law.**

3. The induction coil produces a high p.d. in the secondary when a low p.d. source is connected to the primary. A make-and-break device creates a very rapid change in the lines of force at the break.

4. The simple dynamo has a coil rotating in a magnetic field. An alternating current (A.C.) is obtained by connecting the respective ends of the coil to two slip-rings; a direct current (D.C.) is obtained by connecting the respective ends to the two halves of a split ring commutator.

5. The transformer consists of a primary and a secondary coil wound round soft-iron sheets. The voltage to be changed is connected to the primary.

6. An alternating current produces no electrolysis, and does not make the needle of a moving-coil instrument move. It can be measured by a hot-wire ammeter or by a moving-iron ammeter.

## EXERCISES ON CHAPTER XXXVIII

**1.** State the law giving the *direction* of an induced current in a coil. What is *Fleming's right-hand rule* for the induced current direction in a straight conductor ?

**2.** The south pole of a magnet is moved away from a coil connected to a sensitive galvanometer. Draw a sketch showing the induced current direction. Draw sketches showing the current direction if (i) the south pole is moved towards the coil, (ii) a north pole is moved away from the coil, (iii) a north pole is moved towards the coil.

**3.** Two coils, a battery, connecting wire, and a key are on the table. Draw a sketch showing how you would arrange them to obtain an induced current, and *explain* the results of your demonstration.

**4.** Name the essential features of a *transformer*. What are the uses of a transformer ? A 220-volt a.c. supply is to be changed to a 4-volt a.c. supply. Draw a sketch of the transformer arrangement, and state the ratio of the turns used.

**5.** Draw a sketch of a simple a.c. dynamo, and explain its action. How can the voltage be increased?

**6.** Draw a sketch of a dynamo supplying direct current, and explain its action.

**7.** Describe an experiment to show the production of an electric current by electromagnetic induction. What factors determine the magnitude of the current ? (*L.*)

**8.** With the help of a diagram describe *either* a Ruhmkorff induction coil *or* the coil ignition system of an internal combustion engine. Explain in detail what happens at "make" and at "break." (*N.*)

**9.** Describe Faraday's experiments to demonstrate the phenomenon of electromagnetic induction, and state the laws of electromagnetic induction. Briefly explain the principle of action of *either* (*a*) a sparking (or induction) coil, *or* (*b*) a transformer. (*C.W.B.*)

**10.** Explain the action of *two* of the following, giving in each case a clear diagram: (*a*) An induction coil, (*b*) a d.c. dynamo, (*c*) a dry cell, (*d*) a moving-iron ammeter. (*L.*)

**11.** A straight vertical wire of thick copper is pivoted at its upper end and is free to swing about the pivot in any direction. It hangs between the poles of an electromagnet arranged to produce a uniform horizontal field. Describe and explain, with the help of a diagram, what is observed when an electric current passes through the wire. (The conductors joining the wire to the battery which supplies the current are light and very flexible.)  How could you produce a momentary current in the wire without the aid of a battery ?  Give an explanation.  (*N.*)

**12.** Describe an experiment to illustrate electromagnetic induction, and another to demonstrate Lenz's law.  Explain the construction and mode of action of *either* (*a*) a simple induction coil, *or* (*b*) a step-up transformer.  (*C.W.B.*)

**13.** Describe, with a clearly labelled diagram, the construction of a simple form of direct current dynamo.  Explain the working of the machine and name *four* factors on which the average voltage at the terminals depends.  (*N.*)

**14.** Describe the construction, principles and mode of action of a simple A.C. generator.  What are the advantages of conveying electrical energy at high rather than at low voltages ?  (*C.*)

**15.** Describe, with the aid of a labelled diagram, a simple type of dynamo.  State whether the type you describe gives alternating current or direct current and explain the action of the dynamo when in use. A dynamo is driven by a falling weight suitably geared to it.  When the dynamo is supplying a current of 1·44 amps. to a test-lamp the potential difference across the lamp terminals is 15 volts and the weight of 36 lb. falls at a steady rate of 6 in. per sec.  Calculate the power used by the lamp and the efficiency of the arrangement.  (1 h.p. = 550 ft.-lb. per sec. = $\frac{3}{4}$ kilowatt.)  (*N.*)

**16.** Explain the term *induced current* and specify the conditions essential for the production of induced current.  State the laws of electromagnetic induction and describe *two* experiments which illustrate them  (*O. & C.*)

**17.** State the laws governing the production of *induced currents*. Describe the construction of a transformer and explain how it illustrates the laws of electromagnetic induction.  (*O. & C.*)

**18.** Describe how a galvanometer, a coil of wire and a bar magnet may be used to demonstrate the production of induced currents.  What further experiments could be made with this apparatus to discover the magnitude and direction of such currents ?  State the results you would expect to obtain.  (*L.*)

# SI UNITS

In electrical measurements and formulae, three kinds of units may be met. They are:—(1) *electrostatic units* (*e.s.u.*), which are based on the static effect of electricity; (2) *electromagnetic units* (*e.m.u.*), which are based on the magnetic effect of current electricity; and (3) *practical units*, such as the ampere and volt, which are related to electromagnetic units and used by electrical engineers. In 1901 an Italian scientist, Giorgi, suggested the *metre-kilogramme-second* (M.K.S.) system of units, because, by adjustment of certain constants, only practical units would then need to be used in electrical formulae. The system of units has now been extended to cover all branches of physics. It is known as SI (Système International) units and will be adopted in Great Britain for science and technology.

The unit of length in the SI system is the "metre". Thus a length of 80 cm. is recorded at "0·8 metre" when the SI system is used. The unit of mass on the SI system is the "kilogramme"; thus a mass of 250 grammes is recorded as "0·25 kg.". The unit of time on the SI system is retained as the "second". Only one set of electrical units is used in the SI system, and they are practical units—ampere, volt, ohm, coulomb, farad and so on—which is one advantage of the system.

**The newton.** The unit of force on the SI system is called the *newton*; it is the force acting on a mass of 1 kilogramme which gives it an acceleration of 1 metre per sec.² The dyne, it will be recalled, is the scientific unit of force used previously in the book; it is the force acting on a mass of 1 gramme which gives it an acceleration of 1 cm. per sec.² From $P = mf$, it follows that

$$P(1 \text{ newton}) = 1 \text{ kilogramme} \times 1 \text{ metre per sec.}^2$$

$$= 1000 \text{ g.} \times 100 \text{ cm. per sec.}^2$$

$$\therefore 1 \text{ } newton = 100{,}000 \text{ or } 10^5 \text{ } dynes.$$

614

**Energy and power.** The unit of work or energy on the C.G.S. system, the erg, is the work done when a force of 1 dyne moves through 1 cm. in its own direction. The unit of work or energy on the SI system is the work done when a force of 1 newton moves through 1 metre in its own direction. Now work = force × distance in same direction as the force.

$$\therefore \text{ work done} = 1 \text{ newton} \times 1 \text{ metre}$$
$$= 10^5 \text{ dynes} \times 100 \text{ cm.} = 10^7 \text{ ergs.}$$

But 1 joule = $10^7$ ergs, by definition. Hence

*the unit of work or energy on the SI system is the joule.*

This is an advantage, as the joule is the practical unit of energy.

The unit of *power* on the SI system is thus the rate of working at one joule per second, which, by definition is the *watt*.

**Temperature and Heat.** The SI unit of temperature is the degree Kelvin. The symbol is "K", without the degree sign (see p. 220). A *temperature change* of 1 deg C = a temperature change of 1 K. Thus in SI units, the coefficient of expansion of a gas is about 1/273 per K (in place of 1/273 per deg C).

The SI unit of quantity of heat is the joule, symbol J. This was chosen as the unit since heat is a form of energy. In SI units, the thermal capacity of a calorimeter may be 80 J per K (in place of about 20 cal per deg C). In SI units, the specific heat of water is about 4·2 J per g per K, or, using the kilogramme (kg), about 4200 J per kg per K.

**Electrical units.** We now turn to the electrical units used on the SI system. The unit of *current* is the "ampere". Since the current-carrying wire $X$ has a magnetic field, a neighbouring current-carrying wire $Y$ has a force exerted on it. The "ampere" is defined as "the current which, flowing in each of two infinitely long parallel straight wires one metre apart in a vacuum, creates a force between them of $2 \times 10^{-7}$ newton per metre length of wire". (This force is about 1/50 milligramme wt.)

The *coulomb* is the SI unit of quantity of electricity; it is the quantity of charge which passes a given point in a wire in 1 second when a current of 1 ampere is flowing. The *volt* is the SI unit of potential difference; it is the p.d. between two points when 1 joule of work is expended in taking 1 coulomb between the points. The *ohm* is the SI unit of resistance; it is the resistance of a conductor when the p.d. between its ends is 1 volt and a current of 1 ampere flows in it. The *farad* is the SI unit of capacitance; it is the capacitance of a capacitor when the charge on it is 1 coulomb and the p.d. across it is 1 volt. *Electric field-strength* is measured in "volt per metre"; *resistivity* is measured in "ohm metre".

The *flux*, $\Phi$, in a magnetic field is the number of lines of force in the field, and is measured in "webers". The *flux-density*, $B$, is the "lines per unit area", and is measured in "tesla (T)" or "weber per sq. metre". When the force on a conductor in a magnetic field, or the couple on a coil in a magnetic field, is calculated, the flux-density $B$ is used.

The Ohm's law formulae, the energy formula

$$W = QV \text{ joules} = IVt \text{ joules},$$

and the power formulae $P = IV = V^2/R = I^2R$ watts, used previously in the book, all remain unchanged on the SI system.

## Some Textbooks in SI units

Further discussion of SI units, and their applications in all branches of physics, may be found in the following textbooks by the author:

*Chatto & Windus* (*Educational*) *Ltd*

Fundamentals of Physics (O-level)
Exercises in Ordinary Level Physics
C.S.E. Physics
Revision Book in C.S.E. Physics
SI Units: An Introduction for A-Level

*Heinemann Educational Books Ltd*

Advanced Level Physics (with P. Parker)
Mechanics & Properties of Matter
Light and Sound
Solutions to O-level and A-level Questions

# ANSWERS TO NUMERICAL EXERCISES

## CHAPTER XXVIII (p. 445)

1. (i) 3·2, (ii) 0·8 dynes.
3. 4 cm.

7. (i) 2, (ii) $8\mu$F.

## CHAPTER XXIX (p. 468)

2. (i) 0·2, (ii) 40, (iii) 10 amp.
3. (i) 60, (ii) 6 volts.
5. (i) 2·4, (ii) 35, (iii) $2\frac{6}{7}$, (iv) 25 ohms.
6. (i) 12 amp., $72V$, $48V$, (ii) 20, 30 amp.; $120V$.
7. 2·4 amp.

8. (i) 5, 10 amp.; $30V$, (ii) 2·07 2·59, 10·34 amp.; $10·34V$.
9. (i) 0·05, (ii) 295 ohms.
11. (i) Series 3600, (ii) parallel $4\frac{4}{9}$ ohms.
12. 4, $1\frac{1}{3}$, $2\frac{2}{3}$ amp.
13. (i) $2\frac{1}{4}$ amp., (ii) $30V$.

## CHAPTER XXX (p. 481)

2. (i) 4000, (ii) 72,000 joules; 40 watts.
3. (i) 0·5 amp., (ii) 2s. 1·2d.
6. 200 ohm lamp.
7. 1143, 1714 cal.
8. 180 ohms.
11. 1,296,000.
12. (a) 0·435 amp., (b) 529 ohms, (c) 2s. 1d.

14. 0·26 amp., 881·7 ohms, 7s. 6d.
15. 4·35 amp., 6·6 m., 0·11d.
16. 300 ohms, 9·07° C. per min.
17. 11·2 min.
18. $\frac{1}{2}d$, $\frac{1}{4}d$ (approx.).
19. Parallel : series = 4 : 1.
20. (a) 1·5 amp., (b) 300,000 cal., (c) 1·2d.

## CHAPTER XXXI (p. 495)

4. 4·02 gm.
5. 0·00033 gm. per coulomb.
6. 12 m. 50 s.
8. Hydrogen 0·0083 gm. Oxygen 0·066 gm.
10. 4 hours.

11. 0·067 gm.
12. 0·02 amp.
13. 0·01 amp.
14. 0·15 gm.
15. (a) 10,000, (b) $\frac{1}{45}$.
16. 0·74 c.c.

## CHAPTER XXXII (p. 507)

12. $55\frac{2}{3}$ ohms, 6d.

15. (a) $\frac{1}{2}$, (b) $\frac{1}{3}$ amp.

## CHAPTER XXXIII (p. 525)

1. (i) 0·15, (ii) 0·1, (iii) 0·0375 amp.
2. 0·8, 4, 4·8 volts.
3. 3 volts.
4. 5, 1⅔, 3⅓ amp.
6. 44 × 10⁻⁶ ohm cm.
7. 4 : 1.
9. 100 cm.

10. (i) 3E, 3b, (ii) E, $\dfrac{b}{3}$ ; 3 ohms,
      2 volts.

14. (a) 0·00011 ohm-cm.
      (b) 10 metres.
15. (a) 1·5 volts, (b) 1·82 volts,
      (c) 0·182 amp.
17. 3·7, 0·6 ohms.
18. 2 ohms.
20. 20 ohms.

## CHAPTER XXXV (p. 563)

1. (i) 8, (ii) 8¾ dynes.
2. 4·68 dynes.
3. 25,000 dynes.
8. 0·195 oersted.

10. 1·95 : 1.
15. 0·31 oersted.
16. 26 c.g.s.

## CHAPTER XXXVI (p. 582)

11. (a) 0·084 oersted, (b) 22° 44′.

## CHAPTER XXXVII (p. 593)

8. 0·18 amp.                    11. 30 milliamp.

## CHAPTER XXXVIII (p. 612)

15. 21·6 watts; 88%.

# MODERN PHYSICS

# CHAPTER XXXIX

# USES OF ELECTRONS. ATOMIC STRUCTURE

TOWARDS the end of the nineteenth century, physicists began to direct their attention towards the flow of electricity through gases. In 1897, SIR J. J. THOMSON, the Professor of Physics at Cambridge, discovered the **electron**. This is a small particle (only about 1/2000 part of the weight of a hydrogen atom) which has a negative electric charge and exists as part of every atom.

**Electronics.** With the discovery of the electron, a new and wide field of technical applications was opened up; many instruments known as *electronic devices* have now been invented which utilize these particles. Three of the most important, which affect the lives of millions of people, are (i) the *thermionic valve*, used in broadcasting, (ii) the *cathode ray tube*, used in radar and television, (iii) the *photo-electric cell*, used in the sound track of cinema films.

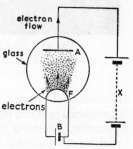

FIG. 430. Diode valve

**The thermionic valve.** Hundreds of thousands of valves are manufactured annually in this country and abroad, to be used in transmitting and receiving sets for both radio and television. The first valve, patented by SIR AMBROSE FLEMING in 1904, enabled easier reception of radio waves to be obtained in the early days of wireless. The **diode valve** consists essentially of (i) a **cathode** or **filament**, *F*, a wire which emits electrons when it is white-hot; (ii) a nickel plate, *A*, called the **anode**, which acts as a "collector" of the electrons; (iii) a glass envelope containing a vacuum (Fig. 430). A battery *B* is required to heat the thin filament, which may then glow. Just as some of a liquid is converted to vapour when its boiling point is reached, so electrons acquire extra energy to escape from the filament when the temperature of the metal is sufficiently high.

A battery $X$ is connected between the anode $A$ and filament $F$ of the valve so that its positive pole is joined to $A$, and the latter has thus a positive potential relative to $F$. Electrons from the filament are thus attracted towards the anode $A$, and flow through the battery $X$ round to $F$, and then across to $A$ again. Hence a current is obtained in the *anode circuit* which is maintained by $X$.

FIG. 431. PART OF A LARGE B.B.C. TRANSMITTER VALVE
showing the leads to the electrodes and the pipe for air-cooling
the glass, which becomes very hot

If, however, the battery $X$ is turned round, so that its negative pole is connected to $A$ and its positive pole to $F$, the anode $A$ has a negative potential relative to $F$, and *no electrons* reach $A$ owing to the repulsive effect exerted round the filament. Thus no current flows in the anode circuit in this case. A valve therefore "conducts" whenever the anode potential is positive relative to $F$, but ceases to conduct whenever its potential is negative relative to $F$. It has therefore a "one-way" action to electricity, and for this reason it was called a "valve."

The emission of electrons from a hot metal was first studied in 1905 by SIR OWEN RICHARDSON, and the phenomenon is hence known as the **Richardson effect.** It is also called the **thermionic effect,** as it concerns heat and electricity, and the valve is known as the *thermionic valve.* Modern valves utilize the heat produced in the cathode by a low alternating p.d. from an A.C. mains transformer, instead of the steady p.d. of the battery *B* in Fig. 430. A high power transmitting valve, used by the B.B.C., is shown in Fig. 431.

**How the valve converts alternating p.d. to steady (direct) p.d.** One of the many uses of a valve is to convert the alternating p.d. from the mains to a steady (direct) p.d. A special diode is used in a radio receiver set when it is used with A.C. mains, so as to obtain a steady p.d. for the valves without using high-tension batteries.

The principle of the action is illustrated in Fig. 432 (*a*), in which *H* represents an alternating p.d. connected to the anode and

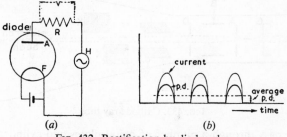

(*a*)                    (*b*)

FIG. 432. Rectification by diode valve

filament of a diode. A resistance *R* is connected in the anode circuit, and electrons flow through the valve and *R* in the anode circuit whenever the potential of *A* is positive relative to *F*. Since *H* is an *alternating* p.d., the potential of *A* relative to *F* is alternately positive and negative on each half of the cycle; hence electrons flow in *R* in the same direction on *one half* only of the cycles. The current variation in *R* is illustrated in Fig. 432 (*b*), and the p.d. obtained across *R* varies in the same way. The p.d. is in one direction, however, and a reading is therefore obtained on a moving-coil voltmeter connected across *R* which is the average value of the p.d. Thus, a p.d. in one direction can be obtained by using a diode valve in series with a resistance *R* and an alternating p.d., a process known as **rectification.**

**The cathode ray tube (C.R.T.).** As we stated at the beginning of the chapter, scientists began to investigate the flow of electricity through gases towards the end of the nineteenth century. When most of the air was pumped out of a glass tube and a high p.d. was applied between two electrodes inside, the glass was observed to fluoresce with a greenish colour. Closer examination showed that the fluorescence was due to the impact of fast-moving electrons on the glass, and from this observation has emerged the *cathode ray tube*, which is used in television and radar (Fig. 435).

Fig. 433 illustrates the internal features of a cathode ray tube. As in the thermionic valve, a heated cathode C emits electrons,

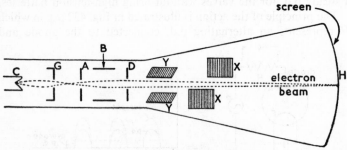

FIG. 433. Cathode ray tube

which pass through a number of metal plates and cylinders contained inside the evacuated tube and strike a glass screen H coated with zinc sulphide. A greenish-blue spot of light is observed on the screen where the electrons strike it. Now a beam of electrons is extremely light, and is therefore very sensitive to electric and magnetic fields imposed on it. Hence, when a varying p.d. V is applied to the Y-plates in Fig. 433, the electron beam moves up and down in response to the variations; a straight line is seen on the screen when the frequency of the p.d. is more than 10 cycles per second, owing to the rapidity of the spot's movement.

In order to show the actual wave-form of V on the screen a p.d. is required which deflects the beam in a *horizontal* direction, thereby making a "time-axis" for V. This p.d. is obtained from a specially designed circuit known as a **time-base**, which is connected to the X-plates shown in Fig. 434. If the frequency of the time-base p.d. is varied until it is exactly the same as the frequency of $V$, a stationary picture of the latter's wave-form is obtained and this can be examined at leisure.

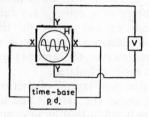

FIG. 434. X- and Y-plates p.d.s

The brightness of the light on the screen $H$ is controlled by the potential of $G$ (Fig. 433). If its potential relative to $C$ is made more negative fewer electrons pass it and strike $H$, as a repelling effect is then produced round the surface of $C$. The potentials of

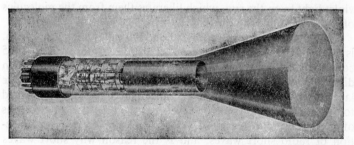

FIG. 435. CATHODE-RAY TUBE

$A$, $B$, $D$ are high and positive relative to $C$, and they create an electric field between them which affects the path taken by the electrons passing through to the screen. $A$, $B$, $D$ are said to act as an **electron lens** in the cathode ray tube, and by altering the potential of $B$ the focal length of the "lens" is altered. In this way the beam of electrons can be focused on to the screen $H$.

**The Kennelly-Heaviside and Appleton layers.** Since 1924 SIR EDWARD APPLETON and other scientists have been exploring the *ionosphere*, a region above the earth containing many electrons and ions. Radio pulses, short bursts of radio signals, are sent skywards from a transmitter, and are reflected back on meeting a layer of electrons of a certain high concentration. The "echo" is received on the screen of a cathode ray tube, and by means of a time axis already on the screen, the time taken for the pulse to the layer of electrons and back is measured. From the known speed of radiowaves, the same as the speed of light, the height of the layer is easily calculated.

Appleton showed that two "layers" of electrons of high concentration existed high above the earth. The lower layer, about 70 miles high, was predicted in 1901 by HEAVISIDE in England and KENNELLY in America, who considered that radio waves sent from one side of the globe were "reflected" at such a layer to the other side of the globe. This belt of electrons is therefore called the *Kennelly-Heaviside* layer. Experiments showed that this layer is effective mainly in the transmission of long radio waves, and that the concentration of electrons in it is due to the action of the ultra-violet light from the sun, which liberates them from the molecules of gas high above the earth.

The upper layer of electrons, about 170 miles above the earth, is known as the *Appleton layer*, after the discoverer, who in 1925 continued his researches well into the night. The sun had then disappeared, and as the electron density of the Kennelly-Heaviside layer, which depends on the ultra-violet light of the sun, diminished to a low value, the Appleton layer was the only belt of electrons of high density in existence. The Appleton layer gives reflection of radio short waves, and is the reason why short waves are used for transmission over long distances, such as for overseas broadcasting.

Meteorological and radio stations all over the world investigate abnormal values of the earth's magnetic field, disturbances during radio communication, and sun-spot activity, which have a connection with each other.

**Radar.** Largely as a result of the work of SIR ROBERT WATSON-WATT, the principles of Appleton's pioneer experiments on the ionosphere were adapted for use in *radar*. This is another example of the industrial applications of experiments in pure science. Very

FIG. 436 (*a*). SHIP RADAR

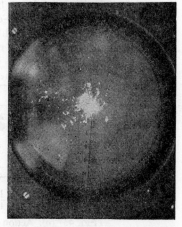

FIG. 436 (*b*). MARINE RADAR SCREEN

briefly, a radio pulse is sent out from a transmitter, and the "echo" from an object such as an aeroplane is received back on a cathode-ray tube with a special time-base design, which enables the operator to know the height of the aeroplane. Its position is located, in conjunction with other ground stations, by using a rotating aerial sending out a narrow beam of radio waves, like a searchlight. During the war of 1939-1945, special transmitters and cathode ray tubes were carried by aeroplanes which enabled a "picture" of the territory over which they were flying to be seen by the pilot.

exhibited in Kensington, London, showing echoes from familiar London landmarks. The scale shown is calibrated in miles.

With the return of peace, radar is applied for the benefit of travellers on the sea and in the air. The "Queen Elizabeth," the largest liner in the world,

21*

and important ferry-boats carry radar sets which permit safe steering in fog and bad weather. Fig. 436 (*a*) illustrates the aerial of a ship sending out radio waves in all directions. If the ship is approaching a cliff, the waves are reflected back, and the appearance on the screen, shown inset, is a wavy line representing the contour of the cliff. A distant ship is represented by a spot of light, and the distance and bearing are read from a scale on the cathode ray screen. See also Fig. 436 (*b*).

**The photo-electric cell.** In 1887 H. HERTZ, a famous German scientist, noticed that light falling on metal electrodes in air enabled the gas to conduct electricity more easily when a high p.d. was connected to the electrodes. The observation was taken up by other scientists, and it was proved that *some electrons are emitted by a metal when light falls on it.* It appears, in fact, that some of the energy in the light is given to the surface electrons in the metal atoms, which are then able to escape completely from the metal if an outward attractive force is exerted on them. This is the principle of one type of *photo-electric cell*, in which, as in the thermionic valve, the electrons liberated from the metal inside are attracted towards another metal maintained at a positive potential.

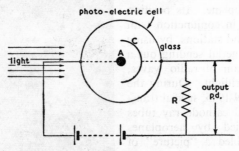

FIG. 437 (*a*). Action of photo-electric cell

The difference, however, is that the electrons are produced when light falls on the metal. There are various other types of photo-electric cells, which enable street lighting to be turned on and off automatically, operate highly efficient alarms, and enable objects of different colours to be automatically distinguished.

Fig. 437 (*a*) illustrates the principle of a photo-electric cell. *C* is a metal of large surface area, known as the cathode, from which electrons are emitted when light falls on it; and *A* is a metal, known as an anode, which has a positive potential relative to *C*. The electrons from *C* are thus attracted to *A*, and a small current flows through a resistance *R* in the circuit. A p.d. is hence obtained across *R*, but because it is small, it has to be amplified by means of a valve (not shown). This p.d. can be used to keep open a device such as a burglar-alarm bell circuit. As soon as the light is cut off from the cell by a person passing between the source of light and the cell itself, the circuit closes and the bell rings.

A device utilizing a photo-electric cell has been installed on the battleship *H.M.S. Vanguard* to assist the stokers in their work. The cell is placed on one side of a funnel just above the place where the smoke issues, and a light, passing through the smoke, illuminates the cell. When too much coal is being burned, the smoke issuing from the funnel is black, and the photo-electric current is practically cut off. Immediately this happens a red lamp on deck is caused to light up, and the observer signals below to the stokers for reduced coal. As long as the smoke is grey, a green lamp is lit up on deck, but the transition stage between grey and black smoke causes an amber lamp to light up and warn the observer.

Fig. 437 (*b*). PHOTO-ELECTRIC CELL

Electrons are emitted when light falls on the V-shaped caesium-coated metal

The modern sound-film carries a "sound-track" near one edge of the celluloid strip on which the pictures are printed. This sound-track has a density to light which varies exactly as the audio-frequency variations of the speech or music to be reproduced. From it are derived all the sounds which accompany the picture. Fig. 438 illustrates the underlying principle. A lamp and

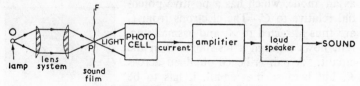

FIG. 438. Film sound-track principle

lens system illuminate each portion of the "sound-track" as it runs past $P$, and the light through the film falls on the photo-cell. Since the intensity of the light incident on the cell is varying, a varying current flows in the cell's circuit (not shown), which contains a resistance $R$ (see Fig. 437 ($a$) ). This current varies at an audio-frequency rate, so that the potential difference across $R$ varies at the same frequency. The potential difference is now amplified by means of a valve, and the audio-frequency current is passed through a loudspeaker, when sound waves of exactly the same frequency are obtained. The variation in density of the "sound-track" thus enables sound waves to be produced with the aid of a photo-electric cell.

**X-rays and X-ray machines.** In 1895 a German physicist, RÖNTGEN, discovered that a penetrating radiation was emitted from a glass tube at any part where fast-moving electrons impinged on it. Many experiments were performed to decide the nature of the radiations, which were called *X-rays* because they were unknown. All the experiments failed until 1912, when it was definitely proved that $X$-rays were radiations of the same nature as light, except that they were invisible. The wavelengths of $X$-rays were accurately measured by SIR WILLIAM BRAGG and his son SIR LAWRENCE BRAGG, and it was shown that their wavelengths are over a thousand times smaller than that of visible light.

It is now known that $X$-rays are always produced when a fast-moving electron impinges on a heavy metal, and that some of the energy of the moving electron is converted into energy which is radiated when the collision occurs. This is an opposite type of phenomenon to the photo-electric effect (p. 630), where some of the energy in light radiation is given to electrons in a metal to enable them to escape from its surface.

Fig. 439 shows the essential features of the interior of a modern X-ray tube. F is a filament which emits electrons when heated by

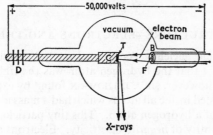

FIG. 439. X-ray tube principle

a small transformer connected to A.C. mains, and T is a tungsten block at the end of a copper rod C maintained at a very high positive potential relative to F, so that the electrons reach A with a very high energy. The whole of the containing tube is evacuated of gas, and X-rays are emitted when electrons strike T. Since C becomes hot in the process, radiating fins, D, are used which dissipate much of the heat conducted along the thick copper rod C. The metal cylinder represented by B is maintained at a negative potential relative to F, and is used to focus the electrons on T.

**Some applications of X-rays.** Since X-rays penetrate the flesh but are stopped by harder substances such as bone, Röntgen immediately realized their importance in medicine. Nowadays, most hospitals are equipped with X-ray machines. A fracture of a bone can easily be detected from a radiograph of the limb concerned, the radiograph being the "picture" obtained when X-rays are incident through the limb on to a photographic plate.

X-ray machines are also used industrially for detecting flaws and defects in steel plates that are invisible to the eye; X-rays pass more easily through the flaws than through the rest of the material.

In recent years, too, X-ray machines have been developed which act as very powerful "microscopes," enabling the arrangement of the molecules of crystalline substances to be discovered. In this way the structure of wool, for example, has been found, and research is now proceeding on new types of fibres to improve the quality of woollen articles. Thus, the discovery of *X*-rays in 1895 will have an impact on society by determining the quality of the clothes which people will wear in future.

## ATOMIC STRUCTURE—ELECTRONS AND THE NUCLEUS

**Electrons and nucleus.** As pointed out previously, until 1897 it was thought that the hydrogen atom was the lightest particle. In that year, however, SIR J. J. THOMSON found by experiment that a particle existed inside all atoms which had a mass about 1/2000th of the mass of a hydrogen atom. This tiny particle, the electron, carried a quantity of *negative* electricity. Electrons are present in all atoms, from the lightest, hydrogen, to the heaviest such as uranium.

Between 1896 and 1912, as a result of investigations by BECQUEREL, the CURIES, LORD RUTHERFORD, BOHR and other eminent scientists (see p. 644), the structure of the atom was gradually made clear. Basically, all atoms have an extremely small central *nucleus* carrying positive electricity, with electrons moving round the nucleus (see also p. 649) Fig. 440. As matter, and therefore the atom, is

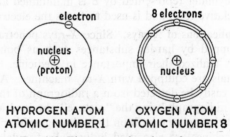

HYDROGEN ATOM        OXYGEN ATOM
ATOMIC NUMBER 1      ATOMIC NUMBER 8

FIG. 440. Electrons and nucleus of atoms

electrically neutral, the quantity of positive electricity on the nucleus is equal to the total quantity of negative electricity on all the electrons. The quantity of electricity on an electron is a constant of nature, and is represented by the symbol *e*. The charge on an electron is thus $-e$, on two electrons it is $-2e$, and so on. The charge $e$ is of the order $10^{-19}$ coulomb, so that when a 60 watt lamp is used on a 240 volt mains and the lamp current is $\frac{1}{4}$ ampere, about $10^{18}$ electrons per second flow past a point in the filament.

There is a simple rule for the number of electrons round the nucleus of nearly every atom. We compile a list of the elements in order of increasing atomic weight, with hydrogen, the lightest element, as number one in the list, the next heavier element helium as number two, and so on, until we reach the heaviest element, uranium, which is number 92. The number of a particular element in the list, apart from a few exceptions, is its *atomic number* (see below), and *the number of electrons round the nucleus of an element is equal to its atomic number*. Thus hydrogen has one electron round its nucleus, helium has two electrons, oxygen has eight electrons, copper has 29 electrons, and uranium has 92 electrons. The atomic number of an atom may be defined accurately as the number of protons in its nucleus (p. 649), and as we shall see later, it is equal to the number of electrons.

**Energy levels.** We have just seen that the nucleus of an atom carries a positive charge, whereas the electrons round the nucleus carry negative charges. Now positive and negative charges, as we saw in electrostatics (p. 411), attract each other. Thus to remove an electron from an atom, an amount of energy would need to be expended against the force of attraction of the nucleus. An electron further away from the nucleus would not require as much energy to remove it from the atom as an electron near the nucleus, where the force of attraction is greater. Thus an electron at a given distance from the nucleus has a characteristic energy. Bohr further suggested that only certain energies were permitted to an electron in an atom. This may be compared to a person restricted to living on definite floors or levels of a building. Those electrons nearest the nucleus are said to occupy the $K$ level or shell of energy, and others the $L, M, N, \ldots$ shells, proceeding further from the nucleus. Thus for atoms with many electrons, groups of electrons can exist which have similar energies, that is, they occupy the same shell. They may be compared to sets of families each living on various floors of a building.

The hydrogen atom has only one electron; it occupies the $K$ shell. The helium atom, atomic number two, has two electrons in the $K$ shell, and this is the maximum number of electrons accommodated in this shell. The next element, lithium, atomic number three, has two electrons in the $K$ shell and one electron in the next, $L$, shell. The $L$ shell can hold up to a maximum of eight electrons, and as the atomic number increases the number of electrons in this shell increases until the element neon, atomic number 10, is reached. The next element is sodium, atomic number 11, which has eleven electrons; two in the $K$ shell, eight

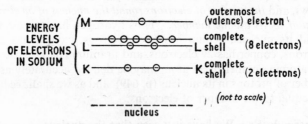

FIG. 441. Energy levels (shells) of electrons in atoms

in the $L$ shell, and one in the $M$ shell (Fig. 441). The element chlorine, atomic number 17, has two electrons in the $K$ shell, eight in the $L$ shell, and seven in the $M$ shell, which can accommodate a maximum of eight electrons. After this, as the atomic number increases, electrons fill other shells up to a maximum number for each shell.

**Chemical activity.** A study of the electrons round the nucleus of atoms shows that they are responsible for many widely different phenomena, such as chemical activity, conduction of electricity, magnetism, and $X$-rays.

Chemical activity can be understood by reference to the number of electrons filling the outermost shells of the atoms concerned. Consider the alkali metal sodium, for example, which has one electron in its outermost or $M$ shell. Since this electron is furthest from the nucleus, the energy with which it is bound to the nucleus is not very large. Now chlorine has seven of eight possible electrons in its outermost or *M* shell, and this atom has therefore a vacancy for one electron. Consequently, when sodium and chlorine atoms are brought together the electron from the outermost shell of the sodium atom is transferred to the outermost shell of the chlorine atom, where it completes the *M* shell and is held very strongly. Thus common salt, sodium chloride, is a very stable compound. We thus see that the sodium atom has an electron to give away, i.e. it is an electron "donor", and that the chlorine atom has an electron deficiency, i.e. it is an electron "acceptor". The electron which takes part in the chemical bonding between the two elements is called a **valence electron**. Lithium and potassium are elements with an outer electronic structure similar to sodium; each of their atoms have one electron outside a closed shell. Fluorine is an element with an outer electron structure similar to chlorine; it has one electron short of a complete shell.

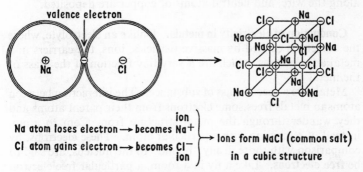

FIG. 442. Electron theory of chemical combination

When a sodium atom gives up an electron, it becomes a sodium **ion** with a positive charge. Similarly, when a chlorine atom takes up an electron, it becomes a negative chlorine ion. Sodium chloride in the solid state consists of sodium and chlorine ions held together by the attraction between positive and negative charges (Fig. 442).

In distinction to sodium and chlorine, elements such as helium and argon, which have a **complete number** of electrons in their outer shells, are chemically inactive.

**Conduction of electricity in electrolytes.** As we have just seen, a salt such as sodium chloride is kept in a solid state by the attraction between sodium and chlorine ions. Copper sulphate crystals are kept in a solid state by the force of attraction between copper and sulphate ions. The medium between the ions is air, but if the medium changes to water, the force of attraction between the charges is considerably weakened. Consequently, when copper sulphate crystals are added to water the solid structure collapses, and copper and sulphate ions drift apart and wander haphazardly in the solution. If a battery is now connected to two plates dipping into the solution, the positive copper ions drift towards one plate, the cathode (p. 486), and the negative sulphate ions drift towards the opposite plate, the anode. When the copper ions reach the cathode their charge is neutralized by electrons flowing along the wire, and neutral atoms of copper are deposited.

**Conduction of electricity in metals.** Unlike an electrolyte, where the current is carried by massive particles, ions, the carriers in a metal are **electrons**, which are a very tiny fraction of the mass of the metal.

Metals are a special class of substance. The interaction between atoms in metals frees some electrons from their parent atoms and they wander through the metal structure from atom to atom, instead of remaining with a particular atom. These electrons, by comparison with those firmly "bound" to the nucleus, are said to be **free electrons**. Like a fly in a room, a particular free electron zig-zags in its motion, darting this way and then in an opposite way, so that, on average, it is not moving in any particular direction through the metal structure. Free electrons occupy a band of energies called the **conduction band**, in which their energy is greater than electrons "bound" to the nucleus.

Suppose, however, a battery is connected to the ends of the metal. The free electrons still move haphazardly, but now they also begin to drift in one direction. This drift is an electric current. When the p.d. of the battery is increased the speed of the drift increases, that is, the current increases. As they move through the metal, the constant collision of electrons with the vibrating atoms of the metal results in a transfer of energy to the atoms, that is, the metal becomes hotter. This is the origin of the heating effect of a current.

### Conduction of electricity in semiconductors. Electrons and holes.

Insulators such as perspex and rubber have no free electrons and hence they do not normally conduct electricity. Their electrical resistance is more than $10^{18}$ times that of a metal conductor such as copper. Germanium and silicon are elements which come between conductors and insulators in electrical resistance, and they are therefore called **semiconductors**. Germanium and silicon have considerable importance today because they are used in transistors (see later).

A germanium atom has four valence electrons (electrons in the outer shell which take part in chemical bonding, p. 637), each of which is shared with surrounding atoms (Fig. 443 $(a)$). At

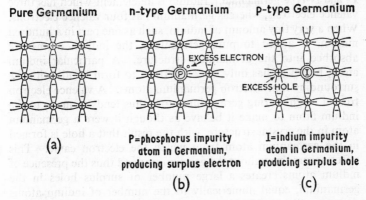

Pure Germanium    n-type Germanium    p-type Germanium

(a)

(b)
P=phosphorus impurity atom in Germanium, producing surplus electron

(c)
I=indium impurity atom in Germanium, producing surplus hole

FIG. 443. Semiconductor and impurities

ordinary temperatures, however, some of the valence electrons have sufficient energy to reach the conduction band or level of energy and then become free electrons. When a valence electron leaves an atom $A$, it leaves a deficiency or **hole** in the energy levels of $A$, and a valence electron in a neighbouring atom $B$ may then move to $A$. The atom $A$ thus becomes complete again, but a hole is now left in the atom $B$. A valence electron in a neighbouring atom $C$ may now move to $B$, thus leaving a hole in $C$. In this way a hole can move through the structure of the semiconductor. The movement of a valence electron from one atom to another is equivalent to a hole moving in the opposite direction; and since the electron is a negative charge, the hole movement is equivalent to the movement of a **positive** charge equal to that on an electron. Although a hole movement may be surprising, it was shown to exist in experiments by the American physicist SHOCKLEY. He observed the movement of positive charges in germanium.

When a battery is connected to a semiconductor, the free electrons are urged to drift in one direction and the holes (behaving as positive charges) in the opposite direction. The current through the semiconductor is therefore carried by both electrons and holes, whereas in a metal it is carried only by electrons.

**P- and N-type germanium**. Indium is an element which has three valence electrons, whereas germanium has four valence electrons. When a very tiny amount of indium, such as one part in a hundred million, is added to pure germanium, the indium atoms are absorbed into the germanium structure. A particular indium atom, however, has only three electrons to form bonds with the surrounding four-electron germanium atoms. A valence electron from a neighbouring germanium atom thus tends to move to the indium atom to make it behave as though it were a germanium atom in the crystal structure, with the result that a hole is formed in the germanium atom from which the electron came. This effect occurs throughout the germanium; and thus the presence of indium atoms creates a large number of surplus holes in the germanium, equal numerically to the number of indium atoms present. Since a hole movement would be equivalent to a positive ($p$) charge movement, the impure germanium is called $p$-type germanium (see Fig. 443 ($c$)).

Phosphorus is an element which has five valence electrons. When a very tiny amount of phosphorus is added to pure germanium, four of the five valence electrons are absorbed into the germanium structure, leaving one surplus electron. The phosphorus impurity thus creates a large number of surplus negative (*n*) charges equal to the number of phosphorus atoms, and the impure germanium is therefore called *n*-type germanium (see Fig. 443 (*b*)).

**The Transistor.** In 1946 BARDEEN and BRATTAIN, working at the Bell Telephone Laboratories in America, designed a "transistor". One form is a thin layer of *n*-type germanium, sandwiched between two regions of *p*-type germanium, in which case it is called a *p-n-p* transistor (Fig. 444 (*a*)). Another form is a thin layer of

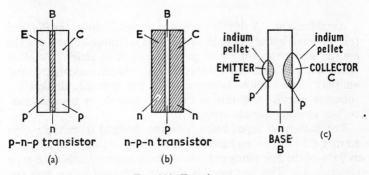

FIG. 444. Transistors

*p*-type germanium sandwiched between *n*-type germanium (Fig. 444 (*b*)). In one method of manufacture, two pellets of indium are placed on a thin plate of *n*-germanium and the whole arrangement is heated. The indium melts and alloys with the germanium, forming *p*-type germanium, and the heating is continued until the alloyed regions are separated by a very thin layer of *n*-type germanium called the **base**, *B* (Fig. 444 (*c*)). This is known as an alloy-junction transistor. One layer of *p*-type germanium, called the **emitter**, *E*, has a smaller junction area with the base than the other layer of *p*-type germanium, called the **collector**, *C*. Fig. 445 (*a*) shows the external appearance of a transistor. The base terminal *B* is between the emitter *E* and collector *C* terminals, the collector terminal *C* being displaced for recognition. Fig. 445 (*b*) shows the conventional representation of a *p-n-p* transistor.

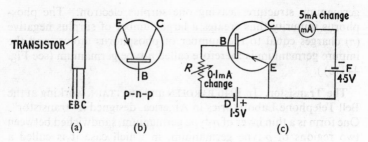

FIG. 445. Transistor and amplifier circuit

**Amplification.** A simple transistor circuit is shown in Fig. **445** (*c*). A small battery *D* such as 1·5 volts is connected between the emitter and base, with a large resistance *R* in series. Another small battery *F* such as 4·5 volts is joined between the collector and emitter. The emitter is common to the base circuit and the collector circuit, and this is therefore known as the **common-emitter circuit.** See also page 670.

The battery *D* urges holes (positive charges) from the *p*-type emitter *E* to the *n*-type base *B*. The base is so thin, that as much as 98% of the holes may pass through the base towards the *p*-type collector *C*. The battery *F* is also so arranged as to urge the holes (positive charges) towards the collector, and consequently a current flows in the collector circuit. In the Mullard transistor OC71, for example, if 1 milliamp. is the current in the emitter circuit, then 0·98 milliamp. flows in the collector circuit and 0·02 milliamp. flows in the base circuit. The amplification is 0·98/0·02 or 49, when the collector current is compared with the base circuit.

If the base current is varied, it will have a similar large effect on the collector current because it will then control, in a sensitive way, the flow of positive holes through itself towards the collector. For example, when a small current change such as 0·1 milliamp. is made in the base circuit, the milliammeter in the collector circuit changes by a much larger current such as 5·0 milliamps. The transistor thus acts as an *amplifier* of current. In contrast, a triode and other radio valves amplify voltages.

On this account transistors are now widely used in radio receivers, tape recorders and computers. They have considerable advantages over radio valves. They are operated by batteries of only a few volts, they absorb little power, they act instantaneously, and they are fairly robust. They have the disadvantage of being sensitive to temperature, and special precautions are necessary to keep their temperature down to a moderate value. Silicon is now replacing germanium as it is much less sensitive to temperature changes.

## THE NUCLEUS

**Radioactivity.** So far we have mainly discussed effects due to the electrons round the nucleus. The first clues of atomic structure and of the nucleus itself were provided by the phenomenon of **radioactivity**, discovered by Becquerel in 1896. He placed a uranium compound on a photographic plate covered with light-proof paper, and found to his surprise that the plate was fogged. He traced this effect to some unknown radiation coming from the uranium compound. The phenomenon was investigated in 1897 by a New Zealand scientist, then Ernest Rutherford and later Lord Rutherford, in whose honour the Rutherford High Energy Laboratory at Didcot, England, is named. This laboratory is part of the National Institute for Research in Nuclear Science, which was opened in 1957.

Using a magnetic field perpendicular to the stream of radiation, it was found that the radiation separated into two parts, called **alpha** ($\alpha$) and **beta** ($\beta$) **rays**. A third part, which was not deflected by the magnetic field, was called **gamma** ($\gamma$) **rays** (Fig. 446). The deflection is the same in principle as that of a conductor carrying a current when situated in a magnetic field (see p. 586), and it shows that the alpha- and beta-rays carry electric charges. By applying Fleming's left-hand rule to the direction of deflection, it was found that $\alpha$-rays carried positive charges and $\beta$-rays carried negative charges. The $\gamma$-rays were found to be more penetrating than $\beta$-rays, which in turn were more penetrating than $\alpha$-rays.

**Nature of $\alpha$- and $\beta$-particles and $\gamma$-rays.** Experiments with $\alpha$-particles carried out by Rutherford and collaborators had shown that they had a mass about four times that of hydrogen atoms and carried a positive charge about twice that on an electron. As its atomic weight is 4, *an α-particle was thus probably a helium atom which has lost two electrons,* or a *helium nucleus.*

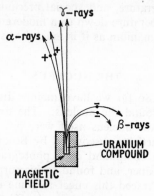

FIG. 446. Separation of α-, β-, γ-rays

In 1909 RUTHERFORD and ROYDS carried out an experiment which showed conclusively that α-particles were helium nuclei. Radon, a gas which disintegrates and gives off α-particles, was collected in a thin-walled tube *A*. After a few days many α-particles had passed through *A* into an evacuated outer tube *B* (Fig. 447). By raising the mercury reservoir the gas collected in *B* was compressed into a capillary tube at the top. The α-particles become neutral atoms by acquiring negative charges from the glass, and a high voltage from an induction coil was applied to the gas. The spectrum of the discharge was then seen to be exactly the same as the characteristic spectrum of helium, thus proving that α-particles are helium nuclei.

**Nature of β-particles and γ-rays.** The nature of β-particles was found from experiments in which they were deflected by electric and magnetic fields. They were shown to be fast-moving electrons.

The nature of γ-rays was found from experiments in which they were incident on crystals. The results showed they were waves and not particles. γ-rays are electromagnetic waves with a wave-length tens of thousands times shorter than that of visible light and shorter even than the wavelength of X-rays (p. 632).

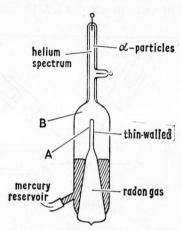

FIG. 447. Identification of α-particles

**Geiger and Marsden's experiment.** At Rutherford's suggestion, GEIGER and MARSDEN in 1909 investigated the effect of α-particles incident on thin films of metal. A radioactive source O was contained in an evacuated vessel V, and emitted α-particles in a narrow beam on to a very thin metal foil of gold or other metal of high atomic weight (Fig. 448). The path of α-particles could be traced by the pin-points of light (scintillations) produced as they struck a fluorescent screen S, in the focal plane of a microscope M. Geiger and Marsden investigated the scattering of α-particles by the atoms of the metal. They observed scintillations not only in the straight-through position M, but also in directions such as FA which were at large angles such as 45° to the incident beam. A few scintillations were also observed in a direction FB, which was opposite to the incident beam. Some α-particles were thus deflected through very large angles, as if they had reached, inside the metal atom, a centre which repelled them violently.

**The Nucleus. Atomic Structure.** Rutherford, with character-istic genius, saw immediately the significance of these obser-vations. He said at the time that the repulsion of α-particles in an opposite direction was as surprising as if a shell fired from a gun had been repelled by tissue paper! Sir J. J. Thomson had thought

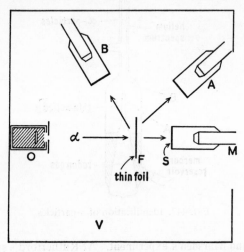

FIG. 448. Principle of Geiger and Marsden's experiment

that an atom might be pictured as a cloud of positive charge, with negative electrons spread uniformly through the cloud like currants in a pudding. Rutherford now proposed, however, that there was a *concentration* of positive electricity in the centre of the atom which repelled the positively-charged α-particles. On this basis he calculated how many α-particles would be scattered through any given angle, and the relationship was verified by Geiger and Marsden in subsequent experiments.

In 1911 Rutherford suggested the first adequate structure of the atom. The atom, he said, consists of a concentration of positive electricity in a very tiny volume in the heart or centre of the atom, called the **nucleus**, with electrons moving round the nucleus. Most of the atom is empty. If we imagine the nucleus magnified to the size of a cricket ball, the furthest part of the atom would be a few miles away, and the space between the ball and the furthest part would be empty except for electrons, about the same size as the ball. The nucleus of the hydrogen is called a **proton**. Since there is one electron round the hydrogen nucleus, the proton carries a charge of $+e$, where $e$ is the numerical value of the charge on the electron.

**Discovery of Protons in Nuclei.** In 1919 Rutherford used energetic α-particles as "bullets" to bombard the atoms of gases such as nitrogen. He found that **protons** were emitted after the collisions. Protons are hydrogen nuclei, that is, hydrogen atoms which have lost an electron.

A simplified diagram of Rutherford's apparatus is shown in Fig. 449. A source of α-particles, $C$, was placed inside a vessel $D$

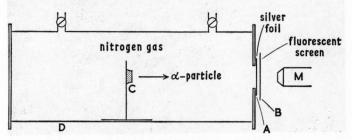

FIG. 449. Artificial disintegration of nitrogen

which was evacuated and then filled with nitrogen. Silver foil, $A$ sufficently thick to absorb the α-particles emitted was placed in front of $C$, and a fluorescent screen $B$ beyond $A$ was observed through a microscope $M$. Scintillations were seen on $B$ when α-particles collided with some nitrogen nuclei. Previous experiments with hydrogen gas showed that the particles emitted from the nuclei were protons, and similar results were obtained with elements other than nitrogen, solid as well as gaseous elements. It was thus established that *protons are in the nucleus of all elements.* These nuclear collisions were also observed by photographing the trail of droplets left behind by the particles in a vessel containing supersaturated water vapour, as shown in Fig. 450. This vessel is called a *cloud chamber*. It was invented by C. T. R. WILSON in 1911, and used for many years to provide valuable information about nuclear structure and nuclear collisions.

FIG. 450. TRANSMUTATION OF NITROGEN ATOM TO OXYGEN ATOM
The streaks are vapour trails made by α-particles moving through nitrogen gas. An α-particle (*right, centre*) has collided and merged with a nitrogen nucleus, which has then broken into an oxygen nucleus (*short track*) and a proton (*long track crossing others*)

**Discovery of the Neutron.** In 1930 α-particles were used by BOTHE and BECKER to bombard the element beryllium. A very penetrating radiation was produced, which was thought to be radiation of a very high energy like γ-rays. The unknown radiation was found by Curie-Joliot in 1932 to produce protons when it was incident on a slab of paraffin-wax, which contained a high percentage of hydrogen. After the laws of conservation of linear momentum and energy were applied to the collision, the calculated energy of the unknown radiation was found to be much less than the observed energy.

The discrepancy was cleared up by SIR JAMES CHADWICK in 1932. He carried out experiments, and showed that the unknown radiation was in fact *a particle which carried no charge* and which had a mass about the same as that of a proton. This particle was called a **neutron**.

**Nucleus of atom.** We now believe that the nucleus of any atom is built up of protons and neutrons. (Other particles have been discovered in the nucleus but this will not concern us.) The particles inside a nucleus are called **nucleons**. The hydrogen nucleus is the proton, which has a positive charge $+e$, numerically equal to the charge $e$ on an electron, but it is nearly 2000 times as heavy as the electron. The next heavy element is helium, whose atomic number is 2; this has therefore a charge of $+2e$ on its nucleus, that is, the nucleus contains 2 protons. The mass of the helium nucleus is 4, and hence the nucleus must also contain 2 neutrons. The symbol for the helium nucleus is $_2\mathrm{He}^4$, the upper number representing the atomic mass or mass number and the lower number representing the atomic number or the number of protons in the nucleus. The proton itself is represented by $_1\mathrm{H}^1$, the neutron by $_0n^1$. Each has a mass number 1, but the neutron has no charge. The nitrogen nucleus is written $_7\mathrm{N}^{14}$; it has an atomic mass of 14 and atomic number 7, that is, there are 7 protons and 7 neutrons in the nucleus. Oxygen, $_8\mathrm{O}^{16}$, has 8 protons and 8 neutrons in its nucleus (see Fig. 451).

|  | Hydrogen | Helium | Oxygen | Iron | Uranium |
|---|---|---|---|---|---|
| Nucleons | $1p$ | $2p+2n$ | $8p+8n$ | $28p+28n$ | $92p+146n$ |
| Charge | $+e$ | $+2e$ | $+8e$ | $+28e$ | $+92e$ |
| Mass No. | 1 | 2 | 16 | 56 | 238 |

FIG. 451 Some atomic nuclei: $p=$ proton, $n=$ neutron

**Natural disintegration of nuclei.** The elements uranium, thorium and actinium all show radioactivity. They emit α- or

$\beta$-particles as their nuclei disintegrate. Between 1902 and 1909 Rutherford and Suddy found that a series of elements were formed from each of the above parent elements. The *uranium series* is given in the table below, together with their *half-life period*. The latter is defined as the time taken for half the number of atoms to disintegrate, so that the half-life period is the time taken for the radioactivity to diminish to one-half of its original intensity.

| Element | Atomic Number | Mass Number | Half-life | Particles emitted |
|---|---|---|---|---|
| Uranium I | 92 | 238 | $4.5 \times 10^9$ years | $\alpha$ |
| Uranium $X_1$ | 90 | 234 | 24 days | $\beta$ |
| Uranium $X_2$ | 91 | 234 | 1.2 min. | $\beta$ |
| Uranium II | 92 | 234 | $2.5 \times 10^5$ years | $\alpha$ |
| Ionium | 90 | 230 | $8 \times 10^4$ years | $\alpha$ |
| Radium | 88 | 226 | $1.6 \times 10^3$ years | $\alpha$ |
| Radon | 86 | 222 | 3.8 days | $\alpha$ |
| Radium A | 84 | 218 | 3 min. | $\alpha$ |
| Radium B | 82 | 214 | 27 min. | $\beta$ |
| Radium C | 83 | 214 | 20 min. | $\alpha$ |
| Radium C' | 84 | 214 | $1.6 \times 10^{-4}$ sec. | $\alpha$ |
| Radium C'' | 81 | 210 | 1.3 min. | $\beta$ |
| Radium D | 82 | 210 | 19 years | $\beta$ |
| Radium E | 83 | 210 | 5 days | $\beta$ |
| Radium F | 84 | 210 | 138 days | $\alpha$ |
| Radium G (Lead) | 82 | 206 | Stable | |

The new elements formed can easily be deduced by considering the number of neutrons and protons in the nucleus. Uranium I, for example, has an atomic weight 238 and an atomic number 92. An $\alpha$-particle, a helium nucleus, has an atomic weight of 4 and an atomic number 2. Uranium I emits $\alpha$-particles. The nucleus of the atom left has therefore an atomic weight 234 and an atomic number 92, that is,

$$_{92}U^{238} \rightarrow _2He^4 + _{90}U^{234}$$

The uranium atom left, $_{90}U^{234}$, is known as Uranium $X_1$, and emits a $\beta$-particle from its nucleus. The $\beta$-particle has negligible mass but it has a negative charge $-e$, where $e$ is the numerical value of the charge on an electron. Consequently the atomic weight or mass number of the new nucleus formed is unaltered, but its charge *increases* from $+90e$ to $+91e$. The atomic number of the new element formed is thus 91, and this is called Uranium X₂. The uranium series ceases at lead, atomic number 82, which is a stable element.

**Artificial disintegration of nuclei. Transmutation.** As we have already mentioned, artificial disintegration was first achieved when energetic α-particles were used by Rutherford to disrupt the nitrogen nucleus, and he found that protons were emitted after the collision. The α-particle is a helium nucleus with 2 protons, the nitrogen nucleus has 7 protons, and hence the total charge concerned is equal to $+9e$. After the disintegration, the products are one proton together with another atom X (Fig. 452). The

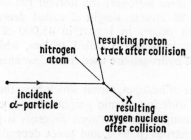

FIG. 452. Transmutation of nitrogen

latter must have 8 protons because the total charge before collision equals the total charge after collision, and hence X is an *oxygen* atom because the atomic number of this element is 8. Thus a nitrogen atom has been transformed into an oxygen atom, a process called *transmutation*.

Also, mass of α-particle + nitrogen nucleus = 4 + 14 = 18,
and mass of proton produced = 1.
∴ mass of oxygen atom produced = 18 − 1 = 17

We can therefore write the nuclear reaction as:

$$_2\text{He}^4 + {}_7\text{N}^{14} \rightarrow {}_8\text{O}^{17} + {}_1\text{H}^1.$$

**Isotopes.** The usual oxygen atom has an atomic number 8 and an atomic mass 16. In the nuclear disintegration just discussed, however, the oxygen atom had an atomic mass 17. We thus have a case of *two* oxygen atoms with the same chemical properties but of different mass, and these are called **isotopes of oxygen.**

When chlorine gas is collected in a laboratory experiment, its atomic weight is found to be 35·5, and this is the *average* mass of all the atoms in the gas. There is no chemical method of telling whether the individual atoms of chlorine are identical. Using magnetic and electric fields, however, Sir J. J. Thomson showed in 1911 that the individual atoms of all elements do not have the same mass. Chlorine, for example, consists of a high percentage of atoms of mass 35, together with atoms of mass 37. The average atomic weight is 35·5. Neon gas has as many as ten isotopes. Hydrogen has three isotopes; the normal gas of atomic weight 1, an isotope of atomic weight 2 called **deuterium** or **heavy hydrogen** which occurs in 1 part in 45,000 of water, and an isotope of atomic weight 3 called **tritium**, which is unstable. The isotopes of hydrogen are used in nuclear energy experiments (p. 659).

The existence of isotopes, atoms which have the same chemical properties but differing atomic weights, brings out very clearly that chemical combination between elements is governed by an exchange of electrons (p. 637), and hence depends on the atomic numbers of the elements. The atomic weight, however, is the weight of the nucleus, and this is different for the isotopes of a given element.

If elements are subjected to neutron bombardment long enough they tend to become radioactive, that is, some of the atomic nuclei are changed into unstable isotopes of the elements and emit alpha, beta or gamma rays. Radioactive isotopes are now widely used in medicine and industry, and are exported abroad by Harwell. The passage of food through the human body, for example, can be studied by adding a little of a harmless radioactive isotope to it, and noting the path taken by means of electrical instruments capable of detecting radioactive particles, such as a Geiger counter.

**Einstein's Mass-Energy Relation.** In 1905, Einstein showed, from his *Theory of Relativity*, that when a decrease of mass $m$ occurs, an amount of energy $E$ is produced which is given by

$$E = mc^2,$$

where $E$ is ergs when $m$ is in grams and $c$ is $3 \times 10^{10}$, the numerical value of the velocity of light in centimetres per second (see p. 34). A decrease in mass of 1 *milligram* would thus produce an amount of energy $E$ given by

$$E = 0 \cdot 001 \times (3 \times 10^{10})^2 \text{ ergs}$$
$$= 9 \times 10^{17} \text{ ergs} = 9 \times 10^{10} \text{ joules}$$
$$= 90,000 \text{ million joules}.$$

Now a 100 watt lamp uses 100 joules of energy per second, and would therefore use $0 \cdot 36$ million joules of energy in 1 hour. Hence 90,000 million joules would keep 250,000 100 watt lamps burning for 1 hour, and this large amount of energy would be obtained from a decrease of mass of only 1 milligram. When coal burns, the energy produced comes from interchange of electrons because there is a chemical change, and this represents only a very, very small part of the masses of the coal and oxygen. Thus combustion is a slow process of producing energy. As we shall see later, if a nucleus disintegrates a far greater amount of energy is liberated, because the masses changed are much greater.

Einstein's mass-energy law was shown to be true in experiments on radioactivity, where energetic particles such as alpha particles are emitted when a nucleus disintegrates, and the decrease of mass was found to be numerically equal to the energy produced. The deep significance of Einstein's law is that *mass is a form of energy*. When we say that the masses of the substances in a reaction are conserved, we also mean that the total energy is constant. The atomic mass of the carbon isotope $_6C^{12}$ is taken as $12 \cdot 00$ atomic mass units (a.m.u.), and since mass and energy can be changed from one to the other by Einstein's relation, energy changes in nuclear reactions are often expressed in terms of a.m.u.

**Binding energy.** The particles of the nucleus, the protons and neutrons, are called **nucleons**. The nucleons are kept together in a

22

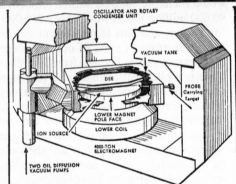

FIG. 453. SYNCHRO-
CYCLOTRON AT BERKELEY,
CALIFORNIA

In this giant atom-smashing machine, the magnet and oscillator cause ions to whirl round in a vacuum chamber with ever-increasing speed on a spiral path, till they hit a target within the chamber or are projected in a high-speed beam to bombard an external target.

**OSCILLATOR AND ROTARY CONDENSER UNIT**

**VACUUM TANK**

**DEE**

**PROBE Carrying Target**

**LOWER MAGNET POLE FACE**

**LOWER COIL**

**ION SOURCE**

**4000-TON ELECTROMAGNET**

**TWO OIL DIFFUSION VACUUM PUMPS**

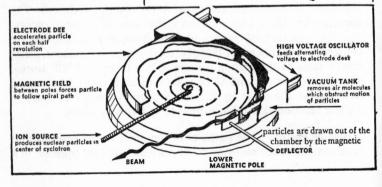

**ELECTRODE DEE** accelerates particle on each half revolution

**HIGH VOLTAGE OSCILLATOR** feeds alternating voltage to electrode dees

**MAGNETIC FIELD** between poles forces particle to follow spiral path

**VACUUM TANK** removes air molecules which obstruct motion of particles

**ION SOURCE** produces nuclear particles in center of cyclotron

particles are drawn out of the chamber by the magnetic **DEFLECTOR**

**BEAM**

**LOWER MAGNETIC POLE**

very tiny volume by powerful forces whose nature is not yet known. To take the nucleons apart requires energy, and this amount of energy is known as the **binding energy** of the nucleus. Fig. 453 shows one of the atom-smashing machines used to break up the nucleus. The mass of the nucleus should be less than the total mass of the separated nucleons; the difference in mass, from Einstein's law, represents the amount of binding energy.

As an illustration, we know by measurements that:

$$\text{mass of proton} = 1.0076 \text{ a.m.u.,}$$

and

$$\text{mass of neutron} = 1.0090 \text{ a.m.u.,}$$

Now the carbon nucleus, $_6C^{12}$, has 6 protons and 6 neutrons, and measurement of the mass of the nucleus shows it is 12·0038 a.m.u.

But mass of 6 protons $= 6 \times 1.0076 = 6.0456$ a.m.u.,

and mass of 6 neutrons $= 6 \times 1.0090 = 6.0540$ a.m.u.

∴ total mass $= 12.0996$ a.m.u.

∴ mass of nucleus is *less* than the mass of its nucleons by 12·0996 − 12·0038, or 0·0958 a.m.u.

∴ binding energy of carbon nucleus $= 0.0958$ a.m.u.

The binding energy of a uranium nucleus, which has 92 protons and 146 neutrons, is much greater than that of a carbon nucleus. If a heavy nucleus can be partly disintegrated, the masses of the particles produced are less than the masses of the particles when they were locked together in the nucleus, and the difference in mass is converted into energy. This will now be discussed.

**Nuclear fission.** As we saw on p. 647, energetic α-particles can disrupt an atomic nucleus. Neutrons, however, can penetrate charge, and are therefore not repelled by the positive charge on the nucleus as the α-particles (positive charges) are. From 1934, Fermi and others began to use neutrons as "bullets" to fire at atomic nuclei. Usually only a small fragment was "chipped" from the nucleus and consequently only a small amount of energy was released. In 1938, however, neutrons were fired at the heaviest nucleus, uranium. It was then found that the nucleus had split into two large nuclei, such as barium, Ba, and Krypton, Kr, with a consequent release of a relatively large amount of energy (Fig. 454). The break-up of the uranium nucleus into two large nuclei is called **nuclear fission**. The total mass of the uranium nucleus and incident neutron was less than the total mass of the products of the nuclear "explosion", as explained before, and calculation showed that if all the atoms of 1 lb. of uranium under-

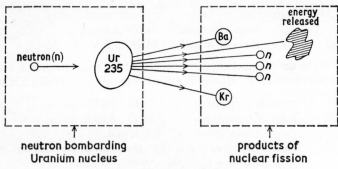

FIG. 454. Nuclear fission

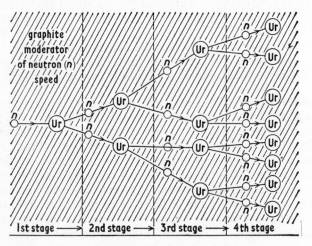

FIG. 455. Principle of chain reaction

go fission, the energy released would be as much, theoretically, as that released by burning three million lb. of coal!

The practical possibility of achieving this enormous energy was due to the release of several neutrons when nuclear fission of a uranium atom took place. Under control, these neutrons could each produce fission of another atom, with the release of more neutrons, which in turn would produce fission of more atoms (Fig. 455). The fission of millions of uranium atoms would thus take place very rapidly throughout the uranium and an enormous amount of energy would be produced in a short time. The speed of the neutrons is a critical factor in nuclear fission, and nuclear reactors such as those at Harwell in England contain material to modify the speed. If the neutrons are too fast for example, fission does not take place.

**Nuclear reactor principle.** Natural uranium contains about 1 part by weight of atoms of atomic weight 235 and 139 parts by weight of atoms of atomic weight 238. Nuclear fission occurs with the atoms of atomic weight 235 when uranium is used in nuclear reactors.

Fission takes place only if the bombarding neutron is slow. Graphite is therefore used round the uranium to moderate the speed of the neutron, so that the chain reaction is prevented from dying out. If there are too many neutrons the reaction will proceed too fast and get out of hand, and this is controlled by neutron-absorbing boron steel rods. These rods are moved in and out of the reactor from the control room by small electric motors, which raise them through the floor to a suitable position. In the event of an electrical failure the rods would fall and shut off the reactor automatically.

The reactor is started by a charge machine, which lowers uranium rods, each about 1 yard long and 1 inch in diameter, into about 1700 fuel channels in the graphite core (Fig. 456). The boron steel rods are then raised slowly, and in a certain position the chain reaction proceeds at the desired rate. The reactor is now said to have "gone critical", and heat is produced steadily.

Carbon dioxide gas is blown through the fuel elements under pressure, and the hot gas is led into heat exchangers outside the reactor. Here it heats up water which flows through an independent pipe system in the exchanger, and this is converted into high pressure and low pressure steam. The steam is used to drive the turbines which turn the electrical generators, and the electricity obtained is fed to the national grid system to supplement the electrical energy produced by burning coal at conventional power

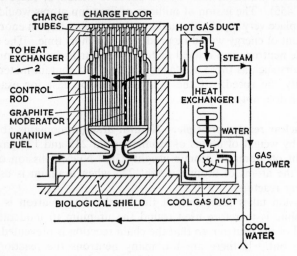

FIG. 456. Principle of nuclear reactor

stations. The nuclear reactor can work night and day for two or three years with the same uranium fuel. The reactor core is shielded by a welded steel pressure vessel and a concrete biological shield, which protects the operators outside from the intense radiation inside the core.

**Nuclear fusion.** We have seen that a considerable amount of energy is released when a heavy nucleus such as uranium disintegrates into two large parts; this is nuclear fission. Experiments show that a considerable release of energy is also obtained when the nuclei of the lightest elements are **fused** to form a heavier nucleus, and this is known as **nuclear fusion.**

One transformation of matter into energy is the fusion of deuterons, $_1H^2$, the nuclei of heavy hydrogen, into helium nuclei. As the sun contains a considerable amount of hydrogen, it was suggested in 1939 that the energy of the sun was basically due to nuclear fusion. It has been estimated that, theoretically, the energy released from the fusion of all the atoms in 1 lb. of deuterium, heavy hydrogen, gas is equivalent to that released by burning nearly 3 million lb. of coal.

In order to obtain nuclear fusion, the nuclei must approach near enough to each other to overcome the repulsion of their like charges. The lightest elements have the smallest nuclear charge, and hence, the chance of nuclear fusion is greatest for such elements. The most practical way of achieving nuclear fusion is to raise the temperature of the deuterium gas to millions of degrees, which is the temperature in the heart of the sun. At these very high temperatures deuterons fuse together, a process known as **thermonuclear fusion.** Very heavy electrical discharges are sent through the gas to heat it to such enormously high temperatures, and researches into methods of retaining the energy in the gas are at present being made in Great Britain and abroad. If methods for nuclear fusion were successful, a cheap source of power would be available. Heavy hydrogen forms about 1 part in 45,000 of water, and if heavy hydrogen were extracted from seawater it could provide limitless power at a very economic price.

## CONCLUSION

We have now come to the end of our journey through the pathways of elementary physics. Throughout it, the reader will have noted how experiments in the laboratory, performed by scientists in their efforts to find out how things work, have led to worldwide applications affecting our everyday lives. Sir Isaac Newton laid the foundations of the science of mechanics, with its practical applications in the design of bridges, buildings, ships and aeroplanes. Faraday discovered the laws of electromagnetic induction, and this made possible the dynamo and the distribution of electrical power for lighting and heating in our homes and factories.

Oersted's discovery of the magnetic effort of a current led to the invention of the telephone and the telegraph; Ampère's experiments on the force on a conductor in a magnetic field led to the electric motor used for driving vehicles and machines; Snell's discovery of the law of refraction led to the design of efficient telescopes and microscopes; and Appleton's experiments on the ionosphere led to the invention of radar, which permits safe navigation by sea and air. Sir J. J. Thomson's discovery of the electron led to the invention of the cathode-ray tube, used both in radar and in television. Rutherford's discovery of the nucleus of the atom led to the development of nuclear reactors.

Today, thousands of scientists of different nationalities are engaged in laboratories all over the world in the quest for knowledge. Will that knowledge be used to increase human wealth, health and happiness? Or will it be used for purposes of destruction and war? The choice will not be made by scientists alone; every citizen of the world must determine to work for peace and plenty.

## Summary

1. The radio valve has a filament (cathode) which emits electrons. When the anode of the valve is positive in potential relative to the filament of the valve conducts.

2. The cathode ray tube has a filament emitting electrons, which produce light on striking a screen at the other end of the tube. The electron beam can be focused, and the light intensity controlled, by altering the potentials of metal cylinders in the tube.

3. The photo-electric cell has a metal inside it which emits electrons when illuminated, and a current then flows through the cell. The magnitude of the current is proportional to the light intensity.

4. X-rays are penetrating rays of very short wavelength, produced when fast-moving electrons strike tungsten, for example.

5. An atom has a tiny nucleus at its centre carrying positive electricity. It is surrounded by electrons which have defined energy levels or shells. The $K$ shell, nearest to the nucleus, can have up to a maximum of 2 electrons, the $L$ and $M$ shells up to a maximum of 8 electrons.

6. The chemical activity of an atom depends on the number of electrons in its outermost shell. A chlorine atom is an electron acceptor, a sodium atom is an electron donor.

7. Electric current is carried through electrolytes by ions, through metals by electrons, and through semiconductors by electrons and holes. Hole movement is equivalent to the movement of a positive charge equal to that on an electron.

8. A $p$-$n$-$p$ transistor consists of a very thin layer of $n$-type semiconductor (one with a large excess number of electrons) sandwiched between two layers of $p$-type semiconductor (one with a large excess number of holes).

9. A radioactive atom emits $\alpha$-particles (helium nuclei) or $\beta$-particles (electrons) and $\gamma$-rays (extremely short electromagnetic waves).

10. The nucleus contains protons (hydrogen nuclei) and neutrons (particles with no charge and mass about the same as a proton). The atomic mass is the mass of all the protons and neutrons and surrounding electrons; the atomic number is the number of protons, or electrons surrounding the nucleus.

11. Nuclear fission occurs when a neutron strikes a uranium nucleus, and two large nuclei, and several neutrons, are obtained. Energy is then released owing to mass change.

12. Nuclear fusion occurs when the nuclei of light elements such as hydrogen fuse together. Energy is then released.

13. Thermonuclear fusion is obtained by heating hydrogen gas to very high temperatures. The energy of the sun comes from a thermonuclear fusion process.

## EXERCISES ON CHAPTER XXXIX

**1.** Describe a *diode valve*. Draw a sketch of its characteristic ($I$-$V$) curve. Explain how a diode valve changes A.C. to D.C. voltage.

**2.** Explain the term *atomic number*. What relationship is there between the electrons in an atom and atomic number ? Explain briefly, using the idea of electron shells, why sodium chloride is a stable compound.

**3.** What are the carriers of electric current in a semiconductor and in a metal ? Describe a *transistor*.

**4.** Describe briefly the structure of an atom, referring in your answer to the hydrogen atom and to one other atom such as uranium.

**5.** Describe an experiment which led to the discovery of the nucleus.

**6.** What are the nucleons (particles) in the helium nucleus, $_2He^4$, and the uranium nucleus, $_{92}U^{238}$? If the mass of a neutron is $1 \cdot 009$ a.m.u. and that of a proton is $1 \cdot 008$ a.m.u., and that of a helium nucleus is $4 \cdot 030$ a.m.u., calculate the binding energy of the helium nucleus.

**7.** What is the difference between *nuclear fission* and *nuclear fusion* ? Describe how neutrons are able to produce a chain reaction leading to a large release of energy.

# MODERN PHYSICS EXPERIMENTS

The conclusions on modern physics discussed in the last chapter were reached as a result of many experiments. These were often devised to detect or measure effects due to ions and electrons, and some of them are described in outline in this section.

When $\alpha$-particles are emitted by a radioactive substance they ionize the air. It is not an easy matter to detect the ions. If an ion carries a charge of the order $e$, $10^{-19}$ coulomb approximately (p. 635), a million ions flowing from one point to another in one second is equivalent to a current of $10^{-19} \times 10^6$ or $10^{-13}$ amperes (approx.). This is less than the smallest current observable with the most sensitive moving-coil galvanometer (p. 588), so that special techniques are necessary to detect this current.

**Pulse (Wulf) electroscope.** In electrostatics we saw that high voltages were concerned but the charges were very small. A gold-leaf electroscope indicates such small charges (p. 414). It has a capacitance of the order of $10^{-12}$ farad, so if voltages of a few hundred volts, such as on a rubbed ebonite rod, produce a deflection of the leaf, this means that the instrument can detect a charge of the order $Q$ given by $Q = CV = 10^{-12} \times 100 = 10^{-10}$ coulomb. If the charge flows away from the leaf in a time $t$ of 10-100 seconds, the current detected would be $Q/t$ or $10^{-11} - 10^{-12}$ amperes.

Special forms of sensitive electroscope with very low capacitances were therefore developed, and much of the fundamental work on radioactivity was done with them. A form suitable for school work is the Pulse or Wulf electroscope, and we shall describe the electroscope made by Griffin & George Limited, London, England, together with experiments to be performed with it. The additional apparatus is also available from Griffin & George. The electroscope consists of a movable side electrode $S$

inside a case $C$ which can be earthed (Fig. 457). The moving part of the instrument is a light metal flag, mounted on a taut phosporbronze suspension. A top electrode $T$ is connected to the flag, and $T$ is used to test the existence and nature of electrical particles, as we shall see.

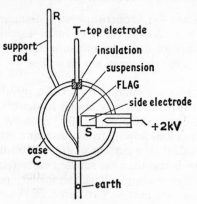

FIG. 457. A pulse electroscope

The side electrode $S$ is kept at a potential of about 2000 volts (2 kV), and it is moved towards the flag until it is about 1 to 2 mm. away. In this position the flag is attracted towards $S$, and after it touches $S$ and is charged to the same potential, the flag breaks away from $S$. If the flag is discharged at regular intervals, for example by opposite charges moving towards the electrode $T$ connected to the flag, the latter is attracted towards $S$ again and flies back again, at a rate depending on the number of charges per second reaching $T$. The flag thus "beats" at a definite rate. *Ions* produced between $R$ and $T$ hence cause the flag to "beat". When the distance between $S$ and the flag is about 1 mm. the rate of beating is about once per second.

**Demonstrations of ions produced by flame.** The heat energy of a flame is sufficient to remove electrons from some gas atoms in the flame, thus creating ions. These ions can be detected by the pulse electroscope.

The side electrode $S$ is maintained at a potential of $+2$ kV (2000 volts), and the case $C$ is earthed (Fig. 457). $S$ is moved near to the flag until it beats once, that is, the potential of the flag and the electrode $T$ is the same as $S$. A lighted match is now held between $R$ and $T$, and the flag is observed to beat at a fast rate. This shows the existence of negative ions, which drift towards $T$ and cause the flag to discharge at a rate depending on the number of ions per second reaching the electrode. When the side potential is changed negatively to $-1.5$ kV, the flame again causes the flag to beat faster, showing the presence of positive ions as well as negative ions.

**Demonstration of α-particles.** The α-particles emitted by radioactive substances produce ionization of air or gas molecules. Radioactive sources can therefore be detected by using a pulse electroscope. To demonstrate the emission of α-particles, a weak radium source is taken out of its container using a lifting tool, and clamped on $R$ by the handle (Fig. 458). The potential of the side

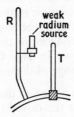

FIG. 458. Demonstration of α-particles

electrode is made $+2$ kV, and then $-1.5$ kV, and the leaf is observed to beat at a faster rate on each occasion; thus showing that the α-particles emitted by the radium source produce both positive and negative ions in air.

**Measurement of range of α-particles in air.** To measure the range of α-particles emitted by a radium source, an ionization chamber is used (Fig. 459 (a)). This chamber is clipped on to the support $R$, with its gauze window uppermost, and is slid down $R$ as far as possible. The weak radioactive source $S$ is now raised by the lifting tool, and fastened to the support rod $R$ so that $S$ faces downward, directly above the gauze window.

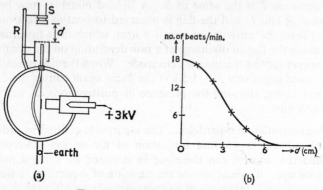

FIG. 459. Range of α-particles in air

The rate of beating of the leaf is observed at different distances $d$ of the source $S$ from the window of the ionization chamber. As $d$ increases, the rate of beating of the flag diminishes, showing that less ions per second are produced in the ionization chamber. A graph similar to that shown in Fig. 459 (b) is obtained, from which it follows that the range of the α-particles is about 6 cm.

**Measurement of half-life period of thorium emanation.** Radioactivity is generally due to the disintegration of the nuclei of a number of atoms in the material concerned. The number of radioactive atoms diminish as time goes on, and the *half-life period* of a radioactive material is defined as the time for the number of radioactive atoms to diminish by half, or as the time taken for the intensity of the radioactivity to diminish by half.

The half-life period of thorium gas can be measured with the aid of a chamber attached to a plastic bottle containing thorium hydroxide. The chamber is clipped on to the support rod, $R$, and slid down the rod so that the top electrode of the electroscope enters the centre of the chamber as far as possible. The side electrode is kept at a potential of $+3$ kV, and the gap between the side electrode and the flag is made 1-2 mm.

The bottle is shaken to release any thorium gas trapped, and then squeezed once or twice, keeping it upright. The thorium gas is transferred to the chamber, where it decays with the emission of α-particles and produces an ionization current. The flag should begin to beat at the rate of 1 beat or more per 10 seconds. The time of successive beats is noted, and this time lengthens as the radioactive gas decays.

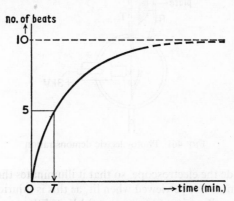

FIG. 460. Half-life period measurement

The actual number of the beats is plotted against the time (Fig. 460). A smooth curve is drawn through the points and extended to level off horizontally. If the number of the beats corresponding to the horizontal part of the curve is ten, as shown, the half-life period, $T$, is the time to reach the fifth beat. This time is read from the graph.

**Photo-electric effect.** Electrons can be produced by ultra-violet light shining on the surface of a suitable metal plate. The rate of emission of electrons is low, so the current is very small, and the pulse electroscope can be used to measure it.

To demonstrate the emission of electrons by an illuminated zinc plate, (a) an ultra-violet lamp, such as Phillip's TUV-6 watt, is required, (b) the zinc plate must first be prepared. For the preparation, a small globule of mercury is placed on the zinc plate and wiped over the plate with cotton wool moistened with dilute sulphuric acid. The plate is wiped dry with clean tissue and should be mirror-bright.

The zinc plate is clipped on to the support rod so that its bright surface faces the rod $T$ (Fig. 461). The ultra-violet lamp is then

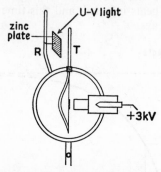

FIG. 461. Photo-electric demonstration

placed beside the electroscope, so that it illuminates the zinc plate. The lamp must not be viewed when lit, as this is injurious to sight.

The high voltage is set at about $+3$ kV and the gap to 1-2 mm. The flag then beats at a steady rate. The ultra-violet lamp is now switched on, and the flag is observed to beat at a different rate. When the high voltage is reduced to $+1.5$ kV, the rate of beating is reduced. When the high voltage is changed to $-1.5$ kV, no change occurs in the rate of beating. Thus the carriers of the current between the zinc plate and $T$ are *negative* particles. They are, in fact, electrons, emitted from the zinc.

When glass is inserted between the lamp and the zinc plate, no change is observed in the rate of beating for a high voltage of $+3$ kV. Thus glass does not transmit ultra-violet rays. When quartz is inserted the rate of beating alters, showing that quartz transmits ultra-violet rays.

**Diode valve characteristic.** By heating a fine tungsten wire, or warming a cylinder coated with the oxides of barium or strontium, a much greater supply of electrons can be produced than by the photo-electric effect. The current produced in this case is of the order of several milliamperes, so that a moving-coil instrument can be used to measure it. The radio valve uses this method of producing electrons, and we shall describe how a diode valve characteristic, a graph showing how the current varies with the applied p.d., is obtained.

*Apparatus.* Diode valve, suitable low voltage (filament) supply, suitable high tension (anode) supply, milliammeter, connecting wire.

*Method.* Connect the low voltage supply, such as 4 volts, to the filament $F$ (Fig. 462 $(a)$). Then join the positive terminal $T$

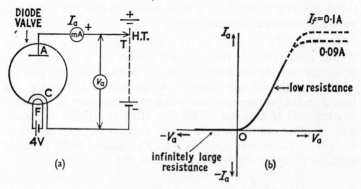

FIG. 462. Diode characteristic

of the high tension supply to the positive terminal of the millammeter, and the negative terminal of the milliammeter to the anode $A$ of the valve. Now join the negative terminal of the high tension supply to the cathode, $C$, of the valve. (If the valve is a directly-heated one, the negative terminal is joined to the negative pole of the filament low-voltage supply.)

Increase the anode voltage $V_a$ from zero, and measure the anode current $I_a$. Reverse the high tension voltage supply so that its positive terminal is now joined to the cathode, and observe that the milliammeter reading is zero whatever the high tension voltage.

*Graph.* Plot the anode current $I_a$ against the anode voltage, $V_a$, for both positive and negative voltages of $V_a$ (Fig. 462 (*b*)).

*Conclusion.* The diode valve conducts when the anode potential relative to the cathode is positive, and has an infinitely-large resistance when the anode potential is negative relative to the cathode.

**Transistor amplification.** As explained on p. 642, transistors can amplify currents. The following experiment demonstrates how the collector current of a transistor varies when its base current is altered.

*Apparatus.* Mullard transistor OC71, 1·5 volt dry battery, 0-500 microammeter, 0-10 milliammeter, 10,000 ohm (10 kΩ) variable resistance, connecting wire.

*Method.* To obtain the transistor action of a common emitter circuit (that is, the emitter $E$ is common in both the base, $B$, circuit and the collector, $C$, circuit, p. 642), join the positive terminal of the 1·5 volt battery $L$ to the emitter terminal, $E$, the base $B$ to the positive terminal of the microammeter $D$, the negative terminal of $D$ to the 10,000 ohm (10 kΩ) variable resistance, and the other end of the latter to the negative terminal of $L$ (Fig. 463 (*a*)). The milliammeter $G$ is then connected to the collector terminal $C$ and to the negative terminal of $L$.

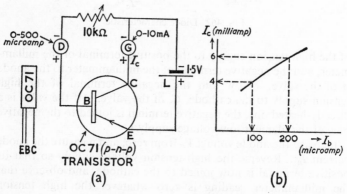

Fig. 463. Transistor amplification

The 10 kΩ resistance is now varied so that the base current $I_b$, read from $D$, increases from a low value, and the reading of the collector current, $I_c$, is observed from $G$.

*Graph.* A graph of collector current $I_c$ against $I_b$ is plotted (Fig. 463 (*b*)).

*Conclusion.* A small change in the base current $I_b$, such as 100 microamp. (0·1 milliamp.), causes a large change, such as 2 milliamp., in the collector current $I_c$. The transistor thus acts as a current amplifier.

## ANSWERS TO NUMERICAL EXERCISES

### CHAPTER XXXIX (p. 662)

**6.** 0·004 a.m.u.

# INDEX

# NOTES

Printed in Great Britain
at Hopetoun Street, Edinburgh,
by T. and A. CONSTABLE LTD.
Printers to the University of Edinburgh

BOOKS BY M. NELKON

\*

*Published by Chatto & Windus*
Fundamentals of Physics (with SI units)
Exercises in Ordinary Level Physics (with SI units)
C.S.E. Physics
Revision Book in C.S.E. Physics
Principles of Physics
SI Units: An introduction for A-level

\*

*Heinemann Educational*
Advanced Level Physics (with SI Units) (*with* P. PARKER)
Advanced Level Practical Physics (*with* J. OGBORN)
Scholarship Physics (with SI units)
Principles of Atomic Physics and Electronics
Light and Sound (with SI units)
Mechanics and Properties of Matter (with SI units)
An Introduction to the Mathematics of Physics
(*with* J. H. AVERY)
Graded Exercises and Worked Examples in Physics
Test Papers in Physics (with SI units)
Revision Notes in Physics
(Book 1. Heat, Light, Sound
Book 2. Electricity, Mechanics, Properties of Matter)
General Science Physics
Electronics and Radio
Elementary Physics, Books 1 and 2 (*with* A. F. ABBOTT)
Elementary Physics (complete edition)
Revision Book in O-Level Physics
Solutions to O-Level Physics Questions (with SI units)
Solutions to A-Level Physics Questions (with SI units)

*Edward Arnold*
Electricity and Magnetism (Scholarship Level)
Electricity (*Advanced Level Course*—SI)

*Blackie*
Heat (for A-Level students)

THE HIGH-TENSION ROOM AT THE CAVENDISH LABORATORY, CAMBRIDGE

In the centre is a 2-million volt Cockcroft and Walton generator, consisting of a series of inter-connected valves and condensers. The high potential terminal is connected to the top of the column on the right containing a source of electrical particles (protons), which are then accelerated down the column to bombard atoms in atomic energy researches.